The Finite Element Method for Elliptic Problems

Books in the Classics in Applied Mathematics series are monographs and textbooks declared out of print by their original publishers, though they are of continued importance and interest to the mathematical community. SIAM publishes this series to ensure that the information presented in these texts is not lost to today's students and researchers.

Classics in Applied Mathematics

C. C. Lin and L. A. Segel, *Mathematics Applied to Deterministic Problems in the Natural Sciences*

Johan G. F. Belinfante and Bernard Kolman, *A Survey of Lie Groups and Lie Algebras with Applications and Computational Methods*

James M. Ortega, *Numerical Analysis: A Second Course*

Anthony V. Fiacco and Garth P. McCormick, *Nonlinear Programming: Sequential Unconstrained Minimization Techniques*

F. H. Clarke, *Optimization and Nonsmooth Analysis*

George F. Carrier and Carl E. Pearson, *Ordinary Differential Equations*

Leo Breiman, *Probability*

R. Bellman and G. M. Wing, *An Introduction to Invariant Imbedding*

Abraham Berman and Robert J. Plemmons, *Nonnegative Matrices in the Mathematical Sciences*

Olvi L. Mangasarian, *Nonlinear Programming*

Carl Friedrich Gauss, *Theory of the Combination of Observations Least Subject to Errors: Part One, Part Two, Supplement*. Translated by G. W. Stewart (*First time in print.*)

Richard Bellman, *Introduction to Matrix Analysis*

U. M. Ascher, R. M. M. Mattheij, and R. D. Russell, *Numerical Solution of Boundary Value Problems for Ordinary Differential Equations*

K. E. Brenan, S. L. Campbell, and L. R. Petzold, *Numerical Solution of Initial-Value Problems in Differential-Algebraic Equations*

Charles L. Lawson and Richard J. Hanson, *Solving Least Squares Problems*

J. E. Dennis, Jr. and Robert B. Schnabel, *Numerical Methods for Unconstrained Optimization and Nonlinear Equations*

Richard E. Barlow and Frank Proschan, *Mathematical Theory of Reliability*

Cornelius Lanczos, *Linear Differential Operators*

Richard Bellman, *Introduction to Matrix Analysis, Second Edition*

Beresford N. Parlett, *The Symmetric Eigenvalue Problem*

Richard Haberman, *Mathematical Models: Mechanical Vibrations, Population Dynamics, and Traffic Flow*

Peter W. M. John, *Statistical Design and Analysis of Experiments*

Tamer Başar and Geert Jan Olsder, *Dynamic Noncooperative Game Theory, Second Edition*

Emanuel Parzen, *Stochastic Processes*

Petar Kokotović, Hassan K. Khalil, and John O'Reilly, *Singular Perturbation Methods in Control: Analysis and Design*

Classics in Applied Mathematics (continued)

Jean Dickinson Gibbons, Ingram Olkin, and Milton Sobel, *Selecting and Ordering Populations: A New Statistical Methodology*

James A. Murdock, *Perturbations: Theory and Methods*

Ivar Ekeland and Roger Témam, *Convex Analysis and Variational Problems*

Ivar Stakgold, *Boundary Value Problems of Mathematical Physics, Volumes I and II*

J. M. Ortega and W. C. Rheinboldt, *Iterative Solution of Nonlinear Equations in Several Variables*

David Kinderlehrer and Guido Stampacchia, *An Introduction to Variational Inequalities and Their Applications*

F. Natterer, *The Mathematics of Computerized Tomography*

Avinash C. Kak and Malcolm Slaney, *Principles of Computerized Tomographic Imaging*

R. Wong, *Asymptotic Approximations of Integrals*

O. Axelsson and V. A. Barker, *Finite Element Solution of Boundary Value Problems: Theory and Computation*

David R. Brillinger, *Time Series: Data Analysis and Theory*

Joel N. Franklin, *Methods of Mathematical Economics: Linear and Nonlinear Programming, Fixed-Point Theorems*

Philip Hartman, *Ordinary Differential Equations, Second Edition*

Michael D. Intriligator, *Mathematical Optimization and Economic Theory*

Philippe G. Ciarlet, *The Finite Element Method for Elliptic Problems*

Jane K. Cullum and Ralph A. Willoughby, *Lanczos Algorithms for Large Symmetric Eigenvalue Computations, Vol. I: Theory*

M. Vidyasagar, *Nonlinear Systems Analysis, Second Edition*

Robert Mattheij and Jaap Molenaar, *Ordinary Differential Equations in Theory and Practice*

Shanti S. Gupta and S. Panchapakesan, *Multiple Decision Procedures: Theory and Methodology of Selecting and Ranking Populations*

Eugene L. Allgower and Kurt Georg, *Introduction to Numerical Continuation Methods*

Leah Edelstein-Keshet, *Mathematical Models in Biology*

Heinz-Otto Kreiss and Jens Lorenz, *Initial-Boundary Value Problems and the Navier-Stokes Equations*

J. L. Hodges, Jr. and E. L. Lehmann, *Basic Concepts of Probability and Statistics, Second Edition*

George F. Carrier, Max Krook, and Carl E. Pearson, *Functions of a Complex Variable: Theory and Technique*

Friedrich Pukelsheim, *Optimal Design of Experiments*

Israel Gohberg, Peter Lancaster, and Leiba Rodman, *Invariant Subspaces of Matrices with Applications*

Lee A. Segel with G. H. Handelman, *Mathematics Applied to Continuum Mechanics*

Rajendra Bhatia, *Perturbation Bounds for Matrix Eigenvalues*

Barry C. Arnold, N. Balakrishnan, and H. N. Nagaraja, *A First Course in Order Statistics*

Charles A. Desoer and M. Vidyasagar, *Feedback Systems: Input-Output Properties*

Stephen L. Campbell and Carl D. Meyer, *Generalized Inverses of Linear Transformations*

Alexander Morgan, *Solving Polynomial Systems Using Continuation for Engineering and Scientific Problems*

I. Gohberg, P. Lancaster, and L. Rodman, *Matrix Polynomials*

Galen R. Shorack and Jon A. Wellner, *Empirical Processes with Applications to Statistics*

Richard W. Cottle, Jong-Shi Pang, and Richard E. Stone, *The Linear Complementarity Problem*

Rabi N. Bhattacharya and Edward C. Waymire, *Stochastic Processes with Applications*

Robert J. Adler, *The Geometry of Random Fields*

The Finite Element Method for Elliptic Problems

Philippe G. Ciarlet
Université Pierre et Marie Curie
Paris, France

siam
Society for Industrial and Applied Mathematics
Philadelphia

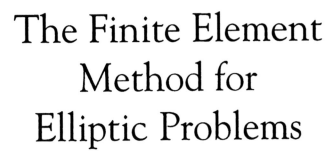

This SIAM edition is an unabridged republication of the work first published by North-Holland, Amsterdam, New York, Oxford, 1978.

10 9 8 7 6 5 4 3

Library of Congress Cataloging-in-Publication Data

Ciarlet, Philippe G.

The finite element method for elliptic problems / Philippe G. Ciarlet.

p. cm. -- (Classics in applied mathematics ; 40)

Includes bibliographical references and index.

ISBN 978-0-898715-14-9 (pbk.)

1. Differential equations, Elliptic--Numerical solutions. 2. Boundary value problems--Numerical solutions. 3. Finite element method. I. Title. II. Series.

QA377 .C53 2002

515'.353--dc21

2002019515

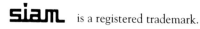

To Monique

TABLE OF CONTENTS

PREFACE TO THE CLASSICS EDITION

Although almost 25 years have elapsed since the manuscript of this book was completed, it is somewhat comforting to see that the content of Chapters 1 to 6, which together could be summarized under the title "The Basic Error Estimates for Elliptic Problems," is still essentially up-to-date. More specifically, the topics covered in these chapters are the following:

- description and mathematical analysis of various problems found in linearized elasticity, such as the membrane and plate equations, the equations of three-dimensional elasticity, and the obstacle problem;
- description of conforming finite elements used for approximating second-order and fourth-order problems, including composite and singular elements;
- derivation of the fundamental error estimates, including those in maximum norm, for conforming finite element methods applied to second-order problems;
- derivation of error estimates for the obstacle problem;
- description of finite element methods with numerical integration for second-order problems and derivation of the corresponding error estimates;
- description of nonconforming finite element methods for second-order and fourth-order problems and derivation of the corresponding error estimates;
- description of the combined use of isoparametric finite elements and isoparametric numerical integration for second-order problems posed over domains with curved boundaries and derivation of the corresponding error estimates;
- derivation of the error estimates for polynomial, composite, and singular finite elements used for solving fourth-order problems.

Otherwise, the topics considered in Chapters 7 and 8 have since undergone considerable progress. Additionally, new topics have emerged that often address the essential issue of the actual implementation of the finite element method. The interested reader may thus wish to consult the following more recent books, the list of which is by no means intended to be exhaustive:

- for further types of error estimates, a posteriori error estimates, locking phenomena, and numerical implementation: Brenner and Scott (1994), Wahlbin (1991, 1995), Lucquin and Pironneau (1998), Apel (1999), Ainsworth and Oden (2000), Bramble and Zhang (2000), Frey and George (2000), Zienkiewicz and Taylor (2000), Babuska and Strouboulis (2001), Braess (2001);
- for mixed and hybrid finite element methods: Girault and Raviart (1986), Brezzi and Fortin (1991), Robert and Thomas (1991);
- for finite element approximations of eigenvalue problems: Babuska and Osborn (1991);
- for finite element approximations of variational inequalities: Glowinski (1984);
- for finite element approximations of shell problems: Bernadou (1995), Bathe (1996);
- for finite element approximations of time-dependent problems: Raviart and Thomas (1983), Thomée (1984), Hughes (1987), Fujita and Suzuki (1991).

Last but not least, it is my pleasure to express my sincere thanks to Sara J. Triller, Arjen Sevenster, and Gilbert Strang, whose friendly cooperation made this reprinting possible.

Philippe G. Ciarlet
October 2001

BIBLIOGRAPHY

AINSWORTH, M.; ODEN, J.T. (2000): *A Posteriori Error Estimation in Finite Element Analysis*, John Wiley, New York.

APEL, T. (1999): *Anisotropic Finite Elements: Local Estimates and Applications*, Teubner, Leipzig.

BABUSKA, I.; OSBORN, J. (1991): Eigenvalue problems, in *Handbook of Numerical Analysis, Volume* II (P.G. Ciarlet & J.L. Lions, Editors), pp. 641–787, North-Holland, Amsterdam.

BABUSKA, I.; STROUBOULIS, T. (2001): *The Finite Element Method and Its Reliability*, Oxford University Press.

BATHE, K.J. (1996): *Finite Element Procedures*, Prentice-Hall, Englewood Cliffs, NJ.

BERNADOU, M. (1995): *Finite Element Methods for Thin Shell Problems*, John Wiley, New York.

BRAMBLE, J.H.; ZHANG, X. (2000): The analysis of multigrid methods, in *Handbook of Numerical Analysis, Volume* VII (P.G. Ciarlet & J.L. Lions, Editors), pp. 173–415, North-Holland, Amsterdam.

BRAESS, D. (2001): *Finite Elements: Theory, Fast Solvers, and Applications in Solid Mechanics*, Second Edition, Cambridge University Press.

BRENNER, S.C.; SCOTT, L.R. (1994): *The Mathematical Theory of Finite Element Methods*, Springer-Verlag, Berlin.

BREZZI, F.; FORTIN, M. (1991): *Mixed and Hybrid Finite Element Methods*, Springer-Verlag, Berlin.

FREY, P.J.; GEORGE, P.L. (2000): *Mesh Generation: Application to Finite Elements*, Hermes Science Publishing, Oxford.

FUJITA, H.; SUZUKI, T. (1991): Evolution problems, in *Handbook of Numerical Analysis, Volume* II (P.G. Ciarlet & J.L. Lions, Editors), pp. 789–928, North-Holland, Amsterdam.

GIRAULT, V.; RAVIART, P.A. (1986): *Finite Element Methods for Navier–Stokes Equations*, Springer-Verlag, Berlin.

GLOWINSKI, R. (1984): *Numerical Methods for Nonlinear Variational Problems*, Springer-Verlag, Berlin.

HUGHES, T.J.R. (1987): *The Finite Element Method: Linear Static and Dynamic Finite Element Analysis*, Prentice-Hall, Englewood Cliffs, NJ.

LUCQUIN, B.; PIRONNEAU, O. (1998): *Introduction to Scientific Computing*, John Wiley, New York.

RAVIART, P.A.; THOMAS, J.M. (1983): *Introduction à l'Analyse Numérique des Equations aux Dérivées Partielles*, Masson, Paris (since 1998: Dunod, Paris).

ROBERT, J.E.; THOMAS, J.M. (1991): Mixed and hybrid methods, in *Handbook of Numerical Analysis, Volume* II (P.G. Ciarlet and J.L. Lions, Editors), pp. 523–639, North-Holland, Amsterdam.

THOMÉE, V. (1984): *Galerkin Finite Element Methods for Parabolic Problems*, Lecture Notes in Mathematics, Vol. 1054, Springer-Verlag, Berlin.

WAHLBIN, L.B. (1991): Local behavior in finite element methods, in *Handbook of Numerical Analysis, Volume* II (P.G. Ciarlet & J.L. Lions, Editors), pp. 353–522, North-Holland, Amsterdam.

WAHLBIN, L.B. (1995): *Superconvergence in Galerkin Finite Element Methods*, Lecture Notes in Mathematics, Vol. 1605, Springer-Verlag, Berlin.

ZIENKIEWICZ, O.C.; TAYLOR, R.L. (2000): *The Finite Element Method, Volume* I: *The Basis*, 5th edition, John Wiley, New York.

PREFACE

The *objective* of this book is to analyze within reasonable limits (it is not a treatise) the basic mathematical aspects of the finite element method. The book should also serve as an introduction to current research on this subject.

On the one hand, it is also intended to be a working textbook for advanced courses in Numerical Analysis, as typically taught in graduate courses in American and French universities. For example, it is the author's experience that a one-semester course (on a three-hour per week basis) can be taught from Chapters 1, 2 and 3 (with the exception of Section 3.3), while another one-semester course can be taught from Chapters 4 and 6.

On the other hand, it is hoped that this book will prove to be useful for researchers interested in advanced aspects of the numerical analysis of the finite element method. In this respect, Section 3.3, Chapters 5, 7 and 8, and the sections on "Additional Bibliography and Comments" should provide many suggestions for conducting seminars.

Although the emphasis is mathematical, it is one of the author's wishes that some parts of the book will be of some value to engineers, whose familiar objects are perhaps seen from a different viewpoint. Indeed, in the selection of topics, we have been careful in considering only actual problems and we have likewise restricted ourselves to finite element methods which are actually used in contemporary engineering applications.

The *prerequisites* consist essentially in a good knowledge of Analysis and Functional Analysis, notably: Hilbert spaces, Sobolev spaces, and Differential Calculus in normed vector spaces. Apart from these preliminaries and some results on elliptic boundary value problems (regularity properties of the solutions, for example), the book is mathematically self-contained.

The *main topics covered* are the following:

Description and mathematical analysis of linear second- and fourth-order boundary value problems which are typically found in elasticity theory: System of equations of two-dimensional and three-dimensional elasticity, problems in the theory of membranes, thin plates, arches, thin shells (Chapters 1 and 8).

Description and mathematical analysis of some nonlinear second-order boundary value problems, such as the obstacle problem (and more generally problems modeled by variational inequalities), the minimal surface problem, problems of monotone type (Chapter 5).

Description of conforming finite element methods for solving second-order or fourth-order problems (Chapter 2).

Analysis of the convergence properties of such methods for second-order problems, including the uniform convergence (Chapter 3), and fourth-order problems (Section 6.1).

Description and convergence analysis of finite element methods with numerical integration (Section 4.1).

Description and convergence analysis of nonconforming finite element methods for second-order problems (Section 4.2) and fourth-order problems (Section 6.2).

Description and interpolation theory for isoparametric finite elements (Section 4.3).

Description and convergence analysis of the combined use of isoparametric finite elements and numerical integration for solving second-order problems over domains with curved boundaries (Section 4.4).

Convergence analysis of finite element approximations of some nonlinear problems (Chapter 5).

Description and convergence analysis of a mixed finite element method for solving the biharmonic problem, with an emphasis on duality theory, especially as regards the solution of the associated discrete problem (Chapter 7).

Description and convergence analysis of finite element methods for arches and shells, including an analysis of the approximation of the geometry by curved and flat elements (Chapter 8).

For more detailed information, the reader should consult the Introductions of the Chapters.

It is also appropriate to comment on some of the *omitted topics*. As suggested by the title, we have restricted ourselves to elliptic problems, and this restriction is obviously responsible for the omission of finite element methods for time-dependent problems, a subject which would require another volume. In fact, for such problems, the content of this book should amply suffice for those aspects of the theory which are directly related to the finite element method. The additional analysis, due to the change in the nature of the partial differential equation, requires functional analytic tools of a different nature.

The main omissions within the realm of elliptic boundary value problems concern the so-called hybrid and equilibrium finite element methods, and also mixed methods other than that described in Chapter 7. There are basically two reasons behind these omissions: First, the basic theory for such methods was not yet in a final form by the time this book was completed. Secondly, these methods form such wide and expanding a topic that their inclusion would have required several additional chapters. Other notable omissions are finite element methods for approximating the solution of particular problems, such as problems on unbounded domains, Stokes and Navier–Stokes problems and eigenvalue problems.

Nevertheless, introductions to, and references for, the topics mentioned in the above paragraph are given in the sections titled "Additional Bibliography and Comments".

As a rule, all topics which would have required further analytic tools (such as nonintegral Sobolev spaces for instance) have been deliberately omitted.

Many results are left as exercises, which is not to say that they should be systematically considered less important than those proved in the text (their inclusion in the text would have meant a much longer book).

The book comprises *eight chapters*. Chapter n, $1 \leq n \leq 8$, contains an *introduction*, several *sections* numbered Section $n.1$, Section $n.2$, etc..., and a section "Bibliography and Comments", sometimes followed by a section "Additional Bibliography and Comments". Theorems, remarks, formulas, figures, and exercises, found in each section are numbered with a three-number system. Thus the second theorem of Section 3.2 is "Theorem 3.3.3", the fourth remark in Section 4.4 is "Remark 4.4.4", the twelfth formula of Section 8.3 is numbered (8.3.12) etc.... The end of a theorem or of a remark is indicated by the symbol □.

Since the sections (which correspond to a logical subdivision of the text) may vary considerably in length, unnumbered subtitles have been added in each section to help the reader (they appear in the table of contents).

The *theorems* are intended to represent important results. Their number have been kept to a minimum, and there are no lemmas, propositions, or corollaries. This is why the proofs of the theorems are sometimes fairly long. In principle, one can skip the *remarks* during a first reading. When a term is defined, it is set in italics. Terms which are only given a loose or intuitive meaning are put between quotation marks. There are very few references in the body of the text. All relevant bibliographical material is instead indicated in the sections "Bibliography and Comments" and "Additional Bibliography and Comments".

Underlying the writing of this book, there has been a deliberate attempt to put an emphasis on *pedagogy*. In particular:

All pertinent prerequisite material is clearly delineated and kept to a minimum. It is introduced only when needed.

Complete proofs are generally given. However, some technical results or proofs which resemble previous proofs are occasionally left to the reader.

The chapters are written in such a way that it should not prove too hard for a reader already reasonably familiar with the finite element method to read a given chapter almost independently of the previous chapters. Of course, this is at the expense of some redundancies, which are purposefully included. For the same reason, the index, the glossary of symbols and the interdependence table should be useful.

It is in particular with an eye towards classroom use and self-study that *exercises* of varying difficulty are included at the end of the sections. Some exercises are easy and are simply intended to help the reader in getting a better understanding of the text. More challenging problems (which are generally provided with hints and/or references) often concern significant extensions of the material of the text (they generally comprise several questions, numbered (i), (ii), . . .).

In most sections, a significant amount of material (generally at the beginning) is devoted to the introductive and descriptive aspects of the topic under consideration.

Many *figures* are included, which hopefully will help the reader. Indeed, it is the author's opinion that one of the most fascinating aspects of the finite element method is that it entails a rehabilitation of old-fashioned "classical" geometry (considered as completely obsolete, it has almost disappeared in the curriculae of French secondary schools).

There was no systematic attempt to compile an exhaustive *bibliography*. In particular, most references before 1970 and/or from the engineering literature and/or from Eastern Europe are not quoted. The interested reader is referred to the bibliography of Whiteman (1975). An effort was made, however, to include the *most recent* references (published or unpublished) of which the author was aware, as of October, 1976.

In attributing proper names to some finite elements and theorems, we have generally simply followed the common usages in French universities, and we hope that these choices will not stir up controversies. Our purpose was not to take issues but rather to give due credit to some of those who are clearly responsible for the invention, or the mathematical justification of, some aspects of the finite element method.

For providing a very stimulating and challenging scientific atmosphere, I wish to thank all my colleagues of the Laboratoire d'Analyse Numérique at the Université Pierre et Marie Curie, particularly Pierre-Arnaud Raviart and Roland Glowinski. Above all, it is my pleasure to express my very deep gratitude to Jacques-Louis Lions, who is responsible for the creation of this atmosphere, and to whom I personally owe so much.

For their respective invitations to Bangalore and Montréal, I express my sincere gratitude to Professor K.G. Ramanathan and to Professor A. Daigneault. Indeed, this book is an outgrowth of Lectures which I was privileged to give in Bangalore as part of the "Applied Mathematics Programme" of the Tata Institute of Fundamental Research, Bombay, and at the University of Montréal, as part of the "Séminaire de Mathématiques Supérieures".

For various improvements, such as shorter proofs and better exposition at various places, I am especially indebted to J. Tinsley Oden, Vidar Thomée, Annie Puech-Raoult and Michel Bernadou, who have been kind enough to entirely read the manuscript.

For kindly providing me with computer graphics and drawings of actual triangulations, I am indebted to Professors J. H. Argyris, C. Felippa, R. Glowinski and O. C. Zienkiewicz, and to the Publishers who authorized the reprinting of these figures.

For their understanding and kind assistance as regards the material realization of this book, sincere thanks are due to Mrs. Damperat, Mrs. Theis and Mr. Riguet.

For their expert, diligent, and especially fast, typing of the entire manuscript, I very sincerely thank Mrs. Bugler and Mrs. Guille.

For a considerable help in proofreading and in the general elaboration of the manuscript, and for a permanent comprehension in spite of a finite, but large, number of lost week-ends and holidays, I deeply thank the one to whom this book is dedicated.

The author welcomes in advance all comments, suggestions, criticisms, etc.

December 1976

Philippe G. Ciarlet

GENERAL PLAN AND INTERDEPENDENCE TABLE

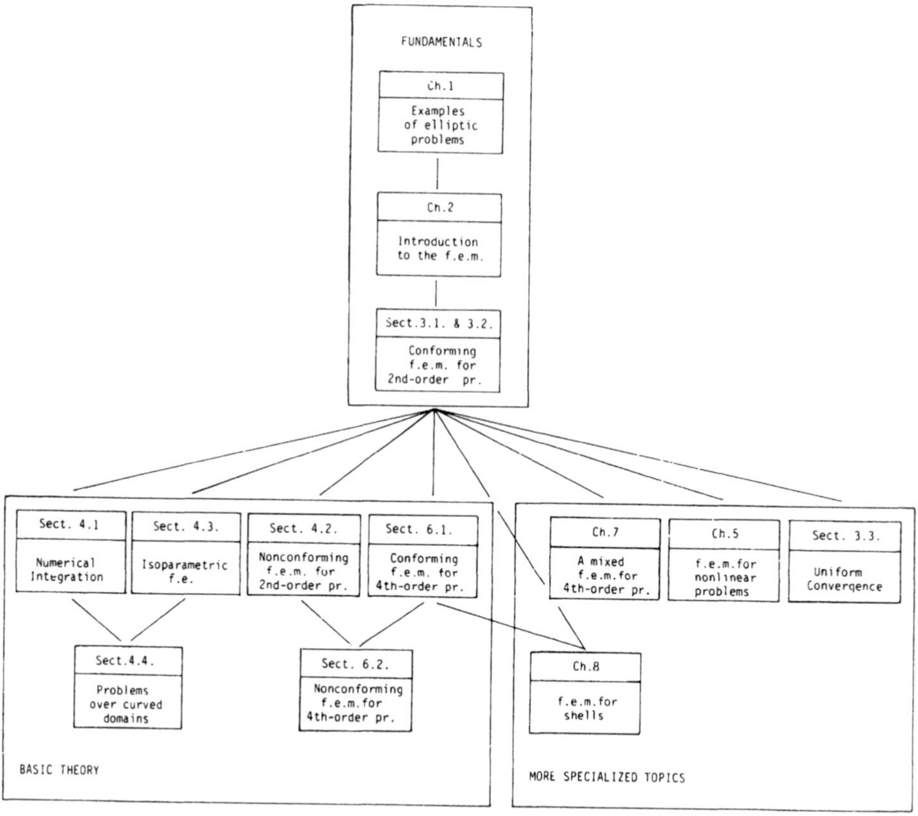

"A mathematician's nightmare is a sequence n_ϵ that tends to 0 as ϵ becomes infinite."

Paul R. HALMOS: *How to Write Mathematics*, A.M.S., 1973.

CHAPTER 1

ELLIPTIC BOUNDARY VALUE PROBLEMS

Introduction

Many problems in elasticity are mathematically represented by the following minimization problem: The unknown u, which is the *displacement* of a *mechanical system*, satisfies

$$u \in U \quad \text{and} \quad J(u) = \inf_{v \in U} J(v),$$

where the set U of *admissible displacements* is a closed convex subset of a Hilbert space V, and the *energy* J of the system takes the form

$$J(v) = \tfrac{1}{2}a(v, v) - f(v),$$

where $a(\cdot, \cdot)$ is a symmetric bilinear form and f is a linear form, both defined and continuous over the space V. In Section 1.1, we first prove a general existence result (Theorem 1.1.1), the main assumptions being the *completeness* of the space V and the *V-ellipticity* of the bilinear form. We also describe other formulations of the same problem (Theorem 1.1.2), known as its *variational formulations*, which, in the absence of the assumption of symmetry for the bilinear form, make up *variational problems* on their own. For such problems, we give an existence theorem when $U = V$ (Theorem 1.1.3), which is the well-known *Lax–Milgram lemma*.

All these problems are called *abstract problems* inasmuch as they represent an "abstract" formulation which is common to many examples, such as those which are examined in Section 1.2.

From the analysis made in Section 1.1, a candidate for the space V must have the following properties: It must be complete on the one hand, and it must be such that the expression $J(v)$ is well-defined for all functions $v \in V$ on the other hand (V is a "space of finite energy"). The *Sobolev spaces* fulfill these requirements. After briefly mentioning some of their properties (other properties will be introduced in later sections, as needed), we examine in Section 1.2 specific examples of the abstract problems of Section 1.1, such as the *membrane problem*, the *clamped plate problem*, and the *system of equations of linear elasticity*, which is by far the most significant example. Indeed, even though throughout this book we will often find it convenient to work with the simpler looking problems described at the beginning of Section 1.2, it must not be forgotten that these are essentially convenient *model problems* for the system of linear elasticity.

Using various *Green's formulas* in Sobolev spaces, we show that when solving these problems, one solves, at least *formally*, elliptic boundary value problems of the second and fourth order posed in the classical way.

1.1. Abstract problems

The symmetric case. Variational inequalities

All functions and vector spaces considered in this book are real.

Let there be given a normed vector space V with norm $\|\cdot\|$, a *continuous* bilinear form $a(\cdot,\cdot): V \times V \to \mathbf{R}$, a *continuous* linear form $f: V \to \mathbf{R}$ and a non empty subset U of the space V. With these data we associate an *abstract minimization problem*: Find an element u such that

$$u \in U \quad \text{and} \quad J(u) = \inf_{v \in U} J(v), \tag{1.1.1}$$

where the functional $J: V \to \mathbf{R}$ is defined by

$$J: v \in V \to J(v) = \tfrac{1}{2}a(v, v) - f(v). \tag{1.1.2}$$

As regards existence and uniqueness properties of the solution of this problem, the following result is essential.

Theorem 1.1.1. *Assume in addition that*

(i) *the space V is complete,*
(ii) *U is a closed convex subset of V,*
(iii) *the bilinear form $a(\cdot,\cdot)$ is symmetric and V-elliptic, in the sense that*

$$\exists \alpha > 0, \quad \forall v \in V, \quad \alpha\|v\|^2 \leqslant a(v, v). \tag{1.1.3}$$

Then the abstract minimization problem (1.1.1) *has one and only one solution.*

Proof. The bilinear form $a(\cdot,\cdot)$ is an inner product over the space V, and the associated norm is equivalent to the given norm $\|\cdot\|$. Thus the space V is a Hilbert space when it is equipped with this inner product. By the Riesz representation theorem, there exists an element $\sigma f \in V$ such that

$$\forall v \in V, \quad f(v) = a(\sigma f, v),$$

so that, taking into account the symmetry of the bilinear form, we may rewrite the functional as

$$J(v) = \tfrac{1}{2}a(v, v) - a(\sigma f, v) = \tfrac{1}{2}a(v - \sigma f, v - \sigma f) - \tfrac{1}{2}a(\sigma f, \sigma f).$$

Hence solving the abstract minimization problem amounts to minimizing the distance between the element σf and the set U, with respect to the norm $\sqrt{a(\cdot,\cdot)}$. Consequently, the solution is simply the projection of the element σf onto the set U, with respect to the inner product $a(\cdot,\cdot)$. By the projection theorem, such a projection exists and is unique, since U is a closed convex subset of the space V. □

Next, we give equivalent formulations of this problem.

Theorem 1.1.2. *An element u is the solution of the abstract minimization problem* (1.1.1) *if and only if it satisfies the relations*

$$u \in U \quad \textit{and} \quad \forall v \in U, \quad a(u, v - u) \geq f(v - u), \tag{1.1.4}$$

in the general case, or

$$u \in U \quad \textit{and} \quad \begin{cases} \forall v \in U, & a(u, v) \geq f(v), \\ & a(u, u) = f(u), \end{cases} \tag{1.1.5}$$

if U is a closed convex cone with vertex 0, *or*

$$u \in U \quad \textit{and} \quad \forall v \in U, \quad a(u, v) = f(v), \tag{1.1.6}$$

if U is a closed subspace.

Proof. The projection u is completely characterized by the relations

$$u \in U \quad \text{and} \quad \forall v \in U, \quad a(\sigma f - u, v - u) \leq 0, \tag{1.1.7}$$

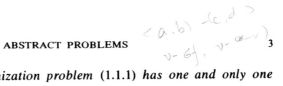

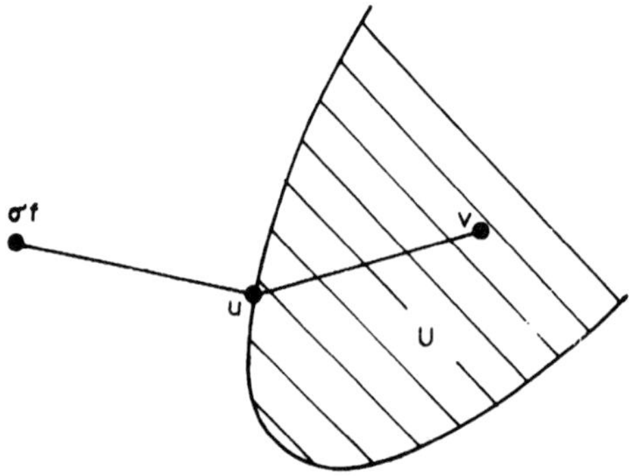

Fig. 1.1.1

the geometrical interpretation of the last inequalities being that the angle between the vectors $(\sigma f - u)$ and $(v - u)$ is obtuse (Fig. 1.1.1) for all $v \in U$. These inequalities may be written as

$$\forall v \in U, \quad a(u, v-u) \geqslant a(\sigma f, v-u) = f(v-u),$$

which proves relations (1.1.4).

Assume next U is a closed convex cone with vertex 0. Then the point $(u+v)$ belongs to the set U whenever the point v belongs to the set U (Fig. 1.1.2).

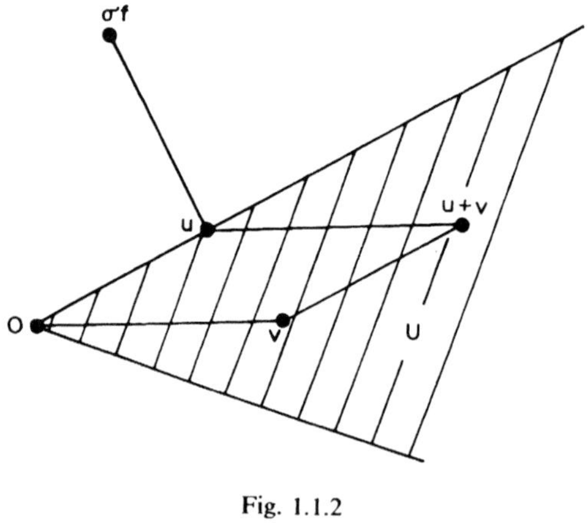

Fig. 1.1.2

Therefore, upon replacing v by $(u+v)$ in inequalities (1.1.4), we obtain the inequalities

$$\forall v \in U, \quad a(u, v) \geqslant f(v),$$

so that, in particular, $a(u, u) \geqslant f(u)$. Letting $v=0$ in (1.1.4), we obtain $a(u, u) \leqslant f(u)$, and thus relations (1.1.5) are proved. The converse is clear.

If U is a subspace (Fig. 1.1.3), then inequalities (1.1.5) written with v and $-v$ yield $a(u, v) \geqslant f(v)$ and $a(u, v) \leqslant f(v)$ for all $v \in U$, from which relations (1.1.6) follow. Again the converse is clear. □

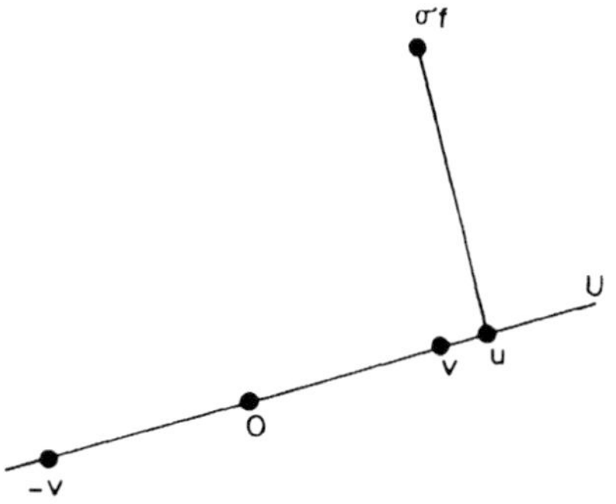

Fig. 1.1.3

The characterizations (1.1.4), (1.1.5) and (1.1.6) are called *variational formulations* of the original minimization problem, the equations (1.1.6) are called *variational equations*, and the inequalities of (1.1.4) and (1.1.5) are referred to as *variational inequalities*. The terminology "variational" will be justified in Remark 1.1.2.

Remark 1.1.1. Since the projection mapping is linear if and only if the subset U is a subspace, it follows that *problems associated with variational inequalities are generally non linear*, the linearity or non linearity being that of the mapping $f \in V' \to u \in V$, where V' is the dual space of V, all other data being fixed. One should not forget, however, that if the resulting problem is linear when one minimizes over a subspace this is also because the functional is *quadratic* i.e., it is of the form (1.1.2). The minimization of more general functionals over a subspace would correspond to nonlinear problems (cf. Section 5.3). □

Remark 1.1.2. *The variational formulations of Theorem* 1.1.2 *may be also interpreted from the point of view of Differential Calculus,* as follows. We first observe that the functional J is differentiable at every point $u \in V$, its (Fréchet) derivative $J'(u) \in V'$ being such that

$$\forall v \in V, \quad J'(u)v = a(u, v) - f(v). \tag{1.1.8}$$

Let then u be the solution of the minimization problem (1.1.1), and let $v = u + w$ be any point of the convex set U. Since the points $(u + \theta w)$ belong to the set U for all $\theta \in [0, 1]$ (Fig. 1.1.4), we have, by definition of

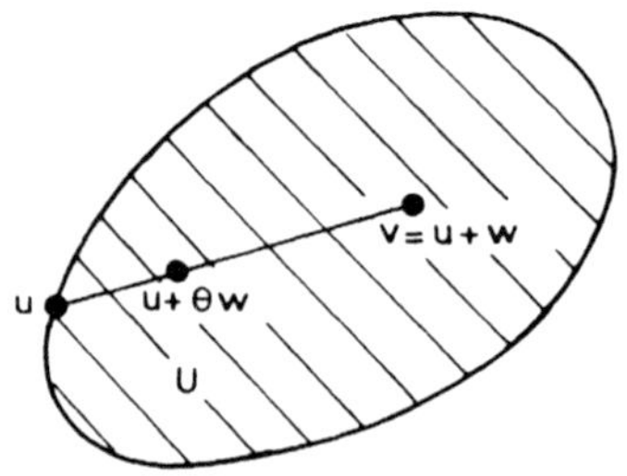

Fig. 1.1.4

the derivative $J'(u)$,

$$0 \leqslant J(u + \theta w) - J(u) = \theta J'(u)w + \theta \|w\| \epsilon(\theta)$$

for all $\theta \in [0, 1]$, with $\lim_{\theta \to 0} \epsilon(\theta) = 0$. As a consequence, we necessarily have

$$J'(u)w \geqslant 0, \tag{1.1.9}$$

since otherwise the difference $J(u + \theta w) - J(u)$ would be strictly negative for θ small enough. Using (1.1.8), inequality (1.1.9) may be rewritten as

$$J'(u)w = J'(u)(v - u) = a(u, v - u) - f(v - u) \geqslant 0,$$

which is precisely (1.1.4). Conversely, assume we have found an element $u \in U$ such that

$$\forall v \in U, \quad J'(u)(v - u) \geqslant 0. \tag{1.1.10}$$

The second derivative $J''(u) \in \mathscr{L}_2(V; \mathbf{R})$ of the functional J is independent of $u \in V$ and it is given by

$$\forall v, w \in V, \quad J''(u)(v, w) = a(v, w). \tag{1.1.11}$$

Therefore, an application of Taylor's formula for any point $v = u + w$ belonging to the set U yields

$$J(u + w) - J(u) = J'(u)(w) + \frac{1}{2}a(w, w) \geq \frac{\alpha}{2}\|w\|^2, \tag{1.1.12}$$

which shows that u is a solution of problem (1.1.1). We have $J(v) - J(u) > 0$ unless $v = u$ so that we see once again the solution is unique.

Arguing as in the proof of Theorem 1.1.2, it is an easy matter to verify that inequalities (1.1.10) are equivalent to the relations

$$\forall v \in U, \quad J'(u)v \geq 0 \quad \text{and} \quad J'(u)u = 0, \tag{1.1.13}$$

when U is a convex cone with vertex 0, alternately,

$$\forall v \in U, \quad J'(u)v = 0, \tag{1.1.14}$$

when U is a subspace. Notice that relations (1.1.13) coincide with relations (1.1.5), while (1.1.14) coincide with (1.1.6).

When $U = V$, relations (1.1.14) reduce to the familiar condition that the *first variation* of the functional J, i.e., the first order term $J'(u)w$ in the Taylor expansion (1.1.12), vanishes for all $w \in V$ when the point u is a minimum of the function $J: V \to \mathbf{R}$, this condition being also sufficient if the function J is convex, as is the case here. Therefore the various relations (1.1.4), (1.1.5) and (1.1.6), through the equivalent relations (1.1.10), (1.1.13) and (1.1.14), appear as generalizations of the previous condition, the expression $a(u, v - u) - f(v - u) = J'(u)(v - u)$, $v \in U$, playing in the present situation the role of the first variation of the functional J *relative to the convex set* U. It is in this sense that the formulations of Theorem 1.1.2 are called "variational". □

The nonsymmetric case. The Lax–Milgram lemma

Without making explicit reference to the functional J, we now define an *abstract variational problem*: Find an element u such that

$$u \in U \quad \text{and} \quad \forall v \in U, \quad a(u, v - u) \geq f(v - u), \tag{1.1.15}$$

or, find an element u such that

$$u \in U \quad \text{and} \quad \begin{cases} \forall v \in U, & a(u, v) \geq f(v), \\ a(u, u) = f(u), \end{cases} \tag{1.1.16}$$

if U is a cone with vertex 0, or, finally, find an element u such that

$$u \in U \quad \text{and} \quad \forall v \in V, \quad a(u, v) = f(v), \tag{1.1.17}$$

if U is a subspace. By Theorem 1.1.1, each such problem has one and only one solution if the space V is complete, the subset U of V is closed and convex, and the bilinear form is V-elliptic, continuous, and symmetric. If the assumption of symmetry of the bilinear form is dropped, the above variational problem still has one and only one solution (LIONS & STAMPACCHIA (1967)) if the space V is a Hilbert space, but there is no longer an associated minimization problem. Here we shall confine ourselves to the case where $U = V$.

Theorem 1.1.3 (*Lax–Milgram lemma*). *Let V be a Hilbert space, let $a(\cdot, \cdot)$: $V \times V \to \mathbf{R}$ be a continuous V-elliptic bilinear form, and let f: $V \to \mathbf{R}$ be a continuous linear form.*

Then the abstract variational problem: Find an element u such that

$$u \in V \quad \text{and} \quad \forall v \in V, \quad a(u, v) = f(v), \tag{1.1.18}$$

has one and only one solution.

Proof. Let M be a constant such that

$$\forall u, v \in V, \quad |a(u, v)| \leq M\|u\| \, \|v\|. \tag{1.1.19}$$

For each $u \in V$, the linear form $v \in V \to a(u, v)$ is continuous and thus there exists a unique element $Au \in V'$ (V' is the dual space of V) such that

$$\forall v \in V, \quad a(u, v) = Au(v). \tag{1.1.20}$$

Denoting by $\|\cdot\|^*$ the norm in the space V', we have

$$\|Au\|^* = \sup_{v \in V} \frac{|Au(v)|}{\|v\|} \leq M\|u\|.$$

Consequently, the linear mapping A: $V \to V'$ is continuous, with

$$\|A\|_{\mathscr{L}(V;\, V')} \leq M. \tag{1.1.21}$$

Let τ: $V' \to V$ denote the Riesz mapping which is such that, by definition,

$$\forall f \in V', \quad \forall v \in V, \quad f(v) = ((\tau f, v)), \tag{1.1.22}$$

$((\cdot,\cdot))$ denoting the inner product in the space V. Then solving the variational problem (1.1.18) is equivalent to solving the equation $\tau A u = \tau f$. We will show that this equation has one and only one solution by showing that, for appropriate values of a parameter $\rho > 0$, the affine mapping

$$v \in V \to v - \rho(\tau A v - \tau f) \in V \tag{1.1.23}$$

is a contraction. To see this, we observe that

$$\begin{aligned}\|v - \rho\tau A v\|^2 &= \|v\|^2 - 2\rho((\tau A v, v)) + \rho^2\|\tau A v\|^2 \\ &\leqslant (1 - 2\rho\alpha + \rho^2 M^2)\|v\|^2,\end{aligned}$$

since, using inequalities (1.1.3) and (1.1.21),

$$\begin{aligned}((\tau A v, v)) = A v(v) = a(v, v) \geqslant \alpha\|v\|^2, \\ \|\tau A v\| = \|A v\|^* \leqslant \|A\|\,\|v\| \leqslant M\|v\|.\end{aligned}$$

Therefore the mapping defined in (1.1.23) is a contraction whenever the number ρ belongs to the interval $]0, 2\alpha/M^2[$ and the proof is complete. □

Remark 1.1.3. It follows from the previous proof that the mapping $A\colon V \to V'$ is onto. Since

$$\alpha\|u\|^2 = a(u, u) = f(u) \leqslant \|f\|^*\|u\|,$$

the mapping A has a continuous inverse A^{-1}, with

$$\|A^{-1}\|_{\mathscr{L}(V';\,V)} \leqslant \frac{1}{\alpha}.$$

Therefore the variational problem (1.1.18) is *well-posed* in the sense that its solution *exists*, *is unique*, and *depends continuously on the data f* (all other data being fixed). □

Exercises

1.1.1. Show that if u_i, $i = 1, 2$, are the solutions of minimization problems (1.1.1) corresponding to linear form $f_i \in V'$, $i = 1, 2$, then

$$\|u_1 - u_2\| \leqslant \frac{1}{\alpha}\|f_1 - f_2\|^*.$$

(i) Give a proof which uses the norm reducing property of the projection operator.

(ii) Give another proof which also applies to the variational problem (1.1.15).

1.1.2. The purpose of this exercise is to give an alternate proof of the Lax–Milgram lemma (Theorem 1.1.3). As in the proof given in the text, one first establishes that the mapping $\mathscr{A} = \tau \cdot A: V \to V$ is continuous with $\|\mathscr{A}\| \leq M$, and that $\alpha\|v\| \leq \|\mathscr{A}v\|$ for all $v \in V$. It remains to show that $\mathscr{A}(V) = V$.

(i) Show that $\mathscr{A}(V)$ is a closed subspace of V.

(ii) Show that the orthogonal complement of $\mathscr{A}(V)$ in the space V is reduced to $\{0\}$.

1.2. Examples of elliptic boundary value problems

The Sobolev spaces $H^m(\Omega)$. Green's formulas

Let us first briefly recall some results from Differential Calculus. Let there be given two normed vector spaces X and Y and a function $v: A \to Y$, where A is a subset of X. If the function is k times differentiable at a point $a \in A$, we shall denote $D^k v(a)$, or simply $Dv(a)$ if $k = 1$, its k-th (*Fréchet*) *derivative*. It is a symmetric element of the space $\mathscr{L}_k(X; Y)$, whose norm is given by

$$\|D^k v(a)\| = \sup_{\substack{\|h_i\| \leq 1 \\ 1 \leq i \leq k}} \|D^k v(a)(h_1, h_2, \ldots, h_k)\|.$$

We shall also use the alternate notations $Dv(a) = v'(a)$ and $D^2 v(a) = v''(a)$.

In the special case where $X = \mathbf{R}^n$ and $Y = \mathbf{R}$, let e_i, $1 \leq i \leq n$, denote the canonical basis vectors of $\mathbf{R}^n$. Then the usual partial derivatives will be denoted by, and are given by, the following:

$$\partial_i v(a) = Dv(a)e_i,$$
$$\partial_{ij} v(a) = D^2 v(a)(e_i, e_j),$$
$$\partial_{ijk} v(a) = D^3 v(a)(e_i, e_j, e_k), \text{ etc.} \ldots$$

Occasionally, we shall use the notation $\nabla v(a)$, or grad $v(a)$, to denote the *gradient* of the function v at the point a, i.e., the vector in $\mathbf{R}^n$ whose components are the partial derivatives $\partial_i v(a)$, $1 \leq i \leq n$.

We shall also use the *multi-index notation*: Given a multi-index $\alpha = (\alpha_1, \alpha_2, \dots, \alpha_n) \in \mathbf{N}^n$, we let $|\alpha| = \Sigma_{i=1}^n \alpha_i$. Then the partial derivative $\partial^\alpha v(a)$ is the result of the application of the $|\alpha|$-th derivative $D^{|\alpha|}v(a)$ to any $|\alpha|$-vector of $(\mathbf{R}^n)^{|\alpha|}$ where each vector e_i occurs α_i times, $1 \leq i \leq n$. For instance, if $n = 3$, we have $\partial_1 v(a) = \partial^{(1,0,0)}v(a)$, $\partial_{123}v(a) = \partial^{(1,1,1)}v(a)$, $\partial_{111}v(a) = \partial^{(3,0,0)}v(a)$, etc. . .

There exist constants $C(m, n)$ such that for any partial derivative $\partial^\alpha v(a)$ with $|\alpha| = m$ and any function v,

$$|\partial^\alpha v(a)| \leq \|D^m v(a)\| \leq C(m, n) \max_{|\alpha|=m} |\partial^\alpha v(a)|,$$

where it is understood that the space $\mathbf{R}^n$ is equipped with the Euclidean norm.

As a rule, we shall represent by symbols such as $D^k v$, v'', $\partial_i v$, $\partial^\alpha v$, etc. . . , the *functions* associated with any derivative or partial derivative.

When $h_1 = h_2 = \cdots = h_k = h$, we shall simply write

$$D^k v(a)(h_1, h_2, \dots, h_k) = D^k v(a)h^k.$$

Thus, given a real-valued function v, *Taylor's formula of order* k is written as

$$v(a + h) = v(a) + \sum_{l=1}^{k} \frac{1}{l!} D^l v(a)h^l + \frac{1}{(k+1)!} D^{k+1}v(a + \theta h)h^{k+1},$$

for some $\theta \in]0, 1[$ (whenever such a formula applies).

Given a bounded open subset Ω in $\mathbf{R}^n$, the space $\mathcal{D}(\Omega)$ consists of all indefinitely differentiable functions $v: \Omega \to \mathbf{R}$ with compact support.

For each integer $m \geq 0$, the *Sobolev space* $H^m(\Omega)$ consists of those functions $v \in L^2(\Omega)$ for which all partial derivatives $\partial^\alpha v$ (in the distribution sense), with $|\alpha| \leq m$, belong to the space $L^2(\Omega)$, i.e., for each multi-index α with $|\alpha| \leq m$, there exists a function $\partial^\alpha v \in L^2(\Omega)$ which satisfies

$$\forall \phi \in \mathcal{D}(\Omega), \quad \int_\Omega \partial^\alpha v\, \phi \,\mathrm{d}x = (-1)^{|\alpha|} \int_\Omega v \partial^\alpha \phi \,\mathrm{d}\chi. \tag{1.2.1}$$

Equipped with the norm

$$\|v\|_{m,\Omega} = \Big(\sum_{|\alpha| \leq m} \int_\Omega |\partial^\alpha v|^2 \,\mathrm{d}x\Big)^{1/2},$$

the space $H^m(\Omega)$ *is a Hilbert space.* We shall also make frequent use of the semi-norm

$$|v|_{m,\Omega} = \Big(\sum_{|\alpha| = m} \int_\Omega |\partial^\alpha v|^2 \,\mathrm{d}x\Big)^{1/2}.$$

We define the *Sobolev space*

$$H_0^m(\Omega) = (\mathcal{D}(\Omega))^-,$$

the closure being understood in the sense of the norm $\|\cdot\|_{m,\Omega}$.

When the set Ω is bounded, there exists a constant $C(\Omega)$ such that

$$\forall v \in H_0^1(\Omega), \quad |v|_{0,\Omega} \leq C(\Omega)|v|_{1,\Omega}, \tag{1.2.2}$$

this inequality being known as the *Poincaré–Friedrichs inequality*.

Therefore, *when the set Ω is bounded, the semi-norm $|\cdot|_{m,\Omega}$ is a norm over the space $H_0^m(\Omega)$, equivalent to the norm $\|\cdot\|_{m,\Omega}$* (another way of reaching the same conclusion is indicated in the proof of Theorem 1.2.1 below).

The next definition will be sufficient for most subsequent purposes whenever some smoothness of the boundary is needed. It allows the consideration of all commonly encountered shapes without cusps. Following NEČAS (1967), we say that an open set Ω has a *Lipschitz-continuous boundary Γ* if the following conditions are fulfilled: There exist constants $\alpha > 0$ and $\beta > 0$, and a finite number of local coordinate systems and local maps a_r, $1 \leq r \leq R$, which are *Lipschitz-continuous on their respective domains of definitions* $\{\hat{x}^r \in \mathbf{R}^{n-1};\ |\hat{x}^r| \leq \alpha\}$, such that (Fig. 1.2.1):

$$\Gamma = \bigcup_{r=1}^{R} \{(x_1^r, \hat{x}^r);\ x_1^r = a_r(\hat{x}^r), \quad |\hat{x}^r| < \alpha\},$$

$$\{(x_1^r, \hat{x}^r); \quad a_r(\hat{x}^r) < x_1^r < a_r(\hat{x}^r) + \beta; \quad |\hat{x}^r| < \alpha\} \subset \Omega, \quad 1 \leq r \leq R,$$

$$\{(x_1^r, \hat{x}^r); \quad a_r(\hat{x}^r) - \beta < x_1^r < a_r(\hat{x}^r); \quad |\hat{x}^r| < \alpha\} \subset \complement\bar{\Omega}, \quad 1 \leq r \leq R,$$

where $\hat{x}^r = (x_2^r, \ldots, x_n^r)$, and $|\hat{x}^r| < \alpha$ stands for $|x_i^r| < \alpha$, $2 \leq i \leq n$. Notice in passing that *an open set with a Lipschitz-continuous boundary is bounded.*

Occasionally, we shall also need the following definitions: A boundary is *of class $\mathscr{X}$* if the functions a_r: $|\hat{x}^r| \leq \alpha \to \mathbf{R}$ are of class $\mathscr{X}$ (such as $\mathscr{C}^m$ or $\mathscr{C}^{m,\alpha}$), and a boundary is said to be *sufficiently smooth* if it is of class $\mathscr{C}^m$, or $\mathscr{C}^{m,\alpha}$, for sufficient high values of m, or m and α (for a given problem).

In the remaining part of this section, it will be always understood that Ω is an open subset in $\mathbf{R}^n$ with a Lipschitz-continuous boundary. This being the case, a superficial measure, which we shall denote $d\gamma$, can be

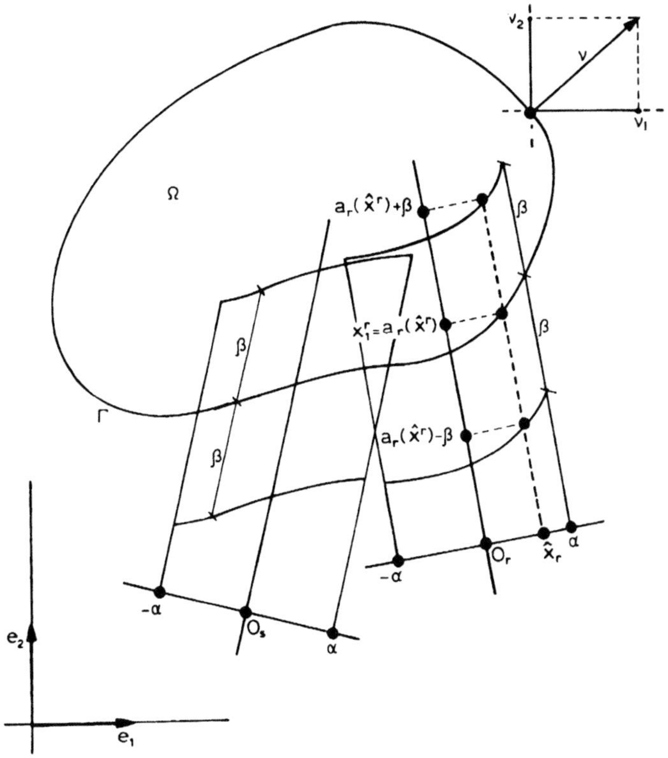

Fig. 1.2.1

defined along the boundary, so that it makes sense to consider the spaces $L^2(\Gamma)$, whose norm shall be denoted $\|\cdot\|_{L^2(\Gamma)}$.

Then it can be proved that there exists a constant $C(\Omega)$ such that

$$\forall v \in \mathscr{C}^\infty(\bar{\Omega}), \quad \|v\|_{L^2(\Gamma)} \leq C(\Omega)\|v\|_{1,\Omega}. \tag{1.2.3}$$

Since in this case $(\mathscr{C}^\infty(\bar{\Omega}))^- = H^1(\Omega)$, the closure being understood in the sense of the norm $\|\cdot\|_{1,\Omega}$, there exists a continuous linear mapping tr: $v \in H^1(\Omega) \to \text{tr } v \in L^2(\Gamma)$, which is called the *trace operator*. However when no confusion should arise, we shall simply write tr $v = v$. The *following characterization holds*:

$$H_0^1(\Omega) = \{v \in H^1(\Omega); \quad v = 0 \quad \text{on} \quad \Gamma\}.$$

Since the unit outer normal $\nu = (\nu_1, \ldots, \nu_n)$ (Fig. 1.2.1) exists almost everywhere along Γ, the (outer) *normal derivative operator*:

$$\partial_\nu = \sum_{i=1}^{n} \nu_i \partial_i$$

is defined almost everywhere along Γ for smooth functions. Extending its definition to $\partial_\nu = \sum_{i=1}^{n} \nu_i \operatorname{tr} \partial_i$ for functions in the space $H^2(\Omega)$, *the following characterization holds*:

$$H_0^2(\Omega) = \{v \in H^2(\Omega);\quad v = \partial_\nu v = 0 \quad \text{on} \quad \Gamma\}.$$

Given two functions $u, v \in H^1(\Omega)$, *the following fundamental Green's formula*

$$\int_\Omega u \partial_i v \,\mathrm{d}x = -\int_\Omega \partial_i u\, v \,\mathrm{d}x + \int_\Gamma u\, v \nu_i \,\mathrm{d}\gamma \tag{1.2.4}$$

holds for any $i \in [1, n]$. From this formula, other *Green's formulas* may be easily deduced. For example, replacing u by $\partial_i u$ and taking the sum from 1 to n, we get

$$\int_\Omega \sum_{i=1}^{n} \partial_i u \partial_i v \,\mathrm{d}x = -\int_\Omega \Delta u\, v \,\mathrm{d}x + \int_\Gamma \partial_\nu u\, v \,\mathrm{d}\gamma \tag{1.2.5}$$

for all $u \in H^2(\Omega)$, $v \in H^1(\Omega)$. As a consequence, we obtain by subtraction:

$$\int_\Omega (u \Delta v - \Delta u\, v) \,\mathrm{d}x = \int_\Gamma (u \partial_\nu v - \partial_\nu u\, v) \,\mathrm{d}\gamma \tag{1.2.6}$$

for all $u, v \in H^2(\Omega)$. Replacing u by Δu in formula (1.2.6), we obtain

$$\int_\Omega \Delta u \Delta v \,\mathrm{d}x = \int_\Omega \Delta^2 u\, v \,\mathrm{d}x - \int_\Gamma \partial_\nu \Delta u\, \nu \,\mathrm{d}\gamma + \int_\Gamma \Delta u \partial_\nu n \,\mathrm{d}\gamma \tag{1.2.7}$$

for all $u \in H^4(\Omega)$, $v \in H^2(\Omega)$. As another application of formula (1.2.4), let us prove the relation

$$\forall v \in H_0^2(\Omega), \quad |\Delta v|_{0,\Omega} = |v|_{2,\Omega}, \tag{1.2.8}$$

which implies that, *over the space* $H_0^2(\Omega)$, *the semi-norm* $v \to |\Delta v|_{0,\Omega}$ *is a norm, equivalent to the norm* $\|\cdot\|_{2,\Omega}$: We have, by definition,

$$|v|_{2,\Omega}^2 = \int_\Omega \Big\{ \sum_i (\partial_{ii} v)^2 + \sum_{i \neq j} (\partial_{ij} v)^2 \Big\} \,\mathrm{d}x,$$

$$|\Delta v|_{0,\Omega}^2 = \int_\Omega \Big\{ \sum_i (\partial_{ii} v)^2 + \sum_{i \neq j} \partial_{ii} v \partial_{jj} v \Big\} \,\mathrm{d}x.$$

Clearly, it suffices to prove relations (1.2.8) for all functions $v \in \mathcal{D}(\Omega)$. For these functions we have

$$\int_\Omega (\partial_{ij}v)^2 \,\mathrm{d}x = -\int_\Omega \partial_i v \partial_{ijj} v \,\mathrm{d}x = \int_\Omega \partial_{ii} v \partial_{jj} v \,\mathrm{d}x,$$

as two applications of Green's formula (1.2.4) show, and thus (1.2.8) is proved.

For $n = 2$, let $\tau = (\tau_1, \tau_2)$ denote the unit tangential vector along the boundary Γ, oriented in the usual way. In addition to the normal derivative operator ∂_ν, we introduce the differential operators ∂_τ, $\partial_{\nu\tau}$, $\partial_{\tau\tau}$ defined by

$$\partial_\tau v(a) = Dv(a)\tau = \sum_{i=1}^{2} \tau_i \partial_i v(a),$$

$$\partial_{\nu\tau} v(a) = D^2 v(a)(\nu, \tau) = \sum_{i,j=1}^{2} \nu_i \tau_j \partial_{ij} v(a),$$

$$\partial_{\tau\tau} v(a) = D^2 v(a)(\tau, \tau) = \sum_{i,j=1}^{2} \tau_i \tau_j \partial_{ij} v(a).$$

Then we shall make use of the following Green's formula, whose proof is left as an exercise (Exercise 1.2.1):

$$\int_\Omega \{2\partial_{12}u\partial_{12}v - \partial_{11}u\partial_{22}v - \partial_{22}u\partial_{11}v\} \,\mathrm{d}x = \int_\Gamma \{-\partial_{\tau\tau}u\partial_\nu v + \partial_{\nu\tau}u\partial_\tau v\} \,\mathrm{d}\gamma. \tag{1.2.9}$$

This relation holds for all functions $u \in H^3(\Omega)$, $v \in H^2(\Omega)$.

First examples of second-order boundary value problems

We next proceed to examine several examples of minimization and variational problems. According to the analysis made in Section 1.1, we need to specify for each example the space V, a subset U of the space V, a bilinear form $a(\cdot, \cdot)$: $V \times V \to \mathbf{R}$, and a linear form f: $V \to \mathbf{R}$. In fact, the examples given in this section correspond to the case where $U = V$, i.e., they all correspond to linear problems (Remark 1.1.1). A non linear problem is considered in Exercise 1.2.5, and another one will be considered in Section 5.1.

The *first example* corresponds to the following data:

$$\begin{cases} V = U = H_0^1(\Omega), \\ a(u, v) = \displaystyle\int_\Omega \Big(\sum_{i=1}^n \partial_i u \partial_i v + auv\Big)\,\mathrm{d}x, \\ f(v) = \displaystyle\int_\Omega fv\,\mathrm{d}x, \end{cases} \tag{1.2.10}$$

and to the following assumptions on the functions a and f:

$$a \in L^\infty(\Omega), \quad a \geq 0 \text{ a.e. on } \Omega, \quad f \in L^2(\Omega). \tag{1.2.11}$$

To begin with, it is clear that the symmetric bilinear form $a(\cdot, \cdot)$ is continuous since for all $u, v \in H^1(\Omega)$,

$$\begin{aligned} |a(u, v)| &\leq \sum_{i=1}^n |\partial_i u|_{0,\Omega} |\partial_i v|_{0,\Omega} + |a|_{0,\infty,\Omega} |u|_{0,\Omega} |v|_{0,\Omega} \\ &\leq \max\{1, |a|_{0,\infty,\Omega}\} \|u\|_{1,\Omega} \|v\|_{1,\Omega}, \end{aligned}$$

where $|\cdot|_{0,\Omega}$ and $|\cdot|_{0,\infty,\Omega}$ denote the norms of the space $L^2(\Omega)$ and $L^\infty(\Omega)$ respectively, and it is $H_0^1(\Omega)$-elliptic since, for all $v \in H^1(\Omega)$,

$$a(v, v) \geq \int_\Omega \sum_{i=1}^n (\partial_i v)^2\,\mathrm{d}x = |v|_{1,\Omega}^2$$

(the semi-norm $|\cdot|_{1,\Omega}$ is a norm over the space $H_0^1(\Omega)$, equivalent to the norm $\|\cdot\|_{1,\Omega}$). Next, the linear form f is continuous since for all $v \in H^1(\Omega)$,

$$|f(v)| \leq |f|_{0,\Omega} |v|_{0,\Omega} \leq |f|_{0,\Omega} \|v\|_{1,\Omega}.$$

Therefore, by Theorem 1.1.1, there exists a unique function $u \in H_0^1(\Omega)$ which minimizes the functional

$$J: v \to J(v) = \frac{1}{2} \int_\Omega \Big\{\sum_{i=1}^n (\partial_i v)^2 + av^2\Big\}\,\mathrm{d}x - \int_\Omega fv\,\mathrm{d}x \tag{1.2.12}$$

over the space $H_0^1(\Omega)$, or equivalently, by Theorem 1.1.2, which satisfies the variational equations

$$\forall v \in H_0^1(\Omega), \quad \int_\Omega \Big\{\sum_{i=1}^n \partial_i u \partial_i v + auv\Big\}\,\mathrm{d}x = \int_\Omega fv\,\mathrm{d}x. \tag{1.2.13}$$

Using these equations, we proceed to show that *we are also solving a partial differential equation in the distributional sense.* More specifically, let $\mathscr{D}'(\Omega)$ denote the *space of distributions over the set* Ω, i.e., the dual space of the space $\mathscr{D}(\Omega)$ equipped with the Schwartz topology, and let $\langle\cdot,\cdot\rangle$ denote the duality pairing between the spaces $\mathscr{D}'(\Omega)$ and $\mathscr{D}(\Omega)$. *If g is a locally integrable function over Ω, we shall identify it with the distribution g: $\phi \in \mathscr{D}(\Omega) \to \int_\Omega g\phi \,\mathrm{d}x$.*

Since the inclusion

$$\mathscr{D}(\Omega) \subset V = H_0^1(\Omega)$$

holds, the variational equations (1.2.13) are satisfied for all functions $v \in \mathscr{D}(\Omega)$. Therefore, by definition of the differentiation for distributions, we may write

$$\forall \phi \in \mathscr{D}(\Omega), \quad a(u,\phi) = \sum_{i=1}^{n} \langle \partial_i u, \partial_i \phi\rangle + \langle au, \phi\rangle = \langle -\Delta u + au, \phi\rangle.$$

Since $f(\phi) = \langle f, \phi\rangle$ for all $\phi \in \mathscr{D}(\Omega)$, it follows from the above relations that u is a solution of the partial differential equation $-\Delta u + au = f$ in $\mathscr{D}'(\Omega)$.

To sum up, the solution u of the minimization (or variational) problem associated with the data (1.2.10) is also a solution of the problem: *Find a distribution $u \in \mathscr{D}'(\Omega)$ such that*

$$u \in H_0^1(\Omega) \quad \textit{and} \quad -\Delta u + au = f \textit{ in } \mathscr{D}'(\Omega), \tag{1.2.14}$$

and conversely, if a distribution u satisfies (1.2.14), it is a solution of the original problem. To see this, we observe that the equalities

$$\forall \phi \in \mathscr{D}(\Omega), \quad a(u,\phi) = \langle -\Delta u + au, \phi\rangle = \langle f, \phi\rangle = f(\phi)$$

hold in fact for all functions $\phi \in H_0^1(\Omega)$ since $\mathscr{D}(\Omega)$ is a dense subspace of the space $H_0^1(\Omega)$.

Remembering that the functions in the space $H_0^1(\Omega)$ have a vanishing trace along Γ, we shall say that we have *formally solved* the *associated boundary value problem*

$$\begin{cases} -\Delta u + au = f & \text{in} \quad \Omega, \\ u = 0 & \text{on} \quad \Gamma. \end{cases} \tag{1.2.15}$$

Problem (1.2.15) is called a *homogeneous Dirichlet problem* for the operator $u \to -\Delta u + au$, since it is formally posed exactly as in the classical sense where, typically, one would seek a solution in the space $\mathscr{C}^0(\bar{\Omega}) \cap \mathscr{C}^2(\Omega)$. Actually, when the data are sufficiently smooth, it can be proved (but this is non trivial) that the solution of (1.2.14) is also a solution of (1.2.15) in the classical sense. Nevertheless, one should keep in mind that, in general, nothing guarantees that the partial differential equation $-\Delta u + au = f$ in Ω can be given a sense otherwise than in the space $\mathscr{D}'(\Omega)$. Likewise the boundary condition $u = 0$ on Γ cannot be understood in general in other than the sense of a vanishing trace, or even in no sense at all if the set Ω were "only" supposed to be bounded.

With $a = 0$ and $n = 2$, the problem under analysis is called the *membrane problem*: It arises when one considers the problem of finding the *equilibrium position of an elastic membrane*, with tension τ, under the action of a "vertical" force, of density $F = \tau f$, and lying in the "horizontal" plane, of coordinates (x_1, x_2), when $f = 0$, as shown in Fig. 1.2.2 (where the vertical scale is considerably distorted if it were to correspond to an actual membrane). More general situations are considered in Exercise 1.2.2 and Section 5.1.

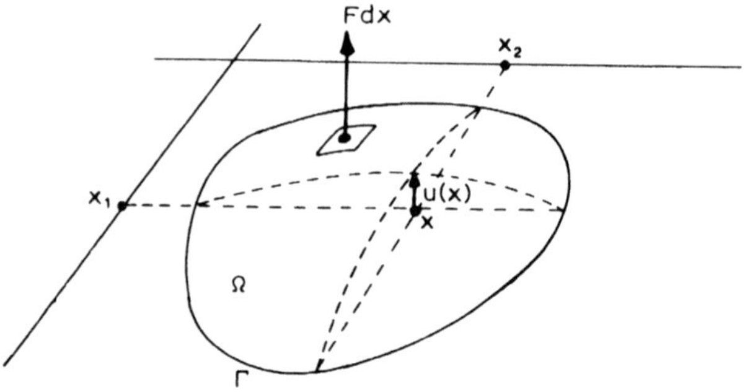

Fig. 1.2.2

The *second example* corresponds to the following data:

$$\left\{\begin{aligned} &V = U = H^1(\Omega),\\ &a(u, v) = \int_\Omega \left(\sum_{i=1}^n \partial_i u \partial_i v + auv\right) \mathrm{d}x,\\ &f(v) = \int_\Omega fv \,\mathrm{d}x + \int_\Gamma gv \,\mathrm{d}\gamma, \end{aligned}\right. \tag{1.2.16}$$

with the following assumptions on the functions a, f and g:

$$a \in L^\infty(\Omega), \quad a \geqslant a_0 > 0 \text{ a.e. on } \Omega, \quad f \in L^2(\Omega), \quad g \in L^2(\Gamma), \tag{1.2.17}$$

for some constant a_0.

The bilinear form is $H^1(\Omega)$-elliptic since $a(v, v) \geqslant \min\{1, a_0\}\|v\|_{1,\Omega}^2$ for all $v \in H^1(\Omega)$ (in Exercise 1.2.3, a case where $a_0 = 0$ is considered). The linear form $v \in H^1(\Omega) \to \int_\Gamma gv \,\mathrm{d}\gamma$ is continuous since by inequality (1.2.3),

$$\left|\int_\Gamma gv \,\mathrm{d}\gamma\right| \leq \|g\|_{L^2(\Gamma)}\|v\|_{L^2(\Gamma)} \leq C(\Omega)\|g\|_{L^2(\Gamma)}\|v\|_{1,\Omega}.$$

Therefore there exists a unique function $u \in H^1(\Omega)$ which minimizes the functional

$$J: v \to J(v) = \frac{1}{2}\int_\Omega \left\{\sum_{i=1}^n (\partial_i v)^2 + av^2\right\} \mathrm{d}x - \int_\Omega fv \,\mathrm{d}x - \int_\Gamma gv \,\mathrm{d}\gamma,$$

over the space $H^1(\Omega)$ or equivalently, such that

$$\forall v \in H^1(\Omega), \int_\Omega \left\{\sum_{i=1}^n \partial_i u \partial_i v + auv\right\} \mathrm{d}x = \int_\Omega fv \,\mathrm{d}x + \int_\Gamma gv \,\mathrm{d}\gamma. \tag{1.2.18}$$

Because $\mathscr{D}(\Omega)$ is a subspace of the space $H^1(\Omega)$, an argument similar to the one used for the first example shows that u is also a solution of the partial differential equation $-\Delta u + au = f$ in $\mathscr{D}'(\Omega)$. Thus we have

$$\forall v \in H^1(\Omega), \int_\Omega (-\Delta u + au)v \,\mathrm{d}x = a(u, v) - \int_\Gamma gv \,\mathrm{d}\gamma.$$

To sum up, the solution u of the minimization (or variational) problem associated with the data (1.2.16) is also a solution of the problem: *Find a distribution $u \in \mathscr{D}'(\Omega)$ such that*

$$\begin{cases} u \in H^1(\Omega), \quad -\Delta u + au = f \quad \text{in} \quad \mathscr{D}'(\Omega), \\ \forall v \in H^1(\Omega), \quad \int_\Omega (-\Delta u + au)v \,\mathrm{d}x = a(u, v) - \int_\Gamma gv \,\mathrm{d}\gamma. \end{cases} \tag{1.2.19}$$

and, conversely, if a distribution u is a solution of problem (1.2.19), it is clearly a solution of the variational equations (1.2.18).

If we assume additional smoothness on the solution, *the second relations* (1.2.19) *can be interpreted as playing the role of boundary conditions*. If the solution u is in the space $H^2(\Omega)$, for example, an application of Green's formula (1.2.5) shows that, for all $v \in H^1(\Omega)$,

$$\begin{aligned} a(u, v) &= \int_\Omega (-\Delta u + au)v \,\mathrm{d}x + \int_\Gamma \partial_\nu u \, v \,\mathrm{d}\gamma, \\ &= \int_\Omega fv \,\mathrm{d}x + \int_\Gamma gv \,\mathrm{d}\gamma. \end{aligned} \tag{1.2.20}$$

Therefore the conjunction of relations (1.2.19) and (1.2.20) implies that

$$\forall v \in H^1(\Omega) \int_\Gamma \partial_\nu u \, v \, \mathrm{d}\gamma = \int_\Gamma g v \, \mathrm{d}\gamma. \tag{1.2.21}$$

From these, one deduces that $\partial_\nu u = g$ on Γ.

Consequently, we shall say that *we have formally solved* the *associated boundary value problem*:

$$\begin{cases} -\Delta u + au = f & \text{in} \quad \Omega, \\ \partial_\nu u = g & \text{on} \quad \Gamma, \end{cases} \tag{1.2.22}$$

which is called a *nonhomogeneous Neumann problem* if $g \neq 0$, or a *homogeneous Neumann problem* if $g = 0$, for the operator $u \to -\Delta u + au$.

Remark 1.2.1. Without using differentiation of distributions, Green's formula (1.2.5) gives another way to obtain the partial differential equation since

$$\forall \phi \in \mathscr{D}(\Omega), \quad a(u, \phi) = \int_\Omega (-\Delta u + au)\phi \, \mathrm{d}x = \langle -\Delta u + au, \phi \rangle.$$

Of course, this is not a coincidence: The definition of differentiation for distributions is precisely based upon the fundamental Green's formula (1.2.4). □

In the *third example*, we shall extend in three directions the previous analysis: First the associated partial differential equation will have non constant coefficients. Secondly, the bilinear form will not be necessarily symmetric so that Theorem 1.1.3 will be needed for the existence analysis, and thirdly, the space V will be "intermediate" between the spaces $H_0^1(\Omega)$ and $H^1(\Omega)$. The data are the following:

$$\begin{cases} V = U = \{v \in H^1(\Omega); \quad v = 0 \quad \text{on} \quad \Gamma_0\}, \\ a(u, v) = \displaystyle\int_\Omega \left\{ \sum_{i,j=1}^n a_{ij} \partial_i u \partial_j v + auv \right\} \mathrm{d}x, \\ f(v) = \displaystyle\int_\Omega f v \, \mathrm{d}x + \int_{\Gamma_1} g v \, \mathrm{d}\gamma, \end{cases} \tag{1.2.23}$$

where $\Gamma_0 = \Gamma - \Gamma_1$ is a $\mathrm{d}\gamma$-measurable subset of the boundary Γ with a strictly positive $\mathrm{d}\gamma$-measure, and the functions a_{ij}, a and f satisfy the following assumptions:

$$\begin{cases} a_{ij} \in L^\infty(\Omega), \quad 1 \leqslant i,j \leqslant n, \quad a \in L^\infty(\Omega), \quad a \geqslant 0 \text{ a.e.} \quad \text{on} \quad \Omega, \\ f \in L^2(\Omega), \quad g \in L^2(\Gamma_1), \end{cases} \tag{1.2.24}$$

$$\exists \beta > 0, \quad \forall \xi_i, \quad 1 \leqslant i \leqslant n, \quad \sum_{i,j=1}^n a_{ij}\xi_i\xi_j \geqslant \beta \sum_{i=1}^n \xi_i^2 \text{ a.e.} \quad \text{on} \quad \Omega. \tag{1.2.25}$$

The V-ellipticity of the bilinear form will be a consequence of the following result.

Theorem 1.2.1. *Let Ω be a connected bounded open subset of $\mathbf{R}^n$. Then the space V defined in* (1.2.23) *is a closed subspace of $H^1(\Omega)$.*

If the dγ*-measure of Γ_0 is strictly positive, the semi-norm $|\cdot|_{1,\Omega}$ is a norm over the space V, equivalent to the norm $\|\cdot\|_{1,\Omega}$.*

Proof. Let (v_k) be a sequence of functions in the space V which converges to an element $v \in H^1(\Omega)$. Since the sequence $(\operatorname{tr} v_k)$ converges to $\operatorname{tr} v$ in the space $L^2(\Gamma)$ (cf. inequalities (1.2.3)), it contains a subsequence which converges almost everywhere to $\operatorname{tr} v$ and thus $\operatorname{tr} v = 0$ a.e. on Γ_0. This implies that the function v belongs to the space V.

Next, let us show that $|\cdot|_{1,\Omega}$ is a norm over the space V. Let v be a function in the space V which satisfies $|v|_{1,\Omega} = 0$. Then it is a constant by virtue of the connectedness of the set Ω and, being as such a smooth function, its trace is the same constant. That this constant is zero follows from the fact that the trace vanishes on the set Γ_0, whose dγ-measure is strictly positive.

Finally, assume that the two norms $|\cdot|_{1,\Omega}$ and $\|\cdot\|_{1,\Omega}$ are not equivalent over the space V. Then there exists a sequence (v_k) of functions $v_k \in V$ such that

$$\begin{cases} \forall k, \quad \|v_k\|_{1,\Omega} = 1, \\ \qquad \lim_{k\to\infty} |v_k|_{1,\Omega} = 0. \end{cases}$$

By *Rellich's theorem*, any bounded sequence in the space $H^1(\Omega)$ contains a subsequence which converges in $L^2(\Omega)$, so that there exists a sequence (v_l) of functions $v_l \in V$ which converges in the space $L^2(\Omega)$ and which is such that $\lim_{l\to\infty} |v_l|_{1,\Omega} = 0$. Thus the sequence (v_l) is a

Cauchy sequence in the complete space V and therefore it converges in the norm $\|\cdot\|_{1,\Omega}$ to an element $v \in V$.

Since $|v|_{1,\Omega} = \lim_{l\to\infty} |v_l|_{1,\Omega} = 0$, we deduce that $v = 0$, which is in contradiction with the equalities $\|v_k\|_{1,\Omega} = 1$ for all k. □

From this theorem, we infer that the bilinear form of (1.2.23) is V-elliptic since we have $a(v, v) \geq \beta |v|^2_{1,\Omega}$ for all $v \in H^1(\Omega)$, as an application of the inequalities of (1.2.24) and (1.2.25) shows.

By the Lax–Milgram lemma (Theorem 1.1.3), there exists a unique function $u \in V$ which satisfies the variational equations

$$\forall v \in V,$$
$$\int_\Omega \left\{\sum_{i,j=1}^n a_{ij}\partial_i u \partial_j v + auv\right\} \mathrm{d}x = \int_\Omega fv \,\mathrm{d}x + \int_{\Gamma_1} gv \,\mathrm{d}\gamma. \tag{1.2.26}$$

Referring once again to formula (1.2.4), we obtain another *Green's formula*:

$$\int_\Omega \sum_{i,j=1}^n a_{ij}\partial_i u \partial_j v \,\mathrm{d}x = -\int_\Omega \sum_{i,j=1}^n \partial_j(a_{ij}\partial_i u) v \,\mathrm{d}x + \int_\Gamma \sum_{i,j=1}^n a_{ij}\partial_i u \; v\nu_j \,\mathrm{d}\gamma, \tag{1.2.27}$$

valid for all functions $u \in H^2(\Omega)$, $v \in H^1(\Omega)$, provided the functions a_{ij} are smooth enough so that the functions $a_{ij}\partial_i u$ belong to the space $H^1(\Omega)$ (for example, $a_{ij} \in \mathscr{C}^1(\bar{\Omega})$). Using (1.2.27), we conclude that *we have formally solved the boundary value problem*

$$\begin{cases} -\sum_{i,j=1}^n \partial_j(a_{ij}\partial_i u) + au = f & \text{in } \Omega, \\ u = 0 & \text{on } \Gamma_0 \\ \sum_{i,j=1}^n a_{ij}\nu_j\partial_i u = g & \text{on } \Gamma_1, \end{cases} \tag{1.2.28}$$

which is called a *homogeneous mixed problem* if $g = 0$, or a *nonhomogeneous mixed problem* if $g \neq 0$, for the operator

$$u \to -\sum_{i,j=1}^n \partial_j(a_{ij}\partial_i u) + au, \tag{1.2.29}$$

assuming in both cases the $\mathrm{d}\gamma$-measures of Γ_0 and Γ_1 are strictly positive. Notice that condition (1.2.25) is the classical *ellipticity con-*

dition for an operator such as that of (1.2.29). The operator

$$u \to \sum_{i,j=1}^{n} a_{ij}\nu_j \partial_i u$$

is called the *conormal derivative operator* associated with the operator of (1.2.29).

If $\Gamma = \Gamma_0$, or $\Gamma = \Gamma_1$, then we have formally solved a *homogeneous Dirichlet problem*, or a *homogeneous* or a *nonhomogeneous Neumann problem*, for the operator of (1.2.29) (in the second case, we would require an inequality such as $a \geqslant a_0 > 0$ a.e. on Ω to get existence).

The elasticity problem

We now come to the *fourth example* which is by far the most significant. Let Ω be a bounded open connected subset of $\mathbf{R}^3$ with a Lipschitz-continuous boundary. We define the space

$$\begin{aligned} \boldsymbol{V} = \boldsymbol{U} = \{\boldsymbol{v} = (v_1, v_2, v_3) \in (H^1(\Omega))^3;& \\ v_i = 0 \quad \text{on} \quad \Gamma_0, \quad 1 \leqslant i \leqslant 3\},& \end{aligned} \tag{1.2.30}$$

where Γ_0 is a dγ-measurable subset of Γ, with a strictly positive dγ-measure. The space $\boldsymbol{V}$ is equipped with the product norm

$$\boldsymbol{v} = (v_1, v_2, v_3) \to \|\boldsymbol{v}\|_{1,\Omega} = \Big(\sum_{i=1}^{3} \|v_i\|_{1,\Omega}^2\Big)^{1/2}.$$

For any $\boldsymbol{v} = (v_1, v_2, v_3) \in (H^1(\Omega))^3$, we let

$$\epsilon_{ij}(\boldsymbol{v}) = \epsilon_{ji}(\boldsymbol{v}) = \tfrac{1}{2}(\partial_j v_i + \partial_i v_j), \quad 1 \leqslant i, j \leqslant 3, \tag{1.2.31}$$

and

$$\sigma_{ij}(\boldsymbol{v}) = \sigma_{ji}(\boldsymbol{v}) = \lambda\Big(\sum_{k=1}^{3} \epsilon_{kk}(\boldsymbol{v})\Big)\delta_{ij} + 2\mu\epsilon_{ij}(\boldsymbol{v}), \quad 1 \leqslant i, j \leqslant 3, \tag{1.2.32}$$

where δ_{ij} is the Kronecker's symbol, and λ and μ are two constants which are assumed to satisfy $\lambda > 0$, $\mu > 0$. We define the bilinear form

$$\begin{aligned} a(\boldsymbol{u}, \boldsymbol{v}) &= \int_\Omega \sum_{i,j=1}^{3} \sigma_{ij}(\boldsymbol{u})\epsilon_{ij}(\boldsymbol{v})\,\mathrm{d}x \\ &= \int_\Omega \Big\{\lambda \operatorname{div} \boldsymbol{u} \operatorname{div} \boldsymbol{v} + 2\mu \sum_{i,j=1}^{3} \epsilon_{ij}(\boldsymbol{u})\epsilon_{ij}(\boldsymbol{v})\Big\}\,\mathrm{d}x, \end{aligned} \tag{1.2.33}$$

and the linear form

$$f(\boldsymbol{v}) = \int_\Omega \boldsymbol{f} \cdot \boldsymbol{v}\, dx + \int_{\Gamma_1} \boldsymbol{g} \cdot \boldsymbol{v}\, d\gamma = \int_\Omega \sum_{i=1}^{3} f_i v_i\, dx + \int_{\Gamma_1} \sum_{i=1}^{3} g_i v_i\, d\gamma, \tag{1.2.34}$$

where $\boldsymbol{f} = (f_1, f_2, f_3) \in (L^2(\Omega))^3$ and $\boldsymbol{g} = (g_1, g_2, g_3) \in (L^2(\Gamma_1))^3$, with $\Gamma_1 = \Gamma - \Gamma_0$ are given functions.

It is clear that these bilinear and linear forms are continuous over the space $\boldsymbol{V}$. To prove the $\boldsymbol{V}$-ellipticity of the bilinear form (see Exercise 1.2.4), one needs *Korn's inequality*: There exists a constant $C(\Omega)$ such that, for all $\boldsymbol{v} = (v_1, v_2, v_3) \in (H^1(\Omega))^3$,

$$\|\boldsymbol{v}\|_{1,\Omega} \leq C(\Omega)\Big(\sum_{i,j=1}^{3} |\epsilon_{ij}(\boldsymbol{v})|_{0,\Omega}^2 + \sum_{i=1}^{3} |v_i|_{0,\Omega}^2\Big)^{1/2}. \tag{1.2.35}$$

This is a nontrivial inequality, whose proof may be found in DUVAUT & LIONS (1972, Chapter 3, §3.3), or in FICHERA (1972, Section 12). From it, one deduces that over the space $\boldsymbol{V}$ defined in (1.2.30) *the mapping*

$$\boldsymbol{v} = (v_1, v_2, v_3) \to |\boldsymbol{v}| = \Big(\sum_{i,j=1}^{3} |\epsilon_{ij}(\boldsymbol{v})|_{0,\Omega}^2\Big)^{1/2}$$

is a norm, equivalent to the product norm, as long as the $d\gamma$-measure of Γ_0 is strictly positive, which is the case here (again the reader is referred to Exercise 1.2.4).

The $\boldsymbol{V}$-ellipticity is therefore a consequence of the inequalities $\lambda > 0$, $\mu > 0$, since by (1.2.33)

$$a(\boldsymbol{v}, \boldsymbol{v}) \geq 2\mu |\boldsymbol{v}|^2.$$

We conclude that there exists a unique function $\boldsymbol{u} \in \boldsymbol{V}$ which minimizes the functional

$$J(\boldsymbol{v}) = \frac{1}{2}\int_\Omega \Big\{\lambda(\operatorname{div} \boldsymbol{v})^2 + 2\mu \sum_{i,j=1}^{3} (\epsilon_{ij}(\boldsymbol{v}))^2\Big\}\, dx - \Big(\int_\Omega \boldsymbol{f} \cdot \boldsymbol{v}\, dx + \int_{\Gamma_1} \boldsymbol{g} \cdot \boldsymbol{v}\, d\gamma\Big) \tag{1.2.36}$$

over the space $\boldsymbol{V}$, or equivalently, which is such that

$$\forall \boldsymbol{v} \in \boldsymbol{V}, \quad \int_\Omega \sum_{i,j=1}^{3} \sigma_{ij}(\boldsymbol{u})\epsilon_{ij}(\boldsymbol{v})\, dx = \int_\Omega \boldsymbol{f} \cdot \boldsymbol{v}\, dx + \int_{\Gamma_1} \boldsymbol{g} \cdot \boldsymbol{v}\, d\gamma. \tag{1.2.37}$$

Since relations (1.2.37) are satisfied by all functions $\boldsymbol{v} \in (\mathcal{D}(\Omega))^3$, they could yield the associated partial differential equation. However, as was pointed out in Remark 1.2.1, it is equivalent to proceed through Green's formulas, which in addition have the advantage of yielding boundary conditions too.

Using Green's formula (1.2.4), we obtain, for all $\boldsymbol{u} \in (H^2(\Omega))^3$ and all $\boldsymbol{v} \in (H^1(\Omega))^3$:

$$\int_\Omega \sigma_{ij}(\boldsymbol{u})\partial_j v_i \,\mathrm{d}x = -\int_\Omega (\partial_j \sigma_{ij}(\boldsymbol{u}))v_i \,\mathrm{d}x + \int_\Gamma \sigma_{ij}(\boldsymbol{u})v_i\nu_j \,\mathrm{d}\gamma,$$

so that, using definitions (1.2.31) and (1.2.32), we have proved that the following *Green's formula* holds:

$$\int_\Omega \sum_{i,j=1}^{3} \sigma_{ij}(\boldsymbol{u})\epsilon_{ij}(\boldsymbol{v}) \,\mathrm{d}x = \int_\Omega \sum_{i=1}^{3} \left(-\sum_{j=1}^{3} \partial_j\sigma_{ij}(\boldsymbol{u})\right) v_i \,\mathrm{d}x + \int_\Gamma \sum_{i=1}^{3} \left(\sum_{j=1}^{3} \sigma_{ij}(\boldsymbol{u})\nu_j\right) v_i \,\mathrm{d}\gamma, \qquad (1.2.38)$$

for all functions $\boldsymbol{u} \in (H^2(\Omega))^3$ and $\boldsymbol{v} \in (H^1(\Omega))^3$.

Arguing as in the previous examples, we find that we are formally solving the equations

$$-\sum_{j=1}^{3} \partial_j\sigma_{ij}(\boldsymbol{u}) = f_i, \quad 1 \leqslant i \leqslant 3. \qquad (1.2.39)$$

It is customary to write these equations in vector form:

$$-\mu\, \Delta\boldsymbol{u} - (\lambda + \mu)\,\mathrm{grad}\,\mathrm{div}\,\boldsymbol{u} = \boldsymbol{f} \quad \text{in} \quad \Omega,$$

which is derived from (1.2.39) simply by using relations (1.2.32).

Taking equations (1.2.39) into account, the variational equations (1.2.37) reduce to

$$\forall \boldsymbol{v} \in V, \quad \int_{\Gamma_1} \sum_{i=1}^{3} \left(\sum_{j=1}^{3} \sigma_{ij}(\boldsymbol{u})\nu_j\right) v_i \,\mathrm{d}\gamma = \int_{\Gamma_1} \sum_{i=1}^{3} g_i v_i \,\mathrm{d}\gamma,$$

since $\boldsymbol{v} = \boldsymbol{0}$ on $\Gamma_0 = \Gamma - \Gamma_1$.

To sum up, *we have formally solved the following associated boundary value problem*:

$$\begin{cases} -\mu\, \Delta\boldsymbol{u} - (\lambda + \mu)\,\mathrm{grad}\,\mathrm{div}\,\boldsymbol{u} = \boldsymbol{f} \quad \text{in} \quad \Omega, \\ \boldsymbol{u} = \boldsymbol{0} \quad \text{on} \quad \Gamma_0, \\ \displaystyle\sum_{j=1}^{3} \sigma_{ij}(\boldsymbol{u})\nu_j = g_i \quad \text{on} \quad \Gamma_1, \quad 1 \leqslant i \leqslant 3, \end{cases} \qquad (1.2.40)$$

which is known as the *system of equations of linear elasticity*. Let us mention that a completely analogous analysis holds in two dimensions, in which case the resulting problem is called the *system of equations of two-dimensional, or plane, elasticity*, the above one being also called by contrast the *system of three-dimensional elasticity*. Accordingly, the variational problem associated with the data (1.2.30), (1.2.33) and (1.2.34) is called the (*three- or two-dimensional*) *elasticity problem*.

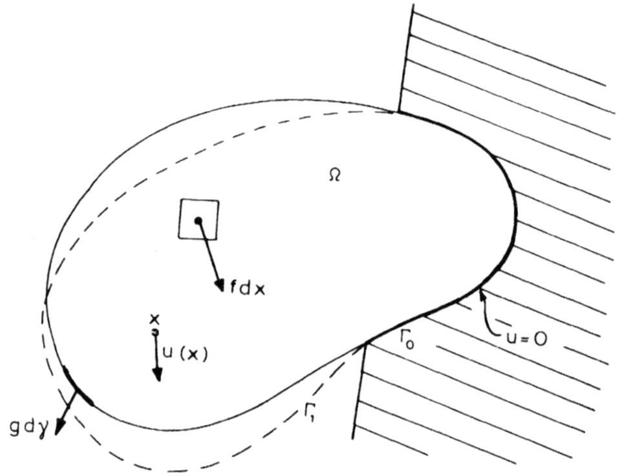

Fig. 1.2.3

Assuming "small" displacements and "small" strains, this system describes the state of a body (Fig. 1.2.3) which occupies the set $\bar{\Omega}$ in the absence of forces, $\boldsymbol{u}$ denoting the displacement of the points of $\bar{\Omega}$ under the influence of given forces (as usual, the scale for the displacements is distorted in the figure).

The body $\bar{\Omega}$ cannot move along Γ_0, and along Γ_1, surface forces of density $\boldsymbol{g}$ are given. In addition, a volumic force, of density $\boldsymbol{f}$, is prescribed inside the body $\bar{\Omega}$.

Then we recognize in $(\epsilon_{ij}(\boldsymbol{u}))$ the *strain tensor* while $(\sigma_{ij}(\boldsymbol{u}))$ is the *stress tensor*, the relationship between the two being given by the linear equations (1.2.32) known in Elasticity as *Hooke's law* for isotropic bodies. The constants λ and μ are the *Lamé coefficients* of the material of which the body is composed.

The variational equations (1.2.37) represent the *principle of virtual work*, valid for all *kinematically admissible displacements* $\boldsymbol{v}$, i.e., which satisfy the boundary condition $\boldsymbol{v} = \mathbf{0}$ on Γ_0.

The functional J of (1.2.36) is the *total potential energy* of the body. It is the sum of the *strain energy*:

$$\frac{1}{2}\int\left\{\lambda(\operatorname{div}\boldsymbol{v})^2 + 2\mu\sum_{i,j=1}^{3}(\epsilon_{ij}(\boldsymbol{v}))^2\right\}\mathrm{d}x,$$

and of the *potential energy of the exterior forces*: $-(\int_\Omega \boldsymbol{f}\cdot\boldsymbol{v}\,\mathrm{d}x + \int_{\Gamma_1}\boldsymbol{g}\cdot\boldsymbol{v}\,\mathrm{d}\gamma)$.

This example is probably the most crucial one, not only because it has obviously many applications, but essentially because its variational formulation, described here, is basically responsible for the invention of the finite element method by engineers.

Remark 1.2.2. It is interesting to notice that the strict positiveness of the $\mathrm{d}\gamma$-measure of Γ_0 has a physical interpretation: It is intuitively clear that in case the $\mathrm{d}\gamma$-measure Γ_0 would vanish, the body would be free and therefore there could not exist an equilibrium position in general. □

Remark 1.2.3. The *membrane problem*, which we have already described, the *plate problem*, which we shall soon describe in this section, and the *shell problem* (Section 8.1), *are derived from the elasticity problem*, through a process which can be briefly described as follows: Because such bodies have a "small" thickness, simplifying *a priori* assumptions can be made (such as linear variations of the stresses over the thickness) which, together with other assumptions (on the constitutive material in the case of membranes, or on the orthogonality of the exterior forces in the case of membranes and plates), allow one to integrate the energy (1.2.36) over the thickness. In this fashion, the problem is reduced to a problem in two variables, and only one function (the "vertical" displacement) in case of membranes and plates. All this is at the expense of a greater mathematical complexity in case of plates and shells however, as we shall see. □

Remark 1.2.4. Since problem (1.2.40) is called system of *linear* elasticity, the linearity being of course that of the mapping $(\boldsymbol{f}, \boldsymbol{g}) \to \boldsymbol{u}$, it is worth saying how this problem might become nonlinear. This may happen in three nonexclusive ways:

(i) Instead of minimizing the energy over the space V, we minimize it over a subset U which is not a subspace. This circumstance, which we already commented upon (Remark 1.1.1) is examined in Exercise 1.2.5 for a simpler model. Another example is treated in Section 5.1.

(ii) Instead of considering the "linearized" strain tensor (1.2.31), the "full" tensor is considered, i.e., we let

$$\epsilon_{ij}(\mathbf{v}) = \frac{1}{2}(\partial_j v_i + \partial_i v_j + \sum_{k=1}^{3} \partial_i v_k \partial_j v_k), \quad 1 \leq i, j \leq 3.$$

Actually, it suffices that for at least one pair (i, j), the above expression be considered. This is the case for instance of the *von Karmann's model of a clamped plate.*

(iii) The linear relation (1.2.32) between the strain tensor and the stress tensor is replaced by a nonlinear relation. □

Examples of fourth-order problems: The biharmonic problem, the plate problem

Whereas in the preceding examples the spaces V were contained in the space $H^1(\Omega)$, we consider in the last examples Sobolev spaces which involve second-order derivatives. We begin with the following data:

$$\begin{cases} V = U = H_0^2(\Omega), \\ a(u, v) = \int_\Omega \Delta u \, \Delta v \, dx, \\ f(v) = \int_\Omega fv \, dx, \quad f \in L^2(\Omega). \end{cases} \tag{1.2.41}$$

Since the mapping $v \to |\Delta v|_{0,\Omega}$ is a norm over the space $H_0^2(\Omega)$, as we showed in (1.2.8), the bilinear form is $H_0^2(\Omega)$-elliptic. Thus there exists a unique function $u \in H_0^2(\Omega)$ which minimizes the functional

$$J: v \to J(v) = \frac{1}{2}\int_\Omega |\Delta v|^2 \, dx - \int_\Omega fv \, dx \tag{1.2.42}$$

over the space $H_0^2(\Omega)$ or, equivalently, which satisfies the variational equations

$$\forall v \in H_0^2(\Omega), \quad \int_\Omega \Delta u \, \Delta v \, dx = \int_\Omega fv \, dx. \tag{1.2.43}$$

Using Green's formula (1.2.7):

$$\int_\Omega \Delta u\, \Delta v\, \mathrm{d}x = \int_\Omega \Delta^2 u\, v\, \mathrm{d}x - \int_\Gamma \partial_\nu \Delta u\, v\, \mathrm{d}\gamma + \int_\Gamma \Delta u \partial_\nu v\, \mathrm{d}\gamma,$$

we find that we have formally solved the following *homogeneous Dirichlet problem for the biharmonic operator* Δ^2:

$$\begin{aligned} &\Delta^2 u = f \quad \text{in} \quad \Omega, \\ &u = \partial_\nu u = 0 \quad \text{on} \quad \Gamma. \end{aligned} \tag{1.2.44}$$

We shall indicate a physical origin of this problem in the section "Additional Bibliography and Comments" of Chapter 4.

As our *last example*, we let, for $n = 2$,

$$\left\{\begin{aligned} &V = U = H_0^2(\Omega), \\ &a(u, v) = \int_\Omega \{\Delta u\, \Delta v + (1-\sigma)(2\partial_{12}u\partial_{12}v - \partial_{11}u\partial_{22}v \\ &\qquad - \partial_{22}u\partial_{11}v)\}\, \mathrm{d}x \\ &\qquad = \int_\Omega \{\sigma \Delta u\, \Delta v + (1-\sigma)(\partial_{11}u\partial_{11}v + \partial_{22}u\partial_{22}v \\ &\qquad + 2\partial_{12}u\partial_{12}v)\}\, \mathrm{d}x, \\ &f(v) = \int_\Omega fv\, \mathrm{d}x, \quad f \in L^2(\Omega). \end{aligned}\right. \tag{1.2.45}$$

These data correspond to the variational formulation of the (*clamped*) *plate problem*, which concerns the equilibrium position of a plate of constant thickness e under the action of a transverse force, of density $F = (Ee^3/12(1-\sigma^2))f$ per unit area. The constants $E = \mu(3\lambda + 2\mu)/(\lambda + \mu)$ and $\sigma = \lambda/2(\lambda + \mu)$ are respectively the *Young's modulus* and the *Poisson's coefficient* of the plate, λ and μ being the Lamé's coefficients of the plate material. When $f = 0$, the plate is in the plane of coordinates (x_1, x_2) (Fig. 1.2.4). The condition $u \in H_0^2(\Omega)$ takes into account the fact that the plate is clamped (see the boundary conditions in (1.2.48) below).

As we pointed out in Remark 1.2.3, the expressions given in (1.2.45) for the bilinear form and the linear form are obtained upon integration over the thickness of the plate of the analogous quantities for the elasticity problem. This integration results in a simpler problem, in that there are now only two independent variables. However, this advantage is compensated by the fact that second partial derivatives are now present

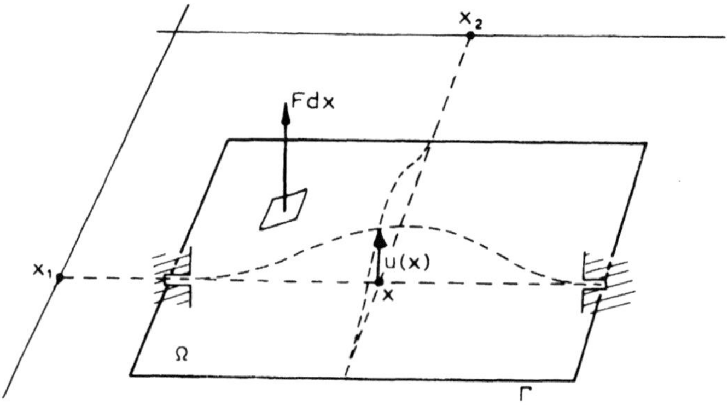

Fig. 1.2.4

in the bilinear form. This will result in a fourth-order partial differential equation. See (1.2.48).

The Poisson's coefficient σ satisfying the inequalities $0<\sigma<\frac{1}{2}$, the bilinear form is $H_0^2(\Omega)$-elliptic, since we have

$$\forall v \in H^2(\Omega), \quad a(v, v) = \sigma|\Delta v|_{0,\Omega}^2 + (1-\sigma)|v|_{2,\Omega}^2.$$

Thus, there exists a unique function $u \in H_0^2(\Omega)$ which minimizes the *total potential energy of the plate*:

$$J(v) = \frac{1}{2}\int_\Omega \{|\Delta v|^2 + 2(1-\sigma)((\partial_{12}v)^2 - \partial_{11}v\partial_{22}v)\}\,\mathrm{d}x - \int_\Omega fv\,\mathrm{d}x, \tag{1.2.46}$$

over the space $H_0^2(\Omega)$ or, equivalently, which is solution of the variational equations

$$\forall v \in H_0^2(\Omega), \quad \int_\Omega \{\Delta u\,\Delta v + (1-\sigma)(2\partial_{12}u\partial_{12}v - \partial_{11}u\partial_{22}v - \partial_{22}u\partial_{11}v)\}\,\mathrm{d}x = \int_\Omega fv\,\mathrm{d}x. \tag{1.2.47}$$

Using Green's formulas (1.2.7) and (1.2.9):

$$\int_\Omega \Delta u\,\Delta v = \int_\Omega \Delta^2 u\, v\,\mathrm{d}x - \int_\Gamma \partial_\nu \Delta u\, v\,\mathrm{d}\gamma + \int_\Gamma \Delta u \partial_\nu v\,\mathrm{d}\gamma,$$

$$\int_\Omega \{2\partial_{12}u\partial_{12}v - \partial_{11}u\partial_{22}v - \partial_{22}u\partial_{11}v\}\,\mathrm{d}x = \int_\Gamma \{-\partial_{\tau\tau}u\partial_\nu v + \partial_{\nu\tau}u\partial_\tau v\}\,\mathrm{d}\gamma,$$

we find that we have again solved, at least formally, the *homogeneous Dirichlet problem for the biharmonic operator* Δ^2:

$$\begin{cases} \Delta^2 u = f & \text{in} \quad \Omega, \\ u = \partial_\nu u = 0 & \text{on} \quad \Gamma. \end{cases} \tag{1.2.48}$$

Therefore, *in spite of a different bilinear form, we eventually find the same problem as in the previous example.* This is so because, in view of the second Green's formula which we used, the contribution of the integral

$$\int_\Omega (1-\sigma)\{2\partial_{12}u\partial_{12}v - \partial_{11}u\partial_{22}v - \partial_{22}u\partial_{11}v\}\,dx$$

is zero when the functions v are in the space $\mathscr{D}(\Omega)$, and consequently in its closure $H_0^2(\Omega)$. Thus, the partial differential equation is still $\Delta^2 u = f$ in Ω. However different boundary conditions might result from another choice for the space V. See Exercise 1.2.7.

To distinguish the two problems, we shall refer to a fourth-order problem corresponding to the functional of (1.2.42) as a *biharmonic problem*, while we shall refer to a fourth-order problem corresponding to the functional of (1.2.46) as a *plate problem*.

In this section, we have examined various minimization or variational problems with each of which is associated a boundary value problem for which the partial differential operator is elliptic (incidentally, this correspondence is not one-to-one, as the last two examples show). This is why, by extension, these minimization or variational problems are themselves called *elliptic boundary value problems*. For the same reasons, such problems are said to be *second-order problems*, or *fourth-order problems*, when the associated partial differential equation is of order two or four, respectively.

Finally, one should recall that even though the association between the two formulations may be formal, it is possible to prove, *under appropriate smoothness assumptions on the data*, that a solution of any of the variational problems considered here is also a solution in the classical sense of the associated boundary value problem.

Remark 1.2.5. In this book, one could conceivably omit all reference to the associated classical boundary value problems, inasmuch as the finite element method is based only on the variational formulations. By contrast, finite difference methods are most often derived from the classical formulations. □

Exercises

1.2.1. Prove Green's formula (1.2.9). The reader should keep in mind that the derivative $\partial_{\tau\tau} v$ generally differs from the second derivative of the function v, considered as a function of the curvilinear abcissa along the boundary.)

1.2.2. Let the space $V = U$ and the bilinear form be as in (1.2.10), and let the linear form be defined by

$$f(v) = \int_\Omega fv \, dx - a(u_0, v),$$

where the functions f and a satisfy assumptions (1.2.11) and u_0 is a given function in the space $H^1(\Omega)$. Show that these data correspond to the formal solution of the *nonhomogeneous Dirichlet problem for the operator* $u \to -\Delta u + au$, i.e.,

$$\begin{cases} -\Delta u + au = f & \text{in} \quad \Omega, \\ u = u_0 & \text{on} \quad \Gamma. \end{cases}$$

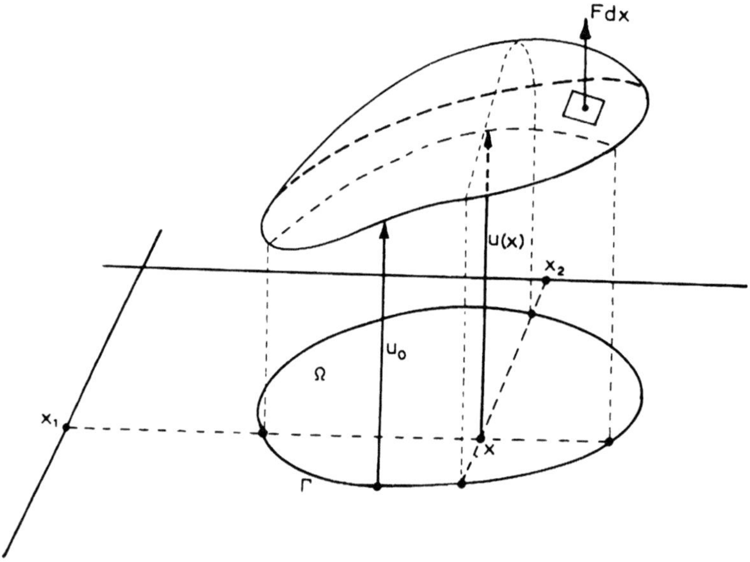

Fig. 1.2.5

Is it equivalent to minimizing the functional (1.2.12) over the subset

$$U = \{v \in H^1(\Omega);\quad (v - u_0) \in H_0^1(\Omega)\}$$

of the space $V = H^1(\Omega)$?

With $a = 0$ and $n = 2$, this is another *membrane problem*. See Fig. 1.2.5, which is self-explanatory.

1.2.3. Find a variational problem which amounts to solving the non-homogeneous Neumann problem for the operator $-\Delta$, i.e., problem (1.2.22) when the function a vanishes identically, and when the equality $\int_\Omega f\,dx + \int_\Gamma g\,d\gamma = 0$ holds. [Hint: Use the fact that over the quotient space $H^1(\Omega)/P_0(\Omega)$, $P_0(\Omega)$: space of constant functions over Ω, the semi-norm $|\cdot|_{1,\Omega}$ is a norm, equivalent to the quotient norm. See Theorem 3.1.1 for a proof.]

1.2.4. Let Ω be a connected open subset of $\mathbf{R}^n$, with $n = 2$ or 3, and let Γ_0 be a $d\gamma$-measurable subset of its boundary Γ, assumed to be Lipschitz-continuous. Let

$$\boldsymbol{V} = \{\boldsymbol{v} = (v_i) \in (H^1(\Omega))^n;\quad v_i = 0 \quad \text{on} \quad \Gamma_0,\quad 1 \leq i \leq n\}$$

(i) Show that $\boldsymbol{V}$ is a closed subspace of the space $(H^1(\Omega))^n$.

(ii) Show that the mapping

$$|\cdot|: \boldsymbol{v} \in \boldsymbol{V} \to |\boldsymbol{v}| = \left(\sum_{i,j=1}^{n} |\epsilon_{ij}(\boldsymbol{v})|_{0,\Omega}^2\right)^{1/2}$$

is a norm over the space $\boldsymbol{V}$, if the $d\gamma$-measure of Γ_0 is strictly positive. [Hint: Show that a function $\boldsymbol{v} \in (H^1(\Omega))^n$ which satisfies $|\boldsymbol{v}| = 0$ is of the form $\boldsymbol{v}(x) = \boldsymbol{a} \times x + \boldsymbol{b}$ for some constant vectors $\boldsymbol{a}$ and $\boldsymbol{b}$, i.e., the displacement $\boldsymbol{v}$ is a *rigid body motion*. Such a result is proved for example in Hlaváček & Nečas (1970, Lemma II.1). See also Section 8.1 for analogous ideas.]

(iii) Using Korn's inequality (1.2.35), show that the norm $|\cdot|$ is equivalent to the norm $\|\cdot\|_{1,\Omega}$. [Hint: Argue as in Theorem 1.2.1.]

1.2.5. Let

$$V = H^1(\Omega),\quad a(u, v) = \int_\Omega \left\{\sum_{i=1}^{n} \partial_i u \partial_i v + auv\right\} dx,\quad f(v) = \int_\Omega fv\,dx,$$

and let

$$U = \{v \in H^1(\Omega);\quad v \geq 0 \text{ a.e.}\quad \text{on}\quad \Gamma\}.$$

Show that U is a closed convex cone with vertex 0. Using characterizations (1.1.5) of Theorem 1.1.2 show that the associated variational problem amounts to formally solving the boundary value problem

$$\begin{cases} -\Delta u + au = f \quad \text{in} \quad \Omega, \\ u \geqslant 0, \quad \partial_\nu u \geqslant 0, \quad u\partial_\nu u = 0 \quad \text{on} \quad \Gamma. \end{cases}$$

This type of nonlinear problem is a model problem for *Signorini problems*, i.e., problems in elasticity for which the boundary conditions are *unilateral constraints* such as the above ones. For extensive discussions of such problems, see DUVAUT & LIONS (1972), FICHERA (1972).

1.2.6. Extend the analysis made for the data (1.2.23) to the case where the bilinear form is given by

$$a(u, v) = \int_\Omega \left\{ \sum_{i,j=1}^n a_{ij}\partial_i u \partial_j v + \sum_{i=1}^n a_i \partial_i u \, v + auv \right\} dx,$$

the functions a_i being in the space $L^\infty(\Omega)$. In particular, find sufficient conditions for the V-ellipticity of the bilinear form.

1.2.7. Let the bilinear form and the linear form be as in (1.2.45), and let

$$V = U = H^2(\Omega) \cap H_0^1(\Omega) = \{v \in H^2(\Omega); \quad v = 0 \quad \text{on} \quad \Gamma\}.$$

This is a mathematical model for a *simply supported plate.* Using the fact that $v \to |\Delta v|_{0,\Omega}$ is again a norm over the space V, equivalent to the norm $\|\cdot\|_{2,\Omega}$, analyze the associated variational problem. What is the associated boundary value problem?

1.2.8. Let

$$\begin{cases} V = U = H^2(\Omega) \cap H_0^1(\Omega) = \{v \in H^2(\Omega); \quad v = 0 \quad \text{on} \quad \Gamma\}, \\ a(u, v) = \displaystyle\int_\Omega \Delta u \, \Delta v \, dx, \\ f(v) = \displaystyle\int_\Omega fv \, dx - \int_\Gamma \lambda \partial_\nu v \, d\gamma, \quad f \in L^2(\Omega), \quad \lambda \in L^2(\Gamma). \end{cases}$$

Using as in Exercise 1.2.7 the fact that $v \to |\Delta v|_{0,\Omega}$ is a norm over the space V, equivalent to the norm $\|\cdot\|_{2,\Omega}$, analyze the associated variational problem. In particular, show that it can be decomposed into two second order problems. What is the associated boundary value problem? Does it share the same property?

Bibliography and comments

1.1. The original reference of the Lax–Milgram lemma is Lax & Milgram (1954). Our proof follows the method of Lions & Stampacchia (1967), where it is applied to the general variational problem (1.1.15), and where the case of semi-positive definite bilinear forms is also considered. Stampacchia (1964) had the original proof. For constructive existence proofs and additional references, see also Glowinski, Lions & Trémolières (1976a).

I. Babuska (Babuska & Aziz (1972, Theorem 5.2.1)) has extended the Lax–Milgram lemma to the case of bilinear forms defined on a product of two distinct Hilbert spaces. This extension turns out to be a useful tool for the analysis of some finite element methods (Babuška (1971b)).

1.2. For treatments of Differential Calculus with Fréchet derivatives, the reader may consult Cartan (1967), Dieudonné (1967), Schwartz (1967). For the theory of distributions and its applications to partial differential equations, see Schwartz (1966). Other references are Trèves (1967), Shilov (1968), Vo-Khac Khoan (1972a, 1972b). The Sobolev spaces are extensively studied in Lions (1962) and Nečas (1967). See also Adams (1975). The original reference is Sobolev (1950).

Thorough treatments of the variational formulations of elliptic boundary value problems are given in Lions (1962), Agmon (1965), Nečas (1967), Lions & Magenes (1968), Vo-Khac Khoan (1972b). Shorter accounts are given in Aubin (1972), Babuška & Aziz (1972), Oden & Reddy (1976a). More specialized treatments, particularly for nonlinear problems, are Ladyženskaja & Ural'ceva (1968), Lions (1969), Ekeland & Témam (1974). For regularity results, see Grisvard (1976), Kondrat'ev (1967).

For more classically oriented treatments, see for example Bers, John & Schechter (1964), Courant & Hilbert (1953, 1962), Miranda (1970), Stakgold (1968).

As an introduction to classical elasticity theory, notably for the elasticity problem, the clamped plate problem, the membrane problem, see for example Landau & Lifschitz (1967). For the variational formulations of problems in elasticity along the lines followed here, consult Duvaut & Lions (1972), Fichera (1972), Oden & Reddy (1976b).

CHAPTER 2

INTRODUCTION TO THE FINITE ELEMENT METHOD

Introduction

The basic scope of this chapter is to introduce the *finite element method* and to give a thorough *description* of the use of this method for approximating the solutions of second-order or fourth-order problems posed in variational form over a space V. A well-known approach for approximating such problems is *Galerkin's method*, which consists in defining similar problems, called *discrete problems*, over finite-dimensional subspaces V_h of the space V. Then the *finite element method in its simplest form* is a Galerkin's method characterized by *three basic aspects* in the construction of the space V_h: First, a *triangulation* $\mathcal{T}_h$ is established over the set $\bar{\Omega}$, i.e., the set $\bar{\Omega}$ is written as a finite union of *finite elements* $K \in \mathcal{T}_h$. Secondly, the function $v_h \in V_h$ are *piecewise polynomials*, in the sense that for each $K \in \mathcal{T}_h$, the spaces $P_K = \{v_{h|K};\ v_h \in V_h\}$ consist of polynomials. Thirdly, there should exist a basis in the space V_h whose functions have *small supports*. These three basic aspects are discussed in Section 2.1, where we also give simple criteria which insure the validity of inclusions such as $V_h \subset H^1(\Omega)$, $V_h \subset H_0^1(\Omega)$, etc... (Theorems 2.1.1 and 2.1.2). We also briefly indicate how the three basic aspects are still present in the more general finite element methods to be subsequently described. In this respect, we shall reserve the terminology *conforming finite element method* for the simplest such method (as described in this chapter).

In Section 2.2, we describe various examples of *finite elements*, which are either n-simplices (*simplicial* finite elements) or n-rectangles (*rectangular* finite elements), in which either all *degrees of freedom* are point values (*Lagrange* finite elements) or some degrees of freedom are *directional derivatives* (*Hermite* finite elements), which yield either the inclusion $X_h \subset H^1(\Omega)$ (finite elements *of class* $\mathscr{C}^0$) or the inclusion

$X_h \subset H^2(\Omega)$ (finite elements *of class* $\mathscr{C}^1$) when they are assembled in a *finite element space* X_h.

Then in Section 2.3, *finite elements* and *finite element spaces* are given general definitions, and we proceed to discuss their various properties. Of particular importance are the notion of an *affine family* of finite elements (where all the finite elements of the family can be obtained as images through affine mappings of a single *reference finite element*) and the notion of the P_K-*interpolation operator* (a basic relationship between these two notions is proved in Theorem 2.3.1). The P_K-interpolation operator and its global counterpart, the X_h-*interpolation operator* both play a fundamental role in the interpolation theory in Sobolev spaces that will be developed in the next chapter. We also show how to impose *boundary conditions* on functions in finite element spaces.

We conclude Section 2.3 by briefly indicating some reasons for which a particular finite element should be preferred to another one in practical computations.

In Section 2.4, we define the *convergence* and the *order of convergence* for a family of discrete problems. In this respect, *Céa's lemma* (Theorem 2.4.1) is crucial: The *error* $\|u - u_h\|$, i.e., the distance (measured in the norm of the space V) between the solution u of the original problem and the solution u_h of the discrete problem, is (up to a constant independent of the space V_h) bounded above by the distance $\inf_{v_h \in V_h} \|u - v_h\|$ between the function u and the subspace V_h. Indeed, all subsequent convergence results will be variations on this theme!

2.1. Basic aspects of the finite element method

The Galerkin and Ritz methods

Consider the linear abstract variational problem: Find $u \in V$ such that

$$\forall v \in V, \quad a(u, v) = f(v), \tag{2.1.1}$$

where the space V, the bilinear form $a(\cdot, \cdot)$, and the linear form f are assumed to satisfy the assumptions of the Lax–Milgram lemma (Theorem 1.1.3). Then the *Galerkin method* for approximating the solution of such a problem consists in defining similar problems in *finite-dimensional subspaces* of the space V. More specifically, with any finite-dimensional subspace V_h of V, we associate the *discrete problem*:

Find $u_h \in V_h$ such that

$$\forall v_h \in V_h, \quad a(u_h, v_h) = f(v_h). \tag{2.1.2}$$

Applying the Lax–Milgram lemma, we infer that such a problem has one and only one solution u_h, which we shall call a *discrete solution.*

Remark 2.1.1. In case the bilinear form is symmetric, the discrete solution is also characterized by the property (Theorem 1.1.2)

$$J(u_h) = \inf_{v_h \in V_h} J(v_h), \tag{2.1.3}$$

where the functional J is given by $J(v) = \frac{1}{2}a(v, v) - f(v)$. This alternate definition of the discrete solution is known as the *Ritz method.* □

The three basic aspects of the finite element method. Conforming finite element methods

Let us henceforth assume that the abstract variational problem (2.1.1) corresponds to a second-order or to a fourth-order elliptic boundary value problem posed over an open subset Ω of $\mathbf{R}^n$, with a Lipschitz-continuous boundary Γ. Typical examples of such problems have been studied in Section 1.2.

In order to apply Galerkin method, we face, by definition, the problem of constructing finite-dimensional subspaces V_h of spaces V such as $H_0^1(\Omega)$, $H^1(\Omega)$, $H_0^2(\Omega)$, etc...

The finite element method, in its simplest form, is a specific process of constructing subspaces V_h, which shall be called *finite element spaces.* This construction is characterized by *three basic aspects*, which for convenience shall be recorded as (FEM 1), (FEM 2) and (FEM 3), respectively, and which shall be described in this section.

(FEM 1) *The first aspect, and certainly the most characteristic, is that a triangulation $\mathcal{T}_h$ is established over the set $\bar{\Omega}$, i.e., the set $\bar{\Omega}$ is subdivided into a finite number of subsets K, called finite elements, in such a way that the following properties are satisfied:*

($\mathcal{T}_h 1$) $\bar{\Omega} = \cup_{K \in \mathcal{T}_h} K$.

($\mathcal{T}_h 2$) For each $K \in \mathcal{T}_h$, the set K is closed and the interior $\mathring{K}$ is non empty.

($\mathcal{T}_h 3$) For each distinct K_1, $K_2 \in \mathcal{T}_h$, one has $\mathring{K}_1 \cap \mathring{K}_2 = \phi$.

($\mathcal{T}_h 4$) For each $K \in \mathcal{T}_h$, the boundary ∂K is Lipschitz-continuous.

Remark 2.1.2. A fifth condition ($\mathcal{T}_h5$) relating "adjacent" finite elements, will be introduced in the next section. □

Once such a triangulation $\mathcal{T}_h$ is established over the set $\bar{\Omega}$, one defines a *finite element space* X_h through a specific process, which will be illustrated by many examples in the next section and subsequently. We shall simply retain for the moment that X_h is a *finite-dimensional* space of functions defined over the set $\bar{\Omega}$ (we shall deliberately ignore at this stage instances of finite element spaces whose "functions" may have two definitions across "adjacent" finite elements; see Section 2.3).

Given a finite element space X_h, we define the (finite-dimensional) spaces

$$P_K = \{v_{h|K};\quad v_h \in X_h\}$$

spanned by the restrictions $v_{h|K}$ of the functions $v_h \in X_h$ to the finite elements $K \in \mathcal{T}_h$. Without specific assumptions concerning the spaces P_K, $K \in \mathcal{T}_h$, there is no reason for an inclusion such as $X_h \subset H^1(\Omega)$ – let alone an inclusion such as $X_h \subset H^2(\Omega)$ – to hold.

In order to obtain such inclusions, we need additional conditions of a particularly simple nature, as we show in the next theorems (converses of these results hold, as we shall show in Theorems 4.2.1 and 6.2.1).

Remark 2.1.3. Here and subsequently, we shall comply with the use of the notation $H^m(K)$, in lieu of $H^m(\mathring{K})$. □

Theorem 2.1.1. *Assume that the inclusions $P_K \subset H^1(K)$ for all $K \in \mathcal{T}_h$ and $X_h \subset \mathscr{C}^0(\bar{\Omega})$ hold. Then the inclusions*

$$X_h \subset H^1(\Omega),$$
$$X_{oh} = \{v_h \in X_h;\quad v_h = 0 \quad \text{on} \quad \Gamma\} \subset H_0^1(\Omega),$$

hold.

Proof. Let a function $v \in X_h$ be given. We already know that it is in the space $L^2(\Omega)$. Therefore, by definition of the space $H^1(\Omega)$, we must find for each $i = 1, \ldots, n$, a function $v_i \in L^2(\Omega)$ such that

$$\forall \phi \in \mathscr{D}(\Omega), \int_\Omega v_i \phi \, dx = -\int_\Omega v \partial_i \phi \, dx.$$

For each i, a natural candidate is the function whose restriction to each finite element K is the function $\partial_i(v|_K)$. Since each finite element K has a Lipschitz-continuous boundary ∂K, we may apply Green's formula (1.2.4): For each $K \in \mathcal{T}_h$,

$$\int_K \partial_i(v|_K)\phi \, \mathrm{d}x = -\int_K v|_K \partial_i \phi \, \mathrm{d}x + \int_{\partial K} v|_K \phi \nu_{i,K} \, \mathrm{d}\gamma,$$

where $\nu_{i,K}$ is the i-th component of the unit outer normal vector along ∂K. By summing over all finite elements, we obtain

$$\int_\Omega v_i \phi \, \mathrm{d}x = -\int_\Omega v \partial_i \phi \, \mathrm{d}x + \sum_{K \in \mathcal{T}_h} \int_{\partial K} v|_K \phi \nu_{i,K} \, \mathrm{d}\gamma,$$

and the proof follows if we notice that the sum $\sum_{K \in \mathcal{T}_h} \int_{\partial K} v|_K \phi \nu_{i,K} \, \mathrm{d}\gamma$ vanishes: Either a portion of ∂K is a portion of the boundary Γ of Ω in which case $\phi = 0$ along this portion, or the contribution of adjacent elements is zero.

The boundary Γ being Lipschitz-continuous by assumption, the second inclusion follows from the characterization

$$H_0^1(\Omega) = \{v \in H^1(\Omega), \quad v = 0 \quad \text{on} \quad \Gamma\},$$

which was mentioned in Section 1.2. □

Assuming Theorem 2.1.1 applies, we shall therefore use the finite element space $V_h = X_{oh}$ if we are solving a second-order homogeneous Dirichlet problem, or $V_h = X_h$ if we are solving a second-order homogeneous or nonhomogeneous Neumann problem.

The proof of the next theorem is similar to that of Theorem 2.1.1 and, for this reason, is left to the reader as an exercise (Exercise 2.1.1).

Theorem 2.1.2. *Assume that the inclusions $P_K \subset H^2(K)$ for all $K \in \mathcal{T}_h$ and $X_h \subset \mathscr{C}^1(\bar{\Omega})$ hold. Then the inclusions*

$$X_h \subset H^2(\Omega),$$
$$X_{oh} = \{v_h \in X_h; \quad v_h = 0 \quad \text{on} \quad \Gamma\} \subset H^2(\Omega) \cap H_0^1(\Omega),$$
$$X_{ooh} = \{v_h \in X_h; \quad v_h = \partial_\nu v_h = 0 \quad \text{on} \quad \Gamma\} \subset H_0^2(\Omega),$$

hold. □

Thus if we are to solve a simply supported plate problem, or a clamped plate problem, we shall use the finite element space $V_h = X_{oh}$, or the finite element space $V_h = X_{ooh}$, respectively, as given in the previous theorem.

Let us return to the description of the finite element method.

(FEM 2) *The second basic aspect* of the finite element method is that *the spaces P_K, $K \in \mathcal{T}_h$, contain polynomials, or, at least, contain functions which are "close to" polynomials.* At this stage, we cannot be too specific about the underlying reasons for this aspect of the method but at least, we can say that

(i) it is the key to all convergence results as we shall see, and

(ii) it yields simple computations of the coefficients of the resulting linear system (see (2.1.4) below).

Let us now briefly examine how the discrete problem (2.1.2) is solved in practice. Let $(w_k)_{k=1}^M$ be a basis in the space V_h. Then *the solution* $u_h = \sum_{k=1}^M u_k w_k$ *of problem* (2.1.2) *is such that the coefficients* u_k *are solutions of the linear system*

$$\sum_{k=1}^{M} a(w_k, w_l)u_k = f(w_l), \quad 1 \leq l \leq M, \tag{2.1.4}$$

whose matrix is always invertible, since the bilinear form, being assumed to be V-elliptic, is *a fortiori* V_h-elliptic. By reference to the elasticity problem, the matrix $(a(w_k, w_l))$ and the vector $(f(w_l))$ are often called the *stiffness matrix* and the *load vector*, respectively.

In the choice of the basis $(w_k)_{k=1}^M$, it is of paramount importance, *from a numerical standpoint*, that *the resulting matrix possess as many zeros as possible.*

For *all* the examples which were considered in Section 1.2 the coefficients $a(w_k, w_l)$ are *integrals* of a specific form: For instance, in the case of the first examples, one has

$$a(w_k, w_l) = \int_\Omega \left(\sum_{i=1}^n \partial_i w_k \partial_i w_l + a w_k w_l \right) \mathrm{d}x,$$

so that a coefficient $a(w_k, w_l)$ vanishes whenever the $\mathrm{d}x$-measure of the intersection of the supports of the basis functions w_k and w_l is zero.

(FEM 3) As a consequence, we shall consider as *the third basic aspect* of the finite element method that *there exists at least one "canonical" basis in the space V_h whose corresponding basis functions have supports which are as "small" as possible, it being implicitly understood that these basis functions can be easily described.*

Remark 2.1.4. When the bilinear form is symmetric, the matrix $(a(w_k, w_l))$ is symmetric and positive definite, which is an advantage for the numerical solution of the linear system (2.1.4). By contrast, this is not generally the case for standard finite-difference methods, except for rectangular domains.

Assuming again the symmetry of the bilinear form, one could conceivably start out with any given basis, and, using the Gram–Schmidt orthonormalization procedure, construct a new basis $(w_k^*)_{k=1}^M$ which is orthonormal with respect to the inner product $a(\cdot, \cdot)$. This is indeed an efficient way of getting a sparse matrix since the corresponding matrix $(a(w_k^*, w_l^*))$ is the identity matrix! However, this process is not recommended from a *practical* standpoint: For comparable computing times, it yields worse results than the solution by standard methods of the linear system corresponding to the "canonical" basis. □

It was mentioned earlier that the three basic aspects were characteristic of the finite element method *in its simplest form*. Indeed, *there are more general finite element methods*:

(i) One may start out with *more general variational problems*, such as variational inequalities (see Section 5.1) or various nonlinear problems (see Sections 5.2 and 5.3), or different variational formulations (see Chapter 7).

(ii) The space V_h, in which one looks for the discrete solution, may no longer be a subspace of the space V. This may happen when the boundary of the set Ω is curved, for instance. Then it cannot be exactly triangulated in general by standard finite elements and thus it is replaced by an approximate set Ω_h (see Section 4.4). This also happens when the functions in the space V_h lack the proper continuity across adjacent finite elements (see the "nonconforming" methods described in Section 4.2 and Section 6.2).

(iii) Finally, the bilinear form and the linear form may be approximated. This is the case for instance when numerical integration is used for computing the coefficients of the linear system (2.1.4) (see Section 4.1), or for the shell problem (see Section 8.2).

Nevertheless, *it is characteristic of all these more general finite element methods that the three basic aspects are again present.*

To conclude these general considerations, we shall reserve the terminology *conforming finite element methods* for the finite element methods described at the beginning of this section, i.e., for which V_h is a subspace of the space V, and the bilinear form and the linear form of the discrete problem are identical to the original ones.

Exercises

2.1.1. Prove Theorem 2.1.2.

2.1.2. The purpose of this problem is to give another proof of the Lax–Milgram lemma (Theorem 1.1.3; see also Exercise 1.1.2) in case the Hilbert space V is separable. Otherwise the bilinear form and the linear form satisfy the same assumptions as in Theorem 1.1.3.

(i) Let V_h be any finite-dimensional subspace of the space V, and let u_h be the discrete solution of the associated discrete problem (2.1.2). Show that there exists a constant C independent of the subspace V_h such that $\|u_h\| \leqslant C$ (as usual, there is a simpler proof when the bilinear form is symmetric).

(ii) The space V being separable, there exists a nested sequence $(V_\nu)_{\nu\in N}$ of finite-dimensional subspaces such that $(\bigcup_{\nu\in N} V_\nu)^- = V$. Let $(u_\nu)_{\nu\in N}$ be the sequence of associated discrete solutions. Show that there exists a subsequence of the sequence $(u_\nu)_{\nu\in N}$ which weakly converges to a solution u of the original variational problem.

(iii) Show that the whole sequence converges in the norm of V to the solution u.

(iv) Show that the Sobolev spaces $H^m(\Omega)$ are separable.

2.2. Examples of finite elements and finite element spaces

Requirements for finite element spaces

Throughout this section, we assume that we are using a *conforming finite element method* for solving a second-order or a fourth-order boundary value problem. Let us first summarize the various requirements that a finite element space X_h must satisfy, according to the discussion made in the previous section. Such a space is associated with a triangulation $\mathcal{T}_h$ of a set $\bar{\Omega} = \bigcup_{K\in\mathcal{T}_h} K$ (FEM 1), and for each finite element $K \in \mathcal{T}_h$, we define the space

$$P_K = \{v_h|_K;\quad v_h \in X_h\}. \tag{2.2.1}$$

Then the requirements are the following:

(i) For each $K \in \mathcal{T}_h$, the space P_K should consist of functions which are polynomials or "nearly polynomials" (FEM 2).

(ii) By Theorems 2.1.1 and 2.1.2, inclusions such as $X_h \subset \mathscr{C}^0(\bar{\Omega})$ or $X_h \subset \mathscr{C}^1(\bar{\Omega})$ should hold, depending upon whether we are solving a second-order or a fourth-order problem. For the time being, *we shall ignore boundary conditions*, which we shall take into account in the next section only.

(iii) Finally, we must check that there exists one canonical basis in the space X_h, whose functions have "small" supports and are easy to describe (FEM 3).

In this section, we shall describe various finite elements K which are all polyedra in $\mathbf{R}^n$, and which are sometimes called *straight finite elements*. As a consequence, *we have to restrict ourselves to problems which are posed over sets $\bar{\Omega}$ which are themselves polyedra*, in which case we shall say that the set Ω is *polygonal*.

First examples of finite elements for second order problems: n-Simplices of type (k), $(3')$

We begin by examining examples for which the inclusion $X_h \subset \mathscr{C}^0(\bar{\Omega})$ holds, and which are the most commonly used by engineers for solving second-order problems with conforming finite element methods. Inasmuch as such problems are most often found in mechanics of continua, it is clear that the value to be assigned in practice to the dimension n in the forthcoming examples is either 2 or 3 (see the examples given in Section 1.2).

We equip the space $\mathbf{R}^n$ with its canonical basis $(e_i)_{i=1}^n$. For each integer $k \geq 0$, we shall denote by P_k the space of all polynomials of degree $\leq k$ in the variables $x_1, x_2, \ldots, x_n$, i.e., a polynomial $p \in P_k$ is of the form

$$p: x = (x_1, x_2, \ldots, x_n) \in \mathbf{R}^n \to p(x) = \sum_{\sum_{i=1}^n \alpha_i \leq k} \gamma_{\alpha_1 \alpha_2 \cdots \alpha_n} x_1^{\alpha_1} x_2^{\alpha_2} \cdots x_n^{\alpha_n},$$

for appropriate coefficients $\gamma_{\alpha_1 \alpha_2 \cdots \alpha_n}$, or using the multi-index notation,

$$p: x \in \mathbf{R}^n \to p(x) = \sum_{|\alpha| \leq k} \gamma_\alpha x^\alpha.$$

The dimension of the space P_k is given by

$$\dim P_k = \binom{n+k}{k}. \tag{2.2.2}$$

If Φ is a space of functions defined over $\mathbf{R}^n$, and if A is any subset of $\mathbf{R}^n$, we shall denote by $\Phi(A)$ the space formed by the restrictions to the set A of the functions in the space Φ. Thus, for instance, we shall let

$$P_k(A) = \{p|_A; \quad p \in P_k\}. \tag{2.2.3}$$

Notice that the dimension of the space $P_k(A)$ is the same as that of the space $P_k = P_k(\mathbf{R}^n)$ as long as the interior of the set A is not empty.

In $\mathbf{R}^n$, a (nondegenerate) *n-simplex* is the convex hull K of $(n+1)$ points $a_j = (a_{ij})_{i=1}^n \in \mathbf{R}^n$, which are called the *vertices* of the n-simplex, and which are such that the matrix

$$A = \begin{pmatrix} a_{11} & a_{12} \cdots & a_{1,n+1} \\ a_{21} & a_{22} \cdots & a_{2,n+1} \\ \vdots & \vdots & \vdots \\ a_{n1} & a_{n2} \cdots & a_{n,n+1} \\ 1 & 1 \cdots & 1 \end{pmatrix} \tag{2.2.4}$$

is regular (equivalently, the $(n+1)$ points a_j are not contained in a hyperplane). Thus, one has

$$K = \left\{x = \sum_{j=1}^{n+1} \lambda_j a_j; \quad 0 \leq \lambda_j \leq 1, \quad 1 \leq j \leq n+1, \quad \sum_{j=1}^{n+1} \lambda_j = 1\right\}. \tag{2.2.5}$$

Notice that a 2-simplex is a triangle and that a 3-simplex is a tetrahedron.

For any integer m with $0 \leq m \leq n$, an *m-face* of the n-simplex K is any m-simplex whose $(m+1)$ vertices are also vertices of K. In particular, any $(n-1)$-face is simply called a *face*, any 1-face is called an *edge*, or a *side*.

The *barycentric coordinates* $\lambda_j = \lambda_j(x)$, $1 \leq j \leq n+1$, of any point $x \in \mathbf{R}^n$, *with respect to the* $(n+1)$ *points* a_j, are the (unique) solutions of the linear system

$$\begin{cases} \sum_{j=1}^{n+1} a_{ij}\lambda_j = x_i, \quad 1 \leq i \leq n, \\ \sum_{j=1}^{n+1} \lambda_j = 1, \end{cases} \tag{2.2.6}$$

whose matrix is precisely the matrix A of (2.2.4). By inspection of the linear system (2.2.6), one sees that *the barycentric coordinates are affine functions of* $x_1, x_2, \ldots, x_n$ (i.e., they belong to the space P_1):

$$\lambda_i = \sum_{j=1}^{n} b_{ij}x_j + b_{in+1}, \quad 1 \leq i \leq n+1, \tag{2.2.7}$$

where the matrix $B = (b_{ij})$ is the inverse of the matrix A.

The *barycenter*, or *center of gravity*, of an n-simplex K is the point of K whose all barycentric coordinates are equal to $1/(n+1)$.

To describe the first finite element, we need to prove that *a polynomial* $p: x \to \sum_{|\alpha|\leq 1} \gamma_\alpha x^\alpha$ *of degree* 1 *is uniquely determined by its values at the* $(n+1)$ *vertices* a_j *of any* n*-simplex in* $\mathbf{R}^n$. To see this, it suffices to show that the linear system

$$\sum_{|\alpha|\leq 1} \gamma_\alpha (a_j)^\alpha = \mu_j, \quad 1 \leq j \leq n+1,$$

has one and only one solution $(\gamma_\alpha, |\alpha| \leq 1)$ for all right-hand sides μ_j, $1 \leq j \leq n+1$. Since $\dim P = \operatorname{card}(\bigcup_{j=1}^{n+1}\{a_j\}) = n+1$, the matrix of this linear system is *square*, and therefore it suffices to prove either *uniqueness* or *existence*. In this case, the existence is clear: The barycentric coordinates verify $\lambda_i(a_j) = \delta_{ij}$, $1 \leq i, j \leq n+1$, and thus the polynomial

$$x \in \mathbf{R}^n \to \sum_{i=1}^{n+1} \mu_i \lambda_i(x)$$

has the desired property. As a consequence, we have the *identity*

$$\forall p \in P_1, \; p = \sum_{i=1}^{n+1} p(a_i)\lambda_i. \tag{2.2.8}$$

Although we shall not repeat this argument in the sequel, it will be often implicitly used.

A polynomial $p \in P_1$ being completely determined by its values $p(a_i)$, $1 \leq i \leq n+1$, we can now define the simplest finite element, which we shall call *n-simplex of type* (1): The set K is an n-simplex with vertices a_i, $1 \leq i \leq n+1$, the space P_K is the space $P_1(K)$, and the *degrees of freedom of the finite element*, i.e., those parameters which uniquely define a function in the space P_K, consist of the values at the vertices. Denoting by Σ_K the corresponding *set of degrees of freedom*, we shall write symbolically

$$\Sigma_K = \{p(a_i), \; 1 \leq i \leq n+1\}.$$

In Fig. 2.2.1, we have recorded the main characteristics of this finite element for arbitrary n, along with the figures in the special cases $n = 2$ and 3. In case $n = 2$, this element is also known as *Courant's triangle* (see the section "Bibliography and Comments").

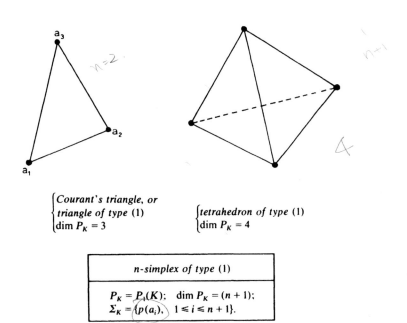

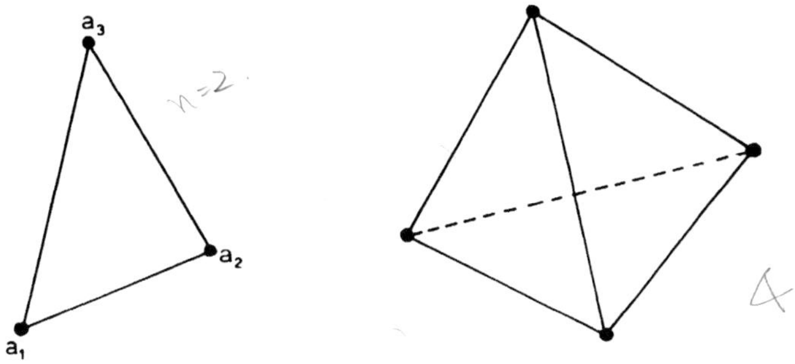

Courant's triangle, or triangle of type (1)
dim $P_K = 3$

tetrahedron of type (1)
dim $P_K = 4$

n-simplex of type (1)
$P_K = P_1(K)$; dim $P_K = (n+1)$; $\Sigma_K = \{p(a_i), \ 1 \leq i \leq n+1\}$.

Fig. 2.2.1

Let us call $a_{ij} = \frac{1}{2}(a_i + a_j)$, $1 \leq i < j \leq n+1$, the mid-points of the edges of the n-simplex K. Since $\lambda_k(a_{ij}) = \frac{1}{2}(\delta_{ki} + \delta_{kj})$, $1 \leq i < j \leq n+1$, $1 \leq k \leq n+1$, we obtain the identity (where, here and subsequently, indices $i, j, k, \ldots$, are always assumed to take all possible values in the set $\{1, 2, \ldots, n\}$ whenever this fact is not specified)

$$\forall p \in P_2, \quad p = \sum_i \lambda_i(2\lambda_i - 1)p(a_i) + \sum_{i<j} 4\lambda_i\lambda_j p(a_{ij}), \tag{2.2.9}$$

which yields the definition of a finite element, called the *n-simplex of type* (2): the space P_K is $P_2(K)$, and the set Σ_K consists of the values at the vertices and at the mid-points of the edges (Fig. 2.2.2).

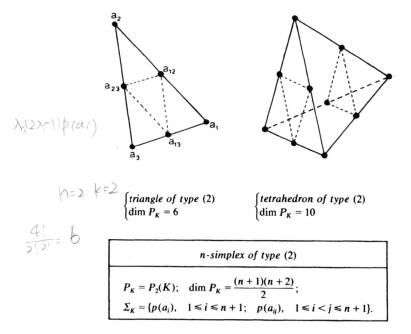

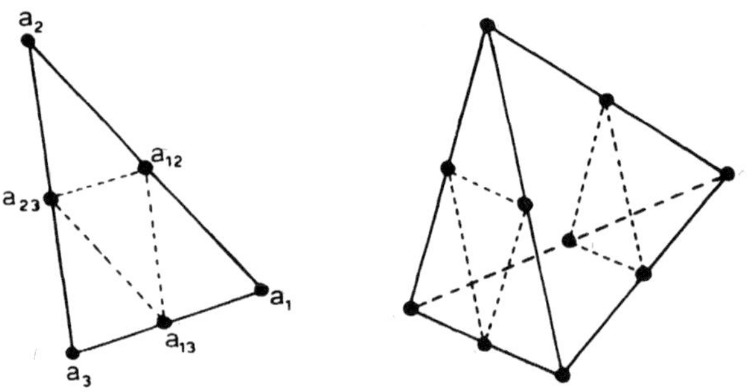

$\begin{cases}\textit{triangle of type } (2)\\ \dim P_K = 6\end{cases}$ $\begin{cases}\textit{tetrahedron of type } (2)\\ \dim P_K = 10\end{cases}$

n-simplex of type (2)
$P_K = P_2(K)$; $\dim P_K = \dfrac{(n+1)(n+2)}{2}$; $\Sigma_K = \{p(a_i),\ 1 \leq i \leq n+1;\ p(a_{ij}),\ 1 \leq i < j \leq n+1\}$.

Fig. 2.2.2

Let $a_{iij} = \frac{1}{3}(2a_i + a_j)$ for $i \neq j$, and $a_{ijk} = \frac{1}{3}(a_i + a_j + a_k)$ for $i < j < k$. From the identity

$$\forall p \in P_3, \quad p = \sum_i \frac{\lambda_i(3\lambda_i - 1)(3\lambda_i - 2)}{2} p(a_i) + \sum_{i \neq j} \frac{9\lambda_i\lambda_j(3\lambda_i - 1)}{2} p(a_{iij}) + \sum_{i<j<k} 27\lambda_i\lambda_j\lambda_k p(a_{ijk}), \tag{2.2.10}$$

we deduce the definition of the *n-simplex of type* (3) (Fig. 2.2.3).

One may define analogous finite elements with polynomials of arbitrary degree, but they are not often used. In this respect, we leave to the reader the proof of the following theorem (Exercise 2.2.2), from which for any integer k, the definition of the *n-simplex of type* (k) can be easily derived.

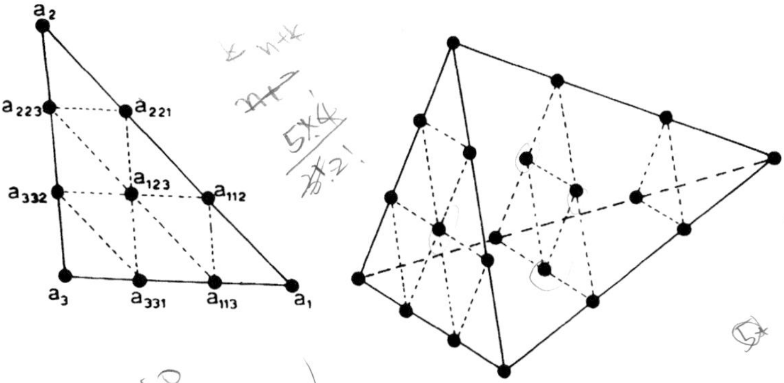

$\begin{cases} \text{triangle of type (3)} \\ \dim P_K = 10 \end{cases}$ $\begin{cases} \text{tetrahedron of type (3)} \\ \dim P_K = 20 \end{cases}$

n-simplex of type (3)
$P_K = P_3(K);\quad \dim P_K = \dfrac{(n+1)(n+2)(n+3)}{6};$ $\Sigma_K = \{p(a_i),\ 1 \le i \le n+1;\ p(a_{iij}),\ 1 \le i, j \le n+1,\ i \ne j;$ $p(a_{ijk}),\ 1 \le i < j < k \le n+1\}.$

Fig. 2.2.3

Theorem 2.2.1. *Let K be an n-simplex with vertices a_j, $1 \le j \le n+1$. Then for a given integer $k \ge 1$, any polynomial $p \in P_k$ is uniquely determined by its values on the set*

$$L_k(K) = \left\{ x = \sum_{j=1}^{n+1} \lambda_j a_j;\ \sum_{j=1}^{n+1} \lambda_j = 1,\right.$$

$$\left. \lambda_j \in \left\{0, \frac{1}{k}, \ldots, \frac{k-1}{k}, 1\right\},\quad 1 \le j \le n+1 \right\}. \tag{2.2.11}$$

□

Let us now examine a modification of the n-simplex of type (3), in which the degrees of freedom $p(a_{ijk})$ are no longer present, and which is often preferred by the engineers to the previous element. To describe the corresponding finite element, we need the following result.

Theorem 2.2.2. *For each triple (i, j, k) with $i<j<k$, let*

$$\phi_{ijk}(p) = 12p(a_{ijk}) + 2 \sum_{l=i,j,k} p(a_l) - 3 \sum_{\substack{l,m=i,j,k \\ l\neq m}} p(a_{llm}). \tag{2.2.12}$$

Then any polynomial in the space

$$P_3' = \{p \in P_3;\ \phi_{ijk}(p) = 0,\quad 1 \leq i<j<k \leq n+1\} \tag{2.2.13}$$

is uniquely determined by its values at the vertices a_i, $1 \leq i \leq n+1$, and at the points a_{iij}, $1 \leq i, j \leq n+1$, $i \neq j$. In addition, the inclusion

$$P_2 \subset P_3' \tag{2.2.14}$$

holds.

Proof. The $\binom{n+1}{3}$ degrees of freedom ϕ_{ijk} are linearly independent (since $\phi_{ijk}(p) = 12p(a_{ijk}) + \cdots$) and thus, the dimension of the space P_3' is

$$\dim P_3' = \dim P_3 - \binom{n+1}{3} = (n+1)^2,$$

i.e., precisely the number of degrees of freedom. Using the identity (2.2.10), and arguing as before, we obtain the identity

$$\begin{aligned} \forall p \in P_3',\quad p = &\sum_i \left(\frac{1}{2}\lambda_i(3\lambda_i - 1)(3\lambda_i - 2) - \frac{9}{2}\lambda_i \sum_{\substack{j<k \\ j,k\neq i}} \lambda_j\lambda_k\right) p(a_i) + \\ &+ \sum_{i\neq j} \left(\frac{9}{2}\lambda_i\lambda_j(3\lambda_i - 1) + \frac{27}{4}\lambda_i\lambda_j \sum_{k\neq i,j} \lambda_k\right) p(a_{iij}), \end{aligned} \tag{2.2.15}$$

which proves the first part of the theorem.

To prove that the inclusion (2.2.14) holds, let p be a polynomial of degree ≤ 2 and let $A \in \mathscr{L}_2(\mathbf{R}^n; \mathbf{R})$ be its second derivative (which is constant). From the expansions

$$p(a_l) = p(a_{ijk}) + Dp(a_{ijk})(a_l - a_{ijk}) + \tfrac{1}{2}A(a_l - a_{ijk})^2, \quad l \in I,$$

valid for any triple $I = \{i, j, k\}$ with $i<j<k$, we deduce

$$\sum_{l\in I} p(a_l) = 3p(a_{ijk}) + \frac{1}{2}\sum_{l\in I} A(a_l - a_{ijk})^2,$$

since $\Sigma_{l\in I}(a_l - a_{ijk}) = 0$. Likewise, from the expansions

$$p(a_{llm}) = p(a_{ijk}) + Dp(a_{ijk})(a_{llm} - a_{ijk}) + \tfrac{1}{2}A(a_{llm} - a_{ijk})^2, \quad l, m \in I, \quad l \neq m,$$

we deduce

$$\sum_{\substack{l,m\in I\\ l\neq m}} p(a_{llm}) = 6p(a_{ijk}) + \frac{1}{2}\sum_{\substack{l,m\in I\\ l\neq m}} A(a_{llm} - a_{ijk})^2,$$

taking into account that $a_{ijk} = \frac{1}{2}(a_{iij} + a_{kkj}) = \frac{1}{2}(a_{jjk} + a_{iik}) = \frac{1}{2}(a_{kki} + a_{jji})$. Because A is a linear mapping, and because

$$a_{llm} - a_{ijk} = \tfrac{1}{3}(2(a_l - a_{ijk}) + (a_m - a_{ijk})),$$

we can write

$$\sum_{l\in I} A(a_l - a_{ijk})^2 - \frac{3}{2}\sum_{\substack{l,m\in I\\ l\neq m}} A(a_{llm} - a_{ijk})^2 = -\frac{2}{3}A\Big(\sum_{l\in I}(a_l - a_{ijk})\Big)^2 = 0,$$

and the proof is complete. □

From Theorem 2.2.2 we deduce the definition of the *n-simplex of type* (3') (Fig. 2.2.4).

Assembly in triangulations. The associated finite element spaces

Next we examine the question of constructing triangulations, using anyone of the finite elements previously described. Being non degenerate n-simplices, these have non empty interiors and Lipschitz-continuous boundaries, and therefore properties ($\mathcal{T}_h2$) and ($\mathcal{T}_h4$) are satisfied. To construct triangulations in the sense understood in Section 2.2, we shall write $\bar{\Omega} = \bigcup_{K\in\mathcal{T}_h} K$ in such a way that the n-simplices have piecewise disjoint interiors (cf. properties ($\mathcal{T}_h1$) and ($\mathcal{T}_h3$)). In order to satisfy inclusions such as $X_h \subset \mathscr{C}^0(\bar{\Omega})$ (and $X_h \subset \mathscr{C}^1(\bar{\Omega})$ later on), we shall impose a fifth condition on a triangulation made up of n-simplices, namely:

($\mathcal{T}_h5$) *Any face of any n-simplex K_1 in the triangulation is either a subset of the boundary Γ, or a face of another n-simplex K_2 in the triangulation.*

In the second case, the n-simplices K_1 and K_2 are said to be *adjacent*.

An example of a triangulation for $n = 2$ is given in Fig. 2.2.5, while

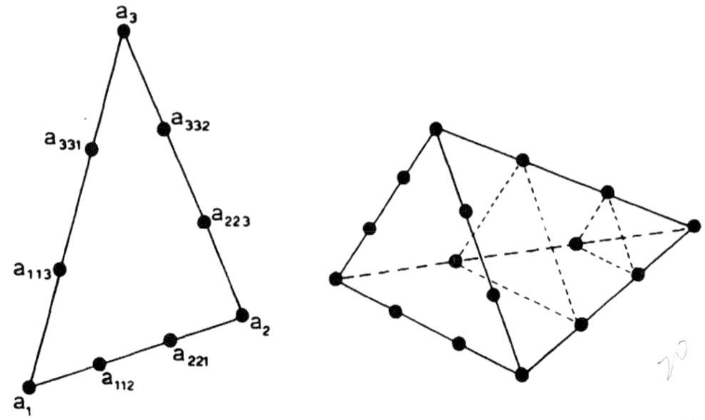

$\begin{cases} \textit{triangle of type } (3') \\ \dim P_K = 9. \end{cases}$ $\begin{cases} \textit{tetrahedron of type } (3') \\ \dim P_K = 16. \end{cases}$

n-simplex of type (3′)
$P_K = P_3'(K)$ (cf. (2.2.13)); $\dim P_K = (n+1)^2$; $\Sigma_K = \{p(a_i),\ 1 \le i \le n+1,\ p(a_{iij}),\ \ 1 \le i, j \le n+1,\ \ i \ne j\}.$

Fig. 2.2.4

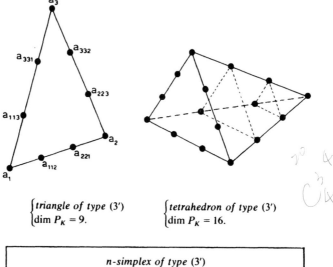

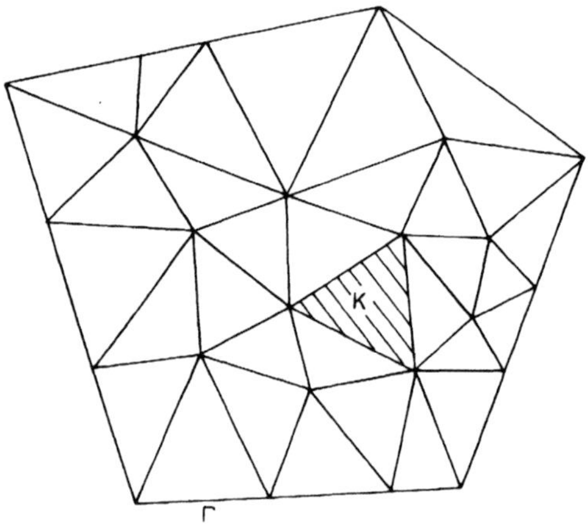

Fig. 2.2.5

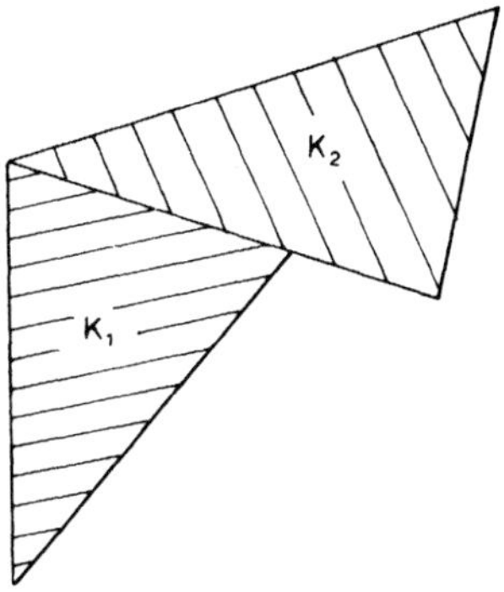

Fig. 2.2.6.

Fig. 2.2.6 shows an example of a "forbidden situation" since the intersection of K_1 and K_2 is not an edge of K_2.

Given a triangulation $\mathcal{T}_h$, we associate in a natural way a finite element space X_h of functions v_h: $\bar{\Omega} \to \mathbf{R}$ with each type of finite element:

With n-simplices of type (1), a function $v_h \in X_h$

(i) is such that each restriction $v_h|_K$ is in the space $P_K = P_1(K)$ for each $K \in \mathcal{T}_h$, and

(ii) is completely determined by its values at all the vertices of the triangulation.

Likewise, with n-simplices of type (2), a function of X_h

(i) is in the space $P_K = P_2(K)$ for each $K \in \mathcal{T}_h$, and

(ii) is completely determined by its values at all the vertices and all the mid-points of the edges of the triangulation.

Similar constructions hold for n-simplices of type (3) or (3′).

In all cases, a function v_h in the space X_h is seen to be determined by *degrees of freedom* which make up a set of the form

$$\Sigma_h = \{v_h(b);\quad b \in \mathcal{N}_h\}, \tag{2.2.16}$$

where $\mathcal{N}_h$ is a finite subset of $\bar{\Omega}$. The set Σ_h is the *set of degrees of freedom of the finite element space* X_h.

One should observe that *if there is no ambiguity in the definition of the degrees of freedom across adjacent finite elements, it is precisely because we have satisfied requirement* ($\mathcal{T}_h5$). This requirement also plays a crucial role in the proof of the following result.

Theorem 2.2.3. *Let X_h be the finite element space associated with n-simplices of type (k) for any integer $k \geqslant 1$ or with n-simplices of type $(3')$. Then the inclusion*

$$X_h \subset \mathscr{C}^0(\bar{\Omega}) \cap H^1(\Omega)$$

holds.

Proof. We shall give the proof in case $n = 2$ and for triangles of type (2), leaving the other cases as a problem (Exercise 2.2.3). Given a function v_h in the space X_h, consider the two functions $v_h|_{K_1}$ and $v_h|_{K_2}$ along the common side $K' = [b_i, b_j]$ of two adjacent triangles K_1 and K_2 (Fig. 2.2.7). Let t denote an abscissa along the axis containing the segment K'. Considered as functions of t, the two functions $v_h|_{K_1}$ and $v_h|_{K_2}$ are quadratic polynomials along K', whose values coincide at the three points b_i, b_j, $b_{ij} = (b_i + b_j)/2$. Therefore these polynomials are identical, and the inclusion $X_h \subset \mathscr{C}^0(\bar{\Omega})$ holds. Finally the inclusion $X_h \subset H^1(\Omega)$ is a consequence of Theorem 2.1.1. □

It remains to verify requirement (FEM 3), i.e., that there is indeed a canonical choice for basis functions with small supports. In each case,

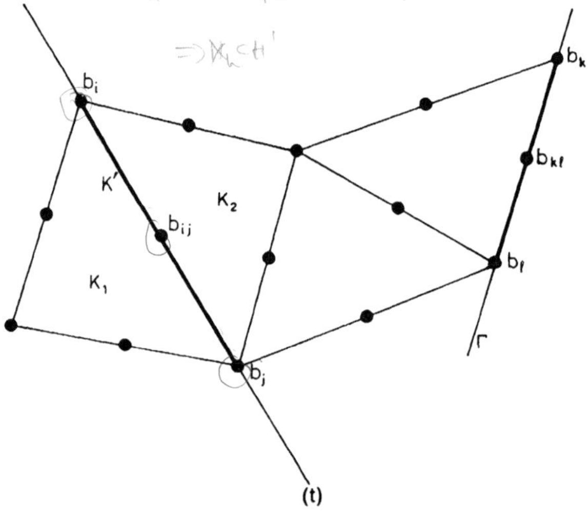

Fig. 2.2.7

the set Σ_h of degrees of freedom of the space is of the form

$$\Sigma_h = \{(b_k); \quad 1 \leqslant k \leqslant M\}. \tag{2.2.17}$$

If we define functions w_k, $1 \leqslant k \leqslant M$, by the conditions

$$w_k \in X_h \quad \text{and} \quad w_k(b_l) = \delta_{kl}, \quad 1 \leqslant k,l \leqslant M, \tag{2.2.18}$$

it is seen that (i) such functions form a basis of the space X_h and that (ii) they have "small" supports. In Fig. 2.2.8, we have represented the three types of supports which are encountered when triangles of type (3) are employed, for instance.

n-Rectangles of type (k). Rectangles of type (2′), (3′). *Assembly in triangulations*

Before we turn to a second category of finite elements, we need a few definitions. For each integer $k \geqslant 0$, we shall denote by Q_k the space of all polynomials which are of degree $\leqslant k$ with respect to each one of the n variables $x_1, x_2, \ldots, x_n$, i.e., a polynomial $p \in Q_k$ is of the form

$$p: x = (x_1, x_2, \ldots, x_n) \in \mathbf{R}^n \to p(x) = \sum_{\substack{\alpha_i \leqslant k, \\ 1 \leqslant i \leqslant n}} \gamma_{\alpha_1\alpha_2\cdots\alpha_n} x_1^{\alpha_1} x_2^{\alpha_2} \cdots x_n^{\alpha_n},$$

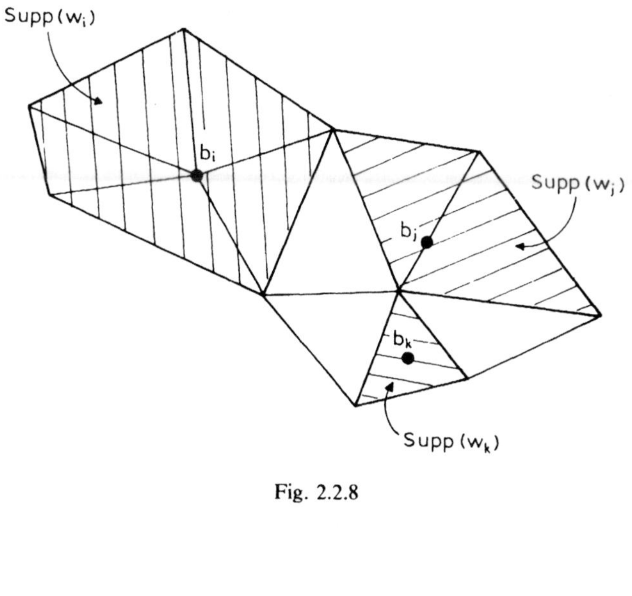

Fig. 2.2.8

for appropriate coefficients $\gamma_{\alpha_1\alpha_2\cdots\alpha_n}$. The dimension of the space Q_k is given by

$$\dim Q_k = (k+1)^n \tag{2.2.19}$$

and the inclusions

$$P_k \subset Q_k \subset P_{nk} \tag{2.2.20}$$

hold.

Notice that the dimension of the space $Q_k(A)$ is the same as that of the space $Q_k = Q_k(\mathbf{R}^n)$ as long as the interior of the set $A \subset \mathbf{R}^n$ is not empty.

Theorem 2.2.4. *A polynomial $p \in Q_k$ is uniquely determined by its values on the set*

$$\hat{M}_k = \left\{x = \left(\frac{i_1}{k}, \frac{i_2}{k}, \ldots, \frac{i_n}{k}\right) \in \mathbf{R}^n;\quad i_j \in \{0, 1, \ldots, k\},\quad 1 \leq j \leq n\right\}. \tag{2.2.21}$$

Proof. It suffices to use the identity

$$\forall p \in Q_k,\quad p = \sum_{\substack{0 \leq i_j \leq k \\ 1 \leq j \leq n}} \prod_{j=1}^{n} \left(\prod_{\substack{i'_j=0 \\ i'_j \neq i_j}}^{k} \frac{kx_j - i'_j}{i_j - i'_j}\right) p\left(\frac{i_1}{k}, \frac{i_2}{k}, \ldots, \frac{i_n}{k}\right). \qquad \square \tag{2.2.22}$$

In $\mathbf{R}^n$, an *n-rectangle*, or simply a *rectangle* if $n = 2$, is a set of the form

$$K = \prod_{i=1}^{n} [a_i, b_i] = \{x = (x_1, x_2, \ldots, x_n);\quad a_i \leq x_i \leq b_i,\quad 1 \leq i \leq n\}, \tag{2.2.23}$$

with $-\infty < a_i < b_i < +\infty$ for each i, i.e., it is a product of compact intervals with non-empty interiors. A *face* of K is any one of the sets

$$\{a_j\} \times \prod_{\substack{i=1 \\ i \neq j}}^{n} [a_i, b_i] \quad \text{or} \quad \{b_j\} \times \prod_{\substack{i=1 \\ i \neq j}}^{n} [a_i, b_i],\quad 1 \leq j \leq n,$$

while an *edge* of K, also called a *side*, is any one of the sets

$$[a_j, b_j] \times \prod_{\substack{i=1 \\ i \neq j}}^{n} \{c_i\},$$

with $c_i = a_i$ or b_i, $1 \leqslant i \leqslant n$, $i \neq j$, $1 \leqslant j \leqslant n$. A *vertex* of K is any point $x = (x_1, x_2, \ldots, x_n)$ of K with $x_i = a_i$ or b_i, $1 \leqslant i \leqslant n$.

Observe that the set $\hat{M}_k$ of (2.2.21) is a subset of a particular n-rectangle, namely the *unit hypercube* $[0, 1]^n$. Then, given any n-rectangle K, we infer that a polynomial $p \in Q_k$ is uniquely determined by its values on the subset

$$M_k(K) = F_K(\hat{M}_k) \tag{2.2.24}$$

of the n-rectangle K, where F_K is a *diagonal affine mapping*, i.e., of the form $F_K: x \in \mathbf{R}^n \to F_K(x) = B_K x + b_K$, with b_K a vector in $\mathbf{R}^n$ and B_K an $n \times n$ diagonal matrix, such that $K = F_K([0, 1]^n)$. From this, we deduce the definition of finite elements, called *n-rectangles of type* (k).

Just as in the case of n-simplices, the values $k = 1, 2$ or 3 are the most commonly encountered. In Fig. 2.2.9, 2.2.10 and 2.2.11, the corresponding elements are represented for $n = 2$ and 3, and the numbering of the points occurring in the sets of degrees of freedom is also indicated for $n = 2$.

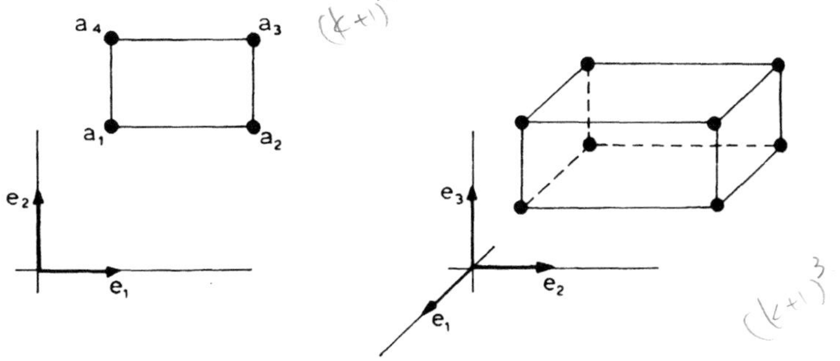

rectangle of type (1)
dim $P_K = 4$

3-rectangle of type (1)
dim $P_K = 8$

n-rectangle of type (1)
$P_K = Q_1(K)$; dim $P_K = 2^n$; $\Sigma_K = \{p(a);\ a \in M_1(K)\}$ (cf. (2.2.24))

Fig. 2.2.9

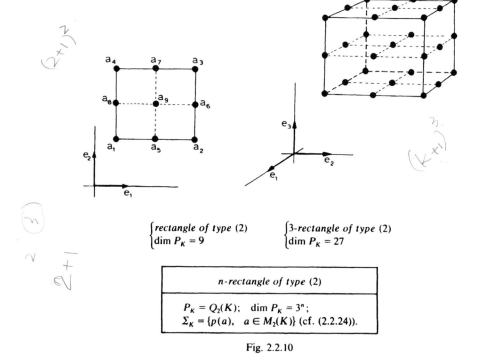

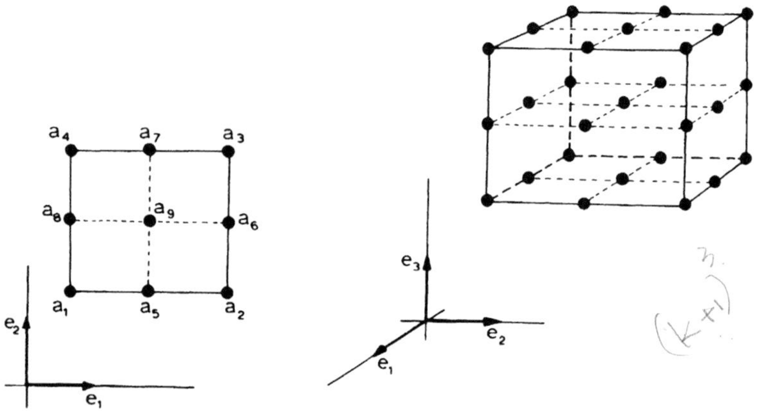

$\begin{cases} \text{rectangle of type (2)} \\ \dim P_K = 9 \end{cases}$ $\qquad$ $\begin{cases} \text{3-rectangle of type (2)} \\ \dim P_K = 27 \end{cases}$

n-rectangle of type (2)
$P_K = Q_2(K)$; $\dim P_K = 3^n$; $\Sigma_K = \{p(a),\ a \in M_2(K)\}$ (cf. (2.2.24)).

Fig. 2.2.10

For the numbering of the nodes when $n = 2$, we have followed this rule: Assuming, without loss of generality, that the set K is the *unit square* $[0, 1]^2$, four points are consecutively numbered if they are the vertices of a square centered at the point $(\frac{1}{2}, \frac{1}{2})$. This rule allows for particularly simple expressions of the corresponding functions p_i appearing in identities of the form

$$\forall p \in Q_k, \quad p = \sum_i p(a_i)p_i,$$

which are special cases (for $k = 1, 2, 3$ and $n = 2$) of the identity (2.2.22). Notice that the coordinates of a given point with respect to the four vertices a_i, $1 \leqslant i \leqslant 4$, of the unit square are

$$(x_1, x_2), \quad (x_2, 1 - x_1), \quad (1 - x_1, 1 - x_2), \quad (1 - x_2, x_1),$$

respectively. Then, if we introduce the variables

$$x_3 = 1 - x_1, \quad x_4 = 1 - x_2, \tag{2.2.25}$$

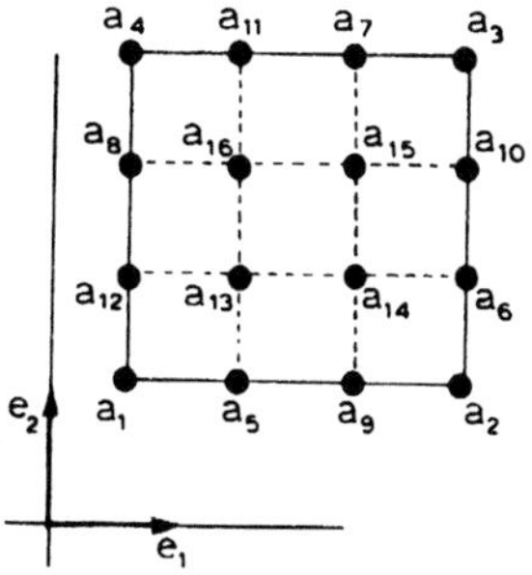

$\begin{cases}\textit{rectangle of type } (3)\\ \dim P_K = 16\end{cases}$

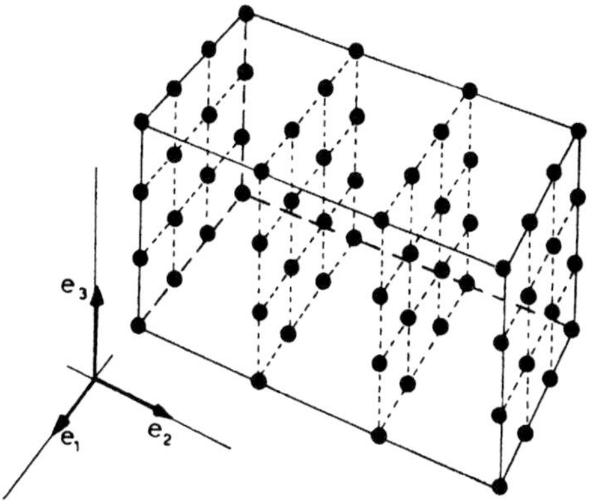

$\begin{cases}\textit{3-rectangle of type } (3)\\ \dim P_K = 64\end{cases}$

n-rectangle of type (3)
$P_K = Q_3(K)$; $\dim P_K = 4^n$: $\Sigma_K = \{p(a);\ \ a \in M_3(K)\}$ (cf. (2.2.24)).

Fig. 2.2.11

the four functions p_i are obtained through circular permutations of the variables x_1, x_2, x_3, x_4 (such permutations correspond to rotations of $+\pi/2$ around the point $(\frac{1}{2}, \frac{1}{2})$).

Corresponding to the unit square of type (1) (recall that $K = [0, 1]^2$), we have the identity

$$\forall p \in Q_1, \quad p = \sum_{i=1}^{4} p(a_i)p_i,$$

with

$$p_1 = (1 - x_1)(1 - x_2), \quad p_2 = x_1(1 - x_2), \quad p_3 = x_1x_2,$$
$$p_4 = (1 - x_1)x_2.$$

We may thus condense these expressions in

$$p_1 = x_3x_4, \ldots . \tag{2.2.26}$$

Likewise, corresponding to the unit square of type (2), we have the identity

$$\forall p \in Q_2, \quad p = \sum_{i=1}^{9} p(a_i)p_i,$$

with

$$\left.\begin{aligned} p_1 &= x_3(2x_3-1)x_4(2x_4-1), \ldots \\ p_5 &= -4x_3(x_3-1)x_4(2x_4-1), \ldots \\ p_9 &= 16x_1x_2x_3x_4, \end{aligned}\right\} \tag{2.2.27}$$

using the above rule. Finally, corresponding to the unit square of type (3), we have

$$\left.\begin{aligned} p_1 &= \frac{1}{4}x_3(3x_3-1)(3x_3-2)x_4(3x_4-1)(3x_4-2), \ldots \\ p_5 &= -\frac{9}{4}x_3(3x_3-1)(x_3-1)x_4(3x_4-1)(3x_4-2), \ldots \\ p_9 &= \frac{9}{4}x_3(3x_3-2)(x_3-1)x_4(3x_4-1)(3x_4-2), \ldots \\ p_{13} &= \frac{81}{4}x_3(3x_3-1)(x_3-1)x_4(3x_4-1)(x_4-1), \ldots . \end{aligned}\right\} \tag{2.2.28}$$

Remark 2.2.1. The inconsistency for the notations a_i, $5 \leq i \leq 9$, between the rectangles of type (2) and (3), avoids the introduction of a new letter. □

In analogy with the n-simplices of type (3′), one can derive two finite elements, in which the "internal" values of the rectangle of type (2) or (3) are no longer degrees of freedom (for simplicity, we shall restrict ourselves to the case $n = 2$). The existence of these finite elements is a consequence of the following two theorems.

Theorem 2.2.5. *Let the points a_i, $1 \leq i \leq 9$, be as in Fig. 2.2.10. Then any polynomial in the space*

$$Q_2' = \left\{p \in Q_2;\ \ 4p(a_9) + \sum_{i=1}^{4} p(a_i) - 2\sum_{i=5}^{8} p(a_i) = 0\right\} \tag{2.2.29}$$

is uniquely determined by its values at the points a_i, $1 \leq i \leq 8$. In addition, *the inclusion*

$$P_2 \subset Q_2' \tag{2.2.30}$$

holds.

Proof. The first part of the proof is similar to the first part of the proof of Theorem 2.2.2. In particular, we have the identity

$$\forall p \in Q_2', \quad p = \sum_{i=1}^{8} p(a_i)p_i,$$

with

$$\left.\begin{aligned} p_1 &= x_3x_4(2x_3 + 2x_4 - 3), \dots \\ p_5 &= -4x_3x_4(x_3 - 1), \dots . \end{aligned}\right\} \tag{2.2.31}$$

To prove the inclusion (2.2.30), let p be a polynomial of degree 2, and let A denote its (constant) second derivative. From the expansions

$$p(a_i) = p(a_9) + Dp(a_9)(a_i - a_9) + \frac{1}{2}A(a_i - a_9)^2, \quad 1 \leqslant i \leqslant 8,$$

we deduce

$$\sum_{i=1}^{4} p(a_i) = 4p(a_9) + \frac{1}{2}\sum_{i=1}^{4} A(a_i - a_9)^2,$$

$$\sum_{i=5}^{8} p(a_i) = 4p(a_9) + \frac{1}{2}\sum_{i=5}^{8} A(a_i - a_9)^2,$$

since

$$\sum_{i=1}^{4} (a_i - a_9) = \sum_{i=5}^{8} (a_i - a_9) = 0.$$

Because the mapping A is bilinear, and because $a_5 = (a_1 + a_2)/2, \dots$, we obtain

$$\sum_{i=5}^{8} A(a_i - a_9)^2 = \frac{1}{2}\sum_{i=1}^{4} A(a_i - a_9)^2.$$

Combining the previous relations, we deduce that

$$4p(a_9) + \sum_{i=1}^{4} p(a_i) - 2\sum_{i=5}^{8} p(a_i) = 0,$$

and the proof is complete. □

Theorem 2.2.6. *Let the points* a_i, $1 \leqslant i \leqslant 16$, *be as in Fig.* 2.2.11. *Define the space*

$$Q_3' = \{p \in Q_3;\quad \psi_i(p) = 0,\quad 1 \leqslant i \leqslant 4\}, \tag{2.2.32}$$

where

$$\begin{aligned}\psi_1(p) = 9p(a_{13}) + 4p(a_1) + 2p(a_2) + p(a_3) + 2p(a_4)\\ - 6p(a_5) - 3p(a_6) - 3p(a_{11}) - 6p(a_{12}),\end{aligned} \tag{2.2.33}$$

and $\psi_2(p)$, $\psi_3(p)$, *and* $\psi_4(p)$ *are derived by circular permutations in the sets* $\bigcup_{i=1}^{4}\{a_i\}$, $\bigcup_{i=5}^{8}\{a_i\}$, $\bigcup_{i=9}^{12}\{a_i\}$ *and* $\bigcup_{i=13}^{16}\{a_i\}$. *Then any polynomial in the space* Q_3' *is uniquely determined by its values at the points* a_i, $1 \leqslant i \leqslant 12$. *In addition, the inclusion*

$$P_3 \subset Q_3' \tag{2.2.34}$$

holds.

Proof. The proof is left as a problem (Exercise 2.2.5). We shall only record the identity

$$\forall p \in Q_3', \quad p = \sum_{i=1}^{12} p(a_i)p_i,$$

with

$$\left.\begin{aligned} p_1 &= x_3x_4\left\{1 + \frac{9}{2}x_3(x_3 - 1) + \frac{9}{2}x_4(x_4 - 1)\right\}, \ldots\\ p_5 &= -\frac{9}{2}x_3(x_3 - 1)(3x_3 - 1)x_4, \ldots\\ p_9 &= \frac{9}{2}x_3(x_3 - 1)(3x_3 - 2)x_4, \ldots\end{aligned}\right\} \tag{2.2.35}$$

□

From these two theorems, we derive the definition of the *rectangle of type* (2′) (Fig. 2.2.12) and of the *rectangle of type* (3′) (Fig. 2.2.13).

If it happens that the set $\bar{\Omega} \subset \mathbf{R}^n$ is *rectangular*, i.e., it is either an n-rectangle or a finite union of n-rectangles, it can be conveniently "triangulated" by finite elements which are themselves n-rectangles: The fifth condition ($\mathcal{T}_h5$) on a triangulation now reads:

($\mathcal{T}_h5$) *Any face of any n-rectangle* K_1 *in the triangulation is either a subset of the boundary* Γ, *or a face of another n-rectangle* K_2 *in the triangulation.*

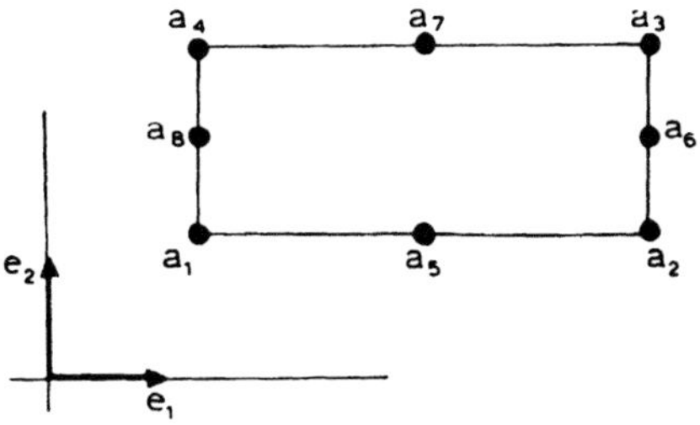

rectangle of type (2′)

$P_K = Q_2'(K)$ (cf. (2.2.29)); dim $P_K = 8$;
$\Sigma_K = \{p(a_i), \quad 1 \leq i \leq 8\}$.

Fig. 2.2.12

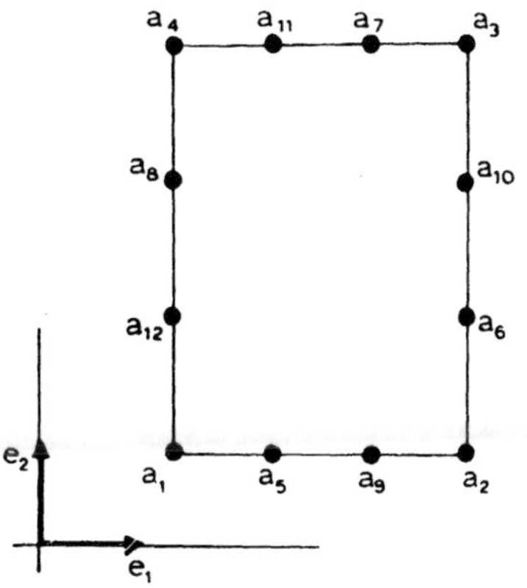

rectangle of type (3′)

$P_K = Q_3'(K)$ (cf. (2.2.32)); dim $P_K = 12$;
$\Sigma_K = \{p(a_i); \quad 1 \leq i \leq 12\}$.

Fig. 2.2.13

In the second case, the n-rectangles K_1 and K_2 are said to be *adjacent*.

An example of a triangulation made up of rectangles is given in Fig. 2.2.14.

With such a triangulation, we may associate in a natural way a finite element space X_h with each type of the rectangular finite elements which we just described. Since the discussion is almost identical to the one concerning n-simplices, we shall be very brief. In particular, one can prove the following analog of Theorem 2.2.3.

Theorem 2.2.7. *Let X_h be the finite element space associated with n-rectangles of type (k) for any integer $k \geq 1$ or with rectangles of type (2′) or (3′). Then the inclusion*

$$X_h \subset \mathscr{C}^0(\bar{\Omega}) \cap H^1(\Omega) \tag{2.2.36}$$

holds. □

Finally, arguing as before, it is easily seen that such finite element spaces possess a basis whose functions have "small" support (FEM 3).

First examples of finite elements with derivatives as degrees of freedom: Hermite n-simplices of type (3), (3′). *Assembly in triangulations*

So far, the degrees of freedom of each finite element K have been "point values", i.e., of the form $p(a)$, for some points $a \in K$. We shall next introduce finite elements in which some degrees of freedom are partial derivatives, or, more generally, *directional derivatives*, i.e.,

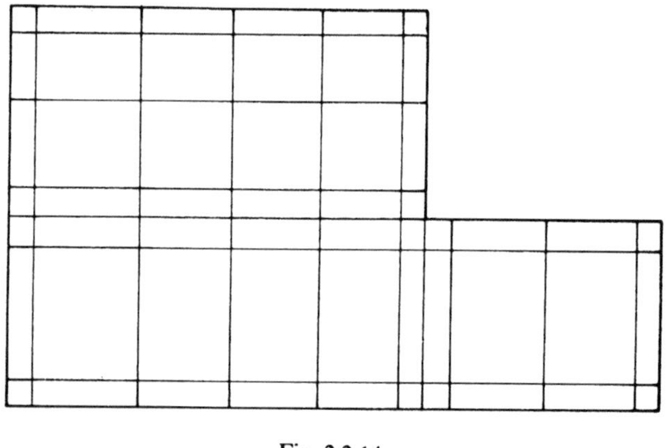

Fig. 2.2.14

expressions such as $Dp(a)b$, $D^2p(a)(b, c)$, etc. . . , where b, c are vectors in $\mathbf{R}^n$.

The first example of this type of finite element is based on the following theorem.

Theorem 2.2.8. *Let K be an n-simplex with vertices a_i, $1 \leq i \leq n+1$, and let $a_{ijk} = \frac{1}{3}(a_i + a_j + a_k)$, $1 \leq i < j < k \leq n+1$. Then any polynomial in the space P_3 is uniquely determined by its values and the values of its n first partial derivatives at the vertices a_i, $1 \leq i \leq n+1$, and its values at the points a_{ijk}, $1 \leq i < j < k \leq n+1$.*

Proof. It suffices to argue as usual so as to obtain the following identity:

$$\forall p \in P_3, \quad p = \sum_i \Big(-2\lambda_i^3 + 3\lambda_i^2 - 7\lambda_i \sum_{\substack{j<k \\ j\neq i,\, k\neq i}} \lambda_j\lambda_k\Big)p(a_i)$$

$$+ 27 \sum_{i<j<k} \lambda_i\lambda_j\lambda_k p(a_{ijk})$$

$$+ \sum_{i\neq j} \lambda_i\lambda_j(2\lambda_i + \lambda_j - 1)Dp(a_i)(a_j - a_i). \tag{2.2.37}$$

The only novelty is that one needs to use the derivatives of the barycentric coordinates in order to show that $Dp(a_i) = D\tilde{p}(a_i)$, $1 \leq i \leq n+1$, denoting momentarily by $\tilde{p}$ the right-hand side of (2.2.37). By differentiating the polynomial $\tilde{p}$, we obtain

$$D\tilde{p}(a_i) = \sum_{j\neq i} \{Dp(a_i)(a_j - a_i)\}D\lambda_j.$$

To show that the above expression is equal to $Dp(a_i)$, it is equivalent to show that

$$D\tilde{p}(a_i)(a_k - a_i) = Dp(a_i)(a_k - a_i), \quad 1 \leq k \leq n+1, \quad k \neq i.$$

These last relations are in turn consequences of the relations

$$D\lambda_j(a_k - a_i) = \delta_{jk} - \lambda_j(a_i), \quad 1 \leq k \leq n+1, \quad k \neq i,$$

which we now establish. Denoting by B the inverse matrix of the matrix A of (2.2.4), we obtain

$$\partial_l\lambda_j = b_{jl}, \quad 1 \leq j \leq n+1, \quad 1 \leq l \leq n$$

(cf. (2.2.7)). Therefore we have

$$D\lambda_j(a_k - x) = \sum_{l=1}^{n} b_{jl}a_{lk} - \sum_{l=1}^{n} b_{jl}x_l = \delta_{jk} - \lambda_j(x)$$

for any $x \in \mathbf{R}^n$, and in particular for $x = a_i$. □

From this theorem, we deduce the definition of a finite element, which is called the *Hermite n-simplex of type* (3) (Fig. 2.2.15), where the directional derivatives $Dp(a_i)(a_j - a_i)$ are degrees of freedom. Of course, the knowledge of these n directional derivatives at a vertex a_i is equivalent to the knowledge of the first derivative $Dp(a_i)$. Such a knowledge is indicated graphically by one small circle, or sphere, centered at the point a_i. Since the first derivative $Dp(a_i)$ is equally well determined by the partial derivatives $\partial_j p(a_i)$, $1 \leq j \leq n$, another possible set of degrees of freedom for this element is the set Σ'_K indicated in Fig. 2.2.15.

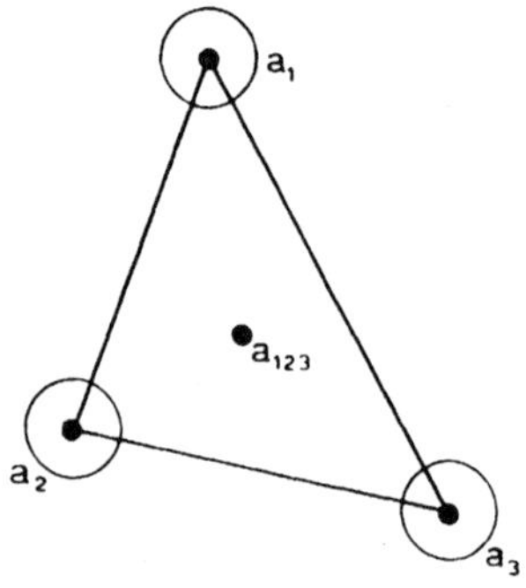

$\left\{\begin{array}{l}\textit{Hermite triangle of type } (3)\\ \dim P_K = 10\end{array}\right.$

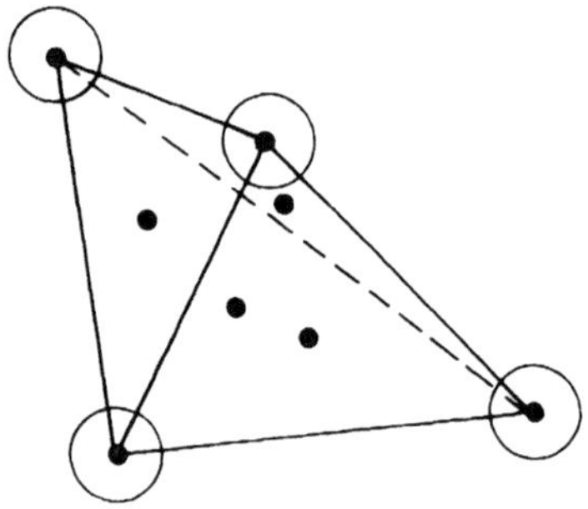

$\left\{\begin{array}{l}\textit{Hermite tetrahedron of type } (3)\\ \dim P_K = 20\end{array}\right.$

Hermite n-simplex of type (3)

$P_K = P_3(K)$; $\dim P_K = \dfrac{(n+1)(n+2)(n+3)}{6}$;

$\Sigma_K = \{p(a_i),\ 1 \leq i \leq n+1;\ p(a_{ijk}),\ 1 \leq i < j < k \leq n+1;$
$Dp(a_i)(a_j - a_i),\ 1 \leq i, j \leq n+1,\ j \neq i\}.$

$\Sigma'_K = \{p(a_i),\ 1 \leq i \leq n+1;\ p(a_{ijk}),\ 1 \leq i < j < k \leq n+1;$
$\partial_j p(a_i),\ 1 \leq i \leq n+1,\ 1 \leq j \leq n\}.$

Fig. 2.2.15

The derivation of a related element without the degrees of freedom $p(a_{ijk})$, $i<j<k$, is based on the following theorem, whose proof is left to the reader (Exercise 2.2.6).

Theorem 2.2.9. *For each triple* (i, j, k) *with* $i<j<k$, *let*

$$\psi_{ijk}(p) = 6p(a_{ijk}) - 2\sum_{l=i,j,k} p(a_l) + \sum_{l=i,j,k} Dp(a_l)(a_l - a_{ijk}). \tag{2.2.38}$$

Then any polynomial in the space

$$P_3'' = \{p \in P_3;\quad \psi_{ijk}(p) = 0,\quad 1 \leqslant i<j<k \leqslant n+1\} \tag{2.2.39}$$

is uniquely determined by its values and the values of its n first partial derivatives at the vertices a_i, $1 \leqslant i \leqslant n+1$. *In addition, the inclusion*

$$P_2 \subset P_3'' \tag{2.2.40}$$

holds. □

From this theorem, one deduces the definition of the *Hermite n-simplex of type* (3′), which, in case $n = 2$, is also called the *Zienkiewicz triangle* (Fig. 2.2.16).

Given a triangulation made up of n-simplices, we associate in a natural way a finite element space X_h with either type of finite elements. To be specific, assume we are using Hermite n-simplices of type (3), the case of Hermite n-simplices of type (3′) being quite similar. Then a function v_h is in the space X_h if (i) each restriction $v_h|_K$ is in the space $P_K = P_3(K)$ for each $K \in \mathcal{T}_h$, and (ii) it is defined by its values at all the vertices of the triangulation, its values at the centers of gravity of all triangles found as 2-faces of the n-simplices $K \in \mathcal{T}_h$, and the values of its n first partial derivatives at all the vertices of the triangulation. The corresponding set of degrees of freedom of the space X_h is thus of the form

$$\Sigma_h = \{v_h(b);\quad b \in \mathcal{N}_v \cup \mathcal{N}_c;\quad \partial_j v_h(b),\quad b \in \mathcal{N}_v;\quad 1 \leqslant j \leqslant n\},$$

where $\mathcal{N}_v$ denotes the set of all the vertices of the n-simplices of the triangulation and $\mathcal{N}_c$ denotes the set of all centers of gravity of all 2-faces of the n-simplices found in the triangulation.

When a finite element space is constructed with n-simplices of type (3) or (3′), the sets Σ_K' are preferred to the sets Σ_K (cf. Figs. 2.2.15 and

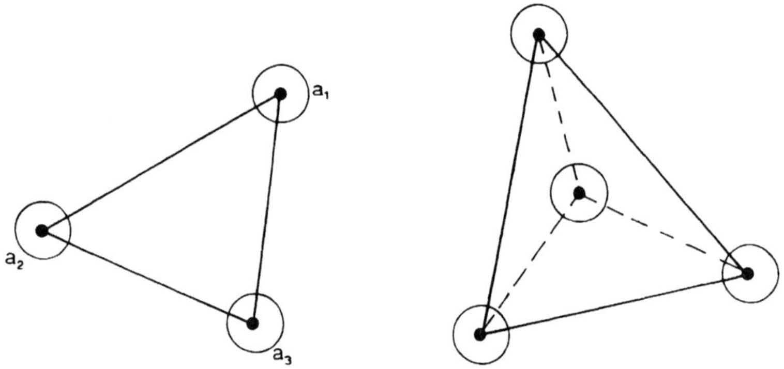

Zienkiewicz triangle, or
Hermite triangle of type (3′)
dim $P_K = 9$

Hermite tetrahedron of type (3′)
dim $P_K = 16$

Hermite n-simplex of type (3′)
$P_K = P_3''(K)$ (cf. (2.2.39)), dim $P_K = (n+1)^2$; $\Sigma_K = \{p(a_i),\ 1 \leqslant i \leqslant n+1;\ Dp(a_i)(a_j - a_i),\ 1 \leqslant i, j \leqslant n+1,\ i \neq j\}$. $\Sigma'_K = \{p(a_i),\ 1 \leqslant i \leqslant n+1;\ \partial_j p(a_i),\ 1 \leqslant i \leqslant n+1,\ 1 \leqslant j \leqslant n\}$.

Fig. 2.2.16

2.2.16) inasmuch as they directly correspond to the set Σ_h, but this observation is of a purely practical nature.

Again, *requirement* $(\mathcal{T}_h5)$ *insures that the degrees of freedom are unambiguously defined across adjacent finite elements*, and it is also the basis for the following theorem.

Theorem 2.2.10. *Let* X_h *be the finite element space associated with Hermite n-simplices of type* (3), *or with Hermite n-simplices of type* (3′). *Then the inclusion*

$$X_h \subset \mathscr{C}^0(\bar{\Omega}) \cap H^1(\Omega) \tag{2.2.41}$$

holds.

Proof. Arguing as in Theorem 2.2.3, it suffices to derive the inclusion $X_h \subset \mathscr{C}^0(\bar{\Omega})$: Along any side common to two adjacent triangles, there is a unique polynomial of degree 3 in one variable which takes on prescribed values and prescribed first derivatives at the end points of the side.

This argument easily extends to the n-dimensional case. □

To verify requirement (FEM 3), let us assume for definiteness that we are considering Hermite triangles of type (3), so that the associated set of degrees of freedom of the space is of the form

$$\begin{aligned}\Sigma_h = \{v(b_k),\quad \partial_1 v(b_k),\quad \partial_2 v(b_k),\quad 1 \leq k \leq J;\\ v(b_k),\quad J+1 \leq k \leq L\}.\end{aligned} \tag{2.2.42}$$

Then if we define functions w_k, w_k^1, $w_k^2 \in X_h$ by the conditions

$$\left.\begin{aligned} w_k(b_l) = \delta_{kl},\quad & 1 \leq k,l \leq L,\quad \partial_1 w_k(b_l) = \partial_2 w_k(b_l) = 0,\\ & 1 \leq k \leq L,\quad 1 \leq l \leq J,\\ w_k^1(b_l) = 0,\quad & 1 \leq k \leq J,\quad 1 \leq l \leq L,\quad \partial_1 w_k^1(b_l) = \delta_{kl},\\ & \partial_2 w_k^1(b_l) = 0,\quad 1 \leq k, l \leq J,\\ w_k^2(b_l) = 0,\quad & 1 \leq k \leq J,\quad 1 \leq l \leq L,\quad \partial_1 w_k^2(b_l) = 0,\\ & \partial_2 w_k^2(b_l) = \delta_{kl},\quad 1 \leq k, l \leq J, \end{aligned}\right\} \tag{2.2.43}$$

it is easily seen that these functions have "small" supports.

First examples of finite elements for fourth-order problems: The Argyris and Bell triangles, the Bogner–Fox–Schmit rectangle. Assembly in triangulations

Finally, we examine some examples of finite elements which yield the inclusion $X_h \subset \mathscr{C}^1(\bar{\Omega})$, and which may therefore be used for solving fourth-order problems. It is legitimate to restrict ourselves to the case where $n = 2$, in view of the examples given in Section 1.2. Our first example is based on the following result.

Theorem 2.2.11. *Let K be a triangle with vertices a_i, $1 \leq i \leq 3$, and let $a_{ij} = \frac{1}{2}(a_i + a_j)$, $1 \leq i < j \leq 3$, denote the mid-points of the sides. Then any polynomial p of degree 5 is uniquely determined by the following set of 21 degrees of freedom:*

$$\Sigma_K = \{\partial^\alpha p(a_i),\quad |\alpha| \leq 2,\ 1 \leq i \leq 3;\quad \partial_\nu p(a_{ij}),\quad 1 \leq i < j \leq 3\}, \tag{2.2.44}$$

where ∂_ν denotes the normal derivative operator along the boundary of K.

Proof. Given a set of degrees of freedom, finding the corresponding polynomial of degree 5 amounts to solving a linear system with a square matrix, for which existence and uniqueness for all right-hand sides are equivalent properties, as we already observed. We shall prove the latter property, i.e., that any polynomial $p \in P_5$ such that

$$\partial^\alpha p(a_i) = 0, \quad |\alpha| \leq 2, \quad 1 \leq i \leq 3, \quad \partial_\nu p(a_{ij}) = 0, \quad 1 \leq i < j \leq 3,$$

is identically zero.

Let t denote an abscissa along the axis which contains the side $K' = [a_1, a_2]$. Then the restriction $p|_{K'}$, considered as a function q of t, is a polynomial of degree 5 which satisfies

$$q(a_1) = q'(a_1) = q''(a_1) = q(a_2) = q'(a_2) = q''(a_2) = 0,$$

since, if τ is a unit vector on the axis containing the side K', we have

$$q'(a_1) = \partial_\tau p(a_1), \quad q''(a_1) = \partial_{\tau\tau} p(a_1), \text{ etc.}\ldots,$$

and thus $q = 0$.

Likewise, considered as a function r of t, the normal derivative $\partial_\nu p$ along K' is a polynomial of degree 4 which satisfies

$$r(a_1) = r'(a_1) = r(a_{12}) = r(a_2) = r'(a_2) = 0,$$

since

$$r(a_1) = \partial_\nu p(a_1), \quad r'(a_1) = \partial_{\nu\tau} p(a_1), \quad r(a_{12}) = \partial_\nu p(a_{12}), \text{ etc.}\ldots,$$

and, thus, $r = 0$.

Since we have $\partial_\tau p = 0$ along K' ($p = 0$ along K'), we have proved that p and its first derivative Dp vanish identically along K'. This implies that the polynomial λ_3^2 is a factor of p, as we now show: After using an appropriate affine mapping if necessary, we may assume without loss of generality that $\lambda_3(x_1, x_2) = x_1$. We can write

$$p(x_1, x_2) = \sum_{i=0}^{5} x_1^i p_i(x_2)$$

where p_i, $0 \leq i \leq 5$, are polynomials of degree $(5 - i)$ in the variable x_2. Therefore

$$\forall x_2 \in \mathbf{R}, \quad p(0, x_2) = p_0(x_2) = 0,$$
$$\forall x_2 \in \mathbf{R}, \quad \partial_1 p(0, x_2) = p_1(x_2) = 0,$$

which proves our assertion.

Similar arguments hold for the other sides, and we find that the polynomial $(\lambda_1^2\lambda_2^2\lambda_3^2)$ is a factor of p. Since the λ_i are polynomials of degree 1 which do not reduce to constants, it necessarily follows that $p = 0$. □

With Theorem 2.2.11, we can define a finite element, the 21-*degree of freedom triangle*, also known as *Argyris triangle* (Fig. 2.2.17).

Fig. 2.2.17 is self-explanatory as regards the graphical symbols used for representing the various degrees of freedom. We observe that at each vertex a_i, the first and second derivatives $Dp(a_i)$ and $D^2p(a_i)$ are known. With this observation in mind, we see that other possible definitions for the set of degrees of freedom are the sets Σ'_K and Σ''_K

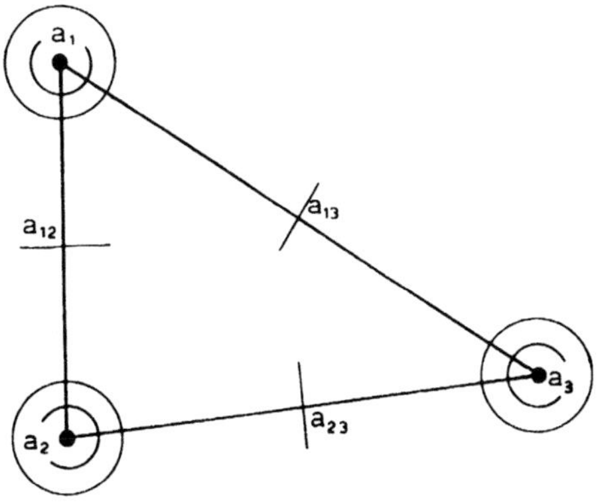

Argyris triangle, or
21-*degree of freedom triangle*

$P_K = P_5(K)$, $\dim P_K = 21$;

$\Sigma_K = \{p(a_i), \partial_1 p(a_i), \partial_2 p(a_i), \partial_{11} p(a_i), \partial_{12} p(a_i), \partial_{22} p(a_i), 1 \leqslant i \leqslant 3;$
$\quad \partial_\nu p(a_{ij}), 1 \leqslant i < j \leqslant 3\}$

$\Sigma'_K = \{p(a_i), 1 \leqslant i \leqslant 3;$
$\quad Dp(a_i)(a_j - a_i), 1 \leqslant i, j \leqslant 3, j \neq i;$
$\quad D^2p(a_i)(a_j - a_i, a_k - a_i), 1 \leqslant i, j, k \leqslant 3, j \neq i, k \neq i,$
$\quad \partial_\nu p(a_{ij}), 1 \leqslant i < j \leqslant 3\}$

$\Sigma''_K = \{p(a_i), \quad Dp(a_i)(a_{i-1} - a_i), \quad Dp(a_i)(a_{i+1} - a_i), \quad 1 \leqslant i \leqslant 3;$
$\quad D^2p(a_i)(a_{j+1} - a_j)^2, \quad 1 \leqslant i, j \leqslant 3;$
$\quad Dp(a_{ij})\nu_k, \quad \{i, j, k\} = \{1, 2, 3\}, \quad i < j\}.$

Fig. 2.2.17

indicated in Fig. 2.2.17. In the expression of the set Σ''_K, the indices are numbered modulo 3, and each vector ν_i, $1 \leq i \leq 3$, is the height originating at the point a_i.

It may be desirable to dispose of the degrees of freedom $\partial_\nu p(a_{ij})$, $1 \leq i < j \leq 3$. This reduction will be a consequence of the following result.

Theorem 2.2.12. *Any polynomial in the space*

$$P'_5(K) = \{p \in P_5(K);\quad \partial_\nu p \in P_3(K') \quad \text{for each side } K' \text{ of } K\} \tag{2.2.45}$$

is uniquely determined by the following set of 18 *degrees of freedom*:

$$\Sigma_K = \{\partial^\alpha p(a_i),\ |\alpha| \leq 2,\ 1 \leq i \leq 3\}. \tag{2.2.46}$$

The space $P'_5(K)$ satisfies the inclusion

$$P_4(K) \subset P'_5(K). \tag{2.2.47}$$

Proof. By writing $\partial_\nu p \in P_3(K')$ in definition (2.2.45), it is of course meant that, considered as a function of an abscissa along an axis containing the side K', the normal derivative $\partial_\nu p$ is a polynomial of degree 3. The inclusion (2.2.47) being obvious, it remains to prove the first part of the theorem.

To begin with, we prove a preliminary result: *Let $K' = [a_i, a_j]$ be a segment in $\mathbf{R}^n$, with mid-point a_{ij}, and let v be a function such that $v|_{K'} \in P_4(K')$. Then we have $v|_{K'} \in P_3(K')$ if and only if $\chi_{ij}(v) = 0$, where*

$$\chi_{ij}(v) = 4(v(a_i) + v(a_j)) - 8v(a_{ij}) + Dv(a_i)(a_j - a_i) \\ + Dv(a_j)(a_i - a_j). \tag{2.2.48}$$

To see this, let, for any $x \in K'$, $\alpha_4 = D^4v(x)\tau^4$, where τ is a unit vector along K', so that α_4 is a constant. Then we have

$$v(a_i) = v(a_{ij}) + Dv(a_{ij})(a_i - a_{ij}) + \frac{1}{2}D^2v(a_{ij})(a_i - a_{ij})^2 \\ + \frac{1}{6}D^3v(a_{ij})(a_i - a_{ij})^3 + \frac{\alpha_4}{24}\|a_i - a_{ij}\|^4,$$

$$v(a_j) = v(a_{ij}) + Dv(a_{ij})(a_j - a_{ij}) + \frac{1}{2}D^2v(a_{ij})(a_j - a_{ij})^2 \\ + \frac{1}{6}D^3v(a_{ij})(a_j - a_{ij})^3 + \frac{\alpha_4}{24}\|a_j - a_{ij}\|^4,$$

from which we deduce ($a_i - a_{ij} = -(a_j - a_{ij})$):

$$v(a_i) + v(a_j) = 2v(a_{ij}) + \frac{1}{2}\{D^2v(a_{ij})(a_i - a_{ij})^2 + D^2v(a_{ij})(a_j - a_{ij})^2\}$$

$$+ \frac{\alpha_4}{24}\{\|a_i - a_{ij}\|^4 + \|a_j - a_{ij}\|^4\}.$$

Likewise,

$$Dv(a_i)(a_i - a_{ij}) = D^2v(a_{ij})(a_i - a_{ij})^2 + \frac{1}{2}D^3v(a_{ij})(a_i - a_{ij})^3$$

$$+ \frac{\alpha_4}{6}\|a_i - a_{ij}\|^4,$$

$$Dv(a_j)(a_j - a_{ij}) = D^2v(a_{ij})(a_j - a_{ij})^2 + \frac{1}{2}D^3v(a_{ij})(a_j - a_{ij})^3$$

$$+ \frac{\alpha_4}{6}\|a_j - a_{ij}\|^4,$$

and therefore,

$$D^2v(a_{ij})(a_i - a_{ij})^2 + D^2v(a_{ij})(a_j - a_{ij})^2$$

$$= Dv(a_i)(a_i - a_{ij}) + Dv(a_j)(a_j - a_{ij})$$

$$- \frac{\alpha_4}{6}\{\|a_i - a_{ij}\|^4 + \|a_j - a_{ij}\|^4\}.$$

Combining our previous relations, we get

$$2v(a_{ij}) = v(a_i) + v(a_j) + \frac{1}{4}\{Dv(a_i)(a_j - a_i) + Dv(a_j)(a_i - a_j)\}$$

$$+ \frac{\alpha_4}{96}\|a_i - a_j\|^4,$$

and the assertion is proved.

As a consequence of this preliminary result, the space $P_5'(K)$ may be also defined as

$$P_5'(K) = \{p \in P_5(K);\quad \chi_{ij}(\partial_\nu p) = 0,\quad 1 \leq i < j \leq 3\}, \tag{2.2.49}$$

i.e., in view of relations (2.2.48), we have characterized the space $P_5'(K)$ by the property that each normal derivative $\partial_\nu p(a_{ij})$ is expressed as a linear combination of the parameters $\partial^\alpha p(a_i)$, $\partial^\alpha p(a_j)$, $|\alpha| = 1, 2$. Then the proof is completed by combining the usual argument with the result of Theorem 2.2.11. □

From Theorem 2.2.12, we deduce the definition of a finite element, called the 18-*degree of freedom triangle*, or, preferably, *Bell's triangle*. See Fig. 2.2.18, where we have indicated three possible sets of degrees of freedom which parallel those of the Argyris triangle.

Given a triangulation made up of triangles, we associate a finite element space X_h with either type of finite elements. We leave it to the reader to derive the associated set of degrees of freedom of the space X_h and to check that the canonical basis is again composed of functions with "small" support. We shall only prove the following result.

Theorem 2.2.13. *Let X_h be the finite element space associated with Argyris triangles or Bell's triangles. Then the inclusion*

$$X_h \subset \mathscr{C}^1(\bar{\Omega}) \cap H^2(\Omega) \tag{2.2.50}$$

holds.

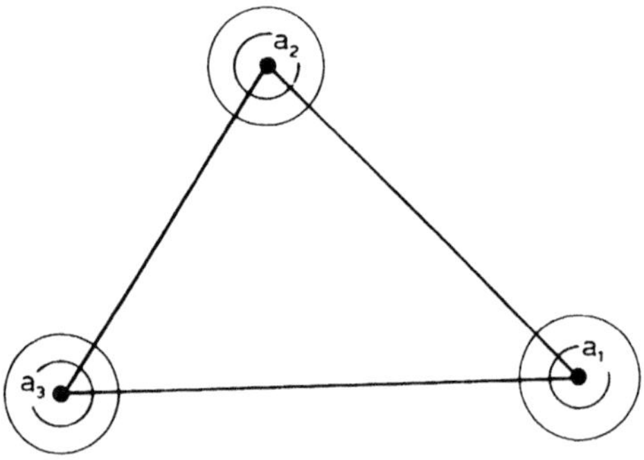

Bell's triangle or
18-degree of freedom triangle

$P_K = P_5'(K)$ (cf. (2.2.45)); $\dim P_K = 18$;

$\Sigma_K = \{p(a_i), \partial_1 p(a_i), \partial_2 p(a_i), \partial_{11} p(a_i), \partial_{12} p(a_i), \partial_{22} p(a_i), 1 \leq i \leq 3\}$

$\Sigma_K' = \{p(a_i), 1 \leq i \leq 3;\ Dp(a_i)(a_j - a_i), 1 \leq i, j \leq 3, j \neq i;$
$D^2 p(a_i)(a_j - a_i, a_k - a_i), 1 \leq i, j, k \leq 3, j \neq i, k \neq i\}$

$\Sigma_K'' = \{p(a_i), Dp(a_i)(a_{i-1} - a_i), Dp(a_i)(a_{i+1} - a_i), 1 \leq i \leq 3;$
$D^2 p(a_i)(a_{j+1} - a_j)^2, 1 \leq i, j \leq 3\}$

Fig. 2.2.18

Proof. By Theorem 2.1.2, it suffices to show that the inclusion $X_h \subset \mathscr{C}^1(\bar{\Omega})$ holds.

Let K_1 and K_2 be two adjacent triangles with a common side $K' = [b_i, b_j]$ (Fig. 2.2.19) and let v_h be a function in the space X_h constructed with Argyris triangles. Considered as functions of an abscissa t along an axis containing the side K', the functions $v_h|_{K_1}$ and $v_h|_{K_2}$ are, along K', polynomials of degree 5 in the variable t. Call these polynomials q_1 and q_2. Since, by definition of the space X_h, we have

$$q(b_i) = q'(b_i) = q''(b_i) = q(b_j) = q'(b_j) = q''(b_j) = 0,$$

with $q = q_1 - q_2$, it follows that $q = 0$ and hence the inclusion $V_h \subset \mathscr{C}^0(\bar{\Omega})$ holds. Likewise, call r_1 and r_2, the restrictions to the side K' of the functions $\partial_\nu v_h|_{K_1}$ and $-\partial_\nu v_h|_{K_2}$. Then both r_1 and r_2 are polynomials of degree 4 in the variable t and, again by definition of the space X_h, we have

$$r(b_i) = r'(b_i) = r(b_{ij}) = r(b_j) = r'(b_j) = 0,$$

with $r = r_1 - r_2$, so that $r = 0$. We have thus proved the continuity of the normal derivative which, combined with the continuity of the tangential derivative ($q = 0$ along K' implies $q' = 0$ along K'), shows that the first derivatives are also continuous on $\bar{\Omega}$.

If the space X_h is constructed with Bell's triangles, the argument is

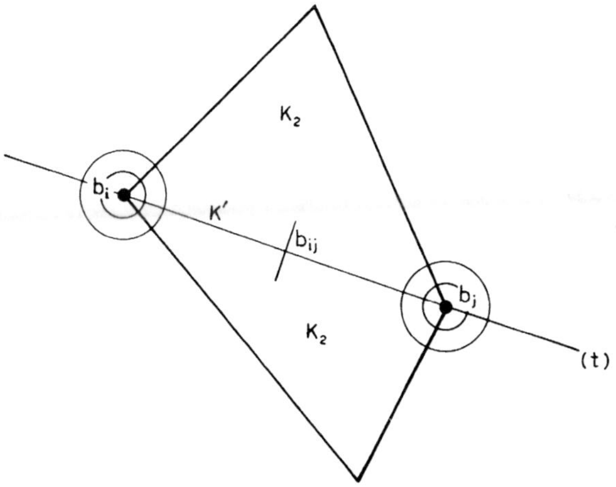

Fig. 2.2.19

identical for the difference $q = q_1 - q_2$. The difference $r = r_1 - r_2$ vanishes because it is a polynomial of degree 3 in the variable t which is such that

$$r(b_i) = r'(b_i) = r(b_j) = r'(b_j) = 0. \qquad \square$$

To conclude, we give one instance of a rectangular finite element which may be used for solving fourth-order problems posed over rectangular domains. Its existence depends upon the following theorem, whose proof is left to the reader (Exercise 2.2.8).

Theorem 2.2.14. *Let K denote a rectangle with vertices a_i, $1 \leqslant i \leqslant 4$. Then a polynomial $p \in Q_3$ is uniquely determined by the following set of degrees of freedom:*

$$\Sigma_K = \{p(a_i),\quad \partial_1 p(a_i),\quad \partial_2 p(a_i),\quad \partial_{12} p(a_i),\quad 1 \leqslant i \leqslant 4\}. \qquad \square \tag{2.2.51}$$

The resulting finite element is the *Bogner–Fox–Schmit rectangle.* See Fig. 2.2.20, which is again self-explanatory for the graphical symbols.

The proof of the next result is also left to the reader (Exercise 2.2.8):

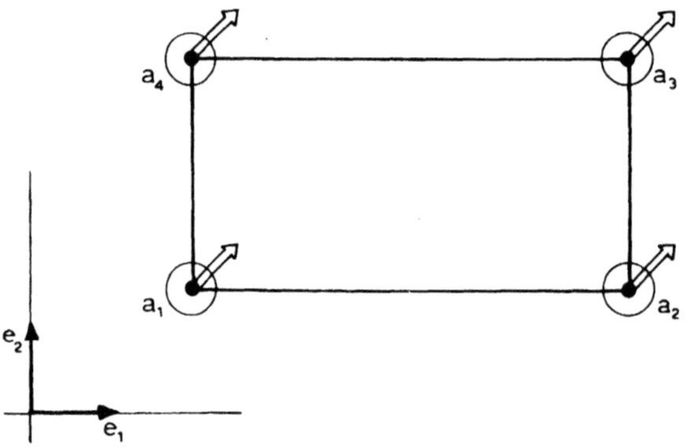

Bogner–Fox–Schmit rectangle
$P_K = Q_3$; $\dim P_K = 16$; $\Sigma_K = \{p(a_i),\ \partial_1 p(a_i),\ \partial_2 p(a_i),\ \partial_{12} p(a_i),\ 1 \leqslant i \leqslant 4\}$

Fig. 2.2.20

Theorem 2.2.15. *Let X_h be the finite element space associated with Bogner–Fox–Schmit rectangles. Then the inclusion*

$$X_h \subset \mathscr{C}^1(\bar{\Omega}) \cap H^2(\Omega) \tag{2.2.52}$$

holds. □

Finally, the reader should check, using the standard construction, that a finite element space constructed with any one of the last three finite elements indeed possesses canonical bases whose functions have "small" supports (FEM 3).

Exercises

2.2.1. (i) Prove that the dimension of the space P_k, resp. Q_k, is $\binom{n+k}{k}$, resp. $(k+1)^n$.

(ii) Prove that $\dim P_k(A) = \dim P_k$, resp. $\dim Q_k(A) = \dim Q_k$, if the interior of the set $A \subset \mathbf{R}^n$ is not empty.

2.2.2. Let K be an n-simplex with vertices a_j, $1 \leq j \leq n+1$. For a given integer $k \geq 1$, show that a polynomial of degree $\leq k$ is uniquely defined by its values on the set $L_k(K)$ defined in (2.2.11) (NICOLAIDES (1972)). The set $L_k(K)$ is called the *principal lattice of order* k of the n-simplex K.

2.2.3. Complete the proof of Theorem 2.2.3 so as to cover all cases.

2.2.4. Give another proof of Theorem 2.2.4 (i.e., without recurring to identity (2.2.22)), by showing that if a polynomial of Q_k vanishes on the set $\hat{M}_k$ defined in (2.2.21), then it is identically zero.

2.2.5. Prove Theorem 2.2.6.

2.2.6. Prove Theorem 2.2.9. Are the spaces P_3' and P_3'' (cf. (2.2.13) and (2.2.39), respectively) identical?

2.2.7. Given a triangle with vertices a_i, $1 \leq i \leq 3$, and mid-points $a_{ij} = \frac{1}{2}(a_i + a_j)$, $1 \leq i < j \leq 3$, show that a polynomial $p \in P_4$ is completely determined by the following degrees of freedom (ŽENÍŠEK (1974)):

$$\left.\begin{array}{l} p(a_i),\quad \partial_{11}p(a_i),\quad \partial_{12}p(a_i),\quad \partial_{22}p(a_i),\quad 1 \leq i \leq 3,\\ p(a_{ij}),\quad 1 \leq i < j \leq 3. \end{array}\right\}$$

Does this element yield the inclusion $X_h \subset \mathscr{C}^0(\bar{\Omega})$, resp. the inclusion $X_h \subset \mathscr{C}^1(\bar{\Omega})$?

2.2.8. (i) Given a rectangle with vertices a_i, $1 \leq i \leq 4$, show that a polynomial $p \in Q_3$ is completely determined by the following degrees of freedom:

$$p(a_i),\quad \partial_1 p(a_i),\quad \partial_2 p(a_i),\quad \partial_{12} p(a_i),\quad 1 \leq i \leq 4.$$

(ii) Show that the corresponding space X_h satisfies the inclusion

$$X_h \subset \mathscr{C}^1(\bar{\Omega}) \cap H^2(\Omega).$$

2.2.9. Consider the finite element space X_h constructed with (Hermite) triangles of type (3) and let w_k^1, $1 \leq k \leq M_1$, be the basis functions of the space X_h associated with the values at the barycenters of all the triangles of the triangulation, so that the discrete solution takes the form

$$u_h = \sum_{k=1}^{M} u_k w_k = \sum_{k=1}^{M_1} u_k^1 w_k^1 + \sum_{k=1}^{M_2} u_k^2 w_k^2.$$

Show that in this case the solution of the linear system (2.1.4) amounts, in fact, to solving a smaller linear system, in the unknowns u_k^2, $1 \leq i \leq M_2$, only.

This process, known as the *static condensation of the degrees of freedom*, is of course to be distinguished from the use of (Hermite) triangles of type (3′).

2.3. General properties of finite elements and finite element spaces

Finite elements as triples (K, P, Σ). Basic definitions. The P-interpolation operator

Let us begin by giving the general definition of a finite element. A *finite element* in $\mathbf{R}^n$ is a triple (K, P, Σ) where:

(i) K is a closed subset of $\mathbf{R}^n$ with a non empty interior and a Lipschitz-continuous boundary,

(ii) P is a space of real-valued functions defined over the set K,

(iii) Σ is a finite set of linearly independent linear forms ϕ_i, $1 \leq i \leq N$, defined over the space P (in order to avoid ambiguities, the forms ϕ_i need to be defined over a larger space; we shall examine this point later; cf. Remark 2.3.3). By definition, it is assumed that the set Σ is P-*unisolvent* in the following sense: given any real scalars α_i, $1 \leq i \leq N$, there exists a unique function $p \in P$ which satisfies

$$\phi_i(p) = \alpha_i,\quad 1 \leq i \leq N. \tag{2.3.1}$$

Consequently, there exist functions $p_i \in P$, $1 \leq i \leq N$, which satisfy

$$\phi_j(p_i) = \delta_{ij}, \quad 1 \leq j \leq N. \tag{2.3.2}$$

Since we have

$$\forall p \in P, \quad p = \sum_{i=1}^{N} \phi_i(p) p_i, \tag{2.3.3}$$

Of course this implies that the space P is finite-dimensional and that $\dim P = N$.

The linear forms ϕ_i, $1 \leq i \leq N$, are called the *degrees of freedom of the finite element*, and the functions p_i, $1 \leq i \leq N$, are called the *basis functions of the finite element*.

Whenever we find it convenient, we shall use the notations P_K, Σ_K, $\phi_{i,K}$ and p_{iK} in lieu of P, Σ, ϕ_i and p_i.

Remark 2.3.1. The set K itself is often called a *finite element*, as we did in the previous section, and as we shall occasionally do in the sequel. □

Remark 2.3.2. The P-unisolvence of the set Σ is equivalent to the fact that the N linear forms ϕ_i form a basis in the dual space of P. As a consequence, one may view the bases $(\phi_i)_{i=1}^N$ and $(p_i)_{i=1}^N$ as being *dual bases*, in the algebraic sense.

In the light of the definition of a finite element, let us briefly review the examples given in the previous section.

We have seen examples for which the set K is either an n-simplex, in which case the finite element is said to be *simplicial*, or *triangular* if $n = 2$, or *tetrahedral* if $n = 3$, or an n-rectangle in $\mathbf{R}^n$, in which case the finite element is said to be *rectangular*. As we already mentioned, these are all special cases of *straight finite elements*, i.e., for which the set K is a polyhedron in $\mathbf{R}^n$. Other polygonal shapes are found in practice, such as *quadrilaterals* (see Section 4.3 and Section 6.1) or "*prismatic*" *finite elements* (see Remark 2.3.6). We shall also describe (Section 4.3) "*curved*" *finite elements*, i.e., whose boundaries are composed of "curved" faces.

The main characteristic of the various spaces P encountered in the examples is that they all contain a "full" polynomial space $P_k(K)$ for *some integer* $k \geq 1$, a property that will be shown in subsequent chapters to be crucial as far as convergence properties are concerned.

In all the examples described previously, the degrees of freedom were of some of the following forms:

$$\left.\begin{aligned} &p \to p(a_i^0), \\ &p \to Dp(a_i^1)\xi_{ik}^1, \\ &p \to D^2p(a_i^2)(\xi_{ik}^2, \xi_{il}^2), \end{aligned}\right\} \tag{2.3.4}$$

where the points a_i^r, $r = 0, 1, 2$, belong to the finite element, and the (non zero) vectors ξ_{ik}^1, ξ_{ik}^2, ξ_{il}^2 are either constructed from the geometry of the finite element (e.g., $Dp(a_i)(a_j - a_i)$, $\partial_\nu p(a_{ij})$, etc. . .) or fixed vectors of $\mathbf{R}^n$ (e.g., $\partial_i p(a_j)$, $\partial_{ij} p(a_k)$). The points a_i^r, $r = 0, 1, 2$, are called the *nodes of the finite element* and make up a set which shall be denoted $\mathcal{N}_K$ in general.

Whereas only directional derivatives of order 1 or 2 occurred in the examples, one could conceivably consider degrees of freedom which would be partial derivatives of arbitrarily high order, but these are seldom used in practice. As we shall see later, however, (Section 4.2 and Section 6.2) there are practical instances of degrees of freedom which are not attached to nodes: They are instead *averages* (over the finite element or over one of its faces) of some partial derivative.

When all the degrees of freedom of a finite element are of the form $p \to p(a_i)$, we shall say that the associated finite element is a *Lagrange finite element* while if at least one directional derivative occurs as a degree of freedom, the associated finite element is said to be a *Hermite finite element*.

As the examples in the previous section have shown, there are essentially two methods for proving that a given set Σ of degrees of freedom is P-unisolvent: *After it has been checked that* $\dim P = \operatorname{card}(\Sigma)$, one either

(i) exhibits the basis functions, or

(ii) shows that if all the degrees of freedom are set equal to zero, then the only corresponding function in the space P is identically zero.

We have used method (i) for all the examples, except for the Argyris triangle where we used method (ii).

Given a finite element (K, P, Σ), and given a function $v = K \to \mathbf{R}$, sufficiently smooth so that the degrees of freedom $\phi_i(v)$, $1 \leq i \leq N$, are well defined, we let

$$\Pi v = \sum_{i=1}^{N} \phi_i(v) p_i \tag{2.3.5}$$

denote the *P-interpolant* of the function v, which is unambiguously defined since the set Σ is P-unisolvent. Indeed, the P-interpolant, also denoted $\Pi_K v$, is equivalently characterized by the conditions

$$\Pi v \in P, \quad \text{and} \quad \phi_i(\Pi v) = \phi_i(v), \quad 1 \leq i \leq N. \tag{2.3.6}$$

Whenever the degrees of freedom are of the form (2.3.4), let s denote the maximal order of derivatives occurring in the definition of the set Σ. Then, for all finite elements of this type described here, the inclusion $P \subset \mathscr{C}^s(K)$ holds. Consequently, we shall usually consider that *the domain* dom Π *of the P-interpolation operator* Π *is the space*

$$\text{dom}\, \Pi = \mathscr{C}^s(K). \tag{2.3.7}$$

This being the case, it follows that *over the space* $P \subset \text{dom}\, \Pi$, *the interpolation operator reduces to the identity*, i.e.,

$$\forall p \in P, \quad \Pi p = p. \tag{2.3.8}$$

Remark 2.3.3. In order that the P-interpolation operator be unambiguously defined, *it is necessary that the forms* ϕ_i *be also defined on the space* $\mathscr{C}^s(K)$, for the following reason. Assume again that the space P is contained in the space $\mathscr{C}^s(K)$. Then if the domain of the operator Π were only the space P, infinitely many extensions to the space $\mathscr{C}^s(K)$ would exist. Let us give one simple example of such a phenomenon: Let K be an n-simplex with barycenter a. Then the linear form $p \in \mathscr{C}^0(K) \rightarrow 1/(\text{meas}\,(K)) \int_K p \, dx$ is one possible extension of the form $p \in P_1(K) \rightarrow p(a)$.

Of course, these considerations are usually omitted inasmuch as when one considers a degree of freedom such as $\partial_i p(a_j)$ for instance, it is implicitly understood that this form is the usual one, i.e., defined over the space $\mathscr{C}^1(K)$, not any one of its possible extensions from the space P to the space $\mathscr{C}^1(K)$. For another illustration of this circumstance, see the description of *Wilson's brick*, in Section 4.2. □

Whereas for a Lagrange finite element, the set of degrees of freedom is unambiguously defined – indeed, it can be conveniently identified with the set of nodes – there are always several possible definitions for the degrees of freedom of a Hermite finite element which correspond to the "same" finite element. More precisely, *we shall say that two finite elements* (K, P, Σ) *and* (L, Q, Ξ) *are equal if we have*

$$K = L, \quad P = Q \quad \text{and} \quad \Pi_K = \Pi_L. \tag{2.3.9}$$

As an example, let us consider the Hermite n-simplex of type (3′) with the two sets of degrees of freedom (cf. Fig. 2.2.16):

$$\Sigma = \{p(a_i), \quad 1 \le i \le n+1; \quad Dp(a_i)(a_j - a_i),$$
$$1 \le i, \; j \le n+1, \quad i \ne j\},$$
$$\Sigma' = \{p(a_i), \quad 1 \le i \le n+1; \quad \partial_k p(a_i), \quad 1 \le i \le n+1,$$
$$1 \le k \le n\}.$$

Let us denote by Π and Π' the corresponding $P_3(K)$-interpolation operators. Then, for any function $v \in \mathscr{C}^1(K) = \operatorname{dom} \Pi = \operatorname{dom} \Pi'$, we have, with obvious notations,

$$\Pi v = \sum_i v(a_i) p_i + \sum_{i,j} Dv(a_i)(a_j - a_i) p_{ij},$$
$$\Pi' v = \sum_i v(a_i) p'_i + \sum_{i,k} \partial_k v(a_i) p'_{ik}.$$

One has, for each pair (i, j), $Dv(a_i)(a_j - a_i) = \Sigma_{k=1}^n \mu_{ijk} \partial_k v(a_i)$ for appropriate coefficients μ_{ijk}. To conclude that $\Pi = \Pi'$, it suffices to observe that for each polynomial $p \in P_K$, one also has $Dp(a_i)(a_j - a_i) = \Sigma_{k=1}^n \mu_{ijk} \partial_k p(a_i)$ with the *same* coefficients μ_{ijk}.

Affine families of finite elements

We now come to an essential idea, which we shall first illustrate by an example.

Suppose we are given a family (K, P_K, Σ_K) of triangles of type (2). Then *our aim is to describe such a family as simply as possible.*

Let $\hat{K}$ be a triangle with vertices $\hat{a}_i$, and mid-points of the sides $\hat{a}_{ij} = (\hat{a}_i + \hat{a}_j)/2$, $1 \le i < j \le 3$, and let

$$\hat{\Sigma} = \{p(\hat{a}_i), \quad 1 \le i \le 3; \qquad p(\hat{a}_{ij}), \quad 1 \le i < j \le 3\},$$

so that the triple $(\hat{K}, \hat{P}, \hat{\Sigma})$ with $\hat{P} = P_2(\hat{K})$ is also a triangle of type (2).

Given any finite element K in the family (Fig. 2.3.1), *there exists a unique invertible affine mapping*

$$F_K: \hat{x} \in \mathbf{R}^2 \to F_K(\hat{x}) = B_K\hat{x} + b_K,$$

i.e., with B_K an invertible 2×2 matrix and b_K a vector of $\mathbf{R}^2$, such that

$$F_K(\hat{a}_i) = a_i, \quad 1 \leq i \leq 3.$$

Then *it automatically follows that*

$$F_K(\hat{a}_{ij}) = a_{ij}, \quad 1 \leq i < j \leq 3,$$

since the property for a point to be the mid-point of a segment is preserved by an affine mapping (likewise, the points which we called a_{iij} or a_{ijk} keep their geometrical definition through an affine mapping).

Once we have established a bijection $\hat{x} \in \hat{K} \to x = F_K(\hat{x}) \in K$ between the points of the sets $\hat{K}$ and K, it is natural to associate the space

$$P_K^* = \{p: K \to \mathbf{R}; \quad p = \hat{p} \cdot F_K^{-1}, \quad \hat{p} \in \hat{P}\}$$

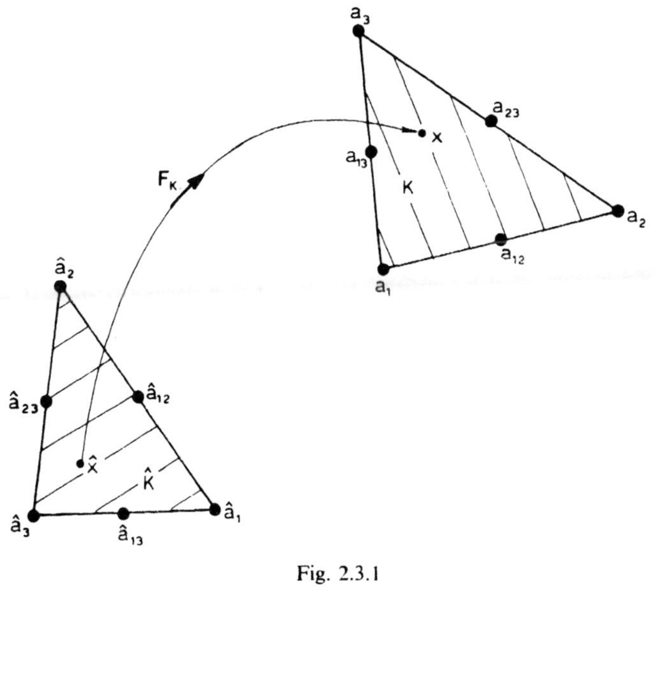

Fig. 2.3.1

with the space P. *Then it follows that*

$$P_K^* = P_2(K) = P_K,$$

because the mapping F_K is affine.

In other words, *rather than prescribing such a family by the data* K, P_K *and* Σ_K, *it suffices to give one reference finite element* $(\hat{K}, \hat{\Sigma}, \hat{P})$ *and the affine mappings* F_K. Then the generic finite element (K, P_K, Σ_K) in the family is such that

$$\left.\begin{aligned} &K = F_K(\hat{K}), \\ &P_K = \{p: K \to \mathbf{R};\quad p = \hat{p} \cdot F^{-1},\quad \hat{p} \in \hat{P}\}, \\ &\Sigma_K = \{p\{F_K(\hat{a}_i)),\quad 1 \leq i \leq 3;\quad p(F_K(\hat{a}_{ij})),\quad 1 \leq i < j \leq 3\}. \end{aligned}\right\}$$

With this example in mind, we are in a position to give the general definition: Two finite elements $(\hat{K}, \hat{P}, \hat{\Sigma})$ and (K, P, Σ), with degrees of freedom of the form (2.3.4), are said to be *affine-equivalent* if there exists an *invertible affine mapping*:

$$F: \hat{x} \in \mathbf{R}^n \to F(\hat{x}) = B\hat{x} + b \in \mathbf{R}^n, \tag{2.3.10}$$

such that the following relations hold:

$$K = F(\hat{K}), \tag{2.3.11}$$

$$P = \{p: K \to \mathbf{R};\quad p = \hat{p} \cdot F^{-1},\quad \hat{p} \in \hat{P}\}, \tag{2.3.12}$$

$$a_i^r = F(\hat{a}_i^r),\quad r = 0, 1, 2, \tag{2.3.13}$$

$$\xi_{ik}^1 = B\hat{\xi}_{ik}^1,\quad \xi_{ik}^2 = B\hat{\xi}_{ik}^2,\quad \xi_{il}^2 = B\hat{\xi}_{il}^2, \tag{2.3.14}$$

whenever the nodes a_i^r, resp. $\hat{a}_i^r$, and vectors ξ_{ik}^1, ξ_{ik}^2, ξ_{il}^2, resp. $\hat{\xi}_{ik}^1$, $\hat{\xi}_{ik}^2$, $\hat{\xi}_{il}^2$, occur in the definition of the set Σ, resp. $\hat{\Sigma}$.

Remark 2.3.4. The justification of the relations (2.3.14) will become apparent in the proof of Theorem 2.3.1. □

With this definition of affine-equivalence, let us return to the examples given in Section 2.2 (the reader should check for oneself the various statements to come).

To begin with, it is clear that two n-simplices of type k for a given integer $k \geq 1$, are affine equivalent, and that this is also the case for n-simplices of type (3′), in view of the definition (2.2.13) of the associated space P_K. Likewise, two n-rectangles of type (k) for a given integer $k \geqslant 1$, or two rectangles of type (2′) or (3′) are affine equivalent through diagonal affine mappings. In other words, *any two identical Lagrange finite elements that we considered are affine-equivalent.*

When we come to Hermite finite elements, the situation is less simple. Consider for example two Hermite n-simplices of type (3) with sets of degrees of freedom in the form Σ_K (Fig. 2.2.15). Then it is clear that they are affine-equivalent because the relations

$$a_j - a_i = F(\hat{a}_j) - F(\hat{a}_i) = B(\hat{a}_j - \hat{a}_i), \quad 1 \leqslant i, j \leqslant n+1, \quad j \neq i,$$

hold, among other things. However, had we taken the sets of degrees of freedom in the form Σ_K', it would not have been clear to decide whether the two finite elements were affine-equivalent, and yet these two sets of degrees of freedom correspond to the *same* finite element, as we already pointed out.

The same analysis and conclusion apply to the Hermite n-simplex of type (3′) or to the Bogner–Fox–Schmit rectangle. In this last case, it suffices to observe that this finite element can also be defined by the following set of degrees of freedom (the index i is counted modulo 4)

$$\Sigma_K' = \{p(a_i), \quad Dp(a_i)(a_{i-1} - a_i), \quad Dp(a_i)(a_{i+1} - a_i),$$
$$D^2p(a_i)(a_{i-1} - a_i, \quad a_{i+1} - a_i), \quad 1 \leqslant i \leqslant 4\}, \tag{2.3.15}$$

for which relations (2.3.14) hold.

There are counter-examples. For instance, consider a finite element where some degrees of freedom are normal derivatives at some nodes. Then two such finite elements are not in general affine equivalent: The property for a vector to be normal to a hyperplane is not in general preserved through an affine mapping. Thus two Argyris triangles are not affine equivalent in general, except for instance if they happen to be both equilateral triangles. The case of Bell's triangles is left as a problem (Exercise 2.3.4).

Let us return to the general case. We shall constantly use the correspondences

$$\hat{x} \in \hat{K} \to x = F(\hat{x}) \in K, \tag{2.3.16}$$

$$\hat{p} \in \hat{P} \to p = \hat{p} \cdot F^{-1} \in P, \tag{2.3.17}$$

between the points $\hat{x} \in \hat{K}$ and $x \in K$, and the functions $\hat{p} \in \hat{P}$ and $p \in P$ corresponding to two affine-equivalent finite elements. As a consequence of the correspondences (2.3.16) and (2.3.17), notice that we have

$$\hat{p}(\hat{x}) = p(x) \quad \text{for all} \quad \hat{x} \in \hat{K}, \quad \hat{p} \in \hat{P}. \tag{2.3.18}$$

We next prove a crucial relationship between the $\hat{P}$-interpolation operator $\hat{\Pi}$ and the P-interpolation operator Π associated with affine-equivalent finite elements. This relationship will be itself a consequence of the fact that the basis functions are also in the correspondence (2.3.17).

Theorem 2.3.1. *Let $(\hat{K}, \hat{P}, \hat{\Sigma})$ and (K, P, Σ) be two affine-equivalent finite elements with degrees of freedom in the form* (2.3.4). *Then if $\hat{p}_i$, $1 \leq i \leq N$, are the basis functions of the finite element $\hat{K}$, the functions p_i, $1 \leq i \leq N$, are the basis functions of the finite element K. The interpolation operators Π and $\hat{\Pi}$ are such that*

$$(\Pi v)\hat{} = \hat{\Pi}\hat{v} \tag{2.3.19}$$

for any functions $\hat{v} \in \operatorname{dom} \hat{\Pi}$ and $v \to \operatorname{dom} \Pi$ associated in the correspondence

$$\hat{v} \in \operatorname{dom} \hat{\Pi} \to v = \hat{v} \cdot F^{-1} \in \operatorname{dom} \pi. \tag{2.3.20}$$

Proof. The P-interpolation operator Π is of the form (with obvious notations):

$$\Pi v = \sum_i v(a_i^0)p_i^0 + \sum_{i,k} \{Dv(a_i^1)\xi_{ik}^1\}p_{ik}^1 + \sum_{i,k,l} \{D^2v(a_i^2)(\xi_{ik}^2, \xi_{il}^2)\}p_{ikl}^2.$$

Using the derivation of composition of functions, we obtain

$$\begin{aligned} Dv(a_i^1)\xi_{ik}^1 &= Dv(F(\hat{a}_i^1))B\hat{\xi}_{ik}^1 = Dv(F(\hat{a}_i^1))DF(\hat{a}_i^1)\hat{\xi}_{ik}^1 \\ &= D(v \cdot F)(\hat{a}_i^1)\hat{\xi}_{ik}^1 = D\hat{v}(\hat{a}_i^1)\hat{\xi}_{ik}^1, \end{aligned}$$

and, taking also into account that $D^2F = 0$,

$$\begin{aligned} D^2v(a_i^2)(\xi_{ik}^2, \xi_{il}^2) &= D^2v(F(\hat{a}_i^2))(B\hat{\xi}_{ik}^2, B\hat{\xi}_{il}^2) \\ &= D^2v(F(\hat{a}_i^2))(DF(\hat{a}_i^2)\hat{\xi}_{ik}^2, DF(\hat{a}_i^2)\hat{\xi}_{il}^2) \\ &= D^2(v \cdot F)(\hat{a}_i^2)(\hat{\xi}_{ik}^2, \hat{\xi}_{il}^2) = D^2\hat{v}(\hat{a}_i^2)(\hat{\xi}_{ik}^2, \hat{\xi}_{il}^2). \end{aligned}$$

Thus we also have

$$\Pi v = \sum_i \hat{v}(\hat{a}_i^0)p_i^0 + \sum_{i,k} \{D\hat{v}(\hat{a}_i^1)\hat{\xi}_{ik}^1\}p_{ik}^1 + \sum_{i,k,l} \{D^2\hat{v}(\hat{a}_i^2)(\hat{\xi}_{ik}^2, \hat{\xi}_{il}^2)\}p_{ikl}^2,$$

from which we deduce, using the correspondence (2.3.17),

$$(\Pi v)\hat{} = \sum_i \hat{v}(\hat{a}_i^0)\hat{p}_i^0 + \sum_{i,k} \{D\hat{v}(\hat{a}_i^1)\hat{\xi}_{ik}^1\}\hat{p}_{ik}^1 + \sum_{i,k,l} \{D^2\hat{v}(\hat{a}_i^2)(\hat{\xi}_{ik}^2, \hat{\xi}_{il}^2)\}\hat{p}_{ikl}^2.$$

If we apply the previous identity to a function $v \in P$, we infer that the functions $\hat{p}_i^0$, $\hat{p}_{ik}^1$, $\hat{p}_{ik}^2$ are the basis functions of the finite element $(\hat{K}, \hat{P}, \hat{\Sigma})$, by virtue of identity (2.3.8). Using this result, we conclude that the function $(\Pi v)\hat{}$ is equal to the function $\hat{\Pi}\hat{v}$, by definition of the $\hat{P}$-interpolation operator $\hat{\Pi}$. □

Remark 2.3.5. To obtain the conclusion of the previous theorem when the sets of degrees of freedom are in the general form $\hat{\Sigma} = \{\hat{\phi}_i;\ 1 \leq i \leq N\}$ and $\Sigma = \{\phi_i;\ 1 \leq i \leq N\}$, it is necessary and sufficient that the degrees of freedom be such that

$$\forall \hat{v} \in \operatorname{dom} \hat{\Pi}, \quad \hat{\phi}_i(\hat{v}) = \phi_i(v), \quad 1 \leq i \leq N, \tag{2.3.21}$$

and, in essence, the proof of Theorem 2.3.1 consisted in showing that the above relations do hold (as consequences of relations (2.3.13)–(2.3.14)) for the type of degrees of freedom heretofore encountered. In Section 4.2, we shall consider a new type of degrees of freedom for which the validity of relations (2.3.21) will be verified. □

A family of finite elements is called an *affine family* if all its finite elements are affine-equivalent to a single finite element, which is called the *reference finite element* of the family (the reference finite element $(\hat{K}, \hat{P}, \hat{\Sigma})$ need not belong itself to the family).

In the case of an affine family of simplicial finite elements, a customary choice for the set $\hat{K}$ is the *unit n-simplex* with vertices

$$\left.\begin{aligned} \hat{a}_1 &= (1, 0, \ldots, 0),\\ \hat{a}_2 &= (0, 1, 0, \ldots, 0),\\ &\ \vdots\\ \hat{a}_n &= (0, \ldots, 0, 1),\\ \hat{a}_{n+1} &= (0, 0, \ldots, 0), \end{aligned}\right\}$$

for which the barycentric coordinates take the simple form

$$\lambda_i = x_i, \quad 1 \leq i \leq n, \quad \text{and} \quad \lambda_{n+1} = 1 - \sum_{i=1}^{n} x_i.$$

In the case of an affine family of rectangular finite elements, a usual choice for the set $\hat{K}$ is either the unit hypercube $[0, 1]^n$ or the hypercube $[-1, +1]^n$.

The concept of an affine family of finite elements is of importance, basically for the following reasons:

(i) *In practical computations, most of the work involved in the computation of the coefficients of the linear system* (2.1.4) *is performed on a reference finite element, not on a generic finite element.* This point will be illustrated in Section 4.1.

(ii) For such affine families, a fairly elegant *interpolation theory* can be developed (Section 3.1), which is, in turn, the basis of most *convergence theorems.*

(iii) Even when a family of finite elements of a given type is not an affine family, it is generally associated in an obvious way with an affine family whose "intermediate" role is essential. For example, when we shall study in Section 6.1 the interpolation properties of the Argyris triangle, an important step will consist in introducing a slightly different finite element (called the "Hermite triangle of type (5)"; cf. Exercise 2.3.5) which *can* be imbedded in an affine family. In the same fashion, we shall consider (Section 4.3) the "isoparametric" families of curved finite elements essentially as perturbations of affine families.

Construction of finite element spaces X_h. Basic definitions. The X_h-interpolation operator

Our next task is to give a precise *description of the construction of a finite element space from the data of finite elements* (K, P_K, Σ_K). For the sake of simplicity however, *we shall restrict ourselves to the case where the finite elements K are all polygonal,* so that the set $\bar{\Omega} = \bigcup_{K \in \mathcal{T}_h} K$ is necessarily *polygonal, and to the case where the finite elements are all of Lagrange type.* These restrictions essentially avoid difficulties (of a purely technical nature) pertaining to appropriate definitions

(i) of "faces" of non polygonal elements in general and

(ii) of compatibility conditions for the degrees of freedom of adjacent finite elements along common faces in the case of Hermite finite elements.

Remark 2.3.6. There are indeed polygonal finite elements which are used in actual computations by engineers and which are neither n-simplices nor n-rectangles. Of course, such finite elements are not just arbitrary polygonal domains. Rather they are adapted to special circumstances: Thus, if the domain $\bar{\Omega}$ is a cylindrical domain in $\mathbf{R}^3$ it might be interesting to use *prismatic finite elements*, an example of which is given in Fig. 2.3.2.

In this case, the space P is the tensor product of the space P_1 in the variables x_1, x_2 by the space P_1 in the variable x_3, i.e., a function p in the space P is of the form

$$p(x_1, x_2, x_3) = \gamma_1 + \gamma_2 x_1 + \gamma_3 x_2 + \gamma_4 x_3 + \gamma_5 x_1 x_3 + \gamma_6 x_2 x_3. \qquad \square$$

We shall assume that each polygonal set K has a non-empty interior and that the interiors of the sets K are pairwise disjoint, so that requirements $(\mathcal{T}_h i)$, $1 \leq i \leq 4$, are satisfied (a polygonal domain has a Lipschitz-continuous boundary). A portion K' of the boundary of a polygonal finite element K is a *face* of K if it is a maximal connected subset of an affine hyperplane $\mathcal{H}$ of $\mathbf{R}^n$ with a nonempty interior relatively to $\mathcal{H}$.

In order to unambiguously define the functions of the finite element

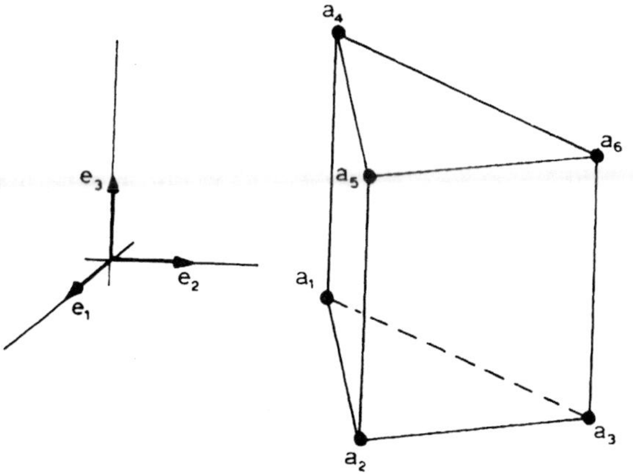

Fig. 2.3.2

space (see below), we need the following obvious extension of the condition ($\mathcal{T}_h5$) already seen for n-simplices and n-rectangles:

($\mathcal{T}_h5$) *Any face of a finite element K_1 is either a face of another finite element K_2, in which case the finite elements K_1 and K_2 are said to be adjacent, or a portion of the boundary Γ of the set Ω.*

Finally, the sets of degrees of freedom of adjacent finite elements shall be related as follows: *Whenever $(K_l, P_{K_l}, \Sigma_{K_l})$ with $\Sigma_{K_l} = \{p(a_i^l), 1 \leq i \leq N_l\}$, $l = 1, 2$, are two adjacent finite elements, then*

$$\left(\bigcup_{i=1}^{N_1}\{a_i^1\}\right) \cap K_2 = \left(\bigcup_{i=1}^{N_2}\{a_i^2\}\right) \cap K_1. \tag{2.3.22}$$

We define the set

$$\mathcal{N}_h = \bigcup_{K \in \mathcal{T}_h} \mathcal{N}_K \tag{2.3.23}$$

where, for each finite element $K \in \mathcal{T}_h$, $\mathcal{N}_K$ denotes the set of nodes. For each $b \in \mathcal{N}_h$, we let K_λ, $\lambda \in \Lambda(b)$, denote all those finite elements for which b is a node. Then the associated *finite element space X_h* is the (generally proper) subspace of the product space $\Pi_{K \in \mathscr{C}_h} P_K$ defined by

$$X_h = \Big\{ v = (v_K)_{K \in \mathcal{T}_h} \in \prod_{K \in \mathcal{T}_h} P_K; \quad \forall b \in \mathcal{N}_h, \quad \forall \lambda, \mu \in \Lambda(b),$$
$$v_{K_\lambda}(b) = v_{K_\mu}(b) \Big\}.$$

Therefore a function in the space X_h is uniquely determined by the set

$$\Sigma_h = \{v(b), \quad b \in \mathcal{N}_h\}, \tag{2.3.24}$$

which is called the *set of degrees of freedom of the finite element space.*

It is thus realized that an element $v \in X_h$ is *not* in general a "function" defined over the set $\bar{\Omega}$, since it need not have a unique definition along faces common to adjacent finite elements. Nevertheless, by virtue of assumption (2.3.22), it is customary to say that the "functions" in the space X_h are at least "continuous at all nodes common to adjacent finite elements" (the inclusions $P_K \subset \mathscr{C}^0(K)$, $K \in \mathcal{T}_h$, hold in practice). It is also a usual practice to consider the functions v_K. $K \in \mathcal{T}_h$, as being the *restrictions* to the finite elements K of the function $v \in X_h$, just as if v were an "ordinary" function defined over the set $\bar{\Omega}$. This is why we shall use the alternate notation $v|_K = v_K$.

If it happens, however, that for each function $v \in X_h$, the restrictions $v|_{K_1}$ and $v|_{K_2}$ coincide along the face common to any pair of adjacent finite elements K_1 and K_2, *then the function v can indeed be identified with a function defined over the set $\bar{\Omega}$.*

Remark 2.3.7. Although this last property was true for all the examples of Section 2.2, it is by no means necessary. Following CROUZEIX & RAVIART (1973), let us consider for example the finite element space constructed with the following finite element (K, P, Σ): The set K is an n-simplex with vertices a_j, $1 \leq j \leq n+1$, the space P is the space $P_1(K)$ and the set of degrees of freedom is the set $\Sigma_K = \{p(b_i), 1 \leq i \leq n+1\}$, where for each i the point b_i is the barycenter of the face which does not contain the point a_i, i.e.,

$$b_i = \frac{1}{n} \sum_{\substack{j=1 \\ j \neq i}}^{n+1} a_j, \quad 1 \leq i \leq n+1.$$

In Fig. 2.3.3, we have represented this finite element for $n = 2$ and $n = 3$.

To show that the set Σ_K is $P_1(K)$-unisolvent, it suffices to observe that the points $b_i = (b_{ji})_{j=1}^n$, $1 \leq i \leq n+1$, are also the vertices of a (non-degenerate) n-simplex: If we let B denote the $(n+1) \times (n+1)$ matrix

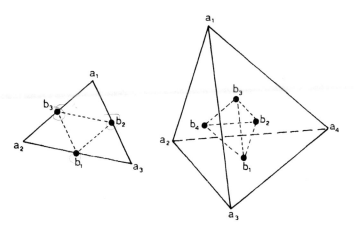

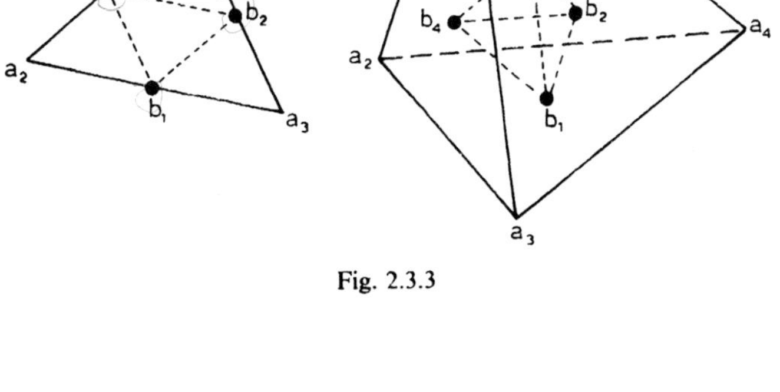

Fig. 2.3.3

defined by

$$B = \begin{pmatrix} b_{11} & b_{12} & \cdots & b_{1,n+1} \\ b_{21} & b_{22} & \cdots & b_{2,n+1} \\ \vdots & & & \vdots \\ b_{n1} & b_{n2} & \cdots & b_{n,n+1} \\ 1 & 1 & \cdots & 1 \end{pmatrix},$$

it is easily verified that $\det B = (-1/n)^n \det A$, where A is the matrix of (2.2.4) and thus $\det B \neq 0$. One may also notice that the functions

$$p_i = 1 - n\lambda_i, \quad 1 \leq i \leq n+1,$$

are the associated basis functions. Then it is clear that the functions of the corresponding finite element space generally have two definitions along faces common to adjacent finite elements, except at the centroids of these faces. □

All the previous considerations can be extended so as to include the case of finite element spaces constructed with Hermite finite elements, and the details are left to the reader (Exercise 2.3.8). We shall simply point out that it is often necessary to choose between various possible sets of degrees of freedom (corresponding to the same finite element) so as to unambiguously define a *set* Σ_h *of degrees of freedom* of the corresponding finite element space. These considerations have been illustrated at various places in Section 2.2.

When the degrees of freedom of all finite elements are of some of the forms (2.3.4), the degrees of freedom of the finite element space are of some of the following forms:

$$\left.\begin{aligned} &v \to v(b_j^0), \\ &v \to Dv(b_j^1)\eta_{jk}^1, \\ &v \to D^2v(b_j^2)(\eta_{jk}^2, \eta_{jl}^2), \end{aligned}\right\} \tag{2.3.25}$$

where the points b_j^r, $r = 0, 1, 2$, called the *nodes of the finite element space*, make up a set which shall be generally denoted $\mathcal{N}_h$.

If we write the set Σ_h as

$$\Sigma_h = \{\phi_{j,h}, \quad 1 \leq j \leq M\}, \tag{2.3.26}$$

then the *basis functions* w_j, $1 \leq j \leq M$, *of the finite element space* are defined by the relations

$$w_j \in X_h \quad \text{and} \quad \phi_{i,h}(w_j) = \delta_{ij}, \quad 1 \leq i \leq M. \tag{2.3.27}$$

We leave it to the reader to verify on each example that *the basis functions of the finite element space are derived from the basis functions of the finite elements*, as follows: Let $\phi_h \in \Sigma_h$ be of one of the form (2.3.25), let b be the associated node, and let K_λ, $\lambda \in \Lambda(b)$, denote all the finite elements of $\mathcal{T}_h$ for which b is a node (see Fig. 2.3.4 in the case of rectangles of type (2)).

For each $\lambda \in \Lambda(b)$, let p_λ denote the basis function of the finite element K_λ associated with the restriction of ϕ_h to K_λ. *Then the function $w \in X_h$ defined by*

$$w = \begin{cases} p_\lambda \text{ over } K_\lambda, & \lambda \in \Lambda(b), \\ 0 \text{ elsewhere,} \end{cases} \tag{2.3.28}$$

is the basis function of the space X_h associated with the degree of freedom ϕ_h.

As a practical consequence, *requirement* (FEM 3) (which was set up in Section 2.1) *is always satisfied in the examples.* The reader should refer to Fig. 2.2.8 where, on an example, it was shown that the basis functions constructed in this fashion have indeed "small" supports. The "worst" case concerns a basis function attached to a vertex, say b, of the triangulation. In this case, the corresponding support is the union of those finite elements which have b as a vertex. In most commonly encountered triangulations in the plane, the number of such finite elements is very low (six or seven, for example).

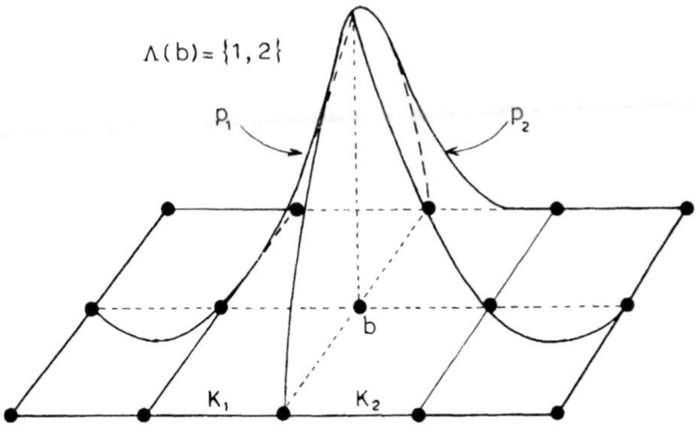

Fig. 2.3.4

Let there be given a finite element space X_h with a set of degrees of freedom of the form (2.3.26). Then with any function $v: \bar{\Omega} \to \mathbf{R}$ sufficiently smooth so that the degrees of freedom $\phi_{j,h}(v)$, $1 \leq j \leq M$, are well defined, we associate the function

$$\Pi_h v = \sum_{j=1}^{M} \phi_{j,h}(v) w_j, \tag{2.3.29}$$

where the functions w_j are the basis functions defined in (2.3.27). The function $\Pi_h v$, called the *X_h-interpolant* of the function v, is equivalently characterized by the conditions

$$\Pi_h v \in X_h \quad \text{and} \quad \phi_{j,h}(\Pi_h v) = \phi_{j,h}(v), \quad 1 \leq j \leq M. \tag{2.3.30}$$

If we let s denote the maximal order of directional derivatives occurring in the finite elements (K, P_K, Σ_K), $K \in \mathcal{T}_h$, we shall usually consider, in view of definition (2.3.7), that *the domain* $\operatorname{dom} \Pi_h$ *of the X_h-interpolation operator* Π_h *is the space*

$$\operatorname{dom} \Pi_h = \mathscr{C}^s(\bar{\Omega}). \tag{2.3.31}$$

It might be helpful to keep in mind the following tableau (Fig. 2.3.5) where we have recapitulated the main "global" (i.e., on $\bar{\Omega}$) versus "local" (i.e., on a generic finite element K) notations, definitions and correspondences.

We next state a relationship of paramount importance between the

"Global" definitions	*"Local" definitions*
$\bar{\Omega}$	K: Finite element
Boundary of the set Ω: Γ	∂K: Boundary of K
	K': Side, or face, of K
Triangulation of the set $\bar{\Omega}$: $\mathcal{T}_h$	
Finite element space: X_h	P or P_K
Generic function of X_h: v or v_h	p or p_K: Generic function of P_K
Set of degrees of freedom of X_h: Σ_h	Σ or Σ_K: Set of degrees of freedom of K
Degrees of freedom of X_h: ϕ_h or $\phi_{j,h}$, $1 \leq j \leq M$	ϕ_i or $\phi_{i,K}$, $1 \leq i \leq N$: Degrees of freedom of K
Basis functions of X_h: w_j, $1 \leq j \leq M$	p_i or $p_{i,K}$, $1 \leq i \leq N$: Basis functions of K
Nodes of X_h: b_j	$a_i, a_{ij}, \ldots$: Nodes of K
X_h-interpolation operator: Π_h	Π or Π_K: P_K-interpolation operator

Fig. 2.3.5

"global" interpolation operator Π_h and the "local" interpolation operators Π_K.

Theorem 2.3.2. *Let v be any function in the space* dom Π_h. *Then the restrictions $v|_K$ belong to the spaces* dom Π_K, *and we have*

$$\forall K \in \mathcal{T}_h, \quad (\Pi_h v)|_K = \Pi_K v|_K. \tag{2.3.32}$$

Proof. The above relations are direct consequences of the way in which the set Σ_h is derived from the sets Σ_K, $K \in \mathcal{T}_h$. □

Finite element of class $\mathscr{C}^0$ and $\mathscr{C}^1$

It has always been assumed thus far that all the finite elements (K, P_K, Σ_K), $K \in \mathcal{T}_h$, which are used in the definition of a finite element space are all *of the same type*: By this, we mean that, for instance, the finite elements are all n-simplices of type (2), or that the finite elements are all Argyris triangles, etc.... If this is the case, we shall say that any finite element (K, P_K, Σ_K), $K \in \mathcal{T}_h$, is the *generic* finite element of the finite element space. We next state two definitions which are of particular importance, in view of Theorems 2.1.1 and 2.1.2.

We shall say that a finite element (K, P_K, Σ_K) is *of class* $\mathscr{C}^0$ if (i) the inclusion $P_K \subset \mathscr{C}^0(K)$ holds and (ii) whenever it is the generic finite element of a triangulation and K_1 and K_2 are two adjacent finite elements, the restrictions $v_h|_{K_1}$ and $v_h|_{K_2}$ coincide along the face common to K_1 and K_2 for any function v_h of the corresponding finite element space. As a consequence, it is legitimate in this case to consider that the inclusion $X_h \subset \mathscr{C}^0(\bar{\Omega})$ holds.

Likewise, we shall say that a finite element (K, P_K, Σ_K) of a given type is *of class* $\mathscr{C}^1$ if (i) the inclusion $P_K \subset \mathscr{C}^1(K)$ holds and (ii) whenever it is the generic finite element of a triangulation and K_1 and K_2 are two adjacent finite elements, for any function v_h in the corresponding finite element space the restrictions $v_h|_{K_1}$ and $v_h|_{K_2}$ coincide along the face K' common to K_1 and K_2 and the outer normal derivatives $\partial_\nu v_h|_{K_1}$ and $\partial_\nu v_h|_{K_2}$ have a zero sum along K'. As a consequence, it is legitimate in this case to consider that the inclusion $X_h \subset \mathscr{C}^1(\bar{\Omega})$ holds.

Thus for instance, *all the finite elements seen in Section* 2.2 *are of class* $\mathscr{C}^0$, *and the Argyris and Bell triangles and the Bogner–Fox–Schmit rectangle are in addition of class* $\mathscr{C}^1$. *There are also finite elements which are not of class* $\mathscr{C}^0$, such as the one that was considered in Remark 2.3.7.

Remark 2.3.8. One may of course use finite elements of *different types* in a triangulation, provided some compatibility conditions are satisfied along faces which are common to adjacent finite elements, in such a way that a function in the space X_h is still unambiguously defined on the one hand, and an inclusion such as $X_h \subset \mathscr{C}^0(\bar{\Omega})$ (for example) holds on the other hand. Thus one may combine triangles of type (k) or (k') with rectangles of type (k) and still obtain the inclusion $X_h \subset \mathscr{C}^0(\bar{\Omega})$, etc. . . . Such an example is considered in Fig. 2.3.6. □

Taking into account boundary conditions. The spaces X_{0h} *and* X_{00h}

The last topic we wish to examine in this section is *the way in which boundary conditions are taken into account in a finite element space.* Again, we shall essentially concentrate on examples.

Let X_h be a finite element space whose generic finite element is any one of the following: n-simplex of type (k), $k \geqslant 1$, or of type $(3')$, n-rectangle of type (k), $k \geqslant 1$, rectangle of type $(2')$ or $(3')$. Then the inclusion $X_h \subset \mathscr{C}^0(\bar{\Omega}) \cap H^1(\Omega)$ holds (Theorems 2.2.3 and 2.2.7) and it follows that the inclusion

$$X_{0h} = \{v_h \in X_h;\quad v_{h|\Gamma} = 0\} \subset H_0^1(\Omega) \tag{2.3.33}$$

holds. In each of the above cases, it is easily verified that a sufficient (and obviously necessary) condition for a function $v_h \in X_h$ to vanish along Γ is that *it vanishes at all the boundary nodes*, i.e., those nodes of the space X_h which are on the boundary Γ. In other words, if we let $\mathscr{N}_h$ denote the set of nodes of the space X_h, the finite element space X_{0h} of (2.3.33) is simply given by

$$X_{0h} = \{v_h \in X_h;\quad \forall b \in \mathscr{N}_h \cap \Gamma,\quad v_h(b) = 0\}. \tag{2.3.34}$$

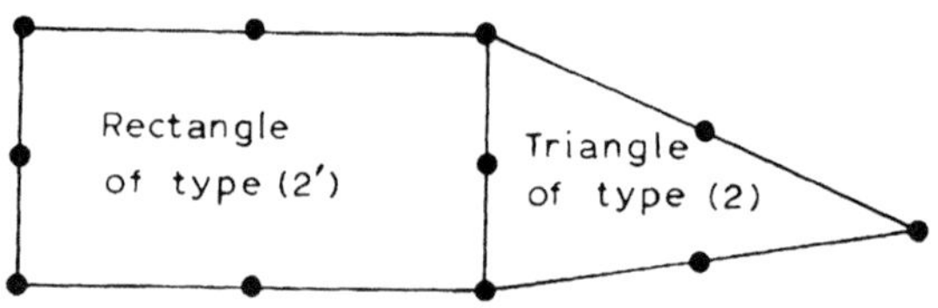

Fig. 2.3.6

When Hermite finite elements are used, the situation is less simple. Let us consider for example the case of Hermite n-simplices of type (3'), in which case the set of nodes of the triangulation coincides with the set of all the vertices of the triangulation. Then for each boundary node $b \in \mathcal{N}_h \cap \Gamma$, we let $\tau_\gamma(b)$, $\gamma \in \Gamma(b)$, denote a maximal set of linearly independent vectors in $\mathbf{R}^n$ with the property that the points $(b + \tau_\gamma(b))$, $\gamma \in \Gamma(b)$, belong to the boundary Γ. Following the notation introduced in Section 1.2, we recall that the differential operator ∂_τ is defined by $\partial_\tau v(a) = Dv(a)\tau$. Then the space X_{0h} of (2.3.33) is given in this case by

$$X_{0h} = \{v_h \in X_h;\quad \forall b \in \mathcal{N}_h \cap \Gamma,\quad v_h(b) = 0,$$
$$\forall b \in \mathcal{N}_h \cap \Gamma,\quad \forall \gamma \in \Gamma(b),\quad \partial_{\tau_\gamma(b)} v_h(b) = 0\}. \qquad (2.3.35)$$

We have indicated in Fig. 2.3.7 the directional derivatives which must be set equal to zero along a specific portion of the boundary of a polygonal set in $\mathbf{R}^2$.

In particular, one should observe that at a corner such as b^*, the directional derivatives $\partial_{\mu_3} v(b)$ and $\partial_{\mu_4} v(b)$ must necessarily vanish.

If we next assume that the inclusion $X_h \subset \mathscr{C}^1(\bar{\Omega}) \cap H^2(\Omega)$ holds, it follows that we have the inclusions

$$X_{0h} = \{v_h \in X_h;\quad v_h|_\Gamma = 0\} \subset H^2(\Omega) \cap H_0^1(\Omega), \qquad (2.3.36)$$

$$X_{00h} = \{v_h \in X_h;\quad v_h|_\Gamma = \partial_\nu v_h = 0\} \subset H_0^2(\Omega), \qquad (2.3.37)$$

so that we are facing the problem of constructing such spaces X_{0h} and X_{00h}. Again they are obtained by canceling appropriate values and directional derivatives at boundary nodes. As an example, we have indicated in Fig. 2.3.8 all the directional derivatives which must be set

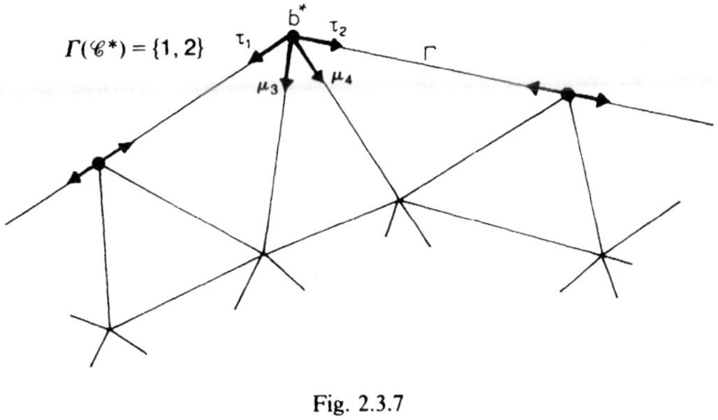

Fig. 2.3.7

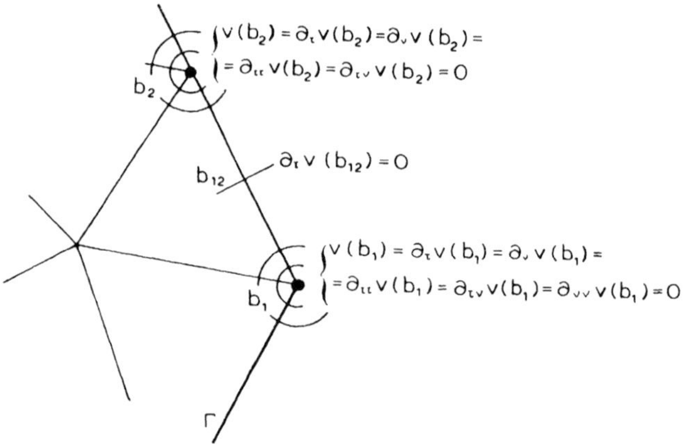

Fig. 2.3.8

equal to zero when Argyris triangles are used and the second inclusion (2.3.37) is needed.

It should be realized that at a boundary node such as b_2, the only "free" degree of freedom is $\partial_{\nu\nu}(b_2)$ while all degrees of freedom are zero at a corner such as b_1.

We shall also record for subsequent uses (particularly in the next chapter) the following crucial properties:

(i) *All finite elements of class* $\mathscr{C}^0$ *and of class* $\mathscr{C}^1$ *described in Section* 2.2 *have the property that*

$$v \in \operatorname{dom} \Pi_h \quad \text{and} \quad v|_\Gamma = 0 \Rightarrow \Pi_h v \in X_{0h}, \tag{2.3.38}$$

where the finite element space X_{0h} is defined as in (2.2.33), or (2.3.36) for finite elements of class $\mathscr{C}^1$.

(ii) *All finite elements of class* $\mathscr{C}^1$ *described in Section* 2.2 *have the property that*

$$v \in \operatorname{dom} \Pi_h \quad \text{and} \quad v|_\Gamma = \partial_\nu v = 0 \Rightarrow \Pi_h v \in X_{00h}, \tag{2.3.39}$$

where the finite element space X_{00h} is defined as in (2.3.37).

Remark 2.3.9. It is clearly possible to extend the previous analyses so as to include the case where boundary conditions are imposed only over a *portion* Γ_0 of the boundary Γ, provided such a portion Γ_0 is exactly the union of some faces of the finite elements found in the triangulation. □

Final comments

Remark 2.3.10. Let us briefly point out how some of the previously studied properties of finite elements and finite element spaces may be in fact derived from a *single "local" property*. For ease of exposition, we shall restrict ourselves to the case of Lagrange finite elements, leaving as a problem (Exercise 2.3.10) the case of Hermite finite elements.

Let (K, P, Σ) be a Lagrange finite element, with $\mathcal{N}$ as its set of nodes, i.e., the set of degrees of freedom is of the form $\Sigma = \{p(a);\ a \in \mathcal{N}\}$. If K' is any face of the set K, we let

$$\Sigma|_{K'} = \{p(a); \quad a \in \mathcal{N} \cap K'\}, \tag{2.3.40}$$

$$P(K') = \{p|_{K'}\colon K' \to \mathbf{R}; \quad p \in P\}. \tag{2.3.41}$$

Then all the Lagrange finite elements heretofore described, with the exception of the finite element of Remark 2.3.7, have the property that *for each one of their faces, say K', the set $\Sigma|_{K'}$ defined in* (2.3.40) *is $P(K')$-unisolvent, where the space $P(K')$ is defined in* (2.3.41). This is a crucial underlying basic property, which has the following easily established consequences:

(i) The P_K-interpolant of a function $v \in \operatorname{dom} \Pi_K$ which vanishes along a face K' is also zero along K'. As a consequence, the global property (2.3.38) holds.

(ii) Let $\phi\colon p \in P \to p(a)$ be any one of the degrees of freedom of the finite element and let p be the associated basis function. Then the function p vanishes along any face which does not contain the node a. From the way the basis functions of the space X_h are constructed from the basis functions of the finite elements (cf. (2.3.28)), we in turn deduce the "global" property that the basis functions of the space X_h have indeed small supports (FEM 3).

(iii) Assume in addition that, for each pair (K_1, K_2) of adjacent finite elements found in a triangulation, one has $P_{K_1}|_{K'} = P_{K_2}|_{K'}$ along the common face K', and that the inclusions $P_K \subset \mathscr{C}^0(K)$, $K \in \mathcal{T}_h$, hold. Then the inclusion $X_h \subset \mathscr{C}^0(\bar{\Omega})$ hold. □

In the *choice* of a finite element for solving a given problem, the following considerations are usually taken into account:

(i) The finite element must be well adapted to the *geometry* of the problem. For example, assembling three-dimensional finite elements is not an easy task. This is especially true for tetrahedra, so that prismatic finite elements (Remark 2.3.6) are usually preferred whenever possible. In this respect, see the discussion in ZIENKIEWICZ (1971, Chapter 6). Geometrical considerations also justify the choice of curved finite elements instead of straight finite elements in the case of "particularly curved" domains.

(ii) The finite element must of course be appropriate for the problem to be solved. For conforming finite element methods, we have seen that this requires the use of finite elements of class $\mathscr{C}^0$ or $\mathscr{C}^1$. In addition, we shall see that a mathematical proof of convergence requires (among other things) the inclusions $P_1(K) \subset P_K$, $K \in \mathscr{T}_h$, for second-order problems and the inclusions $P_2(K) \subset P_K$, $K \in \mathscr{T}_h$, for fourth-order problems. Incidentally, the engineers were well aware of these conditions, which they discovered empirically, long before the mathematicians arrived!

(iii) Once the two previous criteria have been satisfied, it remains to obtain a linear system whose coefficients are easy to compute on the one hand and which is as easy as possible to solve on the other hand. We shall not go very far into this interesting and manifold aspect of finite element methods. However, we shall record two rules which tend to reduce certain computational difficulties:

A *first guideline* is that, if possible, the sets of degrees of freedom associated with a given node in the triangulation be all alike, so as to avoid different instructions depending on the node. This explains why Hermite n-simplices of type (3') may be preferred to Hermite n-simplices of type (3), or why Bell's triangles may be preferred to Argyris triangles, even though there is in both cases a decrease of one in the order of convergence, as we shall see (in addition, such choices slightly reduce the dimension of the resulting linear system).

A *second guideline* is that each node of the space should be common to the greatest number of finite elements. For example, the reader will easily convince himself that *for a given triangulation*, Hermite triangles of type (3) lead to a smaller linear system than triangles of type (3) (with the same order of convergence).

(iv) In addition, miscellaneous questions may be considered. For instance, one may argue in the above example that the use of Hermite triangles of type (3) introduces artificial constraints (the continuity of the first derivatives at the vertices) on the one hand, but on the other hand, this is an advantage if one needs to compute the stresses at the vertices. Likewise, one may argue that the use of Argyris triangles for solving a plate problem introduces artificial constraints (the continuity of the second derivatives at the vertices and "extra" boundary conditions as shown in Fig. 2.3.8) on the one hand, but on the other hand, this is an advantage if one needs to compute the bending moments at the vertices (such moments are obtained from the second partial derivatives of the solution), etc. . . .

To conclude this discussion, we shall simply emphasize the fact that, for all practical purposes, nothing replaces the numerical experience accumulated over the years by the engineers.

Exercises

2.3.1. Let the points a_i, $5 \leq i \leq 8$, be as in Fig. 2.2.10. Is $\{p(a_i), 5 \leq i \leq 8\}$ a Q_1-unisolvent set?

2.3.2. Let there be given a triangle with vertices a_i, $1 \leq i \leq 3$. Among the following sets of degrees of freedom, which ones are P_2-unisolvent?

$$\Sigma = \{p(a_{iij}), \quad 1 \leq i, j \leq 3, \quad i \neq j\},$$

$$\Sigma' = \{p(a_i), \quad 1 \leq i \leq 3; \quad p(a_{iij}), \quad 1 \leq i < j \leq 3\},$$

$$\Sigma'' = \{p(a_i), \quad Dp(a_i)(a_{i+1} - a_i), \quad 1 \leq i \leq 3\}$$

(in Σ'', the indices are counted modulo 3).

2.3.3. Let $(\hat{K}, \hat{P}, \hat{\Sigma})$ be a finite element with degrees of freedom of the form (2.3.4), and let K, P and Σ be defined through relations (2.3.11), (2.3.12), (2.3.13) and (2.3.14), with F any invertible affine mapping.

(i) Show that the triple (K, P, Σ) is a finite element.

(ii) Is a generalization possible so as to include more general (and smooth enough) invertible mappings F?

(iii) Let $(\hat{K}, \hat{P}, \hat{\Sigma})$ be an n-rectangle of type (k). Then the above process, with F affine, allows the derivation of finite elements which are parallelograms for $n = 2$, parallelepipeds for $n = 3$, etc. . . (such finite elements are seldom used in practice however). Describe the corresponding space P.

2.3.4. Are two Bell's triangles affine-equivalent in general?

2.3.5. Let K be a triangle with vertices a_i, $1 \leq i \leq 3$.

(i) Show that the set

$$\Sigma = \{p(a_i), \quad 1 \leq i \leq 3; \quad Dp(a_i)(a_j - a_i), \quad 1 \leq i, j \leq 3, \quad j \neq i;$$
$$D^2p(a_i)(a_j - a_i, \quad a_k - a_i), \quad 1 \leq i, j, k \leq 3, \quad j \neq i, \quad k \neq i;$$
$$Dp(a_{ij})(a_k - a_{ij}), \quad 1 \leq i < j \leq 3, \quad k \neq i, \quad k \neq j\}$$

is $P_5(K)$-unisolvent. The corresponding finite element is called the *Hermite triangle of type* (5).

(ii) Show that this finite element differs in general from the Argyris triangle.

(iii) Is it a finite element of class $\mathscr{C}^1$?

2.3.6. Give a definition and a proof of the statement: "The barycentric coordinates are invariant through an invertible affine mapping". How is this fact reflected when the basis functions of finite elements, such as the n-simplices of type (k), or the Hermite triangle of type (3′), are expressed in terms of barycentric coordinates?

2.3.7. Is a finite element space X_h completely specified by the data of the spaces P_K, $K \in \mathscr{T}_h$, and of the inclusion $X_h \subset \mathscr{C}^0(\bar{\Omega})$?

2.3.8. Give a complete description of the construction of a finite element space made up of Hermite finite elements. In particular, extend condition (2.3.22) so as to include degrees of freedom which involve directional derivatives of the first and second order.

2.3.9. Let K be a triangle with vertices a_i, $1 \leq i \leq 3$. In each one of the following cases, prove the P-unisolvence of the set Σ of degrees of freedom. Which finite elements are of class $\mathscr{C}^0$? (These finite elements have been considered by CROUZEIX & RAVIART (1973), who showed that they could be used for approximating the solution of the Stokes problem):

(i) $P = \mathrm{V}\{\lambda_i^2, \quad 1 \leq i \leq 3; \quad \lambda_i\lambda_j, 1 \leq i < j \leq 3; \quad \lambda_1\lambda_2\lambda_3\}$,

$$\Sigma = \{p(a_i), \quad 1 \leq i \leq 3; \quad p(a_{ij}), 1 \leq i < j \leq 3; \quad p(a_{123})\}.$$

In particular show that the inclusion $P_2 \subset P$ holds.

(ii) $P = \mathrm{V}\{\lambda_i^3, \quad 1 \leq i \leq 3; \quad \lambda_i^2\lambda_j, \lambda_i\lambda_j^2, \quad 1 \leq i < j \leq 3; \quad \lambda_i^2\lambda_{i+1}\lambda_{i+2},$
$1 \leq i \leq 3 \text{ (mod. 3)}\}$,

$$\Sigma = \{p(a_i), \quad 1 \leq i \leq 3; \quad p(a_{iij}), \quad 1 \leq i, j \leq 3, \quad i \neq j;$$
$$p(a_i^*), \quad 1 \leq i \leq 3\},$$

where $a_i^* = \frac{3}{2}(1 - \alpha)a_{123} + \frac{1}{2}(3\alpha - 1)a_i$, $1 \leq i \leq 3$, and α is any number which satisfies $0<\alpha<1$, $\alpha\neq\frac{1}{3}$. In particular show that the inclusion $P_3\subset P$ holds.

(iii) The space P is the same as in (ii),

$$\Sigma=\{p(a_{ij}),\quad 1\leqslant i<j\leqslant 3;\quad p(a_{ij}^*),\quad 1\leqslant i,j\leqslant 3,\ i\neq j;\ p(b_i^*),\\ 1\leqslant i\leqslant 3\},$$

where

$$a_{ij}^*=\gamma_1 a_i+\gamma_2 a_j,\quad 1\leqslant i,j\leqslant 3,\quad i\neq j,$$

$$\gamma_1=\frac{1}{2}\left(1-\sqrt{\frac{3}{5}}\right),\quad \gamma_2=\frac{1}{2}\left(1+\sqrt{\frac{3}{5}}\right)$$

(notice that the points γ_1, $\frac{1}{2}$, γ_2 are the Gaussian quadrature points of the interval $[0,1]$),

$$b_i^*=\alpha a_i+\frac{1-\alpha}{2}(a_{i+1}+a_{i+2}),\quad 1\leqslant i\leqslant 3 \text{ (mod. 3)},$$

and α is any number which satisfies $0<\alpha<1$, $\alpha\neq\frac{1}{3}$.

2.3.10. Extend the content of Remark 2.3.10 so as (i) to cover the cases of Hermite finite elements and (ii) to obtain "local" conditions which imply the inclusion $X_h\subset\mathscr{C}^1(\bar{\Omega})$.

2.3.11. The following finite element which resembles the Hermite triangle of type (3′) has sometimes been used for solving two-dimensional problems: The set K is a triangle with vertices a_i, $1\leqslant i\leqslant 3$, the space P is the space $P_3^*(K)$, where

$$P_3^*=\left\{p: x\in\mathbf{R}^2\rightarrow\sum_{i_1+i_2\leqslant 3}\gamma_{i_1i_2}x_1^{i_1}x_2^{i_2};\quad \gamma_{21}=\gamma_{12}\right\}$$

(obviously the inclusion $P_2(K)\subset P_3^*(K)$ holds), and

$$\Sigma=\{p(a_i),\quad \partial_1 p(a_i),\quad \partial_2 p(a_i),\quad 1\leqslant i\leqslant 3\},$$

so that this finite element is of class $\mathscr{C}^0$.

(i) Is Σ a $P_3^*(K)$-unisolvent set?

(ii) Can this finite element be imbedded in an affine family?

2.4. General considerations on convergence

Convergent family of discrete problems

Whereas up to now, our discussion has been concerned with *one* discrete problem, we shall now consider *families* of discrete problems.

More specifically, assume that we are approximating the solution u of the variational equations

$$\forall v \in V, \quad a(u, v) = f(v), \tag{2.4.1}$$

where the space V, the bilinear form $a(\cdot, \cdot)$ and the linear form f satisfy the assumptions of the Lax–Milgram lemma (Theorem 1.1.3). Confining ourselves to the case of *conforming finite element methods*, we consider a family (V_h) of subspaces of the space V, where it is understood that the *parameter* h (which shall be given a specific meaning in Section 3.2) is the defining parameter of the family and *has limit zero.*

With each finite element space V_h is associated the *discrete solution* u_h which satisfies

$$\forall v_h \in V_h, \quad a(u_h, v_h) = f(v_h). \tag{2.4.2}$$

Then we shall say that *the associated family of discrete problems is convergent*, or equivalently, that *convergence* holds, if, for *any* problem of the form (2.4.1) posed in the space V, one has

$$\lim_{h \to 0} \|u - u_h\| = 0, \tag{2.4.3}$$

where $\|\cdot\|$ denotes the norm in the space V.

Céa's lemma. First consequences. Orders of convergence

We are therefore interested in giving sufficient conditions for convergence and, as a first result in this direction, we have the following basic *abstract error estimate.*

Theorem 2.4.1. (*Céa's lemma*). *There exists a constant C independent upon the subspace V_h such that*

$$\|u - u_h\| \leq C \inf_{v_h \in V_h} \|u - v_h\|. \tag{2.4.4}$$

Consequently, a sufficient condition for convergence is that there exists a family (V_h) of subspaces of the space V such that, for each $u \in V$,

$$\lim_{h \to 0} \inf_{v_h \in V_h} \|u - v_h\| = 0. \tag{2.4.5}$$

Proof. Let w_h be an arbitrary element in V_h: It follows from (2.4.1) and (2.4.2) that $a(u - u_h, w_h) = 0$. Using the same constants α, M as in (1.1.3) and (1.1.19), we have, for any $v_h \in V_h$,

$$\alpha\|u - u_h\|^2 \leq a(u - u_h, u - u_h) = a(u - u_h, u - v_h) \leq M\|u - u_h\| \|v - v_h\|,$$

and the conclusion follows with $C = M/\alpha$. □

Remark 2.4.1. When the bilinear form is *symmetric*, there is a remarkable interpretation of the discrete solution: Since we have $a(u - u_h, w_h) = 0$ for all $w_h \in V_h$, it follows that u_h *is the projection over* V_h *of the exact solution* u, *with respect to the inner product* $a(\cdot, \cdot)$. Therefore, we have in this case

$$a(u - u_h, u - u_h) = \inf_{v_h \in V_h} a(u - v_h, u - v_h).$$

Using the V-ellipticity and the continuity of the bilinear form, we deduce

$$\|u - u_h\| \leq \sqrt{\frac{M}{\alpha}} \inf_{v_h \in V_h} \|u - v_h\|.$$

Thus we have obtained a "better" constant than in the proof of Theorem 2.4.1, since the constant M is necessarily larger than the constant α. □

The simple, yet crucial, inequality (2.4.4) shows that *the problem of estimating the error* $\|u - u_h\|$ *is reduced to a problem in Approximation Theory*: To evaluate the distance $d(u, V_h) = \inf_{v_h \in V_h} \|u - v_h\|$ between a function $u \in V$ and a subspace $V_h \subset V$. This explains why this problem will be a central theme of the next chapter, where we shall essentially prove results of the following type: Assuming appropriate smoothness on the function u, we shall show that the distance $d(u, V_h)$ is itself bounded by a constant (which usually involves norms of higher order derivatives of the function u) times h^β, for some exponent $\beta > 0$. As a consequence, we have the additional information that, *for a given solution* u, there exists a constant $C(u)$ independent of h such that

$$\|u - u_h\| \leq C(u)h^\beta. \tag{2.4.6}$$

If this is the case, we shall say that the *order of convergence* is β, or equivalently, that we have an $0(h^\beta)$ *convergence*, and we shall simply write

$$\|u - u_h\| = 0(h^\beta). \tag{2.4.7}$$

Using more elaborated techniques, we shall also evaluate the difference $(u - u_h)$ in other norms, or semi-norms, than the norm of the space V (which is either the $\|\cdot\|_{1,\Omega}$ or the $\|\cdot\|_{2,\Omega}$ norm), such as the $|\cdot|_{0,\Omega}$ and the $|\cdot|_{0,\infty,\Omega}$ norms (cf. Sections 3.2 and 3.3 respectively), and we shall also call *error* the corresponding norms $|u - u_h|_{0,\Omega}$, $|u - u_h|_{0,\infty,\Omega}$, etc. . . .

Whereas a mathematician is generally satisfied with a sufficient condition for convergence such as that of Theorem 2.4.1, this condition rightly appears as a philosophical matter to many an engineer, who is much more concerned in getting (even rough) estimates of the error *for a given space* V_h: For practical problems, one chooses often one, sometimes two, seldom more, subspaces V_h, but certainly not an infinite family. *In other words, the parameter h never approaches zero in practice*!

Nevertheless, we found it worth examining such questions of convergence because (besides providing the subject of this book . . .) (i) the problem of estimating the error for a given h (i.e., of getting a realistic estimate of the constant $C(u)$ which appears in inequality (2.4.6)) is at the present time not solved in a satisfactory way, and (ii) at least there is a "negative" aspect that few people contest: Presumably, a method should not be used in practice if it were impossible to mathematically prove its convergence. . . .

Bibliography and comments

2.1. The finite element method was first conceived in a paper by COURANT (1943), but the importance of this contribution was ignored at that time. Then the engineers independently re-invented the method in the early fifties: The earliest references generally quoted in the engineering literature are those of ARGYRIS (1954–1955), TURNER, CLOUGH, MARTIN & TOPP (1956). The name of the method was proposed by CLOUGH (1960). Historical accounts on the development of the method, from the engineering point of view, are given in ODEN (1972a), and ZIENKIEWICZ (1973).

It is only in the sixties that mathematicians, notably MIKHLIN (1964, 1971), showed real interest in the analysis of the Galerkin and Ritz methods. Although they were not aware of the engineers contributions, it is interesting to notice that the approximate methods which they studied resembled more and more the finite element method, as exemplified by the basic contributions of CÉA (1964), VARGA (1966) (for the one-dimensional case), BIRKHOFF, SCHULTZ & VARGA (1968) (for the multidimensional – but still tensor-product like – case). Then the outbreak came with the paper of ZLÁMAL (1968), which is generally regarded as the first mathematical error analysis of the "general" finite element method as we know it to-day.

2.2 and 2.3. The finite elements which are described in these sections can be found in the book of ZIENKIEWICZ (1971), where they are sometimes given different names. In this respect, the reader who wishes to look into the Engineering literature may consult the following table, which lists a few correspondences.

Name given in this book	*Name given in Zienkiewicz' book*
Triangle or tetrahedron of type (1), (2), (3)	Linear, quadratic, cubic triangle or tetrahedron
Rectangle of type (1), (2), (3)	Linear, quadratic, cubic rectangle
Rectangle of type (2′), (3′)	Quadratic, cubic rectangle of the serendipity family
3-rectangle	right prism or rectangular prism
barycentric coordinates	area or volume co-ordinates
basis functions	shape functions

Regarding the attribution of names to finite elements, we have tried to follow the most common usages.

In particular, Courant's triangle is named after COURANT (1943). The rectangles of type (2′) and (3′) are also called *serendipity finite elements*, because their discovery required some ingenuity indeed! Other examples of serendipity finite elements may be found in ZIENKIEWICZ (1971, p. 108, p. 121, p. 126), particularly for $n = 3$. We mention that ZLÁMAL (1973d) has given an interesting alternate approach for such serendipity finite elements. The Zienkiewicz triangle is named after BAZELEY, CHEUNG, IRONS & ZIENKIEWICZ (1965). The Argyris triangle is named after ARGYRIS, FRIED & SCHARPF (1968), while Bell's triangle is named after BELL (1969). Although these last two finite elements have appeared in these and several other publications around 1968–1969 (cf. the references given in ZIENKIEWICZ (1971, p. 209)), it was recently brought to the author's attention that they should also be attributed to FELIPPA (1966), where they appeared for the first time.

For the numerical handling of the Argyris triangle (derivation of the basis functions, etc...), the reader is referred to ARGYRIS, FRIED & SCHARPF (1968). See also THOMASSET (1974). Finally the Bogner–Fox–Schmit rectangle is named after BOGNER, FOX & SCHMIT (1965). We also note that Theorem 3 of ZLÁMAL (1968) yields an alternate proof of Theorem 2.2.11.

Whereas it is fairly easy to conceive finite element spaces contained in $\mathscr{C}^0(\bar{\Omega})$, the construction of finite element spaces contained in $\mathscr{C}^1(\bar{\Omega})$ is less obvious, as shown by the last three examples of Section 2.2 (and also by additional examples that will be seen in Section 6.1). In this direction, see the discussion in ZIENKIEWICZ (1971, Section 10.3), whose heuristic considerations have been recently justified by a beautiful *result of* ŽENÍŠEK (1973, 1974), who has proved the following: Let $n = 2$, let X_h be a finite element space for which all finite elements K are triangles, and for which the spaces P_K are spaces of polynomials, i.e., there exists some integer l such that the inclusions $P_K \subset P_l(K)$ hold for all $K \in \mathscr{T}_h$ (therefore finite elements of class $\mathscr{C}^1$ using "singular functions", or of "composite" type, as described in Section 6.1 are excluded from the present analysis). Then for any integer $m \geq 0$, the inclusion $X_h \subset \mathscr{C}^m(\bar{\Omega})$ implies that, at each vertex b of the triangulation, the linear forms $v_h \to \partial^\alpha v_h(b)$ are degrees of freedom of the space X_h for all $|\alpha| \leq 2m$. As a corollary, the inequality $l \geq 4m + 1$ holds (the proof of the corollary is simple, but the proof of the first result is by no means trivial). Thus for instance the particular choice $m = 1$ shows that Bell's triangle is the optimal finite element for fourth-order problems inasmuch as the dimension of its space $P_5'(K)$ is the smallest possible, at least for conforming finite element methods using piecewise polynomial spaces.

ŽENÍŠEK (1972) has also extended his results to the case of higher dimensions, and there has been recently substantial interest in the study of the properties of finite element spaces whose functions are piecewise polynomials and which are contained in $\mathscr{C}^m(\bar{\Omega})$. In this respect, we mention BARNHILL & GREGORY (1975b), DÉLÈZE & GOËL (1976), MORGAN & SCOTT (1975, 1976), SCOTT (1974), STRANG (1973, 1974a).

There is a large literature on the various aspects of the numerical implementation of the finite element method. We shall quote here only a few papers: BIRKHOFF & FIX (1974), BOISSERIE & PLANCHARD (1971), BOSSAVIT (1973), BOSSAVIT & FRÉMOND (1976), DESCLOUX (1972a, 1972b), FIX & LARSEN (1971), FRIED (1971a, 1973a, 1973b), FRÉMOND (1974), GOËL (1968a, 1968b). We also mention BRAUCHLI & ODEN (1971),

ODEN (1973a), ODEN & REDDY (1976a, Section 6.5) for the "conjugate basis functions" approach. The paper of FELIPPA & CLOUGH (1970) is a nice blend of mathematical analysis and practical aspects.

Finally, we mention that the definition of a finite element as a triple (K, P, Σ) is due to CIARLET (1975).

2.4. Céa's lemma (Theorem 2.4.1) appeared in CÉA (1964, Proposition 3.1) in the case of a symmetric bilinear form. It was independently rediscovered by VARGA (1966), and extended to the nonsymmetric case in BIRKHOFF, SCHULTZ & VARGA (1968, Theorem 13).

CHAPTER 3

CONFORMING FINITE ELEMENT METHODS FOR SECOND-ORDER PROBLEMS

Introduction

In this chapter, we consider the problem of determining estimates in various norms of the difference $(u - u_h)$, where $u \in V$ is the solution of a second-order boundary value problem and $u_h \in V_h$ is the discrete solution obtained in a subspace V_h of V.

From Céa's lemma (Theorem 2.4.1), the best error estimate would result in exhibiting the element $\theta_h u \in V_h$ which is such that $\inf_{v_h \in V_h} \|u - v_h\|_{1,\Omega} = \|u - \theta_h u\|_{1,\Omega}$, i.e., the projection of the solution u on the space V_h. However, such a projection is not particularly easy to work with, and it turns out that it is much more convenient to use the *X_h-interpolant $\Pi_h u$ of the solution u*, so that we shall get instead the error estimate $\|u - u_h\|_{1,\Omega} \leq C\|u - \Pi_h u\|_{1,\Omega}$.

Since we shall assume in this chapter that the set $\bar{\Omega}$ is polygonal, it can be written as a union $\bar{\Omega} = \bigcup_{K \in \mathcal{T}_h} K$ of polygonal finite elements K, such as the ones which have been heretofore described. This in turn implies that the corresponding spaces V_h will be contained in the space V (the domains of definition of their functions are identical), i.e., that the corresponding finite element method is indeed *conforming*.

Taking into account that we are using the norm $\|\cdot\|_{1,\Omega}$ and that $(\Pi_h u)|_K = \Pi_K u$ for all $K \in \mathscr{C}_h$ (Theorem 2.3.2), we can write

$$\|u - \Pi_h u\|_{1,\Omega} = \left(\sum_{K \in \mathcal{T}_h} \|u - \Pi_K u\|_{1,K}^2\right)^{1/2}.$$

Therefore, the problem of finding an estimate for the error $\|u - u_h\|_{1,\Omega}$ is reduced to the problem of evaluating quantities such as $\|u - \Pi_K u\|_{1,K}$ and the solution of such "*local*" *interpolation problems* is the object of Section 3.1. In view of other future needs, we shall in fact estimate the difference $(u - \Pi_K u)$ with respect to more general norms and seminorms.

A typical – and crucial – result in this direction is that, for a finite element (K, P_K, Σ_K) *which can be imbedded in an affine family* and whose P_K-interpolation operator leaves invariant the polynomials of degree $\leqslant k$ (equivalently, the inclusions $P_k(K) \subset P_K$ hold), there exists a constant C independent of K such that

$$\forall v \in H^{k+1}(K), \quad |v - \Pi_K v|_{m,K} \leqslant C \frac{h_K^{k+1}}{\rho_K^m} |v|_{k+1,K}, \quad 0 \leqslant m \leqslant k+1,$$

where

h_K = diameter of K,
ρ_K = supremum of the diameters of the spheres inscribed in K.

Such a result is proved (in a more general form) in Theorem 3.1.5.

One key idea in the process of getting such estimates is to go from any finite element K of an affine family to the reference finite element of the family and then back to the finite element K.

Another key is to use *a basic result about Sobolev spaces, due to J. Deny and J.L. Lions, which pervades the mathematical analysis of the finite element method*: Over the quotient space $H^{k+1}(\Omega)/P_k(\Omega)$, the semi-norm $|\cdot|_{k+1,\Omega}$ is a norm equivalent to the quotient norm. This result is proved in Theorem 3.1.1, for the more general *Sobolev spaces* $W^{m,p}(\Omega)$.

In practice, one often considers a *regular family* of finite elements, in the sense that the diameters h_K approach zero, and that there exists a constant σ independent of K such that $h_K \leqslant \sigma\rho_K$. For such a regular family, the previous interpolation error estimate becomes (Theorem 3.1.6)

$$|v - \Pi_K v|_{m,K} = O(h_K^{k+1-m}), \quad 0 \leqslant m \leqslant k+1.$$

Using Céa's lemma, we obtain in Section 3.2 the error estimates (Theorem 3.2.2).

$$\|u - u_h\|_{1,\Omega} \leqslant C \|u - \Pi_h u\|_{1,\Omega} = O(h^k), \quad \text{with } h = \max_{K \in \mathcal{T}_h} h_K,$$

under basically the same assumptions as before about the family of finite elements which make up the finite element spaces. It is worth mentioning here that, although the above error estimate is not the best, it is generally possible to show that the order of convergence *is* the best possible: In other words, it would not be improved by replacing $\Pi_h u$ by $\theta_h u$.

Nevertheless the range of applicability of the above results is limited inasmuch as the X_h-interpolant of the solution u is defined only if some smoothness is assumed on the solution u, and likewise, the above error estimates are obtained provided the solution is sufficiently smooth ($u \in H^{k+1}(\Omega)$). Fortunately, we show in Theorem 3.2.3 that, with the minimal assumptions that the solution u is in the space $H^1(\Omega)$ and that the spaces P_K contain the space $P_1(K)$, $K \in \mathcal{T}_h$, convergence still holds, i.e., one has $\lim_{h\to 0} \|u - u_h\|_{1,\Omega} = 0$.

Next, using a method due to J.P. Aubin and J.A. Nitsche (cf. the Aubin–Nitsche lemma; Theorem 3.2.4), we show that there is in most cases an improvement in the error estimate in the norm $|\cdot|_{0,\Omega}$ in the sense that (Theorem 3.2.5)

$$|u - u_h|_{0,\Omega} = \mathrm{O}(h^{k+1}).$$

Section 3.2 ends up with the so-called *inverse inequalities* (Theorem 3.2.6).

Finally, in Section 3.3, we follow the penetrating *method of weighted norms of J.A. Nitsche*, who has recently shown that, if $u \in W^{k+1,\infty}(\Omega)$,

$$|u - u_h|_{0,\infty,\Omega} = \begin{cases} \mathrm{O}(h^{2-\epsilon}) & \text{for any } \epsilon > 0 \text{ if } k = 1, \\ \mathrm{O}(h^{k+1}) & \text{if } k \geq 2, \end{cases}$$

$$\|u - u_h\|_{1,\infty,\Omega} = \begin{cases} \mathrm{O}(h^{1-\epsilon}) & \text{for any } \epsilon > 0 \text{ if } k = 1, \\ \mathrm{O}(h^{k}) & \text{if } k \geq 2, \end{cases}$$

where $|\cdot|_{0,\infty,\Omega}$ and $\|\cdot\|_{1,\infty,\Omega}$ stand for the norms of the spaces $L^\infty(\Omega)$ and $W^{1,\infty}(\Omega)$, respectively. Restricting ourselves for brevity to the case $k = 1$, the corresponding error estimates are obtained in Theorem 3.3.7.

It is worth pointing out that *all the error estimates found in Section* 3.2 *and* 3.3 *are optimal* in the sense that, with the same regularity assumptions on the function u, one gets the same asymptotic estimates (or "almost" the same for the norms $|\cdot|_{0,\infty,\Omega}$ and $\|\cdot\|_{1,\infty,\Omega}$ when $k = 1$) when the discrete solution $u_h \in V_h$ is replaced by the X_h-interpolant $\Pi_h u \in V_h$.

3.1. Interpolation theory in Sobolev spaces

The Sobolev spaces $W^{m,p}(\Omega)$. The quotient space $W^{k+1,p}(\Omega)/P_k(\Omega)$

We shall consider the *Sobolev space* $W^{m,p}(\Omega)$ which, for any integer $m \geq 0$, and any number p satisfying $1 \leq p \leq \infty$, consists of those functions $v \in L^p(\Omega)$ for which all partial derivatives $\partial^\alpha v$ (in the distribution sense) with $|\alpha| \leq m$ belong to the space $L^p(\Omega)$. *Equipped with the norm*

$$\begin{cases} \|v\|_{m,p,\Omega} = \left(\sum_{|\alpha| \leq m} \int_\Omega |\partial^\alpha v|^p \, dx \right)^{1/p}, & \text{if } 1 \leq p < \infty, \\ \|v\|_{m,\infty,\Omega} = \max_{|\alpha| \leq m} \{ \operatorname{ess.sup}_{x \in \Omega} |\partial^\alpha v(x)| \} & \text{if } p = \infty, \end{cases} \tag{3.1.1}$$

the space $W^{m,p}(\Omega)$ is a Banach space. We shall also use the semi-norms

$$\begin{cases} |v|_{m,p,\Omega} = \left(\sum_{|\alpha| = m} \int_\Omega |\partial^\alpha v|^p \, dx \right)^{1/p}, & \text{if } 1 \leq p < \infty, \\ |v|_{m,\infty,\Omega} = \max_{|\alpha| = m} \left\{ \operatorname{ess.sup}_{x \in \Omega} |\partial^\alpha v(x)| \right\} & \text{if } p = \infty. \end{cases} \tag{3.1.2}$$

The *Sobolev space* $W_0^{m,p}(\Omega)$ is the closure of the space $\mathcal{D}(\Omega)$ in the space $W^{m,p}(\Omega)$.

Given a subset A of $\mathbf{R}^n$ and given a function $v \in \mathscr{C}^m(A)$, the notation $\|v\|_{m,\infty,A}$ and $|v|_{m,\infty,A}$ will also denote the norm $\max_{|\alpha| \leq m} \sup_{x \in A} |\partial^\alpha v(x)|$ and the semi-norm $\max_{|\alpha| = m} \sup_{x \in A} |\partial^\alpha v(x)|$, respectively. Notice that

$$W^{m,2}(\Omega) = H^m(\Omega), \quad W_0^{m,2}(\Omega) = H_0^m(\Omega),$$

$$\|\cdot\|_{m,2,\Omega} = \|\cdot\|_{m,\Omega}, \quad |\cdot|_{m,2,\Omega} = |\cdot|_{m,\Omega}.$$

As usual, *the open sets Ω that will be considered in this section will be assumed to have a Lipschitz-continuous boundary.* In addition they will be assumed to be connected when needed (this assumption is used in the proof of Theorem 3.1.1).

In view of future needs, we shall record here some basic properties of the Sobolev spaces that will be often used. In what follows, the notation $X \hookrightarrow Y$ indicates that the normed linear space X is contained in the normed linear space Y with a continuous injection, and the notation $x \overset{c}{\subset} Y$ indicates in addition the compactness of the injection. Finally, for any integer $m \geq 0$ and any number $\alpha \in]0, 1]$, $\mathscr{C}^{m,\alpha}(\bar{\Omega})$ denotes the space of all functions in $\mathscr{C}^m(\bar{\Omega})$ whose m-th derivatives satisfy a Hölder's condition with exponent α. Equipped with the norm

$$\|v\|_{\mathscr{C}^{m,\alpha}(\bar{\Omega})} = \|v\|_{m,\infty,\bar{\Omega}} + \max_{|\beta| = m} \sup_{\substack{x,y \in \bar{\Omega} \\ x \neq y}} \frac{|\partial^\beta v(x) - \partial^\beta v(y)|}{\|x - y\|^\alpha},$$

where $\|\cdot\|$ denotes the Euclidean norm in $\mathbf{R}^n$, the space $\mathscr{C}^{m,\alpha}(\bar{\Omega})$ is a Banach space.

By the *Sobolev's imbedding theorems*, the following inclusions hold, for all integers $m \geq 0$ and all $1 \leq p \leq \infty$,

$$\left.\begin{array}{lll} W^{m,p}(\Omega) \hookrightarrow L^{p^*}(\Omega) & \text{with } \dfrac{1}{p^*} = \dfrac{1}{p} - \dfrac{m}{n}, & \text{if } m < \dfrac{n}{p}, \\ W^{m,p}(\Omega) \hookrightarrow L^{q}(\Omega) & \text{for all } q \in [1, \infty[, & \text{if } m = \dfrac{n}{p}, \\ W^{m,p}(\Omega) \hookrightarrow \mathscr{C}^{0,m-(n/p)}(\bar{\Omega}), & & \text{if } \dfrac{n}{p} < m < \dfrac{n}{p} + 1, \\ W^{m,p}(\Omega) \hookrightarrow \mathscr{C}^{0,\alpha}(\bar{\Omega}) & \text{for all } 0 < \alpha < 1, & \text{if } m = \dfrac{n}{p} + 1, \\ W^{m,p}(\Omega) \hookrightarrow \mathscr{C}^{0,1}(\bar{\Omega}), & & \text{if } \dfrac{n}{p} + 1 < m. \end{array}\right\} \tag{3.1.3}$$

By the *Kondrasov theorems*, the compact injections

$$\left.\begin{array}{lll} W^{m,p}(\Omega) \overset{c}{\subset} L^{q}(\Omega) & \text{for all } 1 \leq q < p^*, & \\ & \text{with } \dfrac{1}{p^*} = \dfrac{1}{p} - \dfrac{m}{n}, & \text{if } m < \dfrac{n}{p}, \\ W^{m,p}(\Omega) \overset{c}{\subset} L^{q}(\Omega) & \text{for all } q \in [1, \infty[, & \text{if } m = \dfrac{n}{p}, \\ W^{m,p}(\Omega) \overset{c}{\subset} \mathscr{C}^{0}(\bar{\Omega}), & & \text{if } \dfrac{n}{p} < m, \end{array}\right\} \tag{3.1.4}$$

hold for all $1 \leq p \leq \infty$. The compact injection

$$H^1(\Omega) \overset{c}{\subset} L^2(\Omega)$$

(i.e., the special case $p = q = 2$) is known as the *Rellich theorem*.

Of course, analogous inclusion can be derived by "translating" the orders of derivations. Thus for instance, one has $W^{m+r,p}(\Omega) \hookrightarrow W^{r,p^*}(\Omega)$ if $m < \frac{n}{p}$, etc. . .

We also note that, for $1 \leq p < \infty$, one has

$$(\mathscr{C}^{\infty}(\bar{\Omega}))^{-} = W^{m,p}(\Omega).$$

Remark 3.1.1. The assumption that the boundary is Lipschitz-continuous is not always necessary for proving the above properties. For example, one can derive the compact inclusion $W^{1,p}(\Omega) \overset{c}{\subset} L^{q}(\Omega)$ for all $1 \leq q \leq p$, or the above density property, as long as the boundary of Ω is continuous and the set Ω is bounded, etc. . . □

Since an open set Ω with a Lipschitz-continuous boundary is bounded, it makes sense to consider the *quotient space* $W^{k+1,p}(\Omega)/P_k(\Omega)$. This space is a Banach space when it is equipped with the *quotient norm*

$$\dot{v} \in W^{k+1,p}(\Omega)/P_k(\Omega) \to \|\dot{v}\|_{k+1,p,\Omega} = \inf_{p \in P_k(\Omega)} \|v + p\|_{k+1,p,\Omega}, \qquad (3.1.5)$$

where

$$\dot{v} = \{w \in W^{k+1,p}(\Omega); \quad (w - v) \in P_k(\Omega)\} \qquad (3.1.6)$$

denotes the equivalence class of the element $v \in W^{k+1,p}(\Omega)$.

Then the mapping

$$\dot{v} \in W^{k+1,p}(\Omega)/P_k(\Omega) \to |\dot{v}|_{k+1,p,\Omega} = |v|_{k+1,p,\Omega} \qquad (3.1.7)$$

is *a priori* only a semi-norm on the quotient space $W^{k+1,p}(\Omega)/P_k(\Omega)$, which satisfies the inequality

$$\forall \dot{v} \in W^{k+1,p}(\Omega)/P_k(\Omega), \quad |\dot{v}|_{k+1,p,\Omega} \leq \|\dot{v}\|_{k+1,p,\Omega} \qquad (3.1.8)$$

(to see this, observe that, for any polynomial $p \in P_k(\Omega)$,

$$\|v + p\|_{k+1,p,\Omega} = (|v|^p_{k+1,p,\Omega} + \|v + p\|^p_{k,p,\Omega})^{1/p} \geq |v|_{k+1,p,\Omega},$$

with the standard modification for $p = \infty$). It is a fundamental result that *it is in fact a norm over the quotient space, equivalent to the quotient norm* (3.1.5), as we now prove (cf. Exercise 3.1.1 for a generalization).

Theorem 3.1.1. *There exists a constant* $C(\Omega)$ *such that*

$$\forall v \in W^{k+1,p}(\Omega), \quad \inf_{p \in P_k(\Omega)} \|v + p\|_{k+1,p,\Omega} \leq C(\Omega)|v|_{k+1,p,\Omega} \qquad (3.1.9)$$

and consequently, one has

$$\forall \dot{v} \in W^{k+1,p}(\Omega)/P_k(\Omega), \quad \|\dot{v}\|_{k+1,p,\Omega} \leq C(\Omega)|\dot{v}|_{k+1,p,\Omega}. \qquad (3.1.10)$$

Proof. Let $N = \dim P_k(\Omega)$ and let f_i, $1 \leq i \leq N$, be a basis of the dual space of $P_k(\Omega)$. Using the Hahn–Banach extension theorem, there exist continuous linear forms over the space $W^{k+1,p}(\Omega)$, again denoted f_i, $1 \leq i \leq N$, such that for any $p \in P_k(\Omega)$, we have $f_i(p) = 0$, $1 \leq i \leq N$, if and only if $p = 0$. We will show that there exists a constant $C(\Omega)$ such that

$$\forall v \in W^{k+1,p}(\Omega), \ \|v\|_{k+1,p,\Omega} \leq C(\Omega)\Big(|v|_{k+1,p,\Omega} + \sum_{i=1}^{N} |f_i(v)|\Big). \qquad (3.1.11)$$

Inequality (3.1.9) will then be a consequence of inequality (3.1.11): Given any function $v \in W^{k+1,p}(\Omega)$, let $q \in P_k(\Omega)$ be such that $f_i(v+q) = 0$, $1 \leqslant i \leqslant N$. Then, by (3.1.11),

$$\inf_{p \in P_k(\Omega)} \|v + p\|_{k+1,p,\Omega} \leqslant \|v + q\|_{k+1,p,\Omega} \leqslant C(\Omega)|v|_{k+1,p,\Omega},$$

which proves (3.1.9). If inequality (3.1.11) is false, there exists a sequence $(v_l)_{l=1}^{\infty}$ of functions $v_l \in W^{k+1,p}(\Omega)$, such that

$$\forall l \geqslant 1, \quad \|v_l\|_{k+1,p,\Omega} = 1, \quad \text{and} \quad \lim_{l\to\infty}\left(|v_l|_{k+1,p,\Omega} + \sum_{i=1}^{N} |f_i(v_l)|\right) = 0. \tag{3.1.12}$$

Since the sequence (v_l) is bounded in $W^{k+1,p}(\Omega)$, there exists a subsequence, again denoted (v_l), and a function $v \in W^{k,p}(\Omega)$, such that

$$\lim_{l\to\infty} \|v_l - v\|_{k,p,\Omega} = 0 \tag{3.1.13}$$

(this follows from the Kondrasov or Rellich theorems for $1 \leqslant p < \infty$ and from Ascoli's Theorem for $p = \infty$). Since, by (3.1.12),

$$\lim_{l\to\infty} |v_l|_{k+1,p,\Omega} = 0, \tag{3.1.14}$$

and since the space $W^{k+1,p}(\Omega)$ is complete, we conclude from (3.1.13) and (3.1.14) that the sequence (v_l) converges in the space $W^{k+1,p}(\Omega)$. The limit v of this sequence is such that

$$\forall \alpha \quad \text{with} \quad |\alpha| = k+1, \quad |\partial^\alpha v|_{0,p,\Omega} = \lim_{l\to\infty} |\partial^\alpha v_l|_{0,p,\Omega} = 0,$$

and thus $\partial^\alpha v = 0$ for all multi-index α with $|\alpha| = k+1$. With the connectedness of Ω, it follows from distribution theory that the function v is a polynomial of degree $\leqslant k$. Using (3.1.12), we have

$$f_i(v) = \lim_{l\to\infty} f_i(v_l) = 0,$$

so that we conclude that $v = 0$, from the properties of the linear forms f_i. But this contradicts the equality $\|v_l\|_{k+1,p,\Omega} = 1$ for all l. □

Error estimates for polynomial preserving operators

Our main objective in this section is to estimate the *interpolation errors* $|v - \Pi_K v|_{m,q,k}$ and $\|v - \Pi_K v\|_{m,q,k}$, where Π_K is the P_K-interpolation operator associated with some finite element. At other places, however, we shall need similar estimates, but for more general polynomial preserving operators, i.e., not necessarily of interpolation type. This is why we shall develop an approximation theory valid also for such general operators.

To begin with, we need a definition: We shall say that two open subsets Ω and $\hat{\Omega}$ of $\mathbf{R}^n$ are *affine-equivalent* if there exists an invertible affine mapping

$$F: \hat{x} \in \mathbf{R}^n \to F(\hat{x}) = B\hat{x} + b \in \mathbf{R}^n \tag{3.1.15}$$

such that

$$\Omega = F(\hat{\Omega}). \tag{3.1.16}$$

As in the case of affine-equivalent finite elements (compare with (2.3.16)–(2.3.17)), we shall use the correspondences

$$\hat{x} \in \hat{\Omega} \to x = F(\hat{x}) \in \Omega, \tag{3.1.17}$$

$$(\hat{v}: \hat{\Omega} \to \mathbf{R}) \to (v = \hat{v} \cdot F^{-1}: \Omega \to \mathbf{R}), \tag{3.1.18}$$

between the points $\hat{x} \in \hat{\Omega}$ and $x \in \Omega$, and between functions defined over the set $\hat{\Omega}$ and the set Ω. Notice that we have

$$\hat{v}(\hat{x}) = v(x) \tag{3.1.19}$$

for all points $\hat{x}$, x in the correspondence (3.1.17) and all functions $\hat{v}$, v in the correspondence (3.1.18).

Remark 3.1.2. In case the functions v and $\hat{v}$ are defined only almost everywhere (as in the next theorem for instance), it is understood that relation (3.1.19) is to hold for almost all points $\hat{x} \in \hat{\Omega}$, and thus for almost all points $x \in \Omega$. □

We need to know how the Sobolev semi-norms defined in (3.1.2) behave from an open set to an affine-equivalent one. This is the object of the next theorem.

Here and subsequently, $\|\cdot\|$ stands for both the Euclidean norm in $\mathbf{R}^n$ and the associated matrix norm.

Theorem 3.1.2. *Let Ω and $\hat{\Omega}$ be two affine-equivalent open subsets of $\mathbf{R}^n$. If a function v belongs to the space $W^{m,p}(\Omega)$ for some integer $m \geq 0$ and some number $p \in [1, \infty]$, the function $\hat{v} = v \cdot F$ belongs to the space*

$W^{m,p}(\hat{\Omega})$, and in addition, there exists a constant $C = C(m, n)$ such that

$$\forall v \in W^{m,p}(\Omega), \quad |\hat{v}|_{m,p,\hat{\Omega}} \leqslant C\|B\|^m |\det(B)|^{-1/p} |v|_{m,p,\Omega}, \tag{3.1.20}$$

where B is the matrix occurring in the mapping F of (3.1.15).

Analogously, one has

$$\forall \hat{v} \in W^{m,p}(\hat{\Omega}), \quad |v|_{m,p,\Omega} \leqslant C\|B^{-1}\|^m |\det(B)|^{1/p} |\hat{v}|_{m,p,\hat{\Omega}}. \tag{3.1.21}$$

Proof. (i) Let us first assume that the function v belongs to the space $\mathscr{C}^m(\bar{\Omega})$, so that the function $\hat{v}$ belongs to the space $\mathscr{C}^m(\bar{\hat{\Omega}})$.

Since, for any multi-index α with $|\alpha| = m$, one has

$$\partial^\alpha \hat{v}(\hat{x}) = D^m \hat{v}(\hat{x})(e_{\alpha_1}, e_{\alpha_2}, \dots, e_{\alpha_m}),$$

where the vectors e_{α_i}, $1 \leqslant i \leqslant m$, are some of the basis vectors of $\mathbf{R}^n$, we deduce that

$$|\partial^\alpha \hat{v}(\hat{x})| \leqslant \|D^m \hat{v}(\hat{x})\| = \sup_{\substack{\|\xi_i\| \leqslant 1 \\ 1 \leqslant i \leqslant m}} |D^m \hat{v}(\hat{x})(\xi_1, \xi_2, \dots, \xi_m)|.$$

Consequently we obtain

$$\begin{aligned} |\hat{v}|_{m,p,\hat{\Omega}} &= \left(\int_{\hat{\Omega}} \sum_{|\alpha|=m} |\partial^\alpha \hat{v}(\hat{x})|^p \, \mathrm{d}\hat{x} \right)^{1/p} \\ &\leqslant C_1(m, n) \left(\int_{\hat{\Omega}} \|D^m \hat{v}(\hat{x})\|^p \, \mathrm{d}\hat{x} \right)^{1/p}, \end{aligned} \tag{3.1.22}$$

where the constant $C_1(m, n)$ may be chosen as

$$C_1(m, n) = \sup_{1 \leqslant p} (\mathrm{card}\{\alpha \in N^m; \quad |\alpha| = m\})^{1/p}.$$

Using the differentiation rule for composition of functions, we note that for any vectors $\xi_i \in \mathbf{R}^n$, $1 \leqslant i \leqslant m$,

$$D^m \hat{v}(\hat{x})(\xi_1, \xi_2, \dots, \xi_m) = D^m v(x)(B\xi_1, B\xi_2, \dots, B\xi_m),$$

so that

$$\|D^m \hat{v}(\hat{x})\| \leqslant \|D^m v(x)\| \, \|B\|^m,$$

and therefore,

$$\int_{\hat{\Omega}} \|D^m \hat{v}(\hat{x})\|^p \, \mathrm{d}\hat{x} \leqslant \|B\|^{mp} \int_{\hat{\Omega}} \|D^m v(F(\hat{x}))\|^p \, \mathrm{d}\hat{x}. \tag{3.1.23}$$

Using the formula of change of variables in multiple integrals, we get

$$\int_{\hat{\Omega}} \|D^m v(F(\hat{x}))\|^p \, d\hat{x} = |\det(B^{-1})| \int_{\Omega} \|D^m v(x)\|^p \, dx. \tag{3.1.24}$$

Since there exists a constant $C_2(m, n)$ such that

$$\|D^m v(x)\| \leq C_2(m, n) \max_{|\alpha|=m} |\partial^\alpha v(x)|,$$

we obtain

$$\left(\int_{\Omega} \|D^m v(x)\|^p \, dx\right)^{1/p} \leq C_2(m, n)|v|_{m,p,\Omega}. \tag{3.1.25}$$

Inequality (3.1.20) is then a consequence of inequalities (3.1.22), (3.1.23), (3.1.24) and (3.1.25).

(ii) To complete the proof when $p \neq \infty$, it remains to use the continuity of the linear operator ι: $v \in C^m(\bar{\Omega}) \to \hat{v} \in W^{m,p}(\hat{\Omega})$ with respect to the norms $\|\cdot\|_{m,p,\Omega}$ and $\|\cdot\|_{m,p,\hat{\Omega}}$, the denseness of the space $\mathscr{C}^m(\bar{\Omega})$ in the space $W^{m,p}(\Omega)$, and the definition of the (unique) extension of the mapping ι to the space $W^{m,p}(\Omega)$.

(iii) Let us finally consider the case $p = \infty$. A function $v \in W^{m,\infty}(\Omega)$ belongs to the spaces $W^{m,p}(\Omega)$ for all $p < \infty$ (recall that the assumption of Lipschitz-continuity of the boundary implies the boundedness of the set Ω). Therefore, by part (ii), the function $\hat{v}$ belongs to the spaces $W^{m,p}(\hat{\Omega})$ for all $p < \infty$, and there exists a constant $C(m, n)$ such that

$$\forall p \geq 1, \quad \forall \alpha \in \mathbb{N}^m, \quad |\alpha| \leq m,$$
$$|\partial^\alpha \hat{v}|_{0,p,\hat{\Omega}} \leq |\hat{v}|_{|\alpha|,p,\hat{\Omega}} \leq C(m, n)\|B\|^{|\alpha|} \sup_{1 \leq p} |\det(B)|^{-1/p} \|v\|_{m,p,\Omega}.$$

Since the upper bound on the semi-norm $|\partial^\alpha v|_{0,p,\hat{\Omega}}$ is independent of the number p, this shows that, for each $|\alpha| \leq m$, the function $\partial^\alpha \hat{v}$ is in the space $L^\infty(\Omega)$ for each $|\alpha| \leq m$. Consequently, the function $\hat{v}$ belongs to the space $W^{m,\infty}(\hat{\Omega})$. To conclude, it suffices to use inequality (3.1.21) for all $p \geq 1$ in conjunction with the property that for any function $w \in L^\infty(\Omega)$, Ω bounded, one has

$$|w|_{0,\infty,\Omega} = \lim_{p \to \infty} |w|_{0,p,\Omega}.$$

Inequality (3.1.21) is proved in a similar fashion. □

To apply Theorem 3.1.2, it is desirable to evaluate the norms $\|B\|$ and

$\|B^{-1}\|$ in terms of simple geometric quantities. This is the object of the next theorem, where we use the following notations:

$$h = \operatorname{diam}(\Omega),\ \hat{h} = \operatorname{diam}(\hat{\Omega}), \tag{3.1.26}$$

$$\left.\begin{aligned} \rho &= \sup\{\operatorname{diam}(S);\ S \text{ is a ball contained in } \Omega\},\\ \hat{\rho} &= \sup\{\operatorname{diam}(\hat{S});\ \hat{S} \text{ is a ball contained in } \hat{\Omega}\}.\end{aligned}\right\} \tag{3.1.27}$$

Theorem 3.1.3. *Let $\hat{\Omega}$ and $\Omega = F(\hat{\Omega})$ be two affine-equivalent open subsets of $\mathbf{R}^n$, where $F: \hat{x} \in \mathbf{R}^n \to (B\hat{x} + b) \in \mathbf{R}^n$ is an invertible affine mapping. Then the upper bounds*

$$\|B\| \leqslant \frac{h}{\hat{\rho}} \quad \textit{and} \quad \|B^{-1}\| \leqslant \frac{\hat{h}}{\rho} \tag{3.1.28}$$

hold.

Proof. We may write

$$\|B\| = \frac{1}{\hat{\rho}} \sup_{\|\xi\| = \hat{\rho}} \|B\xi\|.$$

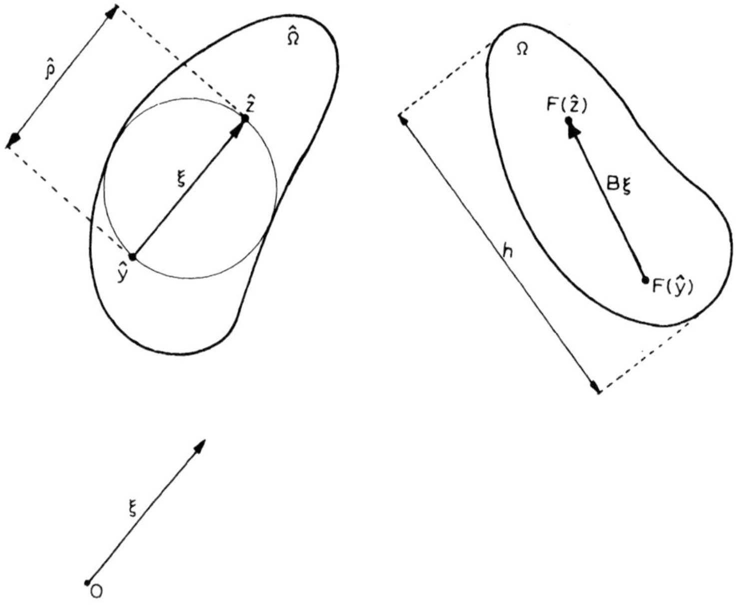

Fig. 3.1.1

Given a vector ξ satisfying $\|\xi\| = \hat{\rho}$, there exist two points $\hat{y}, \hat{z} \in \bar{\hat{\Omega}}$ such that $\hat{y} - \hat{z} = \xi$, by definition of $\hat{\rho}$ (Fig. 3.1.1). Since $B\xi = F(\hat{y}) - F(\hat{z})$ with $F(\hat{y}) \in \bar{\Omega}$, $F(\hat{z}) \in \bar{\Omega}$, we deduce that $\|B\xi\| \leqslant h$, and thus the first inequality (3.1.28) is proved. The other inequality is proved in a similar fashion. □

We are now in a position to prove an important property of *polynomial preserving operators*, i.e., which satisfy a relation of the form (3.1.30) below for some integer $k \geqslant 0$.

Theorem 3.1.4. *For some integers $k \geqslant 0$ and $m \geqslant 0$ and some numbers $p, q \in [1, \infty]$, let $W^{k+1,p}(\hat{\Omega})$ and $W^{m,q}(\hat{\Omega})$ be Sobolev spaces satisfying the inclusion*

$$W^{k+1,p}(\hat{\Omega}) \hookrightarrow W^{m,q}(\hat{\Omega}), \tag{3.1.29}$$

and let $\hat{\Pi} \in \mathscr{L}(W^{k+1,p}(\hat{\Omega});\; W^{m,q}(\hat{\Omega}))$ be a mapping such that

$$\forall \hat{p} \in P_k(\hat{\Omega}), \quad \hat{\Pi}\hat{p} = \hat{p}. \tag{3.1.30}$$

For any open set Ω which is affine-equivalent to the set $\hat{\Omega}$, let the mapping Π_Ω be defined by

$$(\Pi_\Omega v)\hat{} = \hat{\Pi}\hat{v}, \tag{3.1.31}$$

for all functions $\hat{v} \in W^{k+1,p}(\hat{\Omega})$ and $v \in W^{k+1,p}(\Omega)$ in the correspondence (3.1.18). *Then there exists a constant $C(\hat{\Pi}, \hat{\Omega})$ such that, for all affine-equivalent sets Ω,*

$$\forall v \in W^{k+1,p}(\Omega), \quad |v - \Pi_\Omega v|_{m,q,\Omega} \leqslant$$
$$\leqslant C(\hat{\Pi}, \hat{\Omega})(\operatorname{meas}(\Omega))^{(1/q)-(1/p)} \frac{h^{k+1}}{\rho^m} |v|_{k+1,p,\Omega}, \tag{3.1.32}$$

with h and ρ defined as in (3.1.26) *and* (3.1.27) *respectively.*

Proof. Using the polynomial invariance (3.1.30), we obtain the identity

$$\forall \hat{v} \in W^{k+1,p}(\hat{\Omega}), \quad \forall \hat{p} \in P_k(\hat{\Omega}), \quad \hat{v} - \hat{\Pi}\hat{v} = (I - \hat{\Pi})(\hat{v} + \hat{p}),$$

where I, the identity mapping from $W^{k+1,p}(\hat{\Omega})$ into $W^{m,q}(\hat{\Omega})$, is continuous by (3.1.29). From this identity we deduce that

$$|\hat{v} - \hat{\Pi}\hat{v}|_{m,q,\hat{\Omega}} \leq \|I - \hat{\Pi}\|_{\mathscr{L}(W^{k+1,p}(\hat{\Omega});\, W^{m,q}(\hat{\Omega}))} \inf_{\hat{p} \in P_k(\hat{\Omega})} \|\hat{v} + \hat{p}\|_{k+1,p,\hat{\Omega}}$$
$$\leq C(\hat{\Pi}, \hat{\Omega})|\hat{v}|_{k+1,p,\hat{\Omega}}, \tag{3.1.33}$$

by Theorem 3.1.1.

It follows from relation (3.1.31) that

$$\hat{v} - \hat{\Pi}\hat{v} = (v - \Pi_\Omega v)\hat{}\,,$$

and therefore an application of Theorem 3.1.2 yields

$$|v - \Pi_\Omega v|_{m,q,\Omega} \leq C\|B^{-1}\|^m |\det(B)|^{1/q} |\hat{v} - \hat{\Pi}\hat{v}|_{m,q,\hat{\Omega}}. \tag{3.1.34}$$

By the same theorem,

$$|\hat{v}|_{k+1,p,\hat{\Omega}} \leq C\|B\|^{k+1} |\det(B)|^{-1/p} |v|_{k+1,p,\Omega}, \tag{3.1.35}$$

and thus, to obtain inequality (3.1.32), it suffices to combine inequalities (3.1.33), (3.1.34) and (3.1.35), the upper bounds $\|B\| \leq h/\hat{\rho}$ and $\|B^{-1}\| \leq \hat{h}/\rho$ (Theorem 3.1.3), and, finally, to observe that

$$|\det(B)| = \frac{\text{meas}(\Omega)}{\text{meas}(\hat{\Omega})}. \qquad \square$$

Estimates of the interpolation errors $|v - \Pi_K v|_{m,q,K}$ *for affine families of finite elements*

By specializing the above result to finite elements, we obtain *estimates of the interpolation errors* $|v - \Pi_K v|_{m,q,K}$ (for another approach, see Exercise 3.1.2; for a refined analysis of the dependence upon the geometry, see Exercise 3.1.4).

Theorem 3.1.5. *Let* $(\hat{K}, \hat{P}, \hat{\Sigma})$ *be a finite element, for which s denotes the greatest order of partial derivatives occurring in the definition of* $\hat{\Sigma}$. *If the following inclusions hold, for some integers* $m \geq 0$ *and* $k \geq 0$ *and for some numbers* $p, q \in [1, \infty]$,

$$W^{k+1,p}(\hat{K}) \hookrightarrow \mathscr{C}^s(\hat{K}), \tag{3.1.36}$$

$$W^{k+1,p}(\hat{K}) \hookrightarrow W^{m,q}(\hat{K}), \tag{3.1.37}$$

$$P_k(\hat{K}) \subset \hat{P} \subset W^{m,q}(\hat{K}), \tag{3.1.38}$$

there exists a constant $C(\hat{K}, \hat{P}, \hat{\Sigma})$ *such that, for all affine-equivalent finite elements* (K, P, Σ), *and all functions* $v \in W^{k+1,p}(K)$,

$$|v - \Pi_K v|_{m,q,K} \leqslant C(\hat{K}, \hat{P}, \hat{\Sigma})(\text{meas}(K))^{(1/q)-(1/p)} \frac{h_K^{k+1}}{\rho_K^m} |v|_{k+1,p,K}, \tag{3.1.39}$$

where $\Pi_K v$ *denotes the* P_K*-interpolant of the function* v, *and*

$$\left.\begin{aligned} &\text{meas}(K) = \mathrm{d}x - \textit{measure of } K, \\ &h_K = \textit{diam}(K), \\ &\rho_K = \textit{sup}\ \{\textit{diam}(S);\ S \textit{ is a ball contained in } K\}. \end{aligned}\right\} \tag{3.1.40}$$

Proof. The inclusion $P_k(\hat{K}) \subset \hat{P}$ in conjunction with the fact that the $\hat{P}$-interpolation operator $\hat{\Pi}$ reduces to the identity over the space $\hat{P}$ (cf. (2.3.8)) implies that

$$\forall \hat{p} \in P_k(\hat{K}), \quad \hat{\Pi}\hat{p} = \hat{p}. \tag{3.1.41}$$

Let then $\hat{v}$ be a function in the space $W^{k+1,p}(\hat{K})$, so that it belongs to the space dom $\hat{\Pi} = \mathscr{C}^s(\hat{K})$ (cf. (2.3.7)) since the inclusion $W^{k+1,p}(\hat{K}) \subset \mathscr{C}^s(\hat{K})$ holds. For definiteness, let us assume that $s = 2$ (recall that in practice, $s = 0$, 1 or 2) so that the $\hat{P}$-interpolant of the function $\hat{v}$ takes the form

$$\hat{\Pi}\hat{v} = \sum_i \hat{v}(\hat{a}_i^0)\hat{p}_i^0 + \sum_{i,k} \{D\hat{v}(\hat{a}_i^1)\hat{\xi}_{i,k}^1\}\hat{p}_{ik}^1 + \sum_{i,k,l} \{D^2\hat{v}(\hat{a}_i^2)(\hat{\xi}_{ik}^2, \hat{\xi}_{il}^2)\}\hat{p}_{ikl}^2. \tag{3.1.42}$$

We proceed to show that the linear mapping $\hat{\Pi}$: $W^{k+1,p}(\hat{K}) \to W^{m,q}(\hat{K})$ (by (3.1.38), the space $\hat{P}$ is contained in the space $W^{m,q}(\hat{K})$) is continuous: From (3.1.42), we deduce that

$$\begin{aligned} \|\hat{\Pi}\hat{v}\|_{m,q,\hat{K}} \leqslant{}& \sum_i |\hat{v}(\hat{a}_i^0)|\, \|\hat{p}_i^0\|_{m,q,\hat{K}} + \sum_{i,k} |\{D\hat{v}(\hat{a}_i^1)\hat{\xi}_{ik}^1\}|\, \|\hat{p}_{ik}^1\|_{m,q,\hat{K}} \\ &+ \sum_{i,k,l} |\{D^l\hat{v}(\hat{a}_i^2)(\hat{\xi}_{ik}^2, \hat{\xi}_{il}^2)\}|\, \|\hat{p}_{ikl}^2\|_{m,q,\hat{K}} \\ \leqslant{}& C(\|\hat{p}_i^0\|_{m,q,\hat{K}}, \|\hat{\xi}_{ik}^1\|\, \|\hat{p}_{ik}^1\|_{m,q,\hat{K}}, \|\hat{\xi}_{ik}^2\|\, \|\hat{\xi}_{il}^2\|\, \|\hat{p}_{ikl}^2\|_{m,q,\hat{K}})\|\hat{v}\|_{2,\infty,\hat{K}} \\ \leqslant{}& C(\hat{K}, \hat{P}, \hat{\Sigma})\|\hat{v}\|_{k+1,p,\hat{K}}, \end{aligned}$$

where in the last inequality, we have made use of the inclusion (3.1.36).

Since the P_K- and $\hat{P}$-interpolation operators are related through the correspondence

$$(\Pi_K v)\hat{} = \hat{\Pi}\hat{v} \quad \text{for all} \quad v \in \operatorname{dom} \Pi_K$$

(cf. (2.3.19)), we may apply Theorem 3.1.4, and inequality (3.1.39) is just a re-statement of inequality (3.1.32) in the present case. □

Remark 3.1.3. If necessary, the factor $(\operatorname{meas}(K))^{(1/q)-(1/p)}$ may also be expressed in terms of the parameters h_K and ρ_K by means of the inequalities

$$\sigma_n \rho_K^n \leq \operatorname{meas}(K) \leq \sigma_n h_K^n,$$

where σ_n denotes the dx-measure of the unit sphere in $\mathbf{R}^n$. □

It is possible to dispose of the parameter ρ_K in the upper bound (3.1.39) provided we restrict ourselves to finite elements which do not become "flat" in the limit, as we shall show (Theorem 3.1.6). First, we need a definition, of a purely *geometrical* nature.

We shall say that a family of finite elements (K, P_K, Σ_K) is *regular* if the following two conditions are satisfied:

(i) There exists a constant σ such that

$$\forall K, \frac{h_K}{\rho_K} \leq \sigma \tag{3.1.43}$$

(see Exercise 3.1.3 for an equivalent formulation of this condition for triangles).

(ii) The diameters h_K approach zero (in order to avoid introducing new letters, K is viewed as the parameter of the family).

For such families, the interpolation error estimate of Theorem 3.1.5 can be immediately converted into simple estimates of the *norms* $\|v - \Pi_K v\|_{m,q,K}$.

Theorem 3.1.6. *Let there be given a regular affine family of finite elements* (K, P_K, Σ_K) *whose reference finite element* $(\hat{K}, \hat{P}, \hat{\Sigma})$ *satisfies assumptions* (3.1.36), (3.1.37) *and* (3.1.38). *Then there exists a constant* $C(\hat{K}, \hat{P}, \hat{\Sigma})$ *such that, for all finite elements* K *in the family, and all functions* $v \in W^{k+1,p}(K)$,

$$\|v - \Pi_K v\|_{m,q,K} \leq C(\hat{K}, \hat{P}, \hat{\Sigma})(\operatorname{meas}(K))^{(1/q)-(1/p)} h_K^{k+1-m} |v|_{k+1,p,K}. \tag{3.1.44}$$

□

Remark 3.1.4. Only the boundedness of the diameters h_K (implied by condition (ii)) is used in the derivation of the upper bound (3.1.44). □

In order to get a more concrete understanding of such estimates, we have recorded in the next table (Fig. 3.1.2) some interpolation error estimates in the $\|\cdot\|_{m,K}$ norms ($p = q = 2$) for various finite elements which can be imbedded in affine families.

Remark 3.1.5. If the function v lacks the "optimal" regularity, inter-

$\|v - \Pi_K v\|_{m,K}$	$O(h_K^{2-m})$ $0 \leq m \leq 2$ $(k = 1)$	$O(h_K^{3-m})$ $0 \leq m \leq 3$ $(k = 2)$		$O(h_K^{4-m})$ $0 \leq m \leq 4$ $(k = 3)$		
Regularity of the function v	$H^2(K)$	$H^3(K)$		$H^4(K)$		
Upper bound on the dimension n, (to insure that $H^{k+1}(K) \hookrightarrow C^s(K)$)	$n \leq 3$ $(s = 0)$	$n \leq 5$ $(s = 0)$	$n \leq 3$ $(s = 1)$	$n \leq 7$ $(s = 0)$	$n \leq 5$ $(s = 1)$	$n \leq 3$ $(s = 2)$
Simplicial finite elements			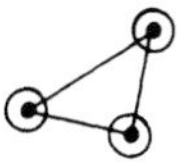		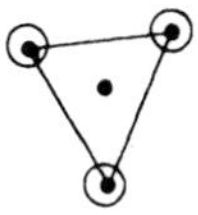	
Rectangular finite elements	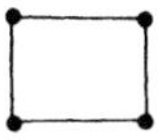			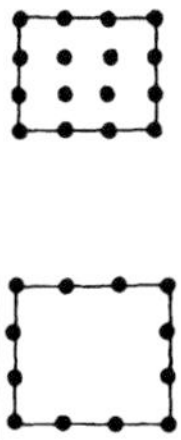		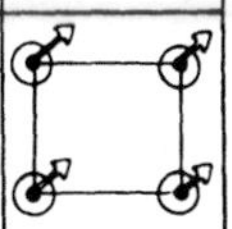

Fig. 3.1.2

polation error estimates may still hold *provided the P_K-interpolant is still defined*, with smaller values of k, however. If we are considering Hermite triangles of type (3) for instance, if the function v is "only" in the space $H^3(K)$ and if $n \leqslant 3$, one has $\|v - \Pi_K v\|_{m,K} = O(h_K^{3-m})$ for $0 \leqslant m \leqslant 3$, and so on... □

Remark 3.1.6. Further conditions may be added in the definition of a regular family made up of a *specific* finite element. For example, this will be the case for the isoparametric n-simplex of type (2) (cf. Section 4.3) and the Hsieh–Clough–Tocher triangle (cf. Section 6.1). □

Exercises

3.1.1. The following abstract generalization of both Theorems 3.1.1 and 3.1.4 is due to L. TARTAR (unpublished) and can be found in BREZZI & MARINI (1975).

Let V be a Banach space and let V_1, V_2 and W be three normed vector spaces. Let $A_i \in \mathscr{L}(V; V_i)$, $i = 1, 2$, be two given mappings, the mapping A_1 being compact. It is assumed that there exists a constant c_0 such that

$$\forall v \in V, \quad \|v\|_V \leqslant c_0(\|A_1 v\|_{V_1} + \|A_2 v\|_{V_2}).$$

Finally, let $L \in \mathscr{L}(V; W)$ be a mapping such that

$$v \in \ker A_2 \Rightarrow Lv = 0.$$

(i) Show that the space $P = \ker A_2$ is finite-dimensional.
(ii) Show that there exists a constant c_1 such that

$$\forall v \in V, \quad \inf_{p \in P} \|v - p\|_V \leqslant c_1 \|A_2 v\|_{V_2}.$$

(iii) Deduce from (ii) that there exists a constant C such that

$$\forall v \in V, \quad \|Lv\|_W \leqslant C \|A_2 v\|_{V_2}.$$

(iv) Let (for simplicity only, we restrict ourselves to $p = q = 2$, $m = k + 1$)

$$V = W = H^{k+1}(\Omega),$$
$$V_1 = H^k(\Omega), \quad A_1 = I,$$
$$V_2 = (L^2(\Omega))^{\Gamma_{k+1}}, \quad A_2\colon v \in H^{k+1}(\Omega) \to (\partial^\alpha v)_{|\alpha|=k+1},$$

where $\Gamma_{k+1} = \operatorname{card}\{\alpha \in \mathbf{N}^n;\ |\alpha| = k+1\}$ and the mapping L is given by

$$L\colon v \in H^{k+1}(\Omega) \to (v - \Pi v) \in H^{k+1}(\Omega),$$

where the mapping $\Pi \in \mathscr{L}(H^{k+1}(\Omega))$ is such that

$$\forall p \in P_k(\Omega),\quad \Pi p = p.$$

Then derive an inequality similar to that of (3.1.33) (we could as well let $V_1 = L^2(\Omega)$; see below).

(v) Let

$$\begin{aligned}
&V = W = H^{k+1}(\Omega),\\
&V_1 = L^2(\Omega),\quad A_1 = I,\\
&V_2 = (L^2(\Omega))^n,\quad A_2\colon v \in H^{k+1}(\Omega)\\
&\qquad \to \{D^{k+1}v(\cdot)(e_i^{k+1})\}_{i=1}^n \in (L^2(\Omega))^n,\\
&L\colon v \in H^{k+1}(\Omega) \to (v - \Pi v) \in H^{k+1}(\Omega),
\end{aligned}$$

where the mapping $\Pi \in \mathscr{L}(H^{k+1}(\Omega))$ is such that

$$\forall p \in Q_k(\Omega),\quad \Pi p = p.$$

Then show that there exists a constant $C(\Pi, \Omega)$ such that

$$\|v - \Pi v\|_{k+1,\Omega} \leq C(\Pi, \Omega)[v]_{k+1,\Omega},$$

where

$$[v]_{k+1,\Omega} = \left(\sum_{i=1}^n \int_\Omega |D^{k+1}v(x)(e_i^{k+1})|^2\,dx\right)^{1/2}.$$

[Hint: Use the following result due to N. Aronszajn and K.T. Smith, and proved in Smith (1961): There exists a constant C such that

$$\forall v \in W^{m,p}(\Omega),\quad \|v\|_{m,p,\Omega} \leq C(|v|_{0,p,\Omega} + [v]_{m,p,\Omega}),$$

where

$$[v]_{m,p,\Omega} = \left(\sum_{i=1}^n \int_\Omega |D^m v(x)(e_i^m)|^p\,dx\right)^{1/2}.]$$

3.1.2. Let (K, P_K, Σ_K) be a Lagrange finite element such that the inclusions

$$P_k(K) \subset P_K \subset \mathscr{C}^k(K)$$

hold for some integer k, and let there be given a function $v\colon K \to \mathbf{R}$ which

will be assumed to be sufficiently smooth for all subsequent purposes.
For any integer $m \geqslant 0$, we let

$$|v|_{m,\infty,K} = \sup_{x \in K} \|D^m v(x)\|_{\mathscr{L}_m(\mathbf{R}^n;\,\mathbf{R})}$$

(we recall that there exists a constant $C(m, n)$ such that

$$|v|_{m,\infty,K} \leqslant |v|_{m,\infty,K} \leqslant C(m, n)|v|_{m,\infty,K}).$$

If a and x are two points in the set K (assumed here to be convex), Taylor's formula of order k reads

$$v(a) = v(x) + Dv(x)(a-x) + \cdots + \frac{1}{k!} D^k v(x)(a-x)^k + \\ + \mathscr{R}_k(v; a, x),$$

where the remainder $\mathscr{R}_k(v; a, x)$ is given by

$$\mathscr{R}_k(v; a, x) = \frac{1}{k!} \int_0^1 (1-t)^k D^{k+1} v(ta + (1-t)x)(a-x)^{k+1}\, dt \\ = \frac{1}{(k+1)!} D^{k+1} v(\theta a + (1-\theta)x)(a-x)^{k+1}.$$

In fact, we shall only need the estimates

$$\mathscr{R}_k(v; a, x)| \leqslant \frac{1}{(k+1)!} |v|_{k+1,\infty,K} \|a - x\|^{k+1}$$

which follow from either expression of the remainder.

(i) Let p_i, $1 \leqslant i \leqslant N$, be the basis functions associated with the set $\Sigma_K = \{p(a_i),\ 1 \leqslant i \leqslant N\}$. Show that, for all $x \in K$ (Ciarlet & Raviart (1972a)).

$$D^m(\Pi_K v - v)(x) = \sum_{i=1}^N \mathscr{R}_k(v; a_i, x) D^m p_i(x), \quad 0 \leqslant m \leqslant k.$$

Notice that for $m = 0$, one obtains a *multi-point Taylor formula* (Ciarlet & Wagschal (1971)):

$$v(x) = \sum_{i=1}^N v(a_i) p_i(x) - \sum_{i=1}^N \mathscr{R}_k(v; a_i, x) p_i(x).$$

(ii) Let $(\hat{K}, \hat{P}, \hat{\Sigma})$ be an affine-equivalent finite element. Show that (with the usual notations)

$$|v - \Pi_K v|_{m,\infty,K} \leqslant C(\hat{K}, \hat{P}, \hat{\Sigma}) \frac{h_K^{k+1}}{\rho_K^m} |v|_{k+1,\infty,K}, \quad 0 \leqslant m \leqslant k,$$

where

$$C(\hat{K}, \hat{P}, \hat{\Sigma}) = \frac{\hat{h}^m}{(k+1)!} \sum_{i=1}^{N} |\hat{p}_i|_{m,\infty,\hat{K}}.$$

Thus, this yields an estimate of the constant which appears in (3.1.39) when $p = q = \infty$.

(iii) In the sequel, p is any number which satisfies $1 \leqslant p \leqslant \infty$ and $k+1 > (n/p)$, so that the inclusion $W^{k+1,p}(K) \hookrightarrow \mathscr{C}^0(K)$ holds. Show that

$$|\mathscr{R}_k(v; a, \cdot)|_{0,p,K} \leqslant \frac{1}{k!\left(k+1-\dfrac{n}{p}\right)} h_K^{k+1} |v|_{k+1,p,K},$$

for all $a \in K$.

(iv) Deduce from (iii) that

$$|v - \Pi_K v|_{m,p,K} \leqslant \frac{1}{k!\left(k+1-\dfrac{n}{p}\right)} \left(\sum_{i=1}^{N} |p_i|_{m,\infty,K}\right) h_K^{k+1} |v|_{k+1,p,K}.$$

(v) Assume that K is an n-simplex and that the basis functions p_i are expressed uniquely in terms of the barycentric coordinates (in this respect, cf. also Exercise 2.3.6):

$$p_i(x) = f_i(\lambda_1(x), \dots, \lambda_{n+1}(x)), \quad 1 \leqslant i \leqslant N,$$

where the (smooth) functions f_i are independent of (K, P_K, Σ_K). Then show that (Arcangéli & Gout (1976))

$$|v - \Pi_K v|_{m,p,K} \leqslant C(\hat{K}, \hat{P}, \hat{\Sigma}) \frac{h_K^{k+1}}{\rho_K^m} |v|_{k+1,p,K},$$

where

$$C(\hat{K}, \hat{P}, \hat{\Sigma}) = \frac{m!}{k!\left(k+1-\dfrac{n}{p}\right)} \times$$

$$\times \sum_{i=1}^{N} \max_{\substack{0 \leqslant \lambda_j \leqslant 1 \\ \Sigma_j \lambda_j = 1}} \sum_{|\alpha|=m} \frac{1}{\alpha_1! \dots \alpha_{n+1}!} |\partial^\alpha f_i(\lambda_1, \dots, \lambda_{n+1})|.$$

Therefore this provides another estimate of the constants which appear in the interpolation error estimate.

[Hint: As in Ciarlet & Wagschal (1971), prove and use the inequalities

$$|\lambda_i|_{1,\infty,K} \leq \frac{1}{\rho_K}, \quad 1 \leq i \leq N.]$$

(vi) Using the result of (v), show that, for triangles of type (1),

$$|v - \Pi_K v|_{m,K} \leq C_m^1 \frac{h_K^2}{\rho_K^m} |v|_{2,K} \quad \text{which} \quad C_0^1 = C_1^1 = 3,$$

and that, for triangles of type (2),

$$|v - \Pi_K v|_{m,K} \leq C_m^2 \frac{h_K^3}{\rho_K^m} |v|_{3,K} \quad \text{with} \quad C_0^2 = 2, \quad C_1^2 = 6, \quad C_2^2 = 9.$$

These estimates can be further improved. See ARCANGELI & GOUT (1976) and GOUT (1976).

3.1.3. Show that for a family of triangular finite elements, condition (3.1.43) is equivalent to *Zlámal's condition* (ZLÁMAL (1968)) that there exists a constant θ_0 such that

$$\forall K, \quad \theta_K \geq \theta_0 > 0,$$

where for each triangle K, θ_K denotes the smallest angle of K.

3.1.4. The object of this problem is to study (in the special case $m = 1$ for simplicity) the improvement of JAMET (1976b) concerning the dependence of the interpolation error estimates upon the geometry of the finite elements.

(i) Let Ω be an open subset of $\mathbf{R}^n$ with $h = \operatorname{diam} \Omega$. In addition, let $\Pi_\Omega \in \mathcal{L}(W^{k(+1),p}(\Omega); W^{1,p}(\Omega))$ be a mapping such that

(a) $\Pi_\Omega p = p$ for all $p \in P_k(\Omega)$ and

(b) there exists a non zero vector $\xi \in \mathbf{R}^n$ such that if $Dv(x)\xi = 0$ for all $x \in \Omega$ for some function $v \in W^{k(+1),p}(\Omega)$, then $D\Pi_\Omega v(x)\xi = 0$ for all $x \in \Omega$.

Show that there exists a mapping $\Phi \in \mathcal{L}(W^{k,p}(\Omega); L^p(\Omega))$ such that

$$\forall v \in W^{k+1,p}(\Omega), \; D\Pi_\Omega v(\cdot)\xi = \Phi(Dv(\cdot)\xi),$$

$$\forall p \in P_{k-1}(\Omega), \quad \Phi p = p.$$

(ii) Let $\hat{\Omega}$ be an open set which is affine-equivalent to Ω and let the mapping $\hat{\Pi}$ be defined in the usual way. Using (i), show that there exists a constant $C(\hat{\Pi}, \hat{\Omega})$ such that

$$\forall v \in W^{k+1,p}(\Omega), \quad |Dv(\cdot)\xi - D\Pi_\Omega v(\cdot)\xi|_{0,p,\Omega} \leq C(\hat{\Pi}, \hat{\Omega}) h^{k(+1)} |v|_{k+1,p,\Omega}.$$

(iii) Assume that the property of (i) is satisfied for n linearly independent vectors ξ_i, $1 \leq i \leq n$. Show that there exists a constant $C(\hat{\Pi}, \hat{\Omega})$ such that (compare with (3.1.32))

$$\forall v \in W^{k+1,p}(\Omega), \quad |v - \Pi_\Omega v|_{1,p,\Omega} \leq C(\hat{\Pi}, \hat{\Omega}) \frac{h^k}{\cos\theta} |v|_{k+1,p,\Omega},$$

where the angle θ (a function of the n vectors ξ_i, $1 \leq i \leq n$) is defined as follows: Given any vector η with $\|\eta\| = 1$, let $\theta_i(\eta) \in [0, \pi/2]$ be the angle between the directions of the vectors η and ξ_i, $1 \leq i \leq n$. Then we let

$$\theta = \max_{\eta \in \mathbf{R}^n} \{\min_{1 \leq i \leq n} \theta_i(\eta)\}.$$

(iv) Assume that $k > (n/p)$. In the case of n-simplices of type (k), show that we may choose for vectors ξ_i, $1 \leq i \leq n$, any n vectors out of the $n(n+1)/2$ vectors of the form $(a_j - a_i)$, $1 \leq i < j \leq n+1$. As a consequence show that in the case of triangles for instance, we may have $\|v - \Pi_K v\|_{1,p,\Omega} = O(h^k)$ even though Zlámal's condition is violated.

(v) Apply the previous analysis to rectangles of type (k).

3.2. Application to second-order problems over polygonal domains

Estimate of the error $\|u - u_h\|_{1,\Omega}$

Let there be given a second-order boundary value problem, posed over a space V which satisfies the usual inclusions $H_0^1(\Omega) \subset V = \bar{V} \subset H^1(\Omega)$. One basic hypothesis will be that the set $\bar{\Omega}$ is *polygonal*, essentially because such an assumption allows us to *exactly* cover the set $\bar{\Omega}$ with polygonal finite elements. Then with any such finite element, we associate a finite element space X_h. Next, we define an appropriate subspace V_h of X_h (this takes into account the boundary conditions contained in the definition of the space V) which is included in the space V, so that we are using a *conforming finite element method.* One main property that we shall assume is that *the space V_h contains the X_h-interpolant of the solution u*: See Section 2.3 where the special cases $V_h = X_h \subset V = H^1(\Omega)$ and $V_h = X_{0h} \subset V = H_0^1(\Omega)$ have been thoroughly discussed. This would also be true of a problem where $V = \{v \in H^1(\Omega);\ v = 0 \text{ on } \Gamma_0\}$ provided the subset Γ_0 of Γ can be written exactly as a union of faces of some finite elements. By contrast, this is not true in general of a nonhomogeneous Dirichlet problem. In this direction, see Exercises 3.2.1 and 3.2.2.

Throughout this section, we shall make the following assumptions, denoted (H 1), (H 2) and (H 3), whose statements will not be repeated.

(H 1) We consider a *regular family of triangulations* $\mathcal{T}_h$ in the following sense:

(i) There exists a constant σ such that

$$\forall K \in \bigcup_h \mathcal{T}_h, \quad \frac{h_K}{\rho_K} \leqslant \sigma. \tag{3.2.1}$$

(ii) The quantity

$$h = \max_{K \in \mathcal{T}_h} h_k \tag{3.2.2}$$

approaches zero.

In other words, *the family formed by the finite elements* (K, P_K, Σ_K), $K \in \bigcup_h \mathcal{T}_h$, *is a regular family of finite elements*, in the sense of Section 3.1.

Remark 3.2.1. There is of course an ambiguity in the meaning of h, which was first considered as a defining parameter of both families $(\mathcal{T}_h)$ and (X_h), and which was next specifically defined in (3.2.2). We have nevertheless conformed to this often followed usage. □

(H 2) All the finite elements (K, P_K, Σ_K), $K \in \bigcup_h \mathcal{T}_h$, are affine-equivalent to a single reference finite element $(\hat{K}, \hat{P}, \hat{\Sigma})$. In other words, *the family* (K, P_K, Σ_K), $K \in \mathcal{T}_h$ *for all* h, *is an affine family of finite elements*, in the sense of Section 2.3.

(H 3) *All the finite elements* (K, P_K, Σ_K), $K \in \bigcup_h \mathcal{T}_h$, *are of class* $\mathscr{C}^0$.

We first prove an approximation property of the family (V_h) (Theorem 3.2.1), from which we derive an estimate for the *error in the norm* $\|\cdot\|_{1,\Omega}$ (Theorem 3.2.2).

In the sequel, C stands for a constant independent of h and of the various functions involved (not necessarily the same at its various occurences).

Theorem 3.2.1. *In addition to* (H 1), (H 2) *and* (H 3), *assume that there exist integers* $k \geqslant 0$ *and* $l \geqslant 0$ *with* $l \leqslant k$, *such that the following inclusions* *are satisfied*:

$$P_k(\hat{K}) \subset \hat{P} \subset H^l(\hat{K}), \tag{3.2.3}$$

$$H^{k+1}(\hat{K}) \hookrightarrow \mathscr{C}^s(\hat{K}), \tag{3.2.4}$$

where s is the maximal order of partial derivatives occurring in the definitions of the set $\hat{\Sigma}$.

Then there exists a constant C independent of h such that, for any function $v \in H^{k+1}(\Omega) \cap V$,

$$\|v - \Pi_h v\|_{m,\Omega} \leq Ch^{k+1-m}|v|_{k+1,\Omega}, \quad 0 \leq m \leq \min\{1, l\}, \tag{3.2.5}$$

$$\Big(\sum_{K \in \mathscr{T}_h} \|v - \Pi_h v\|_{m,K}^2\Big)^{1/2} \leq Ch^{k+1-m}|v|_{k+1,\Omega}, \quad 2 \leq m \leq \min\{k+1, l\} \tag{3.2.6}$$

where $\Pi_h v \in V_h$ *is the* X_h*-interpolant of the function* v.

Proof. Applying Theorem 3.1.6 with $p = q = 2$, we obtain

$$\|v - \Pi_K v\|_{m,K} \leq Ch_K^{k+1-m}|v|_{k+1,K}, \quad 0 \leq m \leq \min\{k+1, l\}.$$

Using the relations $(\Pi_h v)|_K = \Pi_K v$, $K \in \mathscr{T}_h$ (cf. (2.3.32)) and the inequalities $h_K \leq h$, $K \in \mathscr{T}_h$ (cf. (3.2.2)), we get

$$\Big(\sum_{K \in \mathscr{T}_h} \|v - \Pi_h v\|_{m,K}^2\Big)^{1/2} \leq Ch^{k+1-m}\Big(\sum_{K \in \mathscr{T}_h} |v|_{k+1,K}^2\Big)^{1/2}$$

$$= Ch^{k+1-m}|v|_{k+1,\Omega}, \quad 0 \leq m \leq \min\{k+1, l\}.$$

Thus inequalities (3.2.6) are proved and inequalities (3.2.5) follow by observing that

$$\Big(\sum_{K \in \mathscr{T}_h} \|v - \Pi_h v\|_{m,K}^2\Big)^{1/2} = \|v - \Pi_h v\|_{m,\Omega},$$

for $m = 0$ and for $m = 1$ (when $l \geq 1$) since the inclusions $\hat{P} \subset H^1(\hat{K})$ and $X_h \subset \mathscr{C}^0(\bar{\Omega})$ implies $X_h \subset H^1(\Omega)$ (Theorem 2.1.1). □

Remark 3.2.2. Analogous interpolation error estimates hold if the function v is "only" in the spaces $(\mathscr{C}^s(\bar{\Omega}) \cap \Pi_{K \in \mathscr{T}_h} H^{k+1}(K)) \cap V$. It suffices to replace the semi-norm $|v|_{k+1,\Omega}$ by the semi-norm $(\Sigma_{K \in \mathscr{T}_h} |v|_{k+1,K}^2)^{1/2}$ in the right-hand sides of inequalities (3.2.5) and (3.2.6). Such more general estimates are seldom needed. □

Theorem 3.2.2. *In addition to* (H 1), (H 2) *and* (H 3), *assume that there exists an integer* $k \geqslant 1$ *such that the following inclusions are satisfied*:

$$P_k(\hat{K}) \subset \hat{P} \subset H^1(\hat{K}), \tag{3.2.7}$$

$$H^{k+1}(\hat{K}) \hookrightarrow \mathscr{C}^s(\hat{K}), \tag{3.2.8}$$

where s is the maximal order of partial derivatives occurring in the definition of the set $\hat{\Sigma}$.

Then if the solution $u \in V$ *of the variational problem is also in the space* $H^{k+1}(\Omega)$, *there exists a constant C independent of h such that*

$$\|u - u_h\|_{1,\Omega} \leqslant Ch^k |u|_{k+1,\Omega}, \tag{3.2.9}$$

where $u_h \in V_h$ *is the discrete solution.*

Proof. It suffices to use inequality (3.2.5) with $v = u$ and $m = 1$, in conjunction with Céa's lemma (Theorem 2.4.1), which yields

$$\|u - u_h\|_{1,\Omega} \leqslant C \inf_{v_h \in V_h} \|u - v_h\|_{1,\Omega} \leqslant C\|u - \Pi_h u\|_{1,\Omega}. \qquad \square$$

Sufficient conditions for $\lim_{h \to 0} \|u - u_h\|_{1,\Omega} = 0$

The previous result has been established under the assumptions that the solution u is *sufficiently smooth* (in $H^{k+1}(\Omega)$ for some $k \geqslant 1$) and that the X_h-interpolant $\Pi_h u$ exists (cf. the inclusion $H^{k+1}(\hat{K}) \hookrightarrow \mathscr{C}^s(\hat{K})$ which is satisfied if $k > (n/2) - 1 + s$). If these hypotheses are not valid, it is still possible to prove the convergence of the method if the solution u belongs to the space $V \cap H^1(\Omega)$ and if the "minimal" assumptions (3.2.10) below hold, using a "density argument" as we now show (one should notice that the assumption $s \leqslant 1$ in the next theorem is not a restriction *in practice* for second-order problems). For a different approach, see Exercise 3.2.3.

Theorem 3.2.3. *In addition to* (H 1), (H 2) *and* (H 3), *assume that the inclusions*

$$P_1(\hat{K}) \subset \hat{P} \subset H^1(\hat{K}) \tag{3.2.10}$$

are satisfied, and that there are no directional derivatives of order $\geqslant 2$ *in the set* $\hat{\Sigma}$.

Then we have

$$\lim_{h \to 0} \|u - u_h\|_{1,\Omega} = 0. \tag{3.2.11}$$

Proof. Define the space

$$\mathcal{V} = W^{2,\infty}(\Omega) \cap V. \tag{3.2.12}$$

Since the inclusions (3.2.10) and

$$W^{2,\infty}(\hat{K}) \hookrightarrow \mathscr{C}^s(\hat{K}), \quad s = 0 \quad \text{or} \quad 1,$$
$$W^{2,\infty}(\hat{K}) \hookrightarrow H^1(\hat{K}),$$

hold, we may apply Theorem 3.1.6 with $k = 1$, $p = \infty$, $m = 1$, $q = 2$: There exists a constant C such that

$$\forall v \in \mathcal{V}, \quad \|v - \Pi_K v\|_{1,K} \leq C(\text{meas}(K))^{1/2} h_K |v|_{2,\infty,K},$$

from which we deduce that

$$\|v - \Pi_h v\|_{1,\Omega} = \Big(\sum_{K \in \mathcal{T}_h} \|v - \Pi_K v\|_{1,K}^2\Big)^{1/2} \leq Ch(\text{meas}(\Omega))^{1/2} |v|_{2,\infty,\Omega},$$

and thus we have proved that

$$\lim_{h \to 0} \|v - \Pi_h v\|_{1,\Omega} = 0. \tag{3.2.13}$$

For all h and all $v \in \mathcal{V}$, we can write

$$\|u - \Pi_h v\|_{1,\Omega} \leq \|u - v\|_{1,\Omega} + \|v - \Pi_h v\|_{1,\Omega}. \tag{3.2.14}$$

Given the solution $u \in V$ and any number $\epsilon > 0$, we first determine a function $v_\epsilon \in \mathcal{V}$ which satisfies the inequality $\|u - v_\epsilon\|_{1,\Omega} \leq \epsilon/2$. This is possible because *the space $\mathcal{V}$ is dense in the space V*. Then by (3.2.13), there exists an $h_0(\epsilon)$ such that $\|v_\epsilon - \Pi_h v_\epsilon\|_{1,\Omega} \leq \epsilon/2$ for all $h \leq h_0(\epsilon)$. In view of inequality (3.2.14), we have therefore shown that

$$\lim_{h \to 0} \inf_{v_h \in V_h} \|u - v_h\| = 0,$$

and the conclusion follows from Céa's lemma (Theorem 2.4.1). □

A close look at the above proof shows that the choice (3.2.12) of the space $\mathcal{V}$ is the result of the following requirements: On the one hand it had to be dense in the space V and on the other hand the value $k = 1$ was needed in order to apply Theorem 3.1.6 so as to obtain property (3.2.13) with the assumption $P_1(\hat{K}) \subset \hat{P}$. Therefore the space $\mathcal{V}$ had to contain derivatives of order ≤ 2 (this condition limits in turn the admissible values of s to 0 and 1) and consequently, one is naturally led to the space of the form (3.2.12). In fact, any space of the form $\mathcal{V} =$

$W^{2,p}(\Omega) \cap V$ with p sufficiently large, would have also been acceptable, as one may verify.

Estimate of the error $|u - u_h|_{0,\Omega}$. *The Aubin–Nitsche lemma*

In theorem 3.2.2, we have given assumptions which insure that $\|u - u_h\|_{1,\Omega} = O(h^k)$ so that the *error in the norm* $|\cdot|_{0,\Omega}$, i.e., the quantity $|u - u_h|_{0,\Omega}$, is at least of the same order. Our next objective is to show that, under mild additional assumptions, one has in fact $|u - u_h|_{0,\Omega} = O(h^{k+1})$.

Let us first define an abstract setting which is well adapted to this type of improved error estimates:

In addition to the space V, with norm $\|\cdot\|$, we are given a *Hilbert space* H, with norm $|\cdot|$ and inner product $(\cdot,\cdot)$, *such that* $\bar{V} = H$ *with a continuous injection* (in the present case, we shall have typically $V = H_0^1(\Omega)$, $H^1(\Omega)$, or any closed space contained in between these two spaces, and $H = L^2(\Omega)$).

Then we shall identify the space H with its dual, so that the space H may be identified with a subspace of the dual space V' of V, as we now show:

Let $f \in H$ be given. Since $V \subset H$ with a continuous injection ι, we have

$$\forall v \in V, \quad |(f, v)| \leq |f|\,|v| \leq \|\iota\|\,|f|\,\|v\|, \tag{3.2.15}$$

and therefore the mapping $v \in V \to (f, v)$ defines an element $\tilde{f} \in V'$. The mapping $f \in H \to \tilde{f} \in V'$ is an injection for if $(f, v) = 0$ for all $v \in V$, then $(f, v) = 0$ for all $v \in H$ since V is dense in H, and thus $f = 0$. *We shall henceforth identify f and $\tilde{f}$*, i.e., we shall write

$$\forall f \in H, \quad \forall v \in V, \quad (f, v) = f(v). \tag{3.2.16}$$

We next prove an *abstract error estimate.* With the same assumptions as for the Lax–Milgram lemma (Theorem 1.1.3), we let as usual $u \in V$ and $u_h \in V_h$ denote the solutions of the variational problems

$$\forall v \in V, \quad a(u, v) = f(v), \tag{3.2.17}$$

$$\forall v_h \in V_h, \quad a(u_h, v_h) = f(v_h), \tag{3.2.18}$$

respectively. We recall that M denotes an upper bound for the norm of the bilinear form $a(\cdot,\cdot)$ (cf. (1.1.19)).

Theorem 3.2.4 (*Aubin–Nitsche lemma*). *Let H be a Hilbert space, with norm $|\cdot|$ and inner product $(\cdot, \cdot)$, such that*

$$\bar{V} = H \quad \text{and} \quad V \hookrightarrow H. \tag{3.2.19}$$

Then one has

$$|u - u_h| \leq M\|u - u_h\|\left(\sup_{g\in H}\left\{\frac{1}{|g|}\inf_{\varphi_h\in V_h}\|\varphi_g - \varphi_h\|\right\}\right), \tag{3.2.20}$$

where, for any $g \in H$, $\varphi_g \in V$ is the unique solution of the variational problem:

$$\forall v \in V, \quad a(v, \varphi_g) = (g, v). \tag{3.2.21}$$

Proof. To estimate $|u - u_h|$, we shall use the characterization

$$|u - u_h| = \sup_{g\in H}\frac{|(g, u - u_h)|}{|g|}. \tag{3.2.22}$$

Using the identification (3.2.16), we can solve problem (3.2.21) for all $g \in H$ (the proof is exactly the same as that of the Lax–Milgram lemma). Since $(u - u_h)$ is an element of the space V, we have in particular

$$a(u - u_h, \varphi_g) = (g, u - u_h),$$

on the one hand, and we have

$$\forall \varphi_h \in V_h, \quad a(u - u_h, \varphi_h) = 0,$$

on the other, which we obtain by subtracting (3.2.17) and (3.2.18). Using the above relations, we obtain

$$\forall \varphi_h \in V_h, \quad (g, u - u_h) = a(u - u_h, \varphi_g - \varphi_h),$$

and therefore,

$$|(g, u - u_h)| \leq M\|u - u_h\| \inf_{\varphi_h\in V_h}\|\varphi_g - \varphi_h\|. \tag{3.2.23}$$

Inequality (3.2.20) is therefore a consequence of the characterization (3.2.22) and inequality (3.2.23). □

A look at the above proof shows that φ_g had to be the solution of problem (3.2.21), i.e., where the arguments are interchanged in the bilinear form. Problem (3.2.21) is a special case of the following variational problem: Given any element $g \in V'$, find an element $\varphi \in V$ such that

$$\forall v \in V, \quad a(v, \varphi) = g(v).$$

Such a problem is called the *adjoint problem* of problem (3.2.17). Of course the two problems coincide if the bilinear form is symmetric. It is easily verified that when the variational problem (3.2.17) corresponds to a second-order boundary value problem (cf. the examples given in Section 1.2), the same is true for its adjoint problem.

As we shall see, the abstract error estimate of Theorem 3.2.4 yields an improvement in the order of convergence for a restricted class of second-order problems, which we now define: A second-order boundary value problem whose variational formulation is (3.2.17), resp. (3.2.21), is said to be *regular* if the following two conditions are satisfied:

(i) For *any* $f \in L^2(\Omega)$, resp. *any* $g \in L^2(\Omega)$, the corresponding solution u_f, resp. u_g, is in the space $H^2(\Omega) \cap V$.

(ii) There exists a constant C such that

$$\forall f \in L^2(\Omega), \quad \|u_f\|_{2,\Omega} \leqslant C|f|_{0,\Omega}, \tag{3.2.24}$$

$$\text{resp. } \forall g \in L^2(\Omega), \quad \|\varphi_g\|_{2,\Omega} \leqslant C|g|_{0,\Omega}. \tag{3.2.25}$$

Remark 3.2.3. Consider for instance problem (3.2.17). Then without the assumption of regularity, we simply know that (use Remark 1.1.3 and the identification (3.2.16)):

$$\forall f \in L^2(\Omega), \quad \alpha\|u_f\|_{1,\Omega} \leqslant \|f\|^* = \sup_{v \in V} \frac{|f(v)|}{\|v\|_{1,\Omega}}$$
$$= \sup_{v \in V} \frac{|\int fv \, dx|}{\|v\|_{1,\Omega}} \leqslant |f|_{0,\Omega}. \qquad \square$$

Indeed, this regularity is not too restrictive a condition: For example the homogeneous Dirichlet problem and homogeneous Neumann problem associated with the data of (1.2.23) (with $g = 0$) are regular if Ω is convex and if the functions a_{ij} and a are sufficiently smooth. However, this would not be the case for the homogeneous mixed problem of (1.2.28).

We are now in a position to estimate the error in the norm $|\cdot|_{0,\Omega}$.

Theorem 3.2.5. *In addition to* (H 1), (H 2) *and* (H 3), *assume that* $s = 0$, *that the dimension* n *is* $\leqslant 3$, *and that there exists an integer* $k \geqslant 1$ *such* *that the solution u is in the space* $H^{k+1}(\Omega)$ *and such that the inclusions*

$$P_k(\hat{K}) \subset \hat{P} \subset H^1(\hat{K}) \tag{3.2.26}$$

hold.

Then if the adjoint problem is regular, there exists a constant C independent of h such that

$$|u - u_h|_{0,\Omega} \leq Ch^{k+1}|u|_{k+1,\Omega}. \tag{3.2.27}$$

Proof. Since $n \leq 3$, the inclusion $H^2(\hat{K}) \hookrightarrow \mathscr{C}^0(\hat{K})$ holds (if $s = 1$, the inclusion $H^2(\hat{K}) \hookrightarrow \mathscr{C}^1(\hat{K})$ holds only if $n = 1$; this is why we have restricted ourselves to the case $s = 0$). Applying Theorem 3.2.1 and inequality (3.2.25), we obtain, for each $g \in H = L^2(\Omega)$,

$$\inf_{\varphi_h \in V_h} \|\varphi_g - \varphi_h\|_{1,\Omega} \leq \|\varphi_g - \Pi_h \varphi_g\|_{1,\Omega} \leq Ch|\varphi_g|_{2,\Omega} \leq Ch|g|_{0,\Omega}.$$

Combining the above inequality with inequality (3.2.20) yields

$$|u - u_h|_{0,\Omega} \leq Ch\|u - u_h\|_{1,\Omega},$$

and it remains to use inequality (3.2.9) of Theorem 3.2.2. □

Concluding remarks. Inverse inequalities

Although we restricted ourselves to the case of a single partial differential equation, it should be clear that *the analysis of this section includes the systems of equations of plane and three-dimensional elasticity* (cf. (1.2.40)) posed over polygonal domains. In this case, the space $\mathbf{V}_h$ is a *product* of two or three identical finite element spaces V_h: With each degree of freedom of the space V_h, one associates two or three unknowns which are the corresponding components of the approximate displacement.

The asymptotic estimates obtained in Theorems 3.2.2 and 3.2.5 are the best one could hope for, inasmuch as *the orders of convergence are the same as if the discrete solution u_h were replaced by the X_h-interpolant of the function u*: Compare (3.2.9) and (3.2.5) with $m = 1$, and (3.2.27) and (3.2.5) with $m = 0$.

Consequently, the table in Fig. 3.1.2 is also useful for getting a practical appraisal of the upper bounds of Theorems 3.2.2 and 3.2.5. For instance, one gets $\|u - u_h\|_{m,\Omega} = O(h^{2-m})$, $m = 0, 1$, with n-simplices or rectangles of type (1), or $\|u - u_h\|_{m,\Omega} = O(h^{3-m})$, $m = 0, 1$, with n-simplices of type (2) or (3') or rectangles of type (2) or (3'), etc... Nevertheless, the higher the order of convergence, the higher the assumed regularity of the solution, and this observation limits considerably the practical value of such estimates. For example, let us assume that we are using n-simplices of type (3) while the solution is "only" in the space $H^2(\Omega)$: Then the application of Theorems 3.2.2 and 3.2.5 with $k = 1$ shows that one gets only $\|u - u_h\|_{m,\Omega} = O(h^{2-m})$, $m = 0, 1$. Therefore, unless the solution is very smooth, the use of polynomials of high degree does not improve the quality of the approximation. Interestingly, the same conclusion was also drawn through purely practical considerations, by the engineers who seldom use polynomials of degree $\geqslant 4$ for approximating the solution of second-order boundary value problems.

To conclude this section, we shall define a simple property of a family of triangulations (of the type described at the beginning of this section), whose value lies essentially (as usual) in the consequences which we shall derive (cf. Theorem 3.2.6). Although we had no immediate need for this property in the present section, it shall be used subsequently at various places, beginning in the next section, so that it seemed appropriate to record it here.

We shall say that a family of triangulations satisfies an *inverse assumption*, in view of the *inverse inequalities* to be established in the next theorem, if there exists a constant ν such that

$$\forall K \in \bigcup_h \mathcal{T}_h, \quad \frac{h}{h_K} \leqslant \nu. \tag{3.2.28}$$

Notice that this is by no means a restrictive condition in practice.

For such families, we are able to estimate the equivalence constants between familiar semi-norms (we remind the reader that σ is the constant which appears in the regularity assumption; cf. (3.2.1)).

Theorem 3.2.6. *Let there be given a family of triangulations which satisfies hypotheses* (H 1), (H 2) *and an inverse assumption, and let there be given two pairs* (l, r) *and* (m, q) *with* $l, m \geqslant 0$ *and* $(r, q) \in [1, \infty]$ *such that*

$$l \leqslant m \quad \text{and} \quad \hat{P} \subset W^{l,r}(\hat{K}) \cap W^{m,q}(\hat{K}). \tag{3.2.29}$$

Then there exists a constant $C = C(\sigma, \nu, l, r, m, q)$ *such that*

$$\forall v_h \in X_h, \quad \Big(\sum_{K \in \mathcal{T}_h} |v_h|^q_{m,q,K}\Big)^{1/q} \leqslant \frac{C}{(h^n)^{\max\{0,(1/r)-(1/q)\}} h^{m-l}} \Big(\sum_{K \in \mathcal{T}_h} |v_h|^r_{l,r,K}\Big)^{1/r} \tag{3.2.30}$$

if $p, q < \infty$, *with*

$$\max_{K \in \mathcal{T}_h} |v_h|_{m,\infty,K} \quad \textit{in lieu of} \quad \Big(\sum_{K \in \mathcal{T}_h} |v_h|_{m,q,K}^q\Big)^{1/q} \quad \textit{if } q = \infty,$$

$$\max_{K \in \mathcal{T}_h} |v_h|_{l,\infty,K} \quad \textit{in lieu of} \quad \Big(\sum_{K \in \mathcal{T}_h} |v_h|_{l,r,K}^r\Big)^{1/r} \quad \textit{if } r = \infty.$$

Proof. Given a function $v_h \in X_h$ and a finite element $K \in \mathcal{T}_h$, we have by Theorem 3.1.2,

$$\begin{aligned} |\hat{v}_K|_{l,r,\hat{K}} &\leq C\|B_K\|^l |\det(B)|^{-1/r} |v_h|_{l,r,K}, \\ |v_h|_{m,q,K} &\leq C\|B_K^{-1}\|^m |\det(B)|^{1/q} |\hat{v}_K|_{m,q,\hat{K}}, \end{aligned} \tag{3.2.31}$$

where the function $\hat{v}_K$ is in the standard correspondence with the function $v_h|_K$.

Define the space

$$\hat{N} = \{\hat{p} \in \hat{P};\ |\hat{p}|_{l,r,\hat{K}} = 0\} = \begin{cases} \{0\} & \text{if } l = 0, \\ \hat{P} \cap P_{l-1}(\hat{K}) & \text{if } l \geq 1. \end{cases}$$

Since $l \leq m$ by assumption, the implication

$$\hat{p} \in \hat{N} \Rightarrow |\hat{p}|_{m,q,\hat{K}} = 0$$

holds and therefore the mapping

$$\dot{\hat{p}} \in \hat{P}/\hat{N} \to \|\dot{\hat{p}}\|_{m,q,\hat{K}} = \inf_{\hat{s} \in \hat{N}} |\hat{p} - \hat{s}|_{m,q,\hat{K}}$$

is a norm over the quotient space $\hat{P}/\hat{N}$. Since this quotient space is finite-dimensional, this norm is equivalent to the quotient norm $\|\cdot\|_{l,r,\hat{K}}$ and therefore there exists a constant $\hat{C} = \hat{C}(l, r, m, q)$ such that

$$\hat{p} \in \hat{P}, \quad |\hat{p}|_{m,q,\hat{K}} = \|\dot{\hat{p}}\|_{m,q,\hat{K}} \leq \hat{C}\|\dot{\hat{p}}\|_{l,r,\hat{K}} = \hat{C}|\hat{p}|_{l,r,\hat{K}}. \tag{3.2.32}$$

Taking into account the regularity hypothesis and the inverse assumption, we obtain from inequalities (3.2.31) and (3.2.32) and Theorem 3.1.3,

$$|v_h|_{m,q,K} \leq C(\sigma, \nu) \frac{(h^n)^{1/q - 1/r}}{h^{m-l}} |v_h|_{l,r,K}. \tag{3.2.33}$$

Assume first that $q = \infty$, so that there exists a finite element $K_0 \in \mathcal{T}_h$ such that, using (3.2.33),

$$\max_{K \in \mathcal{T}_h} |v_h|_{m,\infty,K} = |v_h|_{m,\infty,K_0} \leq C \frac{(h^n)^{-1/r}}{h^{m-l}} |v_h|_{l,r,K_0} \leq C \frac{(h^n)^{-1/r}}{h^{m-l}} |v_h|_{l,r,\Omega}.$$

Assume next that $q < \infty$. We deduce from inequality (3.2.33) that

$$\Big(\sum_{K\in\mathcal{T}_h} |v_h|_{m,q,K}\Big)^{1/q} \leq C\frac{(h^n)^{(1/q)-(1/r)}}{h^{m-l}}\Big(\sum_{K\in\mathcal{T}_h} |v_h|^q_{l,r,K}\Big)^{1/q}.$$

Then we distinguish three cases: Either $r \leq q$, so that

$$\Big(\sum_{K\in\mathcal{T}_h} |v_h|^q_{l,r,K}\Big)^{1/q} \leq \Big(\sum_{K\in\mathcal{T}_h} |v_h|^r_{l,r,K}\Big)^{1/r}$$

by Jensen's inequality, or $q < r < \infty$, so that Hölder's inequality yields

$$\Big(\sum_{K\in\mathcal{T}_h} |v_h|^q_{l,r,K}\Big)^{1/q} \leq \mathcal{K}_h^{(1/q)-(1/r)}\Big(\sum_{K\in\mathcal{T}_h} |v_h|^r_{l,r,K}\Big)^{1/r}$$

with

$$\mathcal{K}_h = \text{card}\,\mathcal{T}_h \leq \frac{C(\sigma,\nu)}{h^n},$$

or, finally, $r = \infty$, in which case we get

$$\Big(\sum_{K\in\mathcal{T}_h} |v_h|^q_{l,\infty,K}\Big)^{1/q} \leq \mathcal{K}_h^{1/q} \max_{K\in\mathcal{T}_h} |v_h|_{l,\infty,K},$$

and inequality (3.2.30) is proved in all cases. □

Inequalities of the form (3.2.30) are of course immediately converted into inequalities involving the semi-norms $|\cdot|_{m,q,\Omega}$ or $|\cdot|_{l,r,\Omega}$ if it so happens that the inclusions $X_h \subset W^{m,q}(\Omega)$ or $X_h \subset W^{l,r}(\Omega)$ holds.

For example, let us assume that hypothesis (H 3) is satisfied and that the inclusion $\hat{P} \subset H^1(\hat{K})$ holds so that the inclusion $X_h \subset \mathscr{C}^0(\bar{\Omega}) \cap H^1(\Omega)$ holds. Then we have

$$\forall v_h \in X_h, \quad |v_h|_{0,\infty,\Omega} \leq \frac{C}{h^{n/2}}|v_h|_{0,\Omega}, \tag{3.2.34}$$

$$\forall v_h \in X_h, \quad |v_h|_{1,\Omega} \leq \frac{C}{h}|v_h|_{0,\Omega}, \quad \text{etc...} \tag{3.2.35}$$

If now hypothesis (H3) is satisfied and if the inclusion $\hat{P} \subset W^{1,\infty}(\hat{K})$ holds, then we get similarly

$$\forall v_h \in X_h, \quad |v_h|_{1,\infty,\Omega} \leq \frac{C}{h}|v_h|_{0,\infty,\Omega}, \text{etc...} \tag{3.2.36}$$

Another observation is that similar inequalities between *norms* can be directly derived from these inequalities. For instance, we obtain

$$\forall v_h \in X_h, \quad \|v_h\|_{1,\Omega} \leq \frac{C}{h}|v_h|_{0,\Omega}. \tag{3.2.37}$$

Remark 3.2.4. Inverse inequalities can be likewise established between the above semi-norms and other semi-norms or norms, such as $\|\cdot\|_{L^p(\Gamma)}$. In this direction, see Exercise 3.2.4. □

Exercises

3.2.1. The object of this problem is to indicate a way of approximating the solution of a *nonhomogeneous Dirichlet problem* (see also the next problem) whose solution $u \in H^1(\Omega)$ satisfies (cf. Exercise 1.2.2)

$$\begin{cases} (u - u_0) \in H_0^1(\Omega), \\ \forall v \in H_0^1(\Omega), \quad a(u, v) = f(v), \end{cases}$$

where u_0 is a given function in the space $H^1(\Omega)$, and the forms $a(\cdot, \cdot)$ and $f(\cdot)$ satisfy the usual assumptions of the Lax–Milgram lemma, the bilinear form being assumed to be symmetric in addition.

Given a finite element space X_h, we let as usual

$$X_{0h} = \{v_h \in X_h; \quad v_h = 0 \quad \text{on} \quad \Gamma\}.$$

(i) Given a function $u_{0h} \in X_h$, show that the discrete problem: Find $u_h \in X_h$ such that

$$\begin{cases} u_h \in u_{0h} + X_{0h} = \{v_h \in X_h; \quad (v_h - u_{0h}) \in X_{0h}\} \\ \forall v_h \in X_{0h}, \quad a(u_h, v_h) = f(v_h), \end{cases}$$

has a unique solution.

(ii) Show that (Strang & Fix (1973, p. 200)):

$$\|u - u_h\|_{1,\Omega} \leqslant \sqrt{\frac{M}{\alpha}} \inf_{v_h \in (u_{0h} + X_{0h})} \|u - v_h\|.$$

(iii) Assume that the spaces X_h are made up of n-simplices of type (k). Indicate how should one choose the function u_{0h} in order that

$$\|u - u_h\|_{1,\Omega} = \mathrm{O}(h^k),$$

assuming the functions u and u_0 are sufficiently smooth.

3.2.2. This problem describes a *penalty method* for approximating the solution of a nonhomogeneous Dirichlet problem, whose solution $u \in H^1(\Omega)$ satisfies (cf. Exercise 1.2.2):

$$\begin{cases} (u - u_0) \in H_0^1(\Omega), \\ \forall v \in H_0^1(\Omega), \quad a(u, v) = f(v), \end{cases}$$

where u_0 is a given function in the space $H^1(\Omega)$ and where for simplicity, we shall assume that

$$a(u, v) = \int_\Omega \sum_{i=1}^n \partial_i u \partial_i v \, \mathrm{d}x, \qquad f(v) = \int_\Omega fv \, \mathrm{d}x, \quad f \in L^2(\Omega).$$

In what follows, we consider a family of finite element spaces X_h. We are also given a family of real numbers $\epsilon(h) > 0$ with $\lim_{h\to 0} \epsilon(h) = 0$.

(i) Show that, for each h, the discrete problem: Find $u_h \in X_h$ such that

$$\forall v_h \in X_h, \quad a(u_h, v_h) + \frac{1}{\epsilon(h)} \int_\Gamma (u_h - u_0) v_h \, \mathrm{d}\gamma = f(v_h),$$

has a unique solution.

(ii) Assume that the solution u is in the space $H^2(\Omega)$. Show that, for all $v_h \in X_h$,

$$\begin{aligned} |u - u_h|_{1,\Omega}^2 + \frac{1}{\epsilon(h)} \|u_h - u_0\|_{L^2(\Gamma)}^2 = {} & a(u_h - u, v_h - u) \\ & + \frac{1}{\epsilon(h)} \int_\Gamma (u_h - u)(v_h - u) \, \mathrm{d}\gamma \\ & - \int_\Gamma \partial_\nu u (u_h - u) \, \mathrm{d}\gamma \\ & - \int_\Gamma \partial_\nu u (u - v_h) \, \mathrm{d}\gamma. \end{aligned}$$

Using this identity and the inequality $ab \le \eta a^2 + (1/\eta) b^2$ valid for all $\eta > 0$, derive the following abstract error estimate: There exists a constant C independent of h such that

$$\begin{aligned} |u - u_h|_{1,\Omega} \le C \inf_{v_h \in X_h} \Big\{ & |u - v_h|_{1,\Omega}^2 + \frac{1}{\epsilon(h)} \|u_0 - v_h\|_{L^2(\Gamma)}^2 \\ & + \epsilon(h) \|\partial_\nu u\|_{L^2(\Gamma)}^2 \Big\}^{1/2}. \end{aligned}$$

(iii) Assume that the spaces X_h are made up of n-simplices of type (k) and that the solution u is sufficiently smooth. Show that, for some constant C independent of h,

$$|u - u_h|_{1,\Omega} \le C \Big(\frac{h^k}{\sqrt{\epsilon(h)}} + \sqrt{\epsilon(h)} \Big) \|u\|_{k+1,\Omega},$$

and thus deduce the optimal choice for $\epsilon(h)$, so as to maximize the order of convergence (therefore, as far as the order of convergence is concerned, the method proposed in the previous exercise is preferable).

3.2.3. The purpose of this problem is to analyze a procedure of CLÉMENT (1975) for defining an operator whose approximation properties are similar to those of the standard interpolation operator but which can be defined in more general situations.

For simplicity, we shall consider finite element spaces X_h made up of triangles of type (1), but the analysis can be extended to triangular finite elements with polynomials of higher degree.

With each vertex b_i, $1 \leq i \leq M$, of the triangulation, we associate a basis function $w_i \in X_h$ in the usual manner, i.e., one has

$$w_i(b_j) = \delta_{ij}, \quad 1 \leq i, j \leq M.$$

For each i, we set

$$S_i = \text{supp } w_i.$$

Given a function $v \in L^2(\Omega)$, we let $P_i v$ denote, for each $i = 1, \dots, M$, the projection of the function v in the space $L^2(S_i)$ over the subspace $P_1(S_i)$, i.e., one has

$$P_i v \in P_1(S_i) \quad \text{and} \quad \forall p \in P_1(S_i), \quad \int_{S_i} (v - P_i v) p \, \mathrm{d}x = 0,$$

and we set

$$r_h v = \sum_{i=1}^{M} P_i v(b_i) w_i.$$

In this fashion, we have defined a mapping

$$r_h \colon L^2(\Omega) \to X_h.$$

In the sequel, we consider a family of spaces X_h associated with a regular family of triangulations.

(i) Show that there exists a constant C independent of h such that

$$\forall i \in [1, M], \quad K \subset S_i \Rightarrow \text{diam } S_i \leq C h_K,$$

and that there exists an integer ν independent of h such that

$$\forall i \in [1, M], \quad \text{card}\{K \in \mathcal{T}_h; \quad K \subset S_i\} \leq \nu.$$

(ii) Show that there exists a constant C independent of h such that

$$\forall i \in [1, M], \quad \forall v \in H^1(S_i),$$

$$|v - P_i v|_{m,S_i} \leq C(\text{diam } S_i)^{l-m}|v|_{l,S_i}, \quad 0 \leq m \leq l \leq 2.$$

(iii) Show that there exists a constant C independent of h such that

$$\forall K \in \mathcal{T}_h, \quad \forall p \in P_1(K),$$
$$|p|_{m,\infty,K} \leq C(\text{meas}(K))^{-1/2} h_K^{-m} |p|_{0,K}, \quad m = 0, 1.$$

(iv) Show that there exists a constant C independent of h such that

$$\forall i \in [1, M], \quad |w_i|_{m,K} \leq C(\text{meas}(K))^{1/2} h_K^{-m}, \quad m = 0, 1.$$

(v) Show that

$$\forall v \in L^2(\Omega), \quad \begin{cases} |v - r_h v|_{0,\Omega} \leq C|v|_{0,\Omega}, \\ \lim\limits_{h \to 0} |v - r_h v|_{0,\Omega} = 0, \end{cases}$$

$$\forall v \in H^1(\Omega), \quad \begin{cases} |v - r_h v|_{m,\Omega} \leq Ch^{1-m}|v|_{1,\Omega}, \quad m = 0, 1, \\ \lim\limits_{h \to 0} |v - r_h v|_{1,\Omega} = 0, \end{cases}$$

$$\forall v \in H^2(\Omega), \quad \begin{cases} |v - r_h v|_{m,\Omega} \leq Ch^{2-m}|v|_{2,\Omega}, \quad m = 0, 1, \\ \left(\sum\limits_{K \in \mathcal{T}_h} |v - r_h v|_{2,K}^2\right)^{1/2} \leq C|v|_{2,\Omega}, \end{cases}$$

where C denotes as usual various constants independent of h.

[Hint: Let $K \in \mathcal{T}_h$ be a triangle with vertices b_i, $1 \leq i \leq 3$. Prove the identity

$$(r_h v - v)_{|K} = (P_1 v - v)_{|K} + \sum_{i=2}^{3} (P_i v(b_i) - P_1 v(b_i)) w_{i|K}$$

and use the previous questions to estimate appropriate semi-norms of the functions $(P_1 v - v)_{|K}$ and $\Sigma_{i=2}^{3} (P_i v(b_i) - P_1 v(b_i)) w_{i|K}$.]

(vi) If the function v belongs to the space $H_0^1(\Omega)$, is the function $r_h v$ in the space $X_{0h} = \{v_h \in X_h;\ v_h = 0 \text{ on } \Gamma\}$?

(vii) Apply the results of question (v) to the approximation of a second-order boundary value problem. Compare with Theorems 3.2.2 and 3.2.5.

3.2.4. Let there be given a family of triangulations which satisfies hypothesis (H 2) and an inverse assumption. It is also assumed that $\hat{P} \subset \mathscr{C}^0(\hat{K})$. Then show that for each $p \in [1, \infty]$, there exists a constant $C = C(p)$ independent of h such that

$$\forall v_h \in X_h, \quad \|v_h\|_{L^p(\Gamma)} \leq \frac{C}{h^{1/p}} |v_h|_{0,p,\Omega}.$$

3.3. Uniform convergence

A model problem. Weighted semi-norms $|\cdot|_{\phi;m,\Omega}$

For ease of exposition, we shall simply consider the homogeneous Dirichlet problem for the operator $-\Delta$, which corresponds to the following data:

$$\begin{cases} V = H_0^1(\Omega), \\ a(u, v) = \int_\Omega \nabla u \cdot \nabla v \, \mathrm{d}x, \\ f(v) = \int_\Omega fv \, \mathrm{d}x, \quad f \in L^2(\Omega). \end{cases} \tag{3.3.1}$$

Assuming that $\bar{\Omega}$ is a *convex polygonal* subset of $\mathbf{R}^2$, we shall restrict ourselves to finite element spaces X_h which are made up of *triangles of type* (1), so that the corresponding discrete problems are posed in the spaces $V_h = \{v_h \in X_h;\ v_h = 0 \text{ on } \Gamma\}$ (results concerning the use of triangles of type (k) and higher dimensions are indicated at the end of this section and in the section "Bibliography and Comments").

We shall assume once and for all that we are given a family of triangulations of the set $\bar{\Omega}$ which

(i) is *regular* and

(ii) satisfies an *inverse assumption*, i.e., there exist two constants σ and ν such that

$$\forall K \in \bigcup_h \mathcal{T}_h, \quad \frac{h_K}{\rho_K} \leqslant \sigma, \quad \text{and} \quad \frac{h}{h_K} \leqslant \nu. \tag{3.3.2}$$

Our main tool in the study of the error in the norms $|\cdot|_{0,\infty,\Omega}$ and $\|\cdot\|_{1,\infty,\Omega}$ will be the consideration of appropriate *weighted norms and semi-norms*. Accordingly, the first part of this section will be devoted to the study of those properties of such semi-norms which are of interest for our subsequent analysis (cf. Theorems 3.3.1 to 3.3.4).

Given a *weight-function* ϕ, i.e., a function which satisfies

$$\phi \in L^\infty(\Omega) \quad \text{and} \quad \phi \geqslant 0 \text{ a.e.} \quad \text{on} \quad \Omega, \tag{3.3.3}$$

we define, for each integer $m \geqslant 0$, the *weighted semi-norms*

$$v \in H^m(\Omega) \to |v|_{\phi;m,\Omega} = \left(\int_\Omega \phi \sum_{|\beta|=m} |\partial^\beta v|^2 \, \mathrm{d}x \right)^{1/2}. \tag{3.3.4}$$

To begin with, we observe that, *if the function ϕ^{-1} exists and is also in the space $L^\infty(\Omega)$*, an application of Cauchy–Schwarz inequality gives

$$\forall \alpha \in \mathbf{R}, \quad \forall u, v \in H^1(\Omega), \quad a(u, v) \leq |u|_{\phi^\alpha;1,\Omega} |v|_{\phi^{-\alpha};1,\Omega}. \tag{3.3.5}$$

Departing from the general case, we shall in fact concentrate our subsequent study on weighted semi-norms of the particular type $|\cdot|_{\phi^\alpha;m,\Omega}$, $\alpha \in \mathbf{R}$, where the function ϕ is of the form (3.3.7) below. Our first task is to extend to such weighted semi-norms the property that there exists a constant c_1, solely dependent upon the set Ω, such that

$$\forall w \in H_0^1(\Omega) \cap H^2(\Omega), \quad |w|_{2,\Omega} \leq c_1 |\Delta w|_{0,\Omega}. \tag{3.3.6}$$

Theorem 3.3.1. *There exists a constant $C_1 = C_1(\Omega)$ such that, for all functions ϕ of the form*

$$\phi : x \in \bar{\Omega} \to \phi(x) = \frac{1}{\|x - \bar{x}\|^2 + \theta^2}, \quad \theta > 0, \quad \bar{x} = (\bar{x}_1, \bar{x}_2) \in \mathbf{R}^2, \tag{3.3.7}$$

we have

$$\forall v \in H_0^1(\Omega) \cap H^2(\Omega), \quad |v|^2_{\phi^{-1};2,\Omega} \leq C_1 (|\Delta v|^2_{\phi^{-1};0,\Omega} + |v|^2_{1,\Omega}). \tag{3.3.8}$$

Proof. Let v be an arbitrary function in the space $H_0^1(\Omega) \cap H^2(\Omega)$. Then the function

$$w = (x_1 - \bar{x}_1) v$$

also belongs to the space $H_0^1(\Omega) \cap H^2(\Omega)$, and

$$\begin{aligned}
(x_1 - \bar{x}_1)\partial_{11} v &= \partial_{11} w - 2\partial_1 v, \\
(x_1 - \bar{x}_1)\partial_{12} v &= \partial_{12} w - \partial_2 v, \\
(x_1 - \bar{x}_1)\partial_{22} v &= \partial_{22} w, \\
\Delta w &= (x_1 - \bar{x}_1)\Delta v + 2\partial_1 v.
\end{aligned}$$

Using these relations and inequality (3.3.6), we find a constant c_2 such that

$$\begin{aligned}
\int_\Omega (x_1 - \bar{x}_1)^2 \sum_{|\beta|=2} |\partial^\beta v|^2 \, dx &\leq 2c_1^2 |\Delta w|^2_{0,\Omega} + 8|v|^2_{1,\Omega} \\
&\leq c_2 \left\{ \int_\Omega (x_1 - \bar{x}_1)^2 (\Delta v)^2 \, dx + |v|^2_{1,\Omega} \right\}.
\end{aligned}$$

Since we have likewise

$$\int_\Omega (x_2-\bar{x}_2)^2 \sum_{|\beta|=2} |\partial^\beta v|^2 \,\mathrm{d}x \leq c_2\left\{\int_\Omega (x_2-\bar{x}_2)^2(\Delta v)^2\,\mathrm{d}x + |v|^2_{1,\Omega}\right\},$$

we eventually obtain

$$|v|^2_{\phi^{-1};2,\Omega} = \int_\Omega ((x_1-\bar{x}_1)^2+(x_2-\bar{x}_2)^2+\theta^2) \sum_{|\beta|=2} |\partial^\beta v|^2\,\mathrm{d}x$$

$$\leq \max\{2c_2, c_1^2\}(|\Delta v|^2_{\phi^{-1};0,\Omega} + |v|^2_{1,\Omega}),$$

and the proof is complete. □

As exemplified by the above computations, we shall depart in this section from our practice of letting the same letter C denote various constants, not necessarily the same in their various occurrences. This is due not only to the unusually large number of such constants which we shall come across, but also – and essentially – to their sometimes intricate interdependence. Therefore, constants will be numbered and, in addition, their dependence on other quantities will be made explicit when necessary. However the possible dependence upon the set Ω and the constants σ and ν of (3.3.2) will be systematically omitted. While we shall use capital letters C_i, $i \geq 1$, for constants occurring in important inequalities, small letters c_i, $i \geq 1$, will rather be reserved for intermediate computations.

In the next two theorems, we examine the relationships between the weighted semi-norms $|\cdot|_{\phi^\alpha;m,\Omega}$ (the function ϕ being as in (3.3.7)) and the standard semi-norms $|\cdot|_{m,\infty,\Omega}$. Such relationships will play a crucial role in the derivation of the eventual error estimates.

Theorem 3.3.2. *For each number $\alpha > 1$ and each integer $m \geq 0$, there exists a constant $C_2(\alpha, m)$ such that, for all functions ϕ of the form*

$$\phi : x \in \bar{\Omega} \to \phi(x) = \frac{1}{\|x-\bar{x}\|^2+\theta^2}, \quad \theta > 0, \quad \bar{x} \in \bar{\Omega}, \tag{3.3.9}$$

we have

$$\forall v \in W^{m,\infty}(\Omega), \quad |v|_{\phi^\alpha;m,\Omega} \leq C_2(\alpha, m)\frac{1}{\theta^{\alpha-1}}|v|_{m,\infty,\Omega}. \tag{3.3.10}$$

For each number $\beta \in]0, 1[$, and each integer $m \geq 0$, there exists a *constant* $C_3(\beta, m)$ *such that for all functions* ϕ *of the form* (3.3.9), *we have*

$$\forall \theta \leqslant \beta, \quad \forall v \in W^{m,\infty}(\Omega), \quad |v|_{\phi;m,\Omega} \leqslant C_3(\beta, m)|\ln \theta|^{1/2}|v|_{m,\infty,\Omega}. \tag{3.3.11}$$

Proof. Clearly, one has

$$|v|_{\phi^\alpha;m,\Omega} \leqslant c_3(m)\Big(\int_\Omega \phi^\alpha \, dx\Big)^{1/2}|v|_{m,\infty,\Omega}.$$

Next let $\delta = \operatorname{diam}(\Omega)$, so that

$$\int_\Omega \phi^\alpha \, dx \leqslant \int_{B(\bar{x};\delta)} \phi^\alpha \, dx = 2\pi \int_0^\delta \frac{\tau \, d\tau}{(\tau^2 + \theta^2)^\alpha}.$$

If $\alpha > 1$, write

$$\int_0^\delta \frac{\tau \, d\tau}{(\tau^2 + \theta^2)^\alpha} \leqslant \int_0^\infty \frac{\tau \, d\tau}{(\tau^2 + \theta^2)^\alpha} = \frac{1}{2(\alpha - 1)\theta^{2(\alpha-1)}},$$

and inequality (3.3.10) is proved with $C_2(\alpha, m) = c_3(m)(\pi/(\alpha - 1))^{1/2}$. If $\alpha = 1$, we have for $\theta \leqslant \beta < 1$,

$$\begin{aligned}\int_0^\delta \frac{\tau \, d\tau}{\tau^2 + \theta^2} &= |\ln \theta| + \frac{1}{2}\ln(\theta^2 + \delta^2) \\ &\leqslant |\ln \theta| + \frac{1}{2}\ln(1 + \delta^2) \leqslant c_4(\beta)|\ln \theta|,\end{aligned}$$

with

$$c_4(\beta) = 1 + \frac{\ln(1 + \delta^2)}{2|\ln \beta|},$$

and inequality (3.3.11) is proved with $C_3(\beta, m) = c_3(m)(2\pi c_4(\beta))^{1/2}$. □

We next obtain inequalities in the opposite direction. In order that they be useful for our subsequent purposes, however, we shall establish these inequalities only for functions in the finite element space X_h, and further, we shall restrict ourselves to weight-functions of the form ϕ or ϕ^2, with ϕ as in (3.3.9), for which (i) the parameter θ cannot approach zero too rapidly when h approaches zero (cf. (3.3.13)), and for which (ii) the points $\bar{x}$ depend upon the particular function $v_h \in X_h$ under consideration (cf. (3.3.15) and (3.3.17)).

Theorem 3.3.3. *For each number $\gamma > 0$, there exist constants $C_4(\gamma)$ and $C_5(\gamma)$ such that, if for each h, we are given a function ϕ_h of the form*

$$\phi_h : x \in \bar{\Omega} \to \phi_h(x) = \frac{1}{\|x - x_h\|^2 + \theta_h^2}, \quad x_h \in \bar{\Omega}, \tag{3.3.12}$$

in such a way that

$$\exists \gamma > 0, \quad \forall h, \theta_h \geqslant \gamma h, \tag{3.3.13}$$

then (i) *we have*

$$\forall v_h \in X_h, \quad |v_h|_{0,\infty,\Omega} \leqslant C_4(\gamma) \frac{\theta_h^2}{h} |v_h|_{\phi_h^2;0,\Omega} \tag{3.3.14}$$

if, for each function $v_h \in X_h$, the point $x_h \in \bar{\Omega}$ in (3.3.12) *is chosen in such a way that*

$$|v_h(x_h)| = |v_h|_{0,\infty,\Omega}, \tag{3.3.15}$$

and (ii) *we have*

$$\forall v_h \in X_h, \quad |v_h|_{1,\infty,\Omega} \leqslant C_5(\gamma) \frac{\theta_h}{h} |v_h|_{\phi_h;1,\Omega} \tag{3.3.16}$$

if, for each function $v_h \in X_h$, the point $x_h \in \bar{\Omega}$ in (3.3.12) *is chosen in such a way that*

$$\max\{|\partial_1 v_h(x_h)|, \quad |\partial_2 v_h(x_h)|\} = |v_h|_{1,\infty,\Omega}. \tag{3.3.17}$$

Proof. (i) Let v_h be an arbitrary function in the space X_h, and let the point x_h be chosen as in (3.3.15). We can write

$$\begin{aligned} \forall x \in \bar{\Omega}, \quad |v_h|_{0,\infty,\Omega} - |v_h(x)| &= |v_h(x_h)| - |v_h(x)| \leqslant |v_h(x_h) - v_h(x)| \\ &\leqslant \sqrt{2} |v_h|_{1,\infty,\Omega} \|x - x_h\| \\ &\leqslant \frac{c_5}{h} |v_h|_{0,\infty,\Omega} \|x - x_h\|, \end{aligned}$$

for some constant c_5 (in the last inequality we have used the fact that the family of triangulations satisfies an inverse assumption; cf. Theorem 3.2.6). In other words,

$$\forall x \in \bar{\Omega}, \quad |v_h(x)| \geqslant \left(1 - \frac{c_5}{h} \|x - x_h\|\right) |v_h|_{0,\infty,\Omega},$$

and consequently ($B(a\,;r) = \{x \in \mathbf{R}^2;\ \|x - a\| \leq r\}$),

$$|v_h|^2_{\phi_h^2;0,\Omega} \geq |v_h|^2_{0,\infty,\Omega} \int_{\bar{\Omega}\cap B(x_h;h/2c_5)} \left(\frac{1 - \frac{c_5}{h}\|x - x_h\|}{\|x - x_h\|^2 + \theta_h^2} \right)^2 \mathrm{d}x.$$

The set $\bar{\Omega}$ being polygonal, there exists a constant c_6 such that

$$\operatorname{meas}\left\{\bar{\Omega} \cap B\left(x_h;\frac{h}{2c_5}\right)\right\} \geq c_6\left(\frac{h}{2c_5}\right)^2.$$

We also have

$$\forall x \in B\left(x_h;\frac{h}{2c_5}\right),\quad 1 - \frac{c_5}{h}\|x - x_h\| \geq \frac{1}{2},$$

and

$$\forall x \in B\left(x_h;\frac{h}{2c_5}\right),\quad \frac{1}{\|x - x_h\|^2 + \theta_h^2} \geq \frac{1}{\frac{h^2}{4c_5^2} + \theta_h^2} \geq \frac{1}{\theta_h^2\left(1 + \frac{1}{4c_5^2\gamma^2}\right)},$$

by assumption (3.3.13). Combining the previous inequalities, we obtain an inequality of the form (3.3.14) with

$$C_4(\gamma) = \frac{4c_5}{\sqrt{c_6}}\left(1 + \frac{1}{4c_5^2\gamma^2}\right).$$

(ii) Let v_h be an arbitrary function in the space X_h, let the point x_h be chosen as in (3.3.17), and let $K_h \in \mathcal{T}_h$ denote a triangle which contains the point x_h. Since the gradient ∇v_h is constant over the set K_h, we deduce

$$|v_h|^2_{\phi_h;1,\Omega} \geq |v_h|^2_{1,\infty,\Omega} \int_{K_h} \frac{\mathrm{d}x}{\|x - x_h\|^2 + \phi_h^2}.$$

With this inequality and the inequalities

$$\operatorname{meas} K_h \geq c_7(\sigma, \nu)h^2,$$

$$\forall x \in K_h,\quad \frac{1}{\|x - x_h\|^2 + \theta_h^2} \geq \frac{1}{\theta_h^2\left(1 + \frac{1}{\gamma^2}\right)},$$

we obtain an inequality of the form (3.3.16) with

$$C_5(\gamma) = \sqrt{\frac{1}{c_7}\left(1 + \frac{1}{\gamma^2}\right)}.$$ □

To conclude this analysis of weighted semi-norms, we examine in the next theorem the *interpolation error estimates* in the semi-norms $|\cdot|_{\phi_h^\alpha;m,\Omega}$ where, for each h, the function ϕ_h is of the form (3.3.12). The conclusion (cf. (3.3.20)) is that *the error estimates are exactly the same as in the case of the usual semi-norms* $|\cdot|_{m,\Omega}$ *provided the parameter* θ_h *does not approach zero too rapidly with* h (cf. (3.3.19)). Notice, however, that if the behavior of the function θ_h can be "at best" linear as in the previous theorem, the constant which appears in inequality (3.3.19) is not arbitrary, by contrast with the constant γ which appeared in inequality (3.3.13). Finally, observe that no restriction will be imposed upon the points x_h.

Theorem 3.3.4. *There exists a constant C_6 and, for each $\alpha \in \mathbf{R}-\{0\}$, there exist constants $C_7(\alpha)>0$ and $C_8(\alpha)$ such that, if for each h, we are given a function ϕ_h of the form*

$$\phi_h: x \in \bar{\Omega} \to \phi_h(x) = \frac{1}{\|x-x_h\|^2+\theta_h^2}, \quad x_h \in \bar{\Omega}, \tag{3.3.18}$$

in such a way that

$$\forall h, \quad \theta_h \geq C_7(\alpha) h, \tag{3.3.19}$$

then (i) *we have*

$$\forall v \in H^2(\Omega), \quad |v - \Pi_h v|_{\phi_h^\alpha;m,\Omega} \leq C_6 h^{2-m} |v|_{\phi_h^\alpha;2,\Omega}, \quad m = 0, 1, \tag{3.3.20}$$

and (ii) *we have*

$$\forall v_h \in X_h, \quad |\phi_h^\alpha v_h - \Pi_h(\phi_h^\alpha v_h)|_{\phi_h^{-\alpha};1,\Omega} \leq$$

$$\leq C_8(\alpha) \frac{h}{\theta_h} \left(|v_h|_{\phi_h^{\alpha+1};0,\Omega} + |v_h|_{\phi_h^\alpha;1,\Omega} \right). \tag{3.3.21}$$

Proof. (i) There exists a constant c_8 such that

$$\forall v \in H^2(\Omega), \ \forall K \in \mathcal{T}_h, \quad |v - \Pi_K v|_{m,K} \leq c_8 h^{2-m} |v|_{2,K}, \quad m = 0, 1.$$

Next, we have

$$|v - \Pi_K v|_{\phi_h^\alpha;m,K} \leq (\phi_h^\alpha(\bar{x}_K))^{1/2} |v - \Pi_K v|_{m,K},$$

$$|v|_{2,K} \leq (\phi_h^\alpha(x_K))^{-1/2} |v|_{\phi_h^\alpha;2,K},$$

where, for each $K \in \mathcal{T}_h$, the points $x_K \in K$ and $\bar{x}_K \in K$ are chosen in such a way that

$$0 < \phi_h^\alpha(x_K) = \inf_{x\in K} \phi_h^\alpha(x), \quad \phi_h^\alpha(\bar{x}_K) = \sup_{x\in K} \phi_h^\alpha(x).$$

Since

$$\frac{\partial_i(\phi_h^\alpha)(x)}{\phi_h^\alpha(x)} = -2\alpha \frac{(x_i - x_{hi})}{\|x - x_h\|^2 + \theta_h^2} \quad i = 1, 2,$$

we obtain

$$\sup_{x\in\bar{\Omega}} \frac{|\partial_i(\phi_h^\alpha)(x)|}{\phi_h^\alpha(x)} \leqslant \sup_{x\in\bar{\Omega}} \frac{\|D(\phi_h^\alpha)(x)\|}{\phi_h^\alpha(x)} \leqslant 2|\alpha| \sup_{x\in\bar{\Omega}} \frac{\|x - x_h\|}{\|x - x_h\|^2 + \theta_h^2} \leqslant \frac{|\alpha|}{\theta_h},$$

and therefore

$$\phi_h^\alpha(\bar{x}_K) \leqslant \phi_h^\alpha(x_K) + |\alpha| \frac{h}{\theta_h} \phi_h^\alpha(\bar{x}_K).$$

Consequently, if we let

$$C_7(\alpha) = 2|\alpha|,$$

so that

$$\theta_h \geqslant C_7(\alpha)h \Rightarrow \forall K \in \mathcal{T}_h, \ \frac{\phi_h^\alpha(\bar{x}_K)}{\phi_h^\alpha(x_K)} \leqslant 2,$$

the conjunction of the above inequalities yields inequality (3.3.20) with $C_6 = \sqrt{2} c_8$.

(ii) Since the function $\phi_h^\alpha v_h$ is in the space $\mathscr{C}^0(\bar{\Omega}) = \operatorname{dom} \Pi_h$ and since the restrictions $\phi_h^\alpha v_h|_K$ belong to the space $H^2(K)$ for all $K \in \mathcal{T}_h$, the same argument as in (i) shows that

$$|\phi_h^\alpha v_h - \Pi_K(\phi_h^\alpha v_h)|_{\phi_h^{-\alpha};1,\Omega} \leqslant C_7 h \Big(\sum_{K\in\mathcal{T}_h} |\phi_h^\alpha v_h|^2_{\phi_h^{-\alpha};2,K} \Big)^{1/2}.$$

We have (recall that $v_{h|K} \in P_1(K)$ for all $K \in \mathcal{T}_h$)

$$\forall K \in \mathcal{T}_h, \quad \partial_{ij}(\phi_h^\alpha v_h) = (\partial_{ij}\phi_h^\alpha) v_h + (\partial_i \phi_h^\alpha)\partial_j v_h + (\partial_j \phi_h^\alpha)\partial_i v_h \quad \text{in} \quad K,$$

and

$$\frac{\partial_{ij}(\phi_h^\alpha)(x)}{\phi_h^\alpha(x)} = 4\alpha(\alpha+1)\frac{(x_i - x_{hi})(x_j - x_{hj})}{(\|x - x_h\|^2 + \theta_h^2)^2} - 2\alpha \frac{\delta_{ij}}{\|x - x_h\|^2 + \theta_h^2}.$$

Using the inequalities

$$\frac{\|x - x_h\|^2}{(\|x - x_h\|^2 + \theta_h^2)^2} \leqslant \frac{\phi_h^{1/2}(x)}{2\theta_h}, \quad \text{and} \quad \frac{1}{\|x - x_h\|^2 + \theta_h^2} \leqslant \frac{\phi_h^{1/2}(x)}{\theta_h},$$

we deduce that

$$\forall x \in \bar{\Omega}, \quad |\partial_{ij}(\phi_h^\alpha)(x)| \leqslant \frac{2}{\theta_h}(|\alpha^2 + \alpha| + |\alpha|)\phi_h^{\alpha+1/2}(x).$$

Using the above inequalities and the inequality (cf. part (i))

$$\forall x \in \bar{\Omega}, \quad |\partial_i(\phi_h^\alpha)(x)| \leqslant \frac{|\alpha|}{\theta_h}\phi_h^\alpha(x),$$

we conclude that there exists a constant $c_9(\alpha)$ such that

$$\forall x \in K, \quad \phi_h^{-\alpha}(x) \sum_{i,j=1}^{2} |\partial_{ij}(\phi_h^\alpha v_h)(x)|^2 \leqslant$$
$$\leqslant \frac{c_9(\alpha)}{\theta_h^2}\Big(\phi_h^{\alpha+1}(x)|v_h(x)|^2 + \phi_h^\alpha(x) \sum_{i=1}^{2} |\partial_i v_h(x)|^2\Big),$$

and thus,

$$\sum_{K \in \mathcal{T}_h} |\phi_h^\alpha v_h|^2_{\phi_h^{-\alpha};2,K} = \sum_{K \in \mathcal{T}_h} \int_K \phi_h^{-\alpha} \sum_{i,j=1}^{2} |\partial_{ij}(\phi_h^\alpha v_h)(x)|^2 \,\mathrm{d}x \leqslant$$
$$\leqslant \frac{c_9(\alpha)}{\theta_h^2}(|v_h|^2_{\phi_h^{\alpha+1};0,\Omega} + |v_h|^2_{\phi_h^\alpha;1,\Omega}).$$

Therefore we have proved inequality (3.3.21), with

$$C_8(\alpha) = C_7\sqrt{c_9(\alpha)}. \qquad \square$$

Uniform boundedness of the mapping $u \to u_h$ with respect to appropriate weighted norms

After the above preliminaries, we now come to the central object of this section, i.e., the estimate of the errors $|u - u_h|_{0,\infty,\Omega}$ and $|u - u_h|_{1,\infty,\Omega}$ via *the method of weighted norms of J.A. Nitsche.* The analysis will comprise *three stages.* In the *first stage* (cf. the next theorem), we consider for each *h the projection operator*

$$P_h: v \in H_0^1(\Omega) \to P_h v \in V_h \tag{3.3.22}$$

associated with the inner product $a(\cdot, \cdot)$ of (3.3.1), and which is therefore defined for each $v \in H_0^1(\Omega$ by the relations

$$P_h v \in V_h \quad \text{and} \quad \forall w_h \in V_h, \quad a(v - P_h v, w_h) = 0. \tag{3.3.23}$$

Thus we have in particular $u_h = P_h u$, where u_h is the discrete solution found in the space V_h and u is the solution of the problem defined in (3.3.1). We shall then show that for an appropriate choice of the parameters θ_h in the functions ϕ_h (cf. (3.3.25) and (3.3.26) below), *the mappings P_h are also bounded independently of h when both spaces $H_0^1(\Omega)$ and V_h are equipped with the weighted norm*

$$v \to (|v|^2_{\phi_h^2;0,\Omega} + |v|^2_{\phi_h;1,\Omega})^{1/2}. \tag{3.3.24}$$

Theorem 3.3.5. *There exist three constants $h_0 \in]0, 1[$, $C_9 > 0$ and C_{10} such that, if for each h, we are given a function ϕ_h of the form*

$$\phi_h : x \in \bar{\Omega} \to \phi_h(x) = \frac{1}{\|x - x_h\|^2 + \theta_h^2}, \quad x_h \in \bar{\Omega}, \tag{3.3.25}$$

in such a way that

$$\forall h, \quad \theta_h = C_9 h |\ln h|^{1/2}, \tag{3.3.26}$$

then

$$\forall h \leq h_0, \quad \forall v \in H^1(\Omega),$$
$$|P_h v|_{\phi_h^2;0,\Omega} + |P_h v|_{\phi_h;1,\Omega} \leq C_{10}(|v|_{\phi_h^2;0,\Omega} + |v|_{\phi_h;1,\Omega}). \tag{3.3.27}$$

Proof. For convenience, the proof will be divided in four steps.

(i) *There exist two constants C_{11} and C_{12} such that, if*

$$\forall h, \quad \theta_h \geq C_{11} h, \tag{3.3.28}$$

then

$$\forall v \in H_0^1(\Omega), \quad |P_h v|^2_{\phi_h;1,\Omega} \leq C_{12}(|P_h v|^2_{\phi_h^2;0,\Omega} + |v|^2_{\phi_h;1,\Omega}). \tag{3.3.29}$$

For brevity, let

$$v_h = P_h v.$$

Since

$$|v_h|^2_{\phi_h;1,\Omega} = a(v_h, \phi_h v_h) + \frac{1}{2}\int_\Omega \Delta\phi_h \, v_h^2 \, dx$$

and

$$\forall x \in \bar{\Omega}, \quad \Delta\phi_h(x) = \frac{4(\|x - x_h\|^2 - \theta_h^2)}{\|x - x_h\|^2 + \theta_h^2}\phi_h^2(x) \leq 4\phi_h^2(x),$$

we deduce that

$$|v_h|^2_{\phi_h;1,\Omega} \leq a(v_h, \phi_h v_h) + 2|v_h|^2_{\phi_h^2;0,\Omega}. \tag{3.3.30}$$

Using relations (3.3.23) we can write

$$a(v_h, \phi_h v_h) = a(v_h - v, \phi_h v_h - \Pi_h(\phi_h v_h)) + a(v, \phi_h v_h). \tag{3.3.31}$$

An application of inequality (3.3.5) with $\alpha = 1$ shows that

$$|a(v_h - v, \phi_h v_h - \Pi_h(\phi_h v_h))| \leq$$

$$\leq (|v|_{\phi_h;1,\Omega} + |v_h|_{\phi_h;1,\Omega})|\phi_h v_h - \Pi_h(\phi_h v_h)|_{\phi_h^{-1};1,\Omega}.$$

By Theorem 3.3.4, we have that, if

$$\forall h, \quad \theta_h \geq c_{10}h, \quad \text{with } c_{10} = C_7(1), \tag{3.3.32}$$

then (cf. inequality (3.3.21) with $\alpha = 1$)

$$|\phi_h v_h - \Pi_h(\phi_h v_h)|_{\phi_h^{-1};1,\Omega} \leq c_{11}\frac{h}{\theta_h}(|v_h|_{\phi_h^2;0,\Omega} + |v_h|_{\phi_h;1,\Omega}),$$

$$\text{with } c_{11} = C_8(1).$$

Combining the previous inequalities, we find that, for $\theta_h \geq c_{10}h$,

$$|a(v_h - v, \phi_h v_h - \Pi_h(\phi_h v_h))| \leq$$

$$\leq c_{11}\frac{h}{\theta_h}(|v|_{\phi_h;1,\Omega} + |v_h|_{\phi_h;1,\Omega})(|v_h|_{\phi_h^2;0,\Omega} + |v_h|_{\phi_h;1,\Omega}). \tag{3.3.33}$$

By another application of inequality (3.3.5) with $\alpha = 1$, we obtain

$$a(v, \phi_h v_h) \leq |v|_{\phi_h;1,\Omega}|\phi_h v_h|_{\phi_h^{-1};1,\Omega}. \tag{3.3.34}$$

Using the inequality

$$\forall x \in \bar{\Omega}, \quad \sum_{i=1}^{2} |\partial_i\phi_h(x)|^2 \leq 4\phi_h^3(x),$$

we find that, for some constant c_{12},

$$|\phi_h v_h|_{\phi_h^{-1};1,\Omega} \leq c_{12}(|v_h|_{\phi_h^2;0,\Omega} + |v_h|_{\phi_h;1,\Omega}). \tag{3.3.35}$$

Combining relations (3.3.30) to (3.3.35), we have found that, for $\theta_h \geq c_{10}h$,

$$|v_h|^2_{\phi_h;1,\Omega} \leq 2|v_h|^2_{\phi_h^2;0,\Omega} + c_{12}|v|_{\phi_h;1,\Omega}(|v_h|_{\phi_h^2;0,\Omega} + |v_h|_{\phi_h;1,\Omega})$$
$$+ c_{11}\frac{h}{\theta_h}(|v|_{\phi_h;1,\Omega} + |v_h|_{\phi_h;1,\Omega})(|v_h|_{\phi_h^2;0,\Omega} + |v_h|_{\phi_h;1,\Omega}),$$

i.e., an inequality of the form

$$A^2 \leq 2C^2 + c_{12}B(A + C) + c_{11}\frac{h}{\theta_h}(A + B)(A + C).$$

Assuming

$$\forall h, \quad \theta_h \geq 2c_{11}h, \tag{3.3.36}$$

we get

$$A^2 \leq 4C^2 + (1 + 2c_{12})BC + A((1 + 2c_{12})B + C)$$
$$\leq 4C^2 + \left(\frac{1}{2} + c_{12}\right)(B^2 + C^2) + \frac{A^2}{2} + (1 + 2c_{12})^2B^2 + C^2,$$

and therefore step (i) is proved with (cf. (3.3.32) and (3.3.36))

$$C_{11} = \max(c_{10}, 2c_{11}) \tag{3.3.37}$$

in relation (3.3.28) and

$$C_{12} = \max\{11 + 2c_{12}, (1 + 2c_{12})(3 + 4c_{12})\} \tag{3.3.38}$$

in relation (3.3.29).

(ii) *There exists a constant* C_{13} *such that, if we assume* $\theta_h \geq C_{11}h$ *(the constant* C_{11} *has been determined in step* (i)), *we have*

$$\forall v \in H_0^1(\Omega), |P_hv|^2_{\phi_h^2;0,\Omega} + |P_hv|^2_{\phi_h;1,\Omega} \leq$$
$$\leq C_{13}(|v|^2_{\phi_h^2;0,\Omega} + |v|^2_{\phi_h;1,\Omega} + h^2|\psi_h|^2_{\phi_h^{-1};2,\Omega}), \tag{3.3.39}$$

where, for each h, $\psi_h = \psi_h(v)$ *is the solution of the variational problem*:

$$\psi_h \in H_0^1(\Omega) \quad \text{and} \quad \forall w \in H_0^1(\Omega),$$
$$\int_\Omega \nabla\psi_h \cdot \nabla w \,\mathrm{d}x = \int_\Omega \phi_h^2(P_hv)w \,\mathrm{d}x. \tag{3.3.40}$$

Notice that because the set Ω is assumed to be convex, the function

ψ_h is in the space $H^2(\Omega)$ and therefore, it is legitimate to consider the semi-norm $|\cdot|_{\phi_h^{-1};2,\Omega}$ in inequality (3.3.39).

Using the definition of the function ψ_h, and letting again $v_h = P_h v$, we can write

$$|v_h|^2_{\phi_h^2;0,\Omega} = a(v_h - v, \psi_h - \Pi_h\psi_h) + \int_\Omega \phi_h^2 v_h v \,\mathrm{d}x. \tag{3.3.41}$$

By applying inequality (3.3.5) with $\alpha = 1$ and inequality (3.3.20) with $\alpha = -1$ (this is possible because we assume $\theta_h \geqslant C_{11}h$ and $C_{11} \geqslant c_{10} = C_7(-1)$; cf. (3.3.32) and (3.3.37)), we obtain

$$\begin{aligned} &|a(v_h - v, \psi_h - \Pi_h\psi_h)| \leqslant \\ &\quad \leqslant C_7 h(|v_h|_{\phi_h;1,\Omega} + |v|_{\phi_h;1,\Omega})|\psi_h|_{\phi_h^{-1};2,\Omega}, \end{aligned} \tag{3.3.42}$$

Next we have

$$\int_\Omega \phi_h^2 v_h v \,\mathrm{d}x \leqslant |v_h|_{\phi_h^2;0,\Omega}|v|_{\phi_h^2;0,\Omega} \leqslant \frac{1}{2}(|v_h|^2_{\phi_h^2;0,\Omega} + |v|^2_{\phi_h^2;0,\Omega}), \tag{3.3.43}$$

so that, by combining relations (3.3.41), (3.3.42) and (3.3.43), we obtain the inequality

$$\begin{aligned} |v_h|^2_{\phi_h^2;0,\Omega} \leqslant\ & C_7 h(|v_h|_{\phi_h;1,\Omega} + |v|_{\phi_h;1,\Omega})|\psi_h|_{\phi_h^{-1};2,\Omega} \\ & + \frac{1}{2}(|v_h|^2_{\phi_h^2;0,\Omega} + |v|^2_{\phi_h^2;0,\Omega}), \end{aligned}$$

which in turn implies the inequality

$$\begin{aligned} \forall \delta > 0, \quad |v_h|^2_{\phi_h^2;0,\Omega} \leqslant\ & \delta|v_h|^2_{\phi_h;1,\Omega} + |v|^2_{\phi_h^2;0,\Omega} + |v|^2_{\phi_h;1,\Omega} \\ & + \left(1 + \frac{1}{\delta}\right)C_7^2 h^2 |\psi_h|^2_{\phi_h^{-1};2,\Omega}. \end{aligned} \tag{3.3.44}$$

Let then $\delta = 1/(3C_{12})$, where C_{12} is the constant appearing in inequality (3.3.29). The corresponding inequality (3.3.44) added to inequality (3.3.29) times the factor $2/(3C_{12})$ yields

$$\begin{aligned} &\frac{1}{3}|v_h|^2_{\phi_h^2;0,\Omega} + \frac{1}{3C_{12}}|v_h|^2_{\phi_h;1,\Omega} \leqslant \\ &\quad \leqslant |v|^2_{\phi_h^2;0,\Omega} + \frac{5}{3}|v|^2_{\phi_h;1,\Omega} + (1 + 3C_{12})C_7^2 h^2 |\psi_h|^2_{\phi_h^{-1};2,\Omega}, \end{aligned}$$

i.e., an inequality of the form (3.3.39).

(iii) *Given any number $\theta_0 \in]0, 1[$, there exists a constant $C_{14}(\theta_0)$ such that*

$$\forall \theta_h \in]0, \theta_0[, \quad |\psi_h|^2_{\phi_h^{-1};2,\Omega} \leq C_{14}(\theta_0)\frac{|\ln \theta_h|}{\theta_h^2}|P_h v|^2_{\phi_h^2;0,\Omega}. \tag{3.3.45}$$

Since $-\Delta\psi_h = \phi_h^2 v_h (v_h = P_h v)$, we have

$$|\Delta\psi_h|_{\phi_h^{-1};0,\Omega} = |v_h|_{\phi_h^3;0,\Omega},$$

and consequently, by Theorem 3.3.1,

$$|\psi_h|^2_{\phi_h^{-1};2,\Omega} \leq C_1(|v_h|^2_{\phi_h^3;0,\Omega} + |\psi_h|^2_{1,\Omega}). \tag{3.3.46}$$

Since $\phi_h(x) \leq 1/\theta_h^2$ for all $x \in \bar{\Omega}$, we first find that

$$|v_h|^2_{\phi_h^3;0,\Omega} \leq \frac{1}{\theta_h^2}|v_h|^2_{\phi_h^2;0,\Omega}. \tag{3.3.47}$$

To take care of the other term which appears in the right-hand side of inequality (3.3.46), we shall prove that *for each number $\theta_0 \in]0, 1[$, there exists a constant $c_{13}(\theta_0)$ such that, for all functions ϕ of the form*

$$\phi : x \in \bar{\Omega} \to \phi(x) = \frac{1}{\|x - \bar{x}\|^2 + \theta^2}, \quad \bar{x} \in \bar{\Omega}, \quad 0 < \theta \leq \theta_0, \tag{3.3.48}$$

we have

$$\forall \psi \in H_0^1(\Omega) \cap H^2(\Omega), \quad |\psi|^2_{1,\Omega} \leq c_{13}(\theta_0)\frac{|\ln \theta|}{\theta^2}|\Delta\psi|^2_{\phi^{-2};0,\Omega}. \tag{3.3.49}$$

Taking into account that

$$|\Delta\psi_h|_{\phi_h^{-2};0,\Omega} = |v_h|_{\phi_h^2;0,\Omega},$$

and applying inequalities (3.3.49) (with $\psi = \psi_h$ and $\phi = \phi_h$), (3.3.46), and (3.3.47), we then find an inequality of the form (3.3.45), with

$$C_{14}(\theta_0) = C_1\left(\frac{1}{|\ln \theta_0|} + c_{13}(\theta_0)\right). \tag{3.3.50}$$

It therefore remains to prove relation (3.3.49) (another method for proving the same relation is suggested in Exercise 3.3.1). Given an arbitrary function $\psi \in H_0^1(\Omega) \cap H^2(\Omega)$, we have

$$\begin{aligned} |\psi|^2_{1,\Omega} &= -\int_\Omega \psi\, \Delta\psi \,\mathrm{d}x \leq |\Delta\psi|_{\phi^{-2};0,\Omega}|\psi|_{\phi^2;0,\Omega} \\ &\leq \frac{1}{2}\frac{|\ln \theta|}{\theta^2}|\Delta\psi|^2_{\phi^{-2};0,\Omega} + \frac{1}{2}\frac{\theta^2}{|\ln \theta|}|\psi|^2_{\phi^2;0,\Omega}. \end{aligned} \tag{3.3.51}$$

Let then G denote the Green's function associated with the operator $-\Delta$ in Ω and the boundary condition $v = 0$ on Ω, so that

$$\begin{aligned}|\psi|^2_{\phi^2;0,\Omega} &= \int_\Omega \phi^2(x)\left|\int_\Omega G(x,\xi)\Delta\psi(\xi)\,\mathrm{d}\xi\right|^2 \mathrm{d}x \\ &\leqslant \int_\Omega \phi^{-2}(\xi)|\Delta\psi(\xi)|^2\left\{\int_\Omega \phi^2(x)G(x,\xi)\times\right. \\ &\quad \times\left.\left\{\int_\Omega \phi^2(\eta)G(x,\eta)\,\mathrm{d}\eta\right\}\mathrm{d}x\right\}\mathrm{d}\xi. \end{aligned} \tag{3.3.52}$$

There exists a constant c_{14} such that (cf. for example STAKGOLD (1968, p. 143))

$$\forall x, y \in \bar{\Omega},\quad x \neq y,\quad 0 \leqslant G(x,y) \leqslant c_{14}|\ln\|x-y\||. \tag{3.3.53}$$

Using this inequality, we proceed to show that for arbitrary points $x, \bar{x} \in \Omega$ and for any number θ with $0 < \theta \leqslant \theta_0 < 1$, there exists a constant $c_{15}(\theta_0)$ such that

$$\int_\Omega \varphi^2(\eta)G(x,\eta)\,\mathrm{d}\eta = \int_\Omega \frac{G(x,\eta)}{(\|\eta-\bar{x}\|^2+\theta^2)^2}\,\mathrm{d}\eta \leqslant c_{15}(\theta_0)\frac{|\ln\theta|}{\theta^2}. \tag{3.3.54}$$

To see this, write

$$\int_\Omega \frac{|\ln\|x-\eta\||}{(\|\eta-\bar{x}\|^2+\theta^2)^2}\,\mathrm{d}\eta = \sum_{\lambda=1}^{3}\int_{\Omega_\lambda}\frac{|\ln\|x-\eta\||}{(\|\eta-\bar{x}\|^2+\theta^2)^2}\,\mathrm{d}\eta,$$

where

$$\Omega_1 = \Omega_1(x,\theta) = \{\eta \in \Omega;\ \|\eta - x\| \leqslant \theta\},$$

$$\Omega_2 = \Omega_2(x,\theta) = \{\eta \in \Omega;\ \theta \leqslant \|\eta - x\| \leqslant 1\},$$

$$\Omega_3 = \Omega_3(x) = \{\eta \in \Omega;\ 1 \leqslant \|\eta - x\|\}.$$

We then obtain the following inequalities (observe that the last two inequalities make sense only if the sets Ω_2 and Ω_3 are not empty, and that we have diam $\Omega \geqslant 1$ if the set Ω_3 is not empty):

$$\begin{aligned}\int_{\Omega_1}\frac{|\ln\|x-\eta\||}{(\|\eta-\bar{x}\|^2+\theta^2)^2}\,\mathrm{d}\eta &\leqslant -\frac{1}{\theta^4}\int_{\Omega_1}\ln\|\eta-x\|\,\mathrm{d}\eta \\ &= -\frac{1}{\theta^4}\int_{B(0;\theta)}\ln\|\xi\|\,\mathrm{d}\xi \\ &= \frac{\pi}{\theta^2}\left(\frac{1}{2}-\ln\theta\right) \leqslant \pi\left(1+\frac{1}{2|\ln\theta_0|}\right)\frac{|\ln\theta|}{\theta^2},\end{aligned}$$

$$\int_{\Omega_2} \frac{|\ln \|x-\eta\||}{(\|\eta-\bar{x}\|^2+\theta^2)^2}\,\mathrm{d}\eta \leqslant |\ln \theta| \int_{\Omega_2} \frac{\mathrm{d}\eta}{(\|\eta-\bar{x}\|^2+\theta^2)^2}$$
$$\leqslant |\ln \theta| \int_{\mathbf{R}^2} \frac{\mathrm{d}\eta}{(\|\eta-\bar{x}\|^2+\theta^2)^2}$$
$$= \frac{|\ln \theta|}{\theta^2} \int_{\mathbf{R}^2} \frac{\mathrm{d}\xi}{(1+\|\xi\|^2)^2} = \pi \frac{|\ln \theta|}{\theta^2},$$
$$\int_{\Omega_3} \frac{|\ln \|x-\eta\||}{(\|\eta-\bar{x}\|^2+\theta^2)^2}\,\mathrm{d}\eta \leqslant \ln(\operatorname{diam} \Omega) \int_{\Omega_3} \frac{\mathrm{d}\eta}{(\|\eta-\bar{x}\|^2+\theta^2)^2}$$
$$\leqslant \ln(\operatorname{diam} \Omega) \int_{\mathbf{R}^2} \frac{\mathrm{d}\eta}{(\|\eta-\bar{x}\|^2+\theta^2)^2}$$
$$= \frac{\pi}{\theta^2} \ln(\operatorname{diam} \Omega) \leqslant \left(\frac{\pi \ln(\operatorname{diam} \Omega)}{|\ln \theta_0|}\right) \frac{|\ln \theta|}{\theta^2}.$$

Consequently, inequality (3.3.54) is proved, with

$$c_{15}(\theta_0) = \pi c_{14}\left\{2 + \frac{1+2\ln(\operatorname{diam} \Omega)}{2|\ln \theta_0|}\right\}.$$

The conjunction of inequalities (3.3.51) to (3.3.54) then implies inequality (3.3.49) with

$$c_{13}(\theta_0) = \frac{1}{2}(1 + c_{15}^2(\theta_0)).$$

(iv) *It remains to combine the results of steps* (ii) *and* (iii): We have determined constants C_{11}, C_{13} and $C_{14}(\theta_0)$ for each $\theta_0 \in]0, 1[$ such that (cf. inequalities (3.3.39) and (3.3.45))

$$C_{11}h \leqslant \theta_h \leqslant C_0 < 1 \Rightarrow |P_h v|^2_{\phi_h^2;0,\Omega} + |P_h v|^2_{\phi_h;1,\Omega} \leqslant$$
$$\leqslant C_{13}(|v|^2_{\phi_h^2;0,\Omega} + |v|^2_{\phi_h;1,\Omega}) + C_{13}C_{14}(\theta_0)\frac{|\ln \theta_h| h^2}{\theta_h^2} |P_h v|^2_{\phi_h^2,0,\Omega}. \tag{3.3.55}$$

Let for example $\theta_0 = \frac{1}{2}$ and let

$$\theta_h = C_9 h |\ln h|^{1/2} \quad \text{with} \quad C_9 = 2\left(C_{13}C_{14}\left(\frac{1}{2}\right)\right)^{1/2}. \tag{3.3.56}$$

Then there exists a number $h_0 \in]0, 1[$ such that

$$h \leqslant h_0 \Rightarrow \begin{cases} C_{11}h \leqslant \theta_h \leqslant \frac{1}{2} = \theta_0, \\ |\ln \theta_h| \leqslant 2|\ln h|. \end{cases} \tag{3.3.57}$$

This being the case, we have found an inequality of type (3.3.27) with

$$C_{10} = 2\sqrt{C_{13}}. \tag{3.3.58}$$

□

Estimates of the errors $|u - u_h|_{0,\infty,\Omega}$ and $|u - u_h|_{1,\infty,\Omega}$. Nitsche's method of weighted norms

We next develop the *second stage* of our analysis. Using the inequalities (cf. Theorems 3.3.2 and 3.3.3) between the semi-norms $|\cdot|_{m,\infty,\Omega}$, $m = 0, 1$, and the weighted semi-norms which appear in inequality (3.3.27), we show in the next theorem that *the projection operators P_h* of (3.3.22), *considered as acting from the subspace $H_0^1(\Omega) \cap W^{1,\infty}(\Omega)$ of the space $H_0^1(\Omega)$ onto the space V_h, are bounded independently of h when the space $H_0^1(\Omega) \cap W^{1,\infty}(\Omega)$ is equipped with the norm*

$$v \to |v|_{0,\infty,\Omega} + h|\ln h|\,|v|_{1,\infty,\Omega} \tag{3.3.59}$$

and the space V_h is equipped with the norm

$$v \to |\ln h|^{-1/2}|v|_{0,\infty,\Omega} + h|v|_{1,\infty,\Omega}. \tag{3.3.60}$$

Remark 3.3.1. Such norms may be viewed as "weighted $W^{1,\infty}(\Omega)$-like" norms. □

Theorem 3.3.6. *There exists a constant C_{15} such that*

$$\begin{aligned} &\forall h \leq h_0, \quad \forall v \in H_0^1(\Omega) \cap W^{1,\infty}(\Omega), \\ &\quad |\ln h|^{-1/2}|P_h v|_{0,\infty,\Omega} + h|P_h v|_{1,\infty,\Omega} \leq \\ &\quad \leq C_{15}(|v|_{0,\infty,\Omega} + h|\ln h|\,|v|_{1,\infty,\Omega}), \end{aligned} \tag{3.3.61}$$

where the constant $h_0 > 0$ has been determined in Theorem 3.3.5.

Proof. Let there be given a function v in the space $H_0^1(\Omega) \cap W^{1,\infty}(\Omega)$. For each $h \leq h_0$, we define the function

$$\phi_{0h}: x \in \bar{\Omega} \to \phi_{0h}(x) = \frac{1}{\|x - x_h^0\|^2 + \theta_h^2}, \tag{3.3.62}$$

with

$$|P_h v(x_h^0)| = |P_h v|_{0,\infty,\Omega} \quad \text{and} \quad \theta_h = C_9 h|\ln h|^{1/2}, \tag{3.3.63}$$

where h_0 and C_9 are the constants determined in Theorem 3.3.5. Since $\theta_h \geqslant C_{11}h$ for $h \leqslant h_0$ (cf. (3.3.57), we may apply inequality (3.3.14)): There exists a constant

$$c_{16} = C_4(C_{11}) \tag{3.3.64}$$

such that

$$|P_h v|_{0,\infty,\Omega} \leqslant c_{16}\frac{\theta_h^2}{h}|P_h v|_{\phi_{0h}^2;0,\Omega}. \tag{3.3.65}$$

By inequality (3.3.27),

$$|P_h v|_{\phi_{0h}^2;0,\Omega} \leqslant C_{10}(|v|_{\phi_{0h}^2;0,\Omega} + |v|_{\phi_{0h};1,\Omega}), \tag{3.3.66}$$

and by inequalities (3.3.10) and (3.3.11), there exists a constant ($\theta_h \leqslant \theta_0 = \frac{1}{2}$ for $h \leqslant h_0$; cf. (3.3.57))

$$c_{17} = \max\{C_2(2, 0), \quad C_3(\tfrac{1}{2}, 1)\} \tag{3.3.67}$$

such that

$$|v|_{\phi_{0h}^2;0,\Omega} + |v|_{\phi_{0h};1,\Omega} \leqslant c_{17}\left(\frac{1}{\theta_h}|v|_{0,\infty,\Omega} + |\ln\,\theta_h|^{1/2}|v|_{1,\infty,\Omega}\right). \tag{3.3.68}$$

Combining inequalities (3.3.65) to (3.3.68), we find that

$$|P_h v|_{0,\infty,\Omega} \leqslant C_{10}c_{16}c_{17}\left(\frac{\theta_h}{h}|v|_{0,\infty,\Omega} + \frac{\theta_h^2|\ln\,\theta_h|^{1/2}}{h}|v|_{1,\infty,\Omega}\right).$$

Using the relation $\theta_h = C_9 h|\ln\,h|^{1/2}$ (cf. (3.3.26)) and the inequality $|\ln\,\theta_h| \leqslant 2|\ln\,h|$ (cf. (3.3.57)), we eventually get, for all $h \leqslant h_0$,

$$|\ln\,h|^{-1/2}|P_h v|_{0,\infty,\Omega} \leqslant c_{18}(|v|_{0,\infty,\Omega} + h|\ln\,h|\,|v|_{1,\infty,\Omega}), \tag{3.3.69}$$

with

$$c_{18} = C_{10}c_{16}c_{17}\max\{C_9, \sqrt{2}\,C_9^2\}. \tag{3.3.70}$$

Likewise, for each $h \leqslant h_0$, define the function

$$\phi_{1h}: x \in \bar{\Omega} \to \phi_{1h}(x) = \frac{1}{\|x - x_h^1\|^2 + \theta_h^2}, \tag{3.3.71}$$

with

$$\max\{|\partial_1 P_h v(x_h^1), \partial_2 P_h v(x_h^1)|\} = |v_h|_{1,\infty,\Omega} \text{ and } \theta_h = C_9 h|\ln\,h|^{1/2}. \tag{3.3.72}$$

Then there exists (cf. inequality (3.3.16)) a constant

$$c_{19} = C_5(C_{11}) \tag{3.3.73}$$

such that

$$|P_h v|_{1,\infty,\Omega} \leq c_{19} \frac{\theta_h}{h} |P_h v|_{\phi_{1h};1,\Omega}, \tag{3.3.74}$$

and, by inequality (3.3.27),

$$|P_h v|_{\phi_{1h};1,\Omega} \leq C_{10}(|v|_{\phi_{1h}^2;0,\Omega} + |v|_{\phi_{1h};1,\Omega}). \tag{3.3.75}$$

Then, arguing as before, we get, for all $h \leq h_0$,

$$h|P_h v|_{1,\infty,\Omega} \leq c_{20}(|v|_{0,\infty,\Omega} + h|\ln h|\, |v|_{1,\infty,\Omega}) \tag{3.3.76}$$

with

$$c_{20} = C_{10} c_{17} c_{19} \max\{1, \sqrt{2}\, C_9\}. \tag{3.3.77}$$

The conjunction of inequalities (3.3.69) and (3.3.76) implies inequality (3.3.61) with

$$C_{15} = c_{18} + c_{20}.$$ □

Remark 3.3.2. In Theorem 3.3.5, the behavior of the function θ_h was somehow "bounded below" by a constant times $(h|\ln h|^{1/2})$. The key to the success of the present argument was that such a function θ_h tends nevertheless sufficiently rapidly vers zero with h so as to produce the right factors (as functions of h) in the inequalities (3.3.69) and (3.3.76). □

In the *third* – and final – *stage* of our study, the uniform boundedness of the projection mappings P_h which we just established allows us in turn to easily derive the desired error estimates (recall that the discrete solution u_h is nothing but the projection $P_h u$ of the solution u).

Theorem 3.3.7. *Assume that the solution $u \in H_0^1(\Omega)$ of the boundary value problem associated with the data (3.3.1) is also in the space $W^{2,\infty}(\Omega)$.*

Then there exists a constant C independent of h such that

$$|u - u_h|_{0,\infty,\Omega} \leq Ch^2|\ln h|^{3/2}|u|_{2,\infty,\Omega}, \tag{3.3.78}$$

$$|u - u_h|_{1,\infty,\Omega} \leq Ch|\ln h|\, |u|_{2,\infty,\Omega}. \tag{3.3.79}$$

Proof. The norm of the identity mapping acting from the space $H_0^1(\Omega) \cap W^{1,\infty}(\Omega)$ equipped with the norm of (3.3.59) into the same space, but equipped with the norm of (3.3.60), is bounded above by $|\ln h_0|^{-1/2}$ for all $h \leq h_0' = \min\{h_0, 1/e\}$.

Next we have the identity

$$\forall v_h \in V_h, \quad u - u_h = u - P_h u = (I - P_h)(u - v_h),$$

so that we infer from Theorem 3.3.6 that, for all $h \leq h_0'$,

$$|\ln h|^{-1/2} |u - u_h|_{0,\infty,\Omega} + h|u - u_h|_{1,\infty,\Omega} \leq$$

$$\leq (|\ln h_0|^{-1/2} + C_{15}) \inf_{v_h \in V_h} (|u - v_h|_{0,\infty,\Omega} + h|\ln h|\, |u - v_h|_{1,\infty,\Omega})$$

Since there exists a constant c_{21} such that

$$\inf_{v_h \in V_h} (|u - v_h|_{0,\infty,\Omega} + h|\ln h|\, |u - v_h|_{1,\infty,\Omega}) \leq c_{21} h^2 |\ln h|\, |u|_{2,\infty,\Omega},$$

inequalities (3.3.78) and (3.3.79) follow with

$$C = c_{21}(|\ln h_0|^{-1/2} + c_{15}).$$

□

In fact, the error estimate of (3.3.78) is not optimal: J.A. NITSCHE (1976b) gets the improved error bound

$$|u - u_h|_{0,\infty,\Omega} \leq C h^2 |\ln h|\, |u|_{2,\infty,\Omega}, \tag{3.3.80}$$

at the expense, however, of a technical refinement in the argument, special to triangles of type (1). At any rate, the discrepancy between (3.3.78) and (3.3.80) is somehow insignificant: Both error estimates (3.3.78) and (3.3.80) show an $O(h^{2-\epsilon})$ convergence for any $\epsilon > 0$.

To conclude, we point out that all the essential features of *Nitsche's method of weighted norms* have been presented: Indeed, the extension to more general cases proceeds along the same lines. In particular, the use of higher-order polynomial spaces (i.e., $P_K = P_k(K)$ for some $k \geq 2$, n arbitrary) yields a simplification in that the "$|\ln h|$" term present for $k = 1$ disappears in the norms then considered. Thus inequality (3.3.61) is replaced by an inequality of the simpler form (cf. NITSCHE (1975))

$$|P_h v|_{0,\infty,\Omega} + h|P_h v|_{1,\infty,\Omega} \leq C(|v|_{0,\infty,\Omega} + h|v|_{1,\infty,\Omega}). \tag{3.3.81}$$

Such inequalities are obtained after inequalities reminiscent of that of (3.3.27) have been established for appropriate weighted norms of the form $|\cdot|_{\phi_h^{\alpha+1};0,\Omega} + |\cdot|_{\phi_h^{\alpha};1,\Omega}$, $(n/2) < \alpha < (n/2) + 1$, with functions ϕ_h again defined as in (3.3.25).

Exercises

3.3.1. Following NITSCHE (1977), the object of this problem is to provide another proof of inequality (3.3.49), i.e., that for any $\theta_0 \in {]0, 1[}$, there exists a constant $c(\theta_0)$ such that, for any function ϕ of the form

$$\phi : x \in \bar{\Omega} \to \phi(x) = \frac{1}{\|x - \bar{x}\|^2 + \theta^2}, \qquad \bar{x} \in \bar{\Omega}, \quad 0 < \theta \leq \theta_0,$$

we have

$$\forall \psi \in H_0^1(\Omega) \cap H^2(\Omega), \quad |\psi|_{1,\Omega}^2 \leq c(\theta_0) \frac{|\ln \theta|}{\theta^2} |\Delta\psi|_{\phi^{-2};0,\Omega}^2.$$

(i) Let

$$\lambda(\Omega) = \inf_{\psi \in H_0^1(\Omega) \cap H^2(\Omega)} \frac{|\Delta\psi|_{\phi^{-2};0,\Omega}^2}{|\psi|_{1,\Omega}^2}$$

and show that $\lambda(\Omega)$ is the smallest eigenvalue of the eigenvalue problem

$$\begin{cases} -\Delta u = \lambda \phi^2 u \text{ in } \Omega, \\ u = 0 \text{ on } \Gamma, \end{cases}$$

so that $\lambda(\Omega)$ is a strictly positive quantity (references about eigenvalue problems can be found in the section "Additional Bibliography and Comments" at the end of Chapter 4).

(ii) Let $\tilde{\Omega} = B(\bar{x}; \operatorname{diam}(\Omega))$ and show by a direct computation that

$$\frac{1}{\lambda(\tilde{\Omega})} \leq c(\theta_0) \frac{|\ln \theta|}{\theta^2}.$$

(iii) Conclude, by using the implication

$$\Omega_1 \subset \Omega_2 \Rightarrow \lambda(\Omega_2) \leq \lambda(\Omega_1).$$

3.3.2. The object of this exercise is to show how an error estimate in the norm $|\cdot|_{0,\infty,\Omega}$ can be quickly derived, once one is willing to accept a poorer order of convergence than that obtained in Theorem 3.3.7. The terminology is the same as in Section 3.2.

In addition to (H1), (H2) and (H3), assume that $s = 0$, that the dimension n is $\leqslant 3$, that the family of triangulations satisfies an inverse assumption, and that the inclusions

$$P_1(\hat{K}) \subset \hat{P} \subset H^1(\hat{K})$$

hold (notice that by hypothesis (H3), we also have $\hat{P} \subset L^\infty(\hat{K})$).

Let then u_h be the corresponding discrete solution which approximates the solution u of a general second order boundary value problem of the type considered in Section 3.2. Show that if the adjoint problem is regular, and if $u \in H^2(\Omega) \cap V$, there exists a constant C independent of h such that

$$|u - u_h|_{0,\infty,\Omega} \leqslant Ch|u|_{2,\Omega} \text{ if } n = 2,$$
$$|u - u_h|_{0,\infty,\Omega} \leqslant C\sqrt{h}\,|u|_{2,\Omega} \text{ if } n = 3.$$

[Hint: write $|u - u_h|_{0,\infty,\Omega} \leqslant |u_h - \Pi_h u|_{0,\infty,\Omega} + |u - \Pi_h u|_{0,\infty,\Omega}$ and use an appropriate inverse inequality for the first term.]

Bibliography and comments

3.1. The content of this section is essentially based on, and slightly improved upon, CIARLET & RAVIART (1972a). In particular, R. Arcangéli suggested the simpler proof of inequality (3.1.33) given here.

For reference about the Sobolev spaces $W^{m,p}(\Omega)$ and their various properties, see ADAMS (1975), LIONS (1962), NEČAS (1967), ODEN & REDDY (1976a, chapter 3). Theorem 3.1.1 was originally proved in DENY & LIONS (1953–1954) for open sets which satisfy the "cone property" (such sets are slightly more general than those with Lipschitz-continuous boundaries). An abstract extension of this lemma is indicated in Exercise 3.1.1.

There has been considerable interest in interpolation theory and approximation theory in several variables during the past decade, one reason behind this recent interest being the need of such theories for studying convergence properties of finite element methods. Special mention must be made however of the pioneering works of PÓLYA (1952) and SYNGE (1957), for what we call here rectangles of type (1) and triangles of type (1), respectively.

The "classical" approach consists in obtaining error estimates in C^m-norms. In this direction, see the contributions of BARNHILL &

GREGORY (1976b), BARNHILL & WHITEMAN (1973), BIRKHOFF (1971, 1972), BIRKHOFF, SCHULTZ & VARGA (1968), CARLSON & HALL (1973), CIARLET & RAVIART (1972a), CIARLET & WAGSCHAL (1971), COATMÉLEC (1966), LEAF & KAPER (1974), NICOLAIDES (1972, 1973), NIELSON (1973), SCHULTZ (1969b, 1973), STRANG (1971, 1972a), ŽENÍŠEK (1970, 1973), ZLÁMAL (1968, 1970). Although in most cases a special role is played by the canonical Cartesian coordinates, a more powerful coordinate-free approach, using Fréchet derivatives, can be developed, such as in CIARLET & RAVIART (1972a), CIARLET & WAGSCHAL (1971), where the interpolation error estimates are obtained as corollaries of multi-point Taylor formulas (Exercise 3.1.2). See also COATMÉLEC (1966). Another frequently used tool is the kernel Theorem of SARD (1963).

Some authors have considered the problem of estimating the constants which appear in the interpolation error estimates. See ARCANGÉLI & GOUT (1976) (cf. Exercise 3.1.2), ATTÉIA (1977), BARNHILL & WHITEMAN (1973), GOUT (1976), MEINGUET (1975), MEINGUET & DESCLOUX (1977).

The approach in Sobolev spaces which has been followed here has been given much attention. In this respect, we quote the fundamental contributions of BRAMBLE & HILBERT (1970, 1971), BRAMBLE & ZLÁMAL (1970). Other relevant references are AUBIN (1967a, 1967b, 1968a, 1968b, 1972), BABUŠKA (1970, 1972b), BIRKHOFF, SCHULTZ & VARGA (1968), BRAMBLE (1970), CIARLET & RAVIART (1972a), FIX & STRANG (1969), DI GUGLIELMO (1970), HEDSTROM & VARGA (1971), KOUKAL (1973), NITSCHE (1969, 1970), SCHULTZ (1969b), VARGA (1971).

Interesting connections with standard spline theory can be found in ATTÉIA (1975), MANSFIELD (1972b), NIELSON (1973) and, especially, DUCHON (1976a, 1976b).

The dependence of the interpolation error estimates upon the geometry of the element (through the parameters h_K and ρ_K) generalize Zlámal's condition, as given in ZLÁMAL (1968, 1970), and the "uniformity condition" of STRANG (1972a). JAMET (1976a) has recently shown (cf. Exercise 3.1.4) that, for some finite elements at least, the regularity condition given in (3.1.43) can be replaced by a less stringent one. In a special case, the same condition has been simultaneously and independently found by BABUŠKA & AZIZ (1976). In essence, it amounts to saying, in case of triangles, that no angle of the triangle should approach π in the limit while by the present analysis no angle should approach 0 in the limit. Incidentally, this was already observed by SYNGE (1957).

3.2. and 3.3. There exists a very large literature on various possible error estimates one can get for conforming finite element method and here we shall merely record several lists of references, depending upon the viewpoints.

We shall first observe that almost all the papers previously referred to in Section 3.1 also contributed to the error analysis in the norm $\|\cdot\|_{1,\Omega}$, inasmuch as this simply requires a straightforward application of Céa's lemma, just as we did in Theorem 3.2.2.

Even though the X_h-interpolation operator cannot be defined for lack of regularity of the function to be approximated, an approximation theory can still be developed, as in CLÉMENT (1975), HILBERT (1973), PINI (1974), STRANG (1972a). See Exercise 3.2.3 where we have indicated the approach of Ph. Clément.

Historically, the first proof of convergence of a finite element method, albeit in a special case, seems to be due to FRIEDRICHS (1962). Early works on convergence in the engineering literature are JOHNSON & MCLAY (1968), MCLAY (1963), OLIVEIRA (1968, 1969).

The reader who wishes to get general introductions to, and surveys on, the various aspects of the convergence of the finite element method may consult BIRKHOFF & FIX (1974), CARLSON & HALL (1971), CIARLET (1973), FELIPPA & CLOUGH (1970), KIKUCHI (1975c), ODEN (1975), STRANG (1972a, 1974b), THOMÉE (1973a), VEIDINGER (1974), ZLÁMAL (1973c).

Using a priori estimates (in various norms) on the solution (cf. NEČAS (1967) and KONDRAT'EV (1967)), it is possible to get error estimates which depend solely on the data of the problem. See BRAMBLE & ZLÁMAL (1970), NITSCHE (1970), OGANESJAN & RUKHOVETS (1969). In the case of the equation $-\Delta u = f$ over a rectangle, BARNHILL & GREGORY (1976a) obtain theoretical values for the constants which appear in the error estimate, which are realistic, as shown in BARNHILL, BROWN, MCQUEEN & MITCHELL (1976).

"Nonuniform" error estimates are obtained in BABUŠKA & KELLOG (1975), HELFRICH (1976). The case of indefinite bilinear forms is considered in CLÉMENT (1974), SCHATZ (1974). DOUGLAS, DUPONT & WHEELER (1974a) give estimates for the flux on the boundary. HOPPE (1973) has suggested the use of piecewise harmonic polynomials, and his idea has been justified by RABIER (1977). See also BABUŠKA (1974a), ROSE (1975).

There are various ways of treating nonhomogeneous Dirichlet boundary conditions. The most straightforward method is suggested in Exercise 3.2.1. See AUBIN (1972), STRANG & FIX (1973, Section 4.4), THOMÉE (1973a). Lagrange multipliers may also be used as in BABUŠKA (1973a), as well as penalty techniques (cf. Exercise 3.2.2) as in BABUŠKA (1973b). See also the section "Bibliography and Comments", Section 4.4.

For domains with corners or, more generally, for problems where the solution presents singularities, see BARNHILL & WHITEMAN (1973, 1975), BABUŠKA (1972a, 1974b, 1976), BABUŠKA & ROSENZWEIG (1972), BARSOUM (1976), CIARLET, NATTERER & VARGA (1970), CROUZEIX & THOMAS (1973), DAILEY & PIERCE (1972), FIX (1969), FIX, GULATI & WAKOFF (1973), FRIED & YANG (1972), HENNART & MUND (1976), NITSCHE (1976a), SCHATZ & WAHLBIN (1976a), SCOTT (1973b), STRANG & FIX (1973, Chapter 8), THATCHER (1976), VEIDINGER (1972), WAIT & MITCHELL (1971). Recent references in the engineering literature are HENSHELL & SHAW (1975), YAMAMOTO & SUMI (1976).

For further results concerning the error estimates in the norm $\|\cdot\|_{1,\Omega}$, see BABUŠKA & AZIZ (1972, Section 6.4) where it is notably discussed whether they are the best possible, using the theory of n-widths.

Many "abstract" finite element methods, or variants thereof, have been considered, by AUBIN (1967b, 1972), BABUŠKA (1970, 1971a, 1971b, 1972b), FIX & STRANG (1969), DI GUGLIELMO (1971), MOCK (1976), STRANG (1971), STRANG & FIX (1971).

The inverse inequalities established in Section 3.2 are found in many places. See notably DESCLOUX (1973).

The technique which yields the error estimate in the norm $|\cdot|_{0,\Omega}$ was developed independently by AUBIN (1967b) and NITSCHE (1968), and also by OGANESJAN & RUKHOVETS (1969). See KIKUCHI (1975c) for a generalization.

The subject of uniform convergence has a (relatively) long story. In one dimension, we mention NITSCHE (1969), CIARLET (1968), CIARLET & VARGA (1970), and the recent contributions of DOUGLAS & DUPONT (1973, 1976b), DOUGLAS, DUPONT & WAHLBIN (1975b), NATTERER (1977). For special types of triangulations in higher dimensions, see BRAMBLE, NITSCHE & SCHATZ (1975), BRAMBLE & SCHATZ (1976), BRAMBLE & THOMÉE (1974), DOUGLAS, DUPONT & WHEELER (1974b), NATTERER (1975b).

The first contribution to the general case is that of NITSCHE (1970). Then CIARLET & RAVIART (1973) improved the analysis of J.A. Nitsche by using a discrete maximum principle introduced in CIARLET (1970). More specifically, CIARLET & RAVIART (1973) have considered finite element approximations of general second-order nonhomogeneous Dirichlet problems posed over polygonal domains in $\mathbf{R}^n$. Then the discrete problem is said to satisfy a *discrete maximum principle* if one has

$$f \leqslant 0 \Rightarrow \max_{x \in \bar{\Omega}} u_h(x) \leqslant \max\{0, \max_{x \in \Gamma} u_h(x)\}.$$

In the case of the operator $(-\Delta u + au)$ with $a \geqslant 0$ and $n = 2$, it is shown that the discrete maximum principle holds for h small enough if there exists $\epsilon > 0$ such that all the angles of all the triangles found in all the triangulations are $\leqslant [(\pi/2) - \epsilon]$ (in case $a = 0$, it suffices that the angles of the triangles be $\leqslant \pi/2$). Returning to the general case, it is shown that when the discrete problems satisfy a maximum principle, one has

$$\lim_{u \to 0} |u - u_h|_{0,\infty,\Omega} = 0 \quad \text{if } u \in W^{1,p}(\Omega) \text{ with } p > n,$$

$$|u - u_h|_{0,\infty,\Omega} = \mathrm{O}(h) \quad \text{if } u \in W^{2,p}(\Omega) \text{ with } 2p > n,$$

i.e., there was still a loss of one in the expected order of convergence.

Recently, NATTERER (1975a), NITSCHE (1975, 1976b, 1977) and SCOTT (1976a) obtained simultaneously optimal (or nearly optimal) orders of convergence. The greatest generality is achieved in the particularly penetrating analysis of J.A. Nitsche, which we have followed in Section 3.3 (the proof of inequality (3.3.49) is that of RANNACHER (1977)).

While weighted Sobolev norms are also introduced by F. Natterer, R. Scott's main tool is a careful analysis of the approximation of the Green's function. The uniform boundedness in appropriate norms of particular Hilbertian projections, on which J.A. Nitsche's argument is essentially based, was also noticed by DOUGLAS, DUPONT & WAHLBIN (1975a) who have proved (albeit through a different approach) the boundedness in the norms $|\cdot|_{0,q,\Omega}$, $1 \leqslant q \leqslant \infty$, of the projections, with respect to the inner-product of the space $L^2(\Omega)$, onto certain finite element spaces.

J.A. Nitsche's technique has since then been successfully extended in several directions, notably to more general second-order boundary value problems by RANNACHER (1976b), to the obstacle problem by J.A. Nitsche himself, to the minimal surface problem and other nonlinear problems by J. Frehse and R. Rannacher (cf. Chapter 5), to plates by R. Rannacher (cf. Chapter 6), to mixed methods by R. Scholz (cf. Chapter 7).

There is currently a wide interest in obtaining various refinements of the error estimates, such as "interior" estimates, superconvergence results, etc.... In addition to the previously quoted reference, we mention BRAMBLE & SCHATZ (1974, 1976), BRAMBLE & THOMÉE (1974), DESCLOUX (1975), DESCLOUX & NASSIF (1977), DOUGLAS & DUPONT (1973, 1976a), NITSCHE (1972a), NITSCHE & SCHATZ (1974), SCHATZ & WAHLBIN (1976, 1977).

A little explored direction of research is that of the optimal choice of triangulation: For a given number of finite elements of a specific type, the problem consists in finding the "best" triangulation so as to minimize the error in some sense. For references in this direction, see CARROLL & BARKER (1973), MCNEICE & MARCAL (1973), PRAGER (1975), RAJAGOPALAN (1976), TURCKE & MCNEICE (1972).

CHAPTER 4

OTHER FINITE ELEMENT METHODS FOR SECOND-ORDER PROBLEMS

Introduction

Up to now, we have considered finite elements methods which are *conforming*, in the sense that the space V_h is a subspace of the space V, and the bilinear form and the linear form which are used in the definition of the discrete problem are identical to those of the original problem.

In this chapter, we shall analyze several ways of violating this "conformity", which are frequently used in everyday computations.

In Section 4.1, we assume, as before, that the domain $\bar{\Omega}$ is polygonal and that the inclusion $V_h \subset V$ still holds, but we consider the use of a *quadrature scheme* for computing the coefficients of the resulting linear system: each such coefficient being of the form

$$\sum_{K \in \mathcal{T}_h} \int_K \varphi(x) \, \mathrm{d}x,$$

the integrals

$$\int_K \varphi(x) \, \mathrm{d}x, \quad K \in \mathcal{T}_h,$$

are approximated by finite sums of the form

$$\sum_{l=1}^{L} \omega_{l,K} \varphi(b_{l,K}),$$

with weights $\omega_{l,K}$ and nodes $b_{l,K} \in K$, which are derived from a single *quadrature formula* defined over a reference finite element. This process results in an *approximate bilinear form* $a_h(\cdot, \cdot)$ and an *approximate linear form* $f_h(\cdot)$ which are defined over the space V_h.

Our study of this approximation follows a general pattern that will also be the same for the two other methods to be described in this chapter:

First, we prove (the *first Strang lemma*; cf. Theorem 4.1.1) an abstract error estimate (which, as such, is intended to be valid in other situations; cf. Section 8.2). It is established under the critical assumption that the approximate bilinear forms are *uniformly V_h-elliptic*, i.e., that there exists a constant $\tilde{\alpha} > 0$, independent of h, such that $a_h(v_h, v_h) \geq \tilde{\alpha}\|v_h\|^2$ for all $v_h \in V_h$. This is why we next examine (Theorem 4.1.2) under which assumptions (on the quadrature scheme over the reference finite element) this property is true.

The abstract error estimate of Theorem 4.1.1 generalizes Céa's lemma: In the right-hand side of the inequality, there appear, in addition to the term $\inf_{v_h \in V_h} \|u - v_h\|$, two *consistency errors* which measure the quality of the approximation of the bilinear form and of the linear form, respectively.

We are then in a position to study the convergence of such methods. More precisely, we shall essentially concentrate on the following problem: *Find sufficient conditions which insure that the order of convergence in the absence of numerical integration is unaltered by the effect of numerical integration.* Restricting ourselves for simplicity to the case where $P_K = P_k(K)$ for all $K \in \mathcal{T}_h$, our main result in this direction (Theorem 4.1.6) is that one still has

$$\|u - u_h\|_{1,\Omega} = \mathrm{O}(h^k),$$

provided the quadrature formula is exact for all polynomials of degree $(2k-2)$. The proof of this result depends, in particular, on the *Bramble-Hilbert lemma* (Theorem 4.1.3), which is a useful tool for handling linear functionals which vanish on polynomial subspaces. In this particular case, it is repeatedly used in the derivation of the consistency error estimates (Theorems 4.1.4 and 4.1.5).

We next consider in Section 4.2 a first type of finite element method for which the spaces V_h are *not* contained in the space V. This violation of the inclusion $V_h \subset V$ results of the use of finite elements which are not of class $\mathscr{C}^0$ (i.e., which are not continuous across adjacent finite elements), so that the inclusion $V_h \subset H^1(\Omega)$ is not satisfied (Theorem 4.2.1). The terminology "*nonconforming finite element method*" is specifically reserved for this type of method (likewise, for fourth-order problems, nonconforming methods result from the use of finite elements which are not of class $\mathscr{C}^1$; cf. Section 6.2).

For definiteness, we assume through Section 4.2 that we are solving a homogeneous Dirichlet problem posed over a polygonal domain $\bar{\Omega}$. Then the discrete problem consists in finding a function $u_h \in V_h$ such that, for all $v_h \in V_h$, $a_h(u_h, v_h) = f(v_h)$, where the *approximate bilinear form* $a_h(.,.)$ is defined by

$$a_h(.,.) = \sum_{K\in\mathcal{T}_h} \int_K \{\ldots\} \, dx,$$

the integrand $\{\ldots\}$ being the same as in the bilinear form which is used in the definition of the original problem. The linear form $f(.)$ need not be approximated since the inclusion $V_h \subset L^2(\Omega)$ holds.

Assuming that the mapping

$$v_h \in V_h \to \|v_h\|_h = \Big(\sum_{K\in\mathcal{T}_h} |v_h|_{1,K}^2\Big)^{1/2}$$

is a norm over the space V_h, we prove an abstract error estimate (the *second Strang lemma*; cf. Theorem 4.2.2) where the expected term $\inf_{v_h\in V_h}\|u - v_h\|_h$ is added a *consistency error.* Just as in the case of numerical integration, this result holds under the assumption that the approximate bilinear forms are *uniformly* V_h-*elliptic*, in the sense that there exists a constant $\tilde{\alpha} > 0$ independent of h such that $a_h(v_h, v_h) \geq \tilde{\alpha}\|v_h\|_h^2$ for all $v_h \in V_h$.

We then proceed to describe a three-dimensional "nonconforming" finite element, known as *Wilson's brick*, which has gained some popularity among engineers for solving the elasticity problem. Apart from being nonconforming, this finite element presents the added theoretical interest that some of its degrees of freedom are of a form not yet encountered. This is why we need to adapt to this finite element the standard interpolation error analysis (Theorem 4.2.3).

Next, using a "*bilinear lemma*" which extends the Bramble-Hilbert lemma to bilinear forms (Theorem 4.2.5), we analyze the consistency error (Theorem 4.2.6). In this fashion we prove that

$$\|u - u_h\|_h = \Big(\sum_{K\in\mathcal{T}_h} |u - u_h|_{1,K}^2\Big)^{1/2} = O(h),$$

if the solution u is in the space $H^2(\Omega)$. In passing, we establish the connection between the convergence of such nonconforming finite element methods and the *patch test* of B. Irons.

Another violation of the inclusion $V_h \subset V$ occurs in the approximation of a boundary value problem posed over a domain $\bar{\Omega}$ with a *curved* boundary Γ (i.e., the set $\bar{\Omega}$ is no longer assumed to be polygonal). In this case, the set $\bar{\Omega}$ is usually approximated by two types of finite elements: The finite elements of the first type are *straight*, i.e., they have plane faces, and they are typically used "inside" $\bar{\Omega}$. The finite elements of the second type have at least one "curved" face, and they are especially used so as to approximate "as well as possible" the boundary Γ.

In Section 4.3, we consider one way of generating finite elements of the second type, the *isoparametric finite elements*, which are often used in actual computations. The key idea underlying their conception is the generalization of the notion of affine-equivalence: Let there be given a Lagrange finite element $(\hat{K}, \hat{P}, \{\hat{p}(\hat{a}_i), 1 \leq i \leq N\})$ in $\mathbf{R}^n$ and let $F : \hat{x} \in \hat{K} \to F(\hat{x}) = (F_i(\hat{x}))_{i=1}^n \in \mathbf{R}^n$ be a mapping such that $F_i \in \hat{P}$, $1 \leq i \leq n$. Then the triple

$$(K = F(\hat{K}), \quad P = \{p = \hat{p} \cdot F^{-1};\ \hat{p} \in \hat{P}\},$$
$$\{p(a_i = F(\hat{a}_i)), \quad 1 \leq i \leq N\})$$

is also a Lagrange finite element (Theorem 4.3.1), and two cases can be distinguished:

(i) The mapping F is *affine* (i.e., $F_i \in P_1(\hat{K})$, $1 \leq i \leq n$) and therefore the finite elements (K, P, Σ) and $(\hat{K}, \hat{P}, \hat{\Sigma})$ are affine-equivalent.

(ii) Otherwise, the finite element (K, P, Σ) is said to be *isoparametric*, and *isoparametrically equivalent* to the finite element $(\hat{K}, \hat{P}, \hat{\Sigma})$. If $(\hat{K}, \hat{P}, \hat{\Sigma})$ is a standard straight finite element, it is easily seen in the second case that the boundary of the set K is curved in general. This fact is illustrated by several examples.

We then consider the problem (particularly in view of Section 4.4) of developing an interpolation theory adapted to this type of finite element. In this analysis, however, we shall restrict ourselves to the isoparametric n-simplex of type (2), so as to simplify the exposition, yet retaining all the characteristic features of a general analysis. For an *isoparametric family* (K, P_K, Σ_K) of n-simplices of type (2), we show (Theorem 4.3.4) that the Π_K-interpolants of a function v satisfy inequalities of the form

$$|v - \Pi_K v|_{m,K} \leq C h_K^{3-m} \|v\|_{3,K}, \quad 0 \leq m \leq 3,$$

where $h_K = \operatorname{diam}(K)$. This result, which is the same as in the case of affine families (cf. Section 3.1) is established under the crucial assumption that the "isoparametric" mappings F_K do not deviate too much from affine mappings (of course the family is also assumed to be regular, in a sense that generalizes the regularity of affine families).

Even if we use isoparametric finite elements $K \in \mathcal{T}_h$ to "triangulate" a set $\bar{\Omega}$, the boundary of the set $\bar{\Omega}_h = \bigcup_{K \in \mathcal{T}_h} K$ is very close to, but not identical to, the boundary Γ. Consequently, since the domain of definition of the functions in the resulting finite element space V_h is the set $\bar{\Omega}_h$, the space V_h is *not* contained in the space V and therefore both the bilinear form and the linear form need to be approximated.

In order to be in as realistic a situation as possible we then study in Section 4.4 the simultaneous effects of such an approximation of the domain $\bar{\Omega}$ and of isoparametric numerical integration. As in Section 4.1, this last approximation amounts to use a quadrature formula over a reference finite element $\hat{K}$ for computing the integrals of the form $\int_K \varphi(x)\,dx$ (which appear in the linear system) via the isoparametric mappings $F_K : \hat{K} \to K$, $K \in \mathcal{T}_h$. Restricting ourselves again to isoparametric n-simplices of type (2) for simplicity, we show (Theorem 4.4.6) that, if the quadrature formula over the set $\hat{K}$ is exact for polynomials of degree 2, one has

$$\|\tilde{u} - u_h\|_{1,\Omega_h} = O(h^2),$$

where $\tilde{u}$ is an extension of the solution of the given boundary value problem to the set Ω_h (in general $\bar{\Omega}_h \not\subset \bar{\Omega}$), and $h = \max_{K \in \mathcal{T}_h} h_K$. This error estimate is obtained through the familiar process: We first prove an *abstract error estimate* (Theorem 4.4.1), under a *uniform V_h-ellipticity assumption* of the approximate bilinear forms. Then we use the interpolation theory developed in Section 4.3 for evaluating the term $\inf_{v_h \in V_h} \|\tilde{u} - v_h\|_{1,\Omega_h}$ (Theorem 4.4.3) and finally, we estimate the two *consistency errors* (Theorems 4.4.4 and 4.4.5; these results largely depend on related results of Section 4.1). It is precisely in these last estimates that a remarkable conclusion arises: *In order to retain the $O(h^2)$ convergence, it is not necessary to use more sophisticated quadrature schemes for approximating the integrals which correspond to isoparametric finite elements than for straight finite elements.*

4.1. The effect of numerical integration

Taking into account numerical integration.
Description of the resulting discrete problem.

Throughout this section, we shall assume that we are solving the second-order boundary value problem which corresponds to the following data:

$$\begin{cases} V = H_0^1(\Omega), \\ a(u, v) = \int_\Omega \sum_{i,j=1}^n a_{ij}\partial_i u \partial_j v \,\mathrm{d}x, \\ f(v) = \int_\Omega fv \,\mathrm{d}x, \end{cases} \tag{4.1.1}$$

where $\bar{\Omega}$ is a polygonal domain in $\mathbf{R}^n$, the functions $a_{ij} \in L^\infty(\Omega)$ and $f \in L^2(\Omega)$ are assumed to be *everywhere* defined over $\bar{\Omega}$. We shall assume that the ellipticity condition is satisfied i.e.,

$$\begin{gathered} \exists \beta > 0, \quad \forall x \in \bar{\Omega}, \quad \forall \xi_i, \quad 1 \leqslant i \leqslant n, \\ \sum_{i,j=1}^n a_{ij}(x)\xi_i\xi_j \geqslant \beta \sum_{i=1}^n \xi_i^2, \end{gathered} \tag{4.1.2}$$

so that the bilinear form of (4.1.1) is $H_0^1(\Omega)$-elliptic.

This problem corresponds (cf. (1.2.28)) to the homogeneous Dirichlet problem for the operator

$$u \to -\sum_{i,j=1}^n \partial_j(a_{ij}\partial_i u),$$

i.e.,

$$\begin{cases} -\sum_{i,j=1}^n \partial_j(a_{ij}\partial_i u) = f \quad \text{in} \quad \Omega, \\ u = 0 \quad \text{on} \quad \Gamma. \end{cases} \tag{4.1.3}$$

The case of a more general operator of the form

$$u \to -\sum_{i,j=1}^n \partial_j(a_{ij}\partial_i u) + au$$

is left as a problem (Exercise 4.1.5).

We consider in the sequel a family of finite element spaces X_h made up of finite elements (K, P_K, Σ_K), $K \in \mathcal{T}_h$, where $\mathcal{T}_h$ are triangulations of the set $\bar{\Omega}$ (because the set $\bar{\Omega}$ is assumed to be polygonal, it can be exactly covered by triangulations). Then we define the spaces $V_h = \{v_h \in X_h;\ v_h = 0 \text{ on } \Gamma\}$.

The assumptions about the triangulations and the finite elements are the same as in Section 3.2. Let us briefly record these assumptions for convenience:

(H 1) The associated family of triangulations is regular.

(H 2) All the finite elements (K, P_K, Σ_K), $K \in \bigcup_h \mathcal{T}_h$, are affine-equivalent to a single reference finite element $(\hat{K}, \hat{P}, \hat{\Sigma})$.

(H 3) All the finite elements (K, P_K, Σ_K), $K \in \bigcup_h \mathcal{T}_h$, are of class $\mathscr{C}^0$.

As a consequence, the inclusions $X_h \subset H^1(\Omega)$ and $V_h \subset H_0^1(\Omega)$ hold, as long as the inclusion $\hat{P} \subset H^1(\hat{K})$ (which we will assume) holds.

Given a space V_h, solving the corresponding discrete problem amounts to finding the coefficients u_k, $1 \leq k \leq M$, of the expansion $u_h = \sum_{k=1}^M u_k w_k$ of the discrete solution u_h over the basis functions w_k, $1 \leq k \leq M$, of the space V_h. These coefficients are solutions of the linear system (cf. (2.1.4))

$$\sum_{k=1}^{M} a(w_k, w_m) u_k = f(w_m), \quad 1 \leq m \leq M, \tag{4.1.4}$$

where, according to (4.1.1),

$$a(w_k, w_m) = \sum_{K \in \mathcal{T}_h} \int_K \sum_{i,j=1}^{n} a_{ij} \partial_i w_k \partial_j w_m \, \mathrm{d}x, \tag{4.1.5}$$

$$f(w_m) = \sum_{K \in \mathcal{T}_h} \int_K f w_m \, \mathrm{d}x. \tag{4.1.6}$$

In practice, even if the functions a_{ij}, f have simple analytical expressions, the integrals $\int_K \ldots \mathrm{d}x$ which appear in (4.1.5) and (4.1.6) are seldom computed exactly. Instead, they are approximated through the process of *numerical integration*, which we now describe:

Consider one of the integrals appearing in (4.1.5) or (4.1.6), let us say $\int_K \varphi(x) \, \mathrm{d}x$, and let

$$F_K : \hat{x} \in \hat{K} \to F_K(\hat{x}) = B_K \hat{x} + b_K$$

be the invertible affine mapping which maps $\hat{K}$ onto K. Assuming, without loss of generality, that the (constant) Jacobian of the mapping F_K is positive, we can write

$$\int_K \varphi(x) \, \mathrm{d}x = \det(B_K) \int_{\hat{K}} \hat{\varphi}(\hat{x}) \, \mathrm{d}\hat{x}, \tag{4.1.7}$$

the functions φ and $\hat{\varphi}$ being in the usual correspondence, i.e., $\varphi(x) = \hat{\varphi}(\hat{x})$ for all $x = F_K(\hat{x})$, $\hat{x} \in \hat{K}$. In other words, *computing the integral* $\int_K \varphi(x) \, \mathrm{d}x$ *amounts to computing the integral* $\int_{\hat{K}} \hat{\varphi}(\hat{x}) \, \mathrm{d}\hat{x}$.

Then a *quadrature scheme* (over the set $\hat{K}$) consists in replacing the integral $\int_{\hat{K}} \hat{\varphi}(\hat{x}) \, \mathrm{d}\hat{x}$ by a finite sum of the form $\sum_{l=1}^{L} \hat{\omega}_l \hat{\varphi}(\hat{b}_l)$, an approximation process which we shall symbolically represent by

$$\int_{\hat{K}} \hat{\varphi}(\hat{x})\,\mathrm{d}\hat{x} \sim \sum_{l=1}^{L} \hat{\omega}_l \hat{\varphi}(\hat{b}_l). \tag{4.1.8}$$

The numbers $\hat{\omega}_l$ are called the *weights* and the points $\hat{b}_l$ are called the *nodes* of the *quadrature formula* $\sum_{l=1}^{L} \hat{\omega}_l \hat{\varphi}(\hat{b}_l)$. For simplicity, we shall consider in the sequel only examples for which *the nodes belong to the set* $\hat{K}$ *and the weights are strictly positive* (nodes outside the set $\hat{K}$ and negative weights are not excluded in principle, but, as expected, they generally result in quadrature schemes which behave poorly in actual computations).

In view of (4.1.7) and (4.1.8), we see that *the quadrature scheme over the set* $\hat{K}$ *automatically induces a quadrature scheme over the set* K, *namely*

$$\int_{K} \varphi(x)\,\mathrm{d}x \sim \sum_{l=1}^{L} \omega_{l,K} \varphi(b_{l,K}), \tag{4.1.9}$$

with *weights* $\omega_{l,K}$ and *nodes* $b_{l,K}$ defined by

$$\omega_{l,K} = \det(B_K)\hat{\omega}_l \quad \text{and} \quad b_{l,K} = F_K(\hat{b}_l), \quad 1 \leqslant l \leqslant L. \tag{4.1.10}$$

Accordingly, we introduce the *quadrature error functionals*

$$E_K(\varphi) = \int_K \varphi(x)\,\mathrm{d}x - \sum_{l=1}^{L} \omega_{l,K} \varphi(b_{l,K}), \tag{4.1.11}$$

$$\hat{E}(\hat{\varphi}) = \int_{\hat{K}} \hat{\varphi}(\hat{x})\,\mathrm{d}\hat{x} - \sum_{l=1}^{L} \hat{\omega}_l \hat{\varphi}(\hat{b}_l), \tag{4.1.12}$$

which are related by

$$E_K(\varphi) = \det(B_K)\hat{E}(\hat{\varphi}). \tag{4.1.13}$$

Remark 4.1.1. It is realized from the previous description that *one needs only a numerical quadrature scheme over the reference finite element*. This is again in accordance with the pervading principle that most of the analysis needs to be done on the reference finite element only, just as was the case for the interpolation theory (Section 3.1). □

Let us now give a few examples of often used quadrature formulas. Notice that *each scheme preserves some space of polynomials* and it is this polynomial invariance that will subsequently play a crucial role in the problem of estimating the error.

More precisely, given a space $\hat{\Phi}$ of functions $\hat{\varphi}$ defined over the set $\hat{K}$, we shall say that the quadrature scheme is *exact for the space* $\hat{\Phi}$, *or exact for the functions* $\hat{\varphi} \in \hat{\Phi}$, if $\hat{E}(\hat{\varphi}) = 0$ for all $\hat{\varphi} \in \hat{\Phi}$.

Let $\hat{K}$ be an n-simplex with barycenter

$$\hat{a} = \frac{1}{(n+1)} \sum_{i=1}^{n+1} \hat{a}_i.$$

(Fig. 4.1.1).

Then *the quadrature scheme*

$$\int_{\hat{K}} \hat{\varphi}(\hat{x})\,\mathrm{d}\hat{x} \sim \operatorname{meas}(\hat{K})\hat{\varphi}(\hat{a}) \tag{4.1.14}$$

is exact for polynomials of degree $\leqslant 1$, i.e.,

$$\forall \hat{\varphi} \in P_1(\hat{K}), \quad \int_{\hat{K}} \hat{\varphi}(\hat{x})\,\mathrm{d}\hat{x} - \operatorname{meas}(\hat{K})\hat{\varphi}(\hat{a}) = 0. \tag{4.1.15}$$

To see this, let

$$\hat{\varphi} = \sum_{i=1}^{n+1} \hat{\varphi}(\hat{a}_i)\hat{\lambda}_i$$

be any polynomial of degree $\leqslant 1$. Then using the equalities

$$\int_{\hat{K}} \hat{\lambda}_i(\hat{x})\,\mathrm{d}\hat{x} = \operatorname{meas}(\hat{K})/(n+1)$$

(Exercise 4.1.1), $1 \leqslant i \leqslant n+1$, we obtain

$$\int_{\hat{K}} \hat{\varphi}(\hat{x})\,\mathrm{d}\hat{x} = \frac{\operatorname{meas}(\hat{K})}{n+1} \sum_{i=1}^{n+1} \hat{\varphi}(\hat{a}_i) = \operatorname{meas}(\hat{K})\hat{\varphi}(\hat{a}).$$

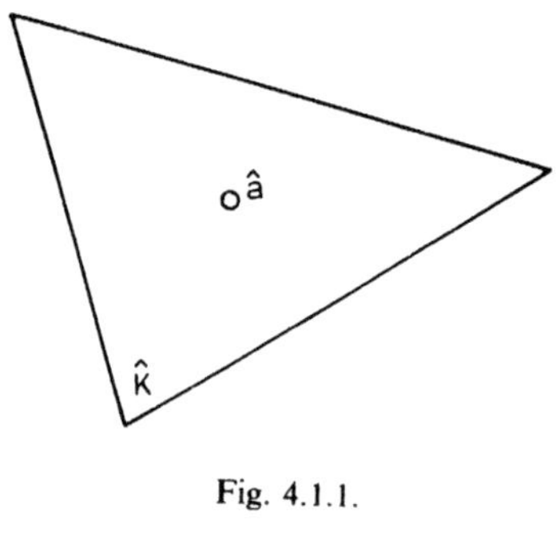

Fig. 4.1.1.

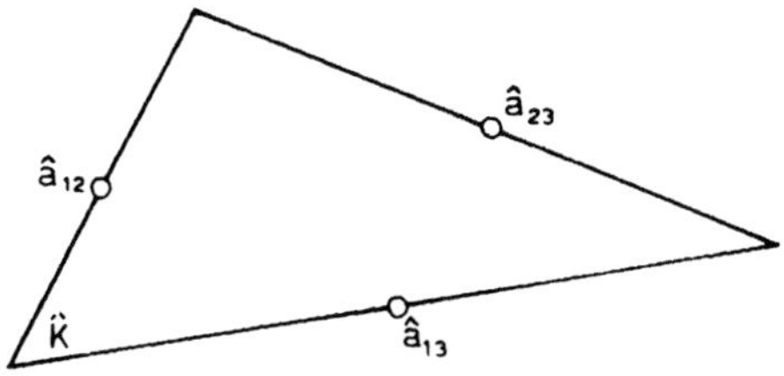

Fig. 4.1.2.

Let $n = 2$ and let $\hat{K}$ be a triangle with mid-points of the sides $\hat{a}_{ij}$, $1 \leq i < j \leq 3$ (Fig. 4.1.2).

Then *the quadrature scheme*

$$\int_{\hat{K}} \hat{\varphi}(\hat{x})\, d\hat{x} \sim \frac{\text{meas}(\hat{K})}{3} \sum_{1 \leq i < j \leq 3} \hat{\varphi}(\hat{a}_{ij}) \tag{4.1.16}$$

is exact for polynomials of degree ≤ 2 (cf. Exercise 4.1.1), i.e.,

$$\forall \hat{\varphi} \in P_2(\hat{K}), \quad \int_{\hat{K}} \hat{\varphi}(\hat{x})\, d\hat{x} - \frac{\text{meas}(\hat{K})}{3} \sum_{1 \leq i < j \leq 3} \hat{\varphi}(\hat{a}_{ij}) = 0. \tag{4.1.17}$$

Let $n = 2$ and let $\hat{K}$ be a triangle with vertices $\hat{a}_i$, $1 \leq i \leq 3$, with mid-points of the sides $\hat{a}_{ij}$, $1 \leq i < j \leq 3$, and with barycenter $\hat{a}_{123}$ (Fig. 4.1.3).

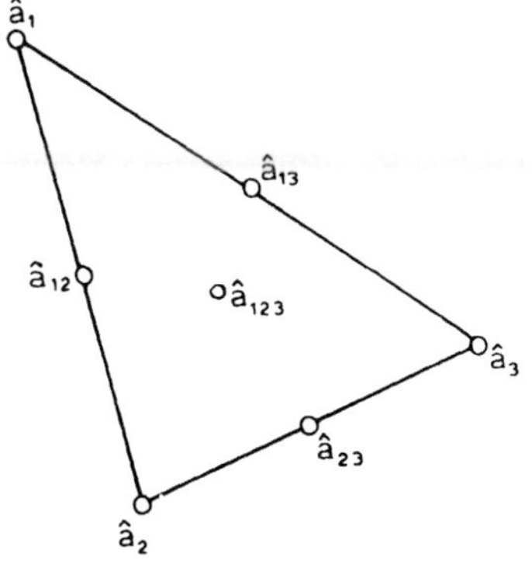

Fig. 4.1.3.

Then *the quadrature scheme*

$$\int_{\hat{K}} \hat{\varphi}(\hat{x})\,\mathrm{d}\hat{x} \sim \frac{\operatorname{meas}(\hat{K})}{60}\Big(3\sum_{i=1}^{3}\hat{\varphi}(\hat{a}_i)+8\sum_{1\leq i<j\leq 3}\hat{\varphi}(\hat{a}_{ij})+27\hat{\varphi}(\hat{a}_{123})\Big) \tag{4.1.18}$$

is exact for polynomials of degree ≤ 3 (cf. Exercise 4.1.1), i.e., $\forall \hat{\varphi} \in P_3(\hat{K})$.

$$\int_{\hat{K}} \hat{\varphi}(\hat{x})\,\mathrm{d}\hat{x} = \frac{\operatorname{meas}(\hat{K})}{60}\Big(3\sum_{i=1}^{3}\hat{\varphi}(\hat{a}_i)+8\sum_{1\leq i<j\leq 3}\hat{\varphi}(\hat{a}_{ij})+27\hat{\varphi}(\hat{a}_{123})\Big). \tag{4.1.19}$$

For examples of numerical quadrature schemes over rectangles, see Exercise 4.1.7.

Let us return to the definition of the discrete problem. Instead of solving the linear system (4.1.4) with the coefficients (4.1.5) and (4.1.6), all integrals $\int_K \ldots \mathrm{d}x$ will be computed using a quadrature scheme given on the set $\hat{K}$. In other words, we are solving the *modified linear system*

$$\sum_{k=1}^{M} a_h(w_k, w_m)u_k = f_h(w_m), \quad 1 \leq m \leq M, \tag{4.1.20}$$

where (compare with (4.1.5) and (4.1.6) respectively)

$$a_h(w_k, w_m) = \sum_{K\in\mathcal{T}_h}\sum_{l=1}^{L}\omega_{l,K}\sum_{i,j=1}^{n}(a_{ij}\partial_i w_k \partial_j w_m)(b_{l,K}), \tag{4.1.21}$$

$$f_h(w_m) = \sum_{K\in\mathcal{T}_h}\sum_{l=1}^{L}\omega_{l,K}(f w_m)(b_{l,K}). \tag{4.1.22}$$

Remark 4.1.2. Conceivably, different quadrature formulas could be used for approximating the coefficients $a(w_k, w_m)$ on the one hand, and the coefficients $f(w_m)$ on the other hand. However, our final result (Theorem 4.1.6) will show that this is not necessary. □

For our subsequent analysis, rather than working with the linear system (4.1.20), it will be more convenient to consider the following equivalent formulation of the *discrete problem*: *We are looking for a discrete solution* $u_h \in V_h$ *which satisfies*

$$\forall v_h \in V_h, \quad a_h(u_h, v_h) = f_h(v_h), \tag{4.1.23}$$

where, for all functions $u_h, v_h \in V_h$, *the bilinear form* a_h *and the linear* *form f_h are respectively given by*

$$a_h(u_h, v_h) = \sum_{K\in\mathcal{T}_h} \sum_{l=1}^{L} \omega_{l,K} \sum_{i,j=1}^{n} (a_{ij}\partial_i u_h \partial_j v_h)(b_{l,K}), \tag{4.1.24}$$

$$f_h(v_h) = \sum_{K\in\mathcal{T}_h} \sum_{l=1}^{L} \omega_{l,K}(f v_h)(b_{l,K}). \tag{4.1.25}$$

Looking at expressions (4.1.24) and (4.1.25), it is understood why the functions a_{ij} and f need to be defined everywhere over the set $\bar{\Omega}$. Also, in order that definition (4.1.21) make sense, it is necessary that, *over each finite element K, the first partial derivatives of the functions in the space X_h be unambiguously defined with unique extensions to the boundary of K*, should some node $b_{l,K}$ be situated on the boundary of K. If this node coincides with a node $b_{l,K^\star}$ corresponding to an adjacent finite element $K^\star$, it should be clear that the values to be assigned to the derivatives $\partial_i v_h(b_{l,K})$ and $\partial_i v_h(b_{l,K^\star})$ are generally different. Notice that, inasmuch as the definition of the discrete problem requires the knowledge of the values of the functions a_{ij} and f only at a finite number of points of $\bar{\Omega}$, it is quite reminiscent of finite-difference methods. In fact, this is true even to the point that *most classical finite difference schemes can be exactly interpreted as finite element methods with specific finite element spaces and specific quadrature schemes*. For results in this direction, see in particular Exercise 4.1.8. Conversely, a finite element method using Lagrange or Hermite finite elements (in which case one may always, at least theoretically, eliminate the unknowns which behave like derivatives) can be viewed as a finite difference method.

Abstract error estimate. The first Strang lemma

To sum up, we started out with a standard variational problem: Find $u \in V$ such that, for all $v \in V$, $a(u, v) = f(v)$, where the space V, the forms $a(\cdot,\cdot)$ and $f(\cdot)$ satisfy the assumptions of the Lax–Milgram lemma. Then given a finite-dimensional subspace V_h of the space V, the discrete problem consists in finding $u_h \in V_h$ such that, for all $v_h \in V_h$ $a_h(u_h, v_h) = f_h(v_h)$, where $a_h(\cdot,\cdot)$ is a bilinear form defined over the space V_h and $f_h(\cdot)$ is a linear form defined over the space V_h.

Notice that, in the present case, *the forms $a_h(\cdot,\cdot)$ and $f_h(\cdot)$ are not defined on the space V* (since the point values are not defined in general for functions in the space $H^1(\Omega)$).

Our first task is to prove an *abstract error estimate* adapted to the above abstract setting, but first we need some definitions.

For convenience, we shall refer to $a_h(\cdot,\cdot)$ as an *approximate bilinear form* and to $f_h(\cdot)$ as an *approximate linear form*. Denoting by $\|\cdot\|$ the norm of the space V, we shall say that approximate bilinear forms $a_h(\cdot,\cdot)$: $V_h \times V_h \to \mathbf{R}$, associated with a family of subspaces V_h of the space V, are *uniformly V_h-elliptic* if

$$\exists \tilde{\alpha} > 0, \quad \forall v_h \in V_h, \quad \tilde{\alpha}\|v_h\|^2 \leq a_h(v_h, v_h), \tag{4.1.26}$$

where the constant $\tilde{\alpha}$ is independent of the subspace V_h. Notice that such an assumption implies the existence of the discrete solutions.

Theorem 4.1.1 (*first Strang lemma*). *Consider a family of discrete problems for which the associated approximate bilinear forms are uniformly V_h-elliptic.*

Then there exists a constant C independent of the space V_h such that

$$\|u - u_h\| \leq C\Big(\inf_{v_h \in V_h} \Big\{\|u - v_h\| + \sup_{w_h \in V_h} \frac{|a(v_h, w_h) - a_h(v_h, w_h)|}{\|w_h\|}\Big\} + \\ + \sup_{w_h \in V_h} \frac{|f(w_h) - f_h(w_h)|}{\|w_h\|}\Big). \tag{4.1.27}$$

Proof. Let v_h be an arbitrary element in the space V_h. With the assumption of uniform V_h-ellipticity, we may write:

$$\begin{aligned}\tilde{\alpha}\|u_h - v_h\|^2 &\leq a_h(u_h - v_h, u_h - v_h) \\ &= a(u - v_h, u_h - v_h) \\ &\quad + \{a(v_h, u_h - v_h) - a_h(v_h, u_h - v_h)\} \\ &\quad + \{f_h(u_h - v_h) - f(u_h - v_h)\},\end{aligned}$$

so that, using the continuity of the bilinear from $a(\cdot,\cdot)$,

$$\begin{aligned}\tilde{\alpha}\|u_h - v_h\| &\leq M\|u - v_h\| + \frac{|a(v_h, u_h - v_h) - a_h(v_h, u_h - v_h)|}{\|u_h - v_h\|} \\ &\quad + \frac{|f_h(u_h - v_h) - f(u_h - v_h)|}{\|u_h - v_h\|} \\ &\leq M\|u - v_h\| + \sup_{w_h \in V_h} \frac{|a(v_h, w_h) - a_h(v_h, w_h)|}{\|w_h\|} \\ &\quad + \sup_{w_h \in V_h} \frac{|f_h(w_h) - f(w_h)|}{\|w_h\|}.\end{aligned}$$

Combining the above inequality with the triangular inequality

$$\|u - u_h\| \leq \|u - v_h\| + \|u_h - v_h\|$$

and taking the infimum with respect to $v_h \in V_h$ yields inequality (4.1.27).

□

Remark 4.1.3. The abstract error estimate (4.1.27) generalizes the abstract error estimate established in Céa's lemma (Theorem 2.4.1) in the case of conforming finite element methods, since, in the absence of numerical integration, we would have $a_h(\cdot,\cdot) = a(\cdot,\cdot)$ and $f_h(\cdot) = f(\cdot)$. □

Sufficient conditions for uniform V_h-ellipticity

We now give sufficient conditions on a quadrature scheme which insure that the approximate bilinear forms are uniformly V_h-elliptic: Notice in particular that in the next theorem *assumptions* (i) *and* (ii) *exhibit the relationship which should exist between the reference finite element* $(\hat{K}, \hat{P}, \hat{\Sigma})$ *and the quadrature scheme defined on* $\hat{K}$ (for the case of negative weights, see Exercise 4.1.2).

Theorem 4.1.2. *Let there be given a quadrature scheme*

$$\int_{\hat{K}} \hat{\varphi}(\hat{x})\,\mathrm{d}\hat{x} \sim \sum_{l=1}^{L} \hat{\omega}_l \hat{\varphi}(\hat{b}_l) \text{ with } \hat{\omega}_l > 0, \quad 1 \leq l \leq L,$$

over the reference finite element $(\hat{K}, \hat{P}, \hat{\Sigma})$, *for which there exists an integer* $k' \geq 1$ *such that*:

(i) *The inclusion* $\hat{P} \subset P_{k'}(\hat{K})$ *holds.*

(ii) *The union* $\bigcup_{l=1}^{L} \{\hat{b}_l\}$ *contains a* $P_{k'-1}(\hat{K})$*-unisolvent subset and/or the quadrature scheme is exact for the space* $P_{2k'-2}(\hat{K})$.

Then there exists a constant $\tilde{\alpha} > 0$ *independent of* h *such that, for all approximate bilinear forms of the form* (4.1.24) *and all spaces* V_h,

$$\forall v_h \in V_h, \quad \tilde{\alpha}\,|v_h|_{1,\Omega}^2 \leq a_h(v_h, v_h). \tag{4.1.28}$$

Proof. (i) Let us first assume that the union $\bigcup_{l=1}^{L} \{\hat{b}_l\}$ contains a $P_{k'-1}(\hat{K})$-unisolvent subset. Using the strict positivity of the weights, we find that

$$\hat{p} \in \hat{P} \quad \text{and} \quad \sum_{l=1}^{L} \hat{\omega}_l \sum_{i=1}^{n} (\partial_i \hat{p}(\hat{b}_l))^2 = 0 \Rightarrow \partial_i \hat{p}(\hat{b}_l) = 0,$$

$$1 \leq i \leq n, \quad 1 \leq l \leq L.$$

For each $i \in [1, n]$, the function $\partial_i\hat{p}$ is in the space $P_{k'-1}(\hat{K})$ by assumption (i), and thus it is identically zero since it vanishes on a $P_{k'-1}(\hat{K})$-unisolvent subset, by assumption (ii). As a consequence, the mapping

$$\hat{p} \to \left(\sum_{l=1}^{L} \hat{\omega}_l \sum_{i=1}^{n} (\partial_i\hat{p}(\hat{b}_l))^2\right)^{1/2}$$

defines a norm over the quotient space $\hat{P}/P_0(\hat{K})$. Since the mapping $\hat{p} \to |\hat{p}|_{1,\hat{K}}$ is also a norm over this space and since this space is finite-dimensional, *there exists a constant* $\hat{C} > 0$ *such that*

$$\forall \hat{p} \in \hat{P}, \quad \hat{C}|\hat{p}|^2_{1,\hat{K}} \leqslant \sum_{l=1}^{L} \hat{\omega}_l \sum_{i=1}^{n} (\partial_i\hat{p}(\hat{b}_l))^2. \tag{4.1.29}$$

If the quadrature scheme is exact for the space $P_{2k'-2}(\hat{K})$, the above inequality becomes an equality with $\hat{C} = 1$, since the function $\Sigma_{i=1}^{n} (\partial_i\hat{p})^2$ belongs to the space $P_{2k'-2}(\hat{K})$ for all $\hat{p} \in \hat{P}$ and since

$$\sum_{l=1}^{L} \hat{\omega}_l \sum_{i=1}^{n} (\partial_i\hat{p}(\hat{b}_l))^2$$

is precisely the quadrature formula which corresponds to the integral

$$\int_{\hat{K}} \sum_{i=1}^{n} (\partial_i\hat{p})^2 \,\mathrm{d}\hat{x} = |\hat{p}|^2_{1,\hat{K}}.$$

(ii) Let us next consider the approximation of one of the integrals

$$\int_{\hat{K}} \sum_{i,j=1}^{n} a_{ij}\partial_i v_h \partial_j v_h \,\mathrm{d}x.$$

Let $v_h|_K = p_K$ and let $\hat{p}_K \in \hat{P}$ be the function associated with p_K through the usual correspondence $\hat{x} \in \hat{K} \to F(\hat{x}) = B_K\hat{x} + b_K = x \in K$. We can write, using the ellipticity condition (4.1.2), and the positivity of the weights,

$$\sum_{l=1}^{L} \omega_{l,K} \sum_{i,j=1}^{n} (a_{ij}\partial_i v_h \partial_j v_h)(b_{l,K}) = \sum_{l=1}^{L} \omega_{l,K} \sum_{i,j=1}^{n} (a_{ij}\partial_i p_K \partial_j p_K)(b_{l,K})$$

$$\geqslant \beta \sum_{l=1}^{L} \omega_{l,K} \sum_{i=1}^{n} (\partial_i p_K(b_{l,K}))^2. \tag{4.1.30}$$

Observe that $\Sigma_{i=1}^{n} \partial_i p_K(b_{l,K}))^2$ is the square of the Euclidean norm $\|\cdot\|$ of the vector $Dp_K(b_{l,K})$. Since $\|D\hat{p}_K(\hat{b}_l)\| \leqslant \|B_K\| \, \|Dp_K(b_{l,K})\|$ (for all $\xi \in \mathbf{R}^n$, we have $D\hat{p}(\hat{b}_l)\xi = Dp(b_{l,K})(B_K\xi)$), we can write, using relations (4.1.10) and (4.1.29),

$$\begin{aligned}\sum_{l=1}^{L} \omega_{l,K} \sum_{i=1}^{n} (\partial_i p_K(b_{l,K}))^2 &\geq \|B_K\|^{-2} \sum_{l=1}^{L} \omega_{l,K} \sum_{i=1}^{n} (\partial_i \hat{p}_K(\hat{b}_l))^2 \\ &= \det(B_K)\|B_K\|^{-2} \sum_{l=1}^{L} \hat{\omega}_l \sum_{i=1}^{n} (\partial_i \hat{p}_K(\hat{b}_l))^2 \\ &\geq \hat{C} \det(B_K)\|B_K\|^{-2} |\hat{p}_K|_{1,\hat{K}}^2 \\ &\geq \hat{C}(\|B_K\| \, \|B_K^{-1}\|)^{-2} |p_K|_{1,K}^2, \end{aligned} \tag{4.1.31}$$

where we have also used Theorem 3.1.2. Since we are considering a regular family of triangulations, we have

$$\|B_K\| \, \|B_K^{-1}\| \leq \frac{\hat{h}}{\hat{\rho}} \frac{h_K}{\rho_K} \leq C, \tag{4.1.32}$$

for some constant C independent of $K \in \mathcal{T}_h$ and h. Combining inequalities (4.1.30), (4.1.31) and (4.1.32), we find that *there exists a constant $\tilde{\alpha} > 0$ independent of $K \in \mathcal{T}_h$ and h such that*

$$\forall v_h \in V_h, \quad \sum_{l=1}^{L} \omega_{l,K} \sum_{i,j=1}^{n} (a_{ij} \partial_i v_h \partial_j v_h)(b_{l,K}) \geq \tilde{\alpha} |v_h|_{1,K}^2. \tag{4.1.33}$$

(iii) It is then easy to conclude: Using inequalities (4.1.33) for all $K \in \mathcal{T}_h$, we obtain

$$\begin{aligned}\forall v_h \in V_h, \quad a_h(v_h, v_h) &= \sum_{K \in \mathcal{T}_h} \sum_{l=1}^{L} \omega_{l,K} \sum_{i,j=1}^{n} (a_{ij} \partial_i v_h \partial_j v_h)(b_{l,K}) \\ &\geq \tilde{\alpha} \sum_{K \in \mathcal{T}_h} |v_h|_{1,K}^2 = \tilde{\alpha} |v_h|_{1,\Omega}^2. \end{aligned}$$ □

Remark 4.1.4. Notice that the expressions

$$\sum_{l=1}^{L} \hat{\omega}_l \sum_{i=1}^{n} (\partial_i \hat{p}_K(\hat{b}_l))^2$$

are exactly the approximations we get when we apply the quadrature scheme to the integrals $|\hat{p}_K|_{1,\hat{K}}^2$, which in turn correspond to the model problem $-\Delta u = f$ in Ω, $u = 0$ on Γ. Therefore it is natural to ask for assumptions (ii) which essentially guarantee that the mapping

$$\hat{p} \to \left(\sum_{l=1}^{L} \hat{\omega}_l \sum_{i=1}^{n} (\partial_i \hat{p}(\hat{b}_l))^2 \right)^{1/2}$$

is a norm over the quotient space $\hat{P}/P_0(\hat{K})$. □

In view of this theorem, let us return to the examples of quadrature schemes given at the beginning of this section.

If $(\hat{K}, \hat{P}, \hat{\Sigma})$ is an n-simplex of type (1) ($\hat{P} = P_1(\hat{K})$ and thus $k' = 1$), we may use the quadrature scheme of (4.1.14) since $\{\hat{a}\}$ is a $P_0(\hat{K})$-unisolvent set.

If $(\hat{K}, \hat{P}, \hat{\Sigma})$ is a triangle of type (2) ($\hat{P} = P_2(\hat{K})$ and thus $k' = 2$), we may use the quadrature scheme of (4.1.16) since $\bigcup_{i<j}\{\hat{a}_{ij}\}$ is a $P_1(\hat{K})$-unisolvent set.

Notice that in both cases, the second assumption of (ii) is also satisfied.

If $(\hat{K}, \hat{P}, \hat{\Sigma})$ is a triangle of type (3) or (3') ($\hat{P} \subset P_3(\hat{K})$ and thus $k' = 3$), we may use the quadrature scheme of (4.1.18) since the set of numerical integration nodes (strictly) contains the $P_2(\hat{K})$-unisolvent subset $(\bigcup_i\{\hat{a}_i\}) \cup (\bigcup_{i<j}\{\hat{a}_{ij}\})$. However the quadrature scheme is not exact for the space $P_4(\hat{K})$ as the second assumption of (ii) would have required.

Consistency error estimates. The Bramble-Hilbert lemma

Now that the question of uniform V_h-ellipticity has been taken care of, we can turn to the problem of estimating the various terms appearing in the right-hand side of inequality (4.1.27). For the sake of clarity, we shall essentially concentrate on one special case (which nevertheless displays all the characteristic properties of the general case), namely the case where

$$\hat{P} = P_k(\hat{K})$$

for some integer $k \geq 1$ (the cases where $P_k(\hat{K}) \subset \hat{P} \subset P_{k'}(\hat{K})$ or where $P_k(\hat{K}) \subset \hat{P} \subset Q_k(\hat{K})$ are left as problems; cf. Exercises 4.1.6 and 4.1.7).

This being the case, if the solution is smooth enough so that it belongs to the space $H^{k+1}(\Omega)$, we have

$$\inf_{v_h \in V_h} \|u - v_h\|_{1,\Omega} \leq \|u - \Pi_h u\|_{1,\Omega} \leq Ch^k |u|_{k+1,\Omega},$$

assuming the X_h-interpolant of the solution u is well-defined, and thus, in the absence of numerical integration, we would have an $O(h^k)$ convergence. Then our basic objective is *to give sufficient conditions on the quadrature scheme which insure that the effect of numerical integration does not decrease this order of convergence.*

Remark 4.1.5. This criterion for appraising the required quality of the quadrature scheme is perhaps arbitrary, but at least it is well-defined. Surprisingly, the results that shall be obtained in this fashion are nevertheless quite similar to the conclusions usually drawn by engineers through purely empirical criteria. □

Let us assume that the approximate bilinear forms are uniformly V_h-elliptic so that we may apply the abstract error estimate (4.1.27) of Theorem 4.1.1. Consequently, our aim is to obtain *consistency error estimates* of the form

$$\sup_{w_h \in V_h} \frac{|a(\Pi_h u, w_h) - a_h(\Pi_h u, w_h)|}{\|w_h\|_{1,\Omega}} \leqslant C(a_{ij}, u) h^k, \tag{4.1.34}$$

$$\sup_{w_h \in V_h} \frac{|f(w_h) - f_h(w_h)|}{\|w_h\|_{1,\Omega}} \leqslant C(f) h^k. \tag{4.1.35}$$

Notice that, in the usual terminology of numerical analysis, the uniform ellipticity condition appears as a *stability condition*, while the conditions (implied by the above error estimates)

$$\lim_{h \to 0} \sup_{w_h \in V_h} \frac{|a(\Pi_h u, w_h) - a_h(\Pi_h u, w_h)|}{\|w_h\|_{1,\Omega}} = 0,$$

$$\lim_{h \to 0} \sup_{w_h \in V_h} \frac{|f(w_h) - f_h(w_h)|}{\|w_h\|_{1,\Omega}} = 0,$$

appear as *consistency conditions*. This is why we call *consistency errors* the two terms of the form $\sup_{w_h \in V_h}(\ldots)$ appearing in the left-hand side of inequalities (4.1.34) and (4.1.35). By definition of the quadrature error functionals $E_K(\cdot)$ (cf.(4.1.11)), we have, for all $w_h \in V_h$,

$$a(\Pi_h u, w_h) - a_h(\Pi_h u, w_h) = \sum_{K \in \mathcal{T}_h} E_K\Big(\sum_{i,j=1}^{n} a_{ij} \partial_i(\Pi_h u) \partial_j w_h\Big), \tag{4.1.36}$$

$$f(w_h) - f_h(w_h) = \sum_{K \in \mathcal{T}_h} E_K(f w_h). \tag{4.1.37}$$

It turns out that we shall obtain (Theorems 4.1.4 and 4.1.5) "local" *quadrature error estimates* of the form

$$\forall p' \in P_K, \quad \forall p \in P_K, \quad |E_K(a_{ij} \partial_i p' \partial_j p)| \leqslant C(a_{ij|K}; \partial_i p') h_K^k |\partial_j p|_{0,K}, \tag{4.1.38}$$

$$\forall p \in P_K, \quad |E_K(fp)| \leqslant C(f_{|K}) h_K^k \|p\|_{1,K}, \tag{4.1.39}$$

from which the "global" consistency error estimates (4.1.34) and (4.1.35) are deduced by an application of the Cauchy-Schwarz inequality (this is possible only because the constants $C(a_{ij|K}; \partial_i p')$ and $C(f_{|K})$ appearing in the above inequalities are of an oppropriate form).

To begin with, we prove a useful preliminary result.

Theorem 4.1.3 (*Bramble-Hilbert lemma*). *Let Ω be an open subset of $\mathbf{R}^n$ with a Lipschitz-continuous boundary. For some integer $k \geq 0$ and some number $p \in [0, \infty]$, let f be a continuous linear form on the space $W^{k+1,p}(\Omega)$ with the property that*

$$\forall p \in P_k(\Omega), \quad f(p) = 0. \tag{4.1.40}$$

Then there exists a constant $C(\Omega)$ such that

$$\forall v \in W^{k+1,p}(\Omega), \quad |f(v)| \leq C(\Omega) \|f\|^{*}_{k+1,p,\Omega} |v|_{k+1,p,\Omega}, \tag{4.1.41}$$

where $\|\cdot\|^{*}_{k+1,p,\Omega}$ is the norm in the dual space of $W^{k+1,p}(\Omega)$.

Proof. Let v be any function in the space $W^{k+1,p}(\Omega)$. Since by assumption, $f(v) = f(v+p)$ for all $p \in P_k(\Omega)$, we may write

$$\forall p \in P_k(\Omega), \quad |f(v)| = |f(v+p)| \leq \|f\|^{*}_{k+1,p,\Omega} \|v+p\|_{k+1,p,\Omega},$$

and thus

$$|f(v)| \leq \|f\|^{*}_{k+1,p,\Omega} \inf_{p \in P_k(\Omega)} \|v+p\|_{k+1,p,\Omega}.$$

The conclusion follows by Theorem 3.1.1. □

In the sequel, we shall often use the following result: *Let the functions $\varphi \in W^{m,q}(\Omega)$ and $w \in W^{m,\infty}(\Omega)$ be given. Then the function φw belongs to the space $W^{m,q}(\Omega)$, and*

$$|\varphi w|_{m,q,\Omega} \leq C \sum_{j=0}^{m} |\varphi|_{m-j,q,\Omega} |w|_{j,\infty,\Omega}, \tag{4.1.42}$$

for some constant C solely dependent upon the integers m and n, i.e., it is in particular independent of the set Ω.

To prove this, use the formula

$$\forall \alpha, \quad |\alpha| = m, \quad \partial^\alpha(\varphi w) = \sum_{j=0}^{m} \sum_{\substack{|\beta| = j \\ \beta + \beta' = \alpha}} \partial^\beta w \partial^{\beta'} \varphi,$$

in conjunction with inequalities of the form

$$\left|\sum_{\lambda=1}^{\Lambda} \alpha_\lambda f_\lambda\right|_{0,q,\Omega} \leq \sum_{\lambda=1}^{\Lambda} |a_\lambda|_{0,\infty,\Omega}|f_\lambda|_{0,q,\Omega}.$$

Theorem 4.1.4. *Assume that, for some integer* $k \geq 1$,

$$\hat{P} = P_k(\hat{K}), \tag{4.1.43}$$

$$\forall \hat{\varphi} \in P_{2k-2}(\hat{K}), \quad \hat{E}(\hat{\varphi}) = 0. \tag{4.1.44}$$

Then there exists a constant C *independent of* $K \in \mathcal{T}_h$ *and* h *such that*

$$\begin{aligned} &\forall a \in W^{k,\infty}(K), \quad \forall p \in P_k(K), \quad \forall p' \in P_k(K), \\ &|E_k(a\partial_i p'\partial_j p)| \leq Ch_K^k\|a\|_{k,\infty,K}\|\partial_i p'\|_{k-1,K}|\partial_j p|_{0,K} \\ &\qquad \leq Ch_K^k\|a\|_{k,\infty,K}\|p'\|_{k,K}|p|_{1,K}. \end{aligned} \tag{4.1.45}$$

Proof. We shall get an error estimate for the expression $E_K(avw)$ for $a \in W^{k,\infty}(K)$, $v \in P_{k-1}(K)$, $w \in P_{k-1}$. From (4.1.13), we infer that

$$E_K(avw) = \det(B_K)\, \hat{E}(\hat{a}\hat{v}\hat{w}), \tag{4.1.46}$$

with $\hat{a} \in W^{k,\infty}(\hat{K})$, $\hat{v} \in P_{k-1}(\hat{K})$, $\hat{w} \in P_{k-1}(\hat{K})$. For a given $\hat{w} \in P_{k-1}(\hat{K})$ and any $\hat{\varphi} \in W^{k,\infty}(\hat{K})$, we have ($W^{k,\infty}(\hat{K}) \hookrightarrow \mathscr{C}^0(\hat{K})$ since $k \geq 1$)

$$\begin{aligned} |\hat{E}(\hat{\varphi}\hat{w})| &= \left|\int_{\hat{K}} \hat{\varphi}\hat{w}\, d\hat{x} - \sum_{l=1}^{L} \hat{\omega}_l(\hat{\varphi}\hat{w})(\hat{b}_l)\right| \\ &\leq \hat{C}|\hat{\varphi}\hat{w}|_{0,\infty,\hat{K}} \leq \hat{C}|\hat{\varphi}|_{0,\infty,\hat{K}}|\hat{w}|_{0,\infty,\hat{K}}, \end{aligned}$$

where, here and subsequently, the letter $\hat{C}$ represents various constants solely dependent upon the reference finite element. Since $|\hat{\varphi}|_{0,\infty,\hat{K}} \leq \|\hat{\varphi}\|_{k,\infty,\hat{K}}$, and since all norms are equivalent on the finite-dimensional space $P_{k-1}(\hat{K})$, we deduce that

$$|\hat{E}(\hat{\varphi}\hat{w})| \leq \hat{C}\|\hat{\varphi}\|_{k,\infty,\hat{K}}|\hat{w}|_{0,\hat{K}}.$$

Thus, *for a given* $\hat{w} \in P_{k-1}(\hat{K})$, *the linear from*

$$\hat{\varphi} \in W^{k,\infty}(\hat{K}) \to \hat{E}(\hat{\varphi}\hat{w})$$

is continuous with norm $\leq \hat{C}|\hat{w}|_{0,\hat{K}}$ on the one hand, and it vanishes over the space $P_{k-1}(\hat{K})$ on the other hand, by assumption (4.1.44). Therefore, using the Bramble-Hilbert lemma, there exists a constant $\hat{C}$

such that

$$\forall \hat{\varphi} \in W^{k,\infty}(\hat{K}), \quad \forall \hat{w} \in P_{k-1}(\hat{K}),$$
$$|\hat{E}(\hat{\varphi}\hat{w})| \leq \hat{C}|\hat{\varphi}|_{k,\infty,\hat{K}}|\hat{w}|_{0,\hat{K}}.$$

Next, let $\hat{\varphi} = \hat{a}\hat{v}$ with $\hat{a} \in W^{k,\infty}(\hat{K})$, $\hat{v} \in P_{k-1}(\hat{K})$. Using (4.1.42) and taking into account that $|\hat{v}|_{k,\infty,\hat{K}} = 0$, we get

$$|\hat{\varphi}|_{k,\infty,\hat{K}} = |\hat{a}\hat{v}|_{k,\infty,\hat{K}} \leq \hat{C}\sum_{j=0}^{k-1} |\hat{a}|_{k-j,\infty,\hat{K}}|\hat{v}|_{j,\infty,\hat{K}} \leq \hat{C}\sum_{j=0}^{k-1} |\hat{a}|_{k-j,\infty,\hat{K}}|\hat{v}|_{j,\hat{K}},$$

where, in the last inequality, we have again used the equivalence of norms over the finite-dimensional space $P_{k-1}(\hat{K})$. Therefore, we obtain

$$\forall \hat{a} \in W^{k,\infty}(\hat{K}), \quad \forall \hat{v} \in P_{k-1}(\hat{K}), \quad \forall \hat{w} \in P_{k-1}(\hat{K}),$$
$$|\hat{E}(\hat{a}\hat{v}\hat{w})| \leq \hat{C}\left(\sum_{j=0}^{k-1} |\hat{a}|_{k-j,\infty,\hat{K}}|\hat{v}|_{j,\hat{K}}\right)|\hat{w}|_{0,\hat{K}}. \tag{4.1.47}$$

Then it suffices to use the inequalities (cf. Theorems 3.1.2 and 3.1.3)

$$|\hat{a}|_{k-j,\infty,\hat{K}} \leq \hat{C}h_K^{k-j}|a|_{k-j,\infty,K}, \quad 0 \leq j \leq k-1,$$
$$|\hat{v}|_{j,\hat{K}} \leq \hat{C}h_K^j(\det(B_K))^{-1/2}|v|_{j,K}, \quad 0 \leq j \leq k-1,$$
$$|\hat{w}|_{0,\hat{K}} \leq \hat{C}(\det(B_K))^{-1/2}|w|_{0,K},$$

in conjunction with relations (4.1.46) and (4.1.47). We obtain in this fashion:

$$\forall a \in W^{k,\infty}(K), \quad \forall v \in P_{k-1}(K), \quad \forall w \in P_{k-1}(K),$$
$$|E_K(avw)| \leq Ch_K^k\left(\sum_{j=0}^{k-1} |a|_{k-j,\infty,K}|v|_{j,K}\right)|w|_{0,K}$$
$$\leq Ch_K^k\|a\|_{k,\infty,K}\|v\|_{k-1,K}|w|_{0,K},$$

and the conclusion follows by replacing v by $\partial_i p'$ and w by $\partial_j p$ in the last inequality. □

Remark 4.1.6. Let us indicate why a *direct* application of the Bramble-Hilbert lemma to the quadrature error functionals $E_K(\cdot)$ (in this direction, see also Exercise 4.1.4) would not have yielded the proper estimate. Let us assume that

$$\forall \hat{\varphi} \in P_l(\hat{K}), \quad \hat{E}(\hat{\varphi}) = 0,$$

for some integer $l \geq 0$, and let $r \in [1, \infty]$ be such that the inclusion

$W^{l+1,r}(\hat{K}) \hookrightarrow \mathscr{C}^0(\hat{K})$ holds, so that we have

$$\forall \hat{\varphi} \in W^{l+1,r}(\hat{K}), \quad |\hat{E}(\hat{\varphi})| \leq \hat{C}|\hat{\varphi}|_{0,\infty,\hat{K}} \leq \hat{C}\|\hat{\varphi}\|_{l+1,r,\hat{K}}.$$

Then assumption (4.1.44), together with the Bramble-Hilbert lemma, implies that

$$\forall \hat{\varphi} \in W^{l+1,r}(\hat{K}), \quad |\hat{E}(\hat{\varphi})| \leq \hat{C}|\hat{\varphi}|_{l+1,r,\hat{K}}.$$

Let us then replace $\hat{\varphi}$ by the product $\hat{a}\hat{v}\hat{w}$, with a sufficiently smooth function $\hat{a}$, $\hat{v} \in P_{k-1}(\hat{K})$, $\hat{w} \in P_{k-1}(\hat{K})$. Using inequalities of the form (4.1.42) and the equivalence of norms over the space $P_{k-1}(\hat{K})$, we would automatically get all the semi-norms $|w|_{j,K}$, $0 \leq j \leq \min\{l+1, k-1\}$, in the right-hand side of the final inequality, whereas only the semi-norm $|w|_{0,K}$ should appear. □

The reader should notice that the ideas involved in the proof of the previous theorem are very reminiscent of those involved in the proof of Theorem 3.1.4. In both cases, the central idea is to apply the fundamental result of Theorem 3.1.1 (in the disguised form of the Bramble-Hilbert lemma in the present case) over the reference finite element and then to use the standard inequalities to go from the finite element K to $\hat{K}$, and back. The same analogies also hold for our next result.

Theorem 4.1.5. *Assume that, for some integer $k \geq 1$,*

$$\hat{P} = P_k(\hat{K}), \tag{4.1.48}$$

$$\forall \hat{\varphi} \in P_{2k-2}(\hat{K}), \quad \hat{E}(\hat{\varphi}) = 0, \tag{4.1.49}$$

and let $q \in [1, \infty]$ be any number which satisfies the inequality

$$k - \frac{n}{q} > 0. \tag{4.1.50}$$

Then there exists a constant C independent of $K \in \mathscr{T}_h$ and h such that

$$\forall f \in W^{k,q}(K), \quad \forall p \in P_k(K), \tag{4.1.51}$$
$$|E_K(fp)| \leq Ch_K^k(\operatorname{meas}(K))^{(1/2)-(1/q)}\|f\|_{k,q,K}\|p\|_{1,K}.$$

Proof. For any $f \in W^{k,q}(K)$ and any $p \in P_k(K)$, we have

$$E_K(fp) = \det(B_K)\hat{E}(\hat{f}\hat{p}), \tag{4.1.52}$$

with $\hat{f} \in W^{k,q}(\hat{K})$, $\hat{p} \in P_k(\hat{K})$. Let us write

$$\hat{E}(\hat{f}\hat{p}) = \hat{E}(\hat{f}\hat{\Pi}\hat{p}) + \hat{E}(\hat{f}(\hat{p} - \hat{\Pi}\hat{p})), \tag{4.1.53}$$

where $\hat{\Pi}$ *is the orthogonal projection in the space* $L^2(\hat{K})$ *onto the subspace* $P_1(\hat{K})$.

(i) *Let us estimate* $\hat{E}(\hat{f}\hat{\Pi}\hat{p})$. For all $\hat{\psi} \in W^{k,q}(\hat{K})$, we have

$$|\hat{E}(\hat{\psi})| \leq \hat{C}|\hat{\psi}|_{0,\infty,\hat{K}} \leq \hat{C}\|\hat{\psi}\|_{k,q,\hat{K}},$$

since inequality (4.1.50) implies that the inclusion $W^{k,q}(\hat{K}) \hookrightarrow \mathscr{C}^0(\hat{K})$ holds, and, in addition, $\hat{E}(\hat{\psi}) = 0$ for all $\hat{\psi} \in P_{k-1}(\hat{K})$, by virtue of assumption (4.1.49) (therefore, this assumption is not fully used at this stage, unless $k = 1$). Using the Bramble-Hilbert lemma, we obtain

$$\forall \hat{\psi} \in W^{k,q}(\hat{K}), \quad |\hat{E}(\hat{\psi})| \leq \hat{C}|\hat{\psi}|_{k,q,\hat{K}}.$$

In particular, let $\hat{\psi} = \hat{f}\hat{\Pi}\hat{p}$ with $\hat{f} \in W^{k,q}(\hat{K})$, $\hat{p} \in P_k(\hat{K})$. Using inequality (4.1.42), we find:

$$|\hat{f}\hat{\Pi}\hat{p}|_{k,q,\hat{K}} \leq \hat{C}(|\hat{f}|_{k,q,\hat{K}}|\hat{\Pi}\hat{p}|_{0,\infty,\hat{K}} + |\hat{f}|_{k-1,q,\hat{K}}|\hat{\Pi}\hat{p}|_{1,\infty,\hat{K}}),$$

since all semi-norms $|\hat{\Pi}\hat{p}|_{l,\infty,\hat{K}}$ are zero for $l \geq 2$ ($\hat{\Pi}\hat{p} \in P_1(\hat{K})$). Using the equivalence of norms over the finite-dimensional space $P_1(\hat{K})$, we get

$$|\hat{f}\hat{\Pi}\hat{p}|_{k,q,\hat{K}} \leq \hat{C}(|\hat{f}|_{k,q,\hat{K}}|\hat{\Pi}\hat{p}|_{0,\hat{K}} + |\hat{f}|_{k-1,q,\hat{K}}|\hat{\Pi}\hat{p}|_{1,\hat{K}}).$$

Further we have

$$|\hat{\Pi}\hat{p}|_{0,\hat{K}} \leq |\hat{p}|_{0,\hat{K}},$$

since $\hat{\Pi}$ is a projection operator, and

$$|\hat{\Pi}\hat{p}|_{1,\hat{K}} \leq |\hat{p} - \hat{\Pi}\hat{p}|_{1,\hat{K}} + |\hat{p}|_{1,\hat{K}}.$$

Applying Theorem 3.1.4 to the operator $\hat{\Pi}$, which leaves the space $P_0(\hat{K})$ invariant, we find, for some constant $\hat{C}$,

$$|\hat{p} - \hat{\Pi}\hat{p}|_{1,\hat{K}} \leq \hat{C}|\hat{p}|_{1,\hat{K}}.$$

Thus, upon combining all our previous inequalities, we have found a constant $\hat{C}$ such that

$$\forall \hat{f} \in W^{k,q}(\hat{K}), \quad \forall \hat{p} \in P_k(\hat{K}), \tag{4.1.54}$$
$$|\hat{E}(\hat{f}\hat{\Pi}\hat{p})| \leq \hat{C}(|\hat{f}|_{k,q,\hat{K}}|\hat{p}|_{0,\hat{K}} + |\hat{f}|_{k-1,q,\hat{K}}|\hat{p}|_{1,\hat{K}}).$$

(ii) *Let us next estimate* $\hat{E}(\hat{f}(\hat{p} - \hat{\Pi}\hat{p}))$. Observe that if $k = 1$, the difference $(\hat{p} - \hat{\Pi}\hat{p})$ vanishes and therefore, we may henceforth assume that $k \geq 2$. This being the case, there exists a number $\rho \in [1, +\infty]$ such that the inclusions

$$W^{k,q}(\hat{K}) \hookrightarrow W^{k-1,\rho}(\hat{K}) \hookrightarrow C^0(\hat{K})$$

hold.

To see this, consider first the case where $1 \leq q < n$, and define a number ρ by letting $(1/\rho) = (1/q) - (1/n)$, so that the inclusion $W^{1,q}(\hat{K}) \hookrightarrow L^\rho(\hat{K})$ (and consequently the inclusion $W^{k,q}(\hat{K}) \hookrightarrow W^{k-1,\rho}(\hat{K})$) holds. Then the inclusion $W^{k-1,\rho}(\hat{K}) \hookrightarrow \mathscr{C}^0(\hat{K})$ also holds because we have $k - 1 - (n/\rho) = k - (n/q) > 0$ by (4.1.50).

Consider next the case where $n \leq q$. Then either $n < q$ and the inclusion $W^{1,q}(\hat{K}) \hookrightarrow L^\rho(\hat{K})$ holds for all $\rho \in [1, \infty]$, or $n = q$ and the same inclusion holds for all (finite) $\rho \geq 1$, so that in both cases the inclusion $W^{k,q}(\hat{K}) \hookrightarrow W^{k-1,\rho}(\hat{K})$ holds for all $\rho \geq 1$. Since in this part (ii) we assume $k \geq 2$, it suffices to choose ρ large enough so that $k - 1 - (n/\rho) > 0$ and then the inclusion $W^{k-1,\rho}(\hat{K}) \hookrightarrow \mathscr{C}^0(\hat{K})$ holds.

Proceeding with the familiar arguments, we eventually find that

$$\begin{aligned} &\forall \hat{f} \in W^{k-1,\rho}(\hat{K}) \hookrightarrow \mathscr{C}^0(\hat{K}),\ \forall \hat{p} \in P_k(\hat{K}),\\ &|\hat{E}(\hat{f}(\hat{p} - \hat{\Pi}\hat{p}))| \leq \hat{C}|\hat{f}(\hat{p} - \hat{\Pi}\hat{p})|_{0,\infty,\hat{K}} \leq \hat{C}|\hat{f}|_{0,\infty,\hat{K}}|\hat{p} - \hat{\Pi}\hat{p}|_{0,\infty,\hat{K}}\\ &\qquad \leq \hat{C}\|\hat{f}\|_{k-1,\rho,\hat{K}}|\hat{p} - \hat{\Pi}\hat{p}|_{0,\infty,\hat{K}}. \end{aligned}$$

Thus for a given $\hat{p} \in P_k(\hat{K})$, the linear form

$$\hat{f} \in W^{k-1,\rho}(\hat{K}) \to \hat{E}(\hat{f}(\hat{p} - \hat{\Pi}\hat{p}))$$

is continuous with norm $\leq \hat{C}|\hat{p} - \hat{\Pi}\hat{p}|_{0,\infty,\hat{K}}$, and it vanishes over the space $P_{k-2}(\hat{K})$ (notice that, contrary to step (i), the "full" assumption (4.1.49) is used here). Another application of the Bramble-Hilbert lemma shows that

$$\begin{aligned} &\forall \hat{f} \in W^{k-1,\rho}(\hat{K}), \quad \forall \hat{p} \in P_k(\hat{K}),\\ &|\hat{E}(\hat{f}(\hat{p} - \hat{\Pi}\hat{p}))| \leq \hat{C}|\hat{f}|_{k-1,\rho,\hat{K}}|\hat{p} - \hat{\Pi}\hat{p}|_{0,\hat{K}}. \end{aligned}$$

Since the operator $\hat{\Pi}$ leaves the space $P_0(\hat{K})$ invariant we have, again by Theorem 3.1.4,

$$|\hat{p} - \hat{\Pi}\hat{p}|_{0,\hat{K}} \leq \hat{C}|\hat{p}|_{1,\hat{K}}.$$

Also, we have

$$\forall \hat{g} \in W^{1,q}(\hat{K}), \quad |\hat{g}|_{0,\rho,\hat{K}} \leq \hat{C}(|\hat{g}|_{0,q,\hat{K}} + |\hat{g}|_{1,q,\hat{K}}),$$

since the inclusion $W^{1,q}(\hat{K}) \hookrightarrow L^p(\hat{K})$ holds, and thus

$$\forall \hat{f} \in W^{k,q}(\hat{K}), \quad |\hat{f}|_{k-1,p,\hat{K}} \leq \hat{C}(|\hat{f}|_{k-1,q,\hat{K}} + |\hat{f}|_{k,q,\hat{K}}).$$

Combining all our previous inequalities, we obtain:

$$\begin{aligned} &\forall \hat{f} \in W^{k,q}(\hat{K}), \quad \forall \hat{p} \in P_k(\hat{K}), \\ &|\hat{E}(\hat{f}(\hat{p} - \hat{\Pi}\hat{p}))| \leq \hat{C}(|\hat{f}|_{k-1,q,\hat{K}} + |\hat{f}|_{k,q,\hat{K}})|\hat{p}|_{1,\hat{K}}. \end{aligned} \tag{4.1.55}$$

(iii) The proof is completed by combining relations (4.1.52), (4.1.53), (4.1.54), (4.1.55), and

$$\begin{aligned} &|\hat{f}|_{k-j,q,\hat{K}} \leq \hat{C} h_K^{k-j} (\det(B_K))^{-1/q} |f|_{k-j,q,K}, \quad j = 0, 1, \\ &|\hat{p}|_{j,\hat{K}} \leq \hat{C} h_K^j (\det(B_K))^{-1/2} |p|_{j,K}, \quad j = 0, 1. \end{aligned}$$

□

Remark 4.1.7. Several comments are in order about this proof.

(i) First, there always exists a number q which satisfies inequality (4.1.50). In particular, the choice $q = \infty$ is possible in all cases.

(ii) Just as in the case of Theorem 4.1.4, a direct application of the Bramble-Hilbert lemma would yield unwanted norms in the right-hand side of the final inequality, which should be of the form $|E_K(fp)| \leq \cdots \|p\|_{1,K}$ (cf. Remark 4.1.6).

(iii) Why did we have to introduce the projection $\hat{\Pi}$? otherwise (arguing as in part (ii) of the proof), we would find either

$$|\hat{E}(\hat{f}\hat{p})| \leq \hat{C}|\hat{f}|_{k-1,p,\hat{K}}|\hat{p}|_{0,\hat{K}}, \quad \text{or} \quad |\hat{E}(\hat{f}\hat{p})| \leq \hat{C}|\hat{f}|_{k-1,p,\hat{K}}\|\hat{p}\|_{1,\hat{K}}.$$

In both cases, there would be a loss of one in the exponent of h_K.

(iv) Since in both steps (i) and (ii) of the proof, only the invariance of the space $P_0(\hat{K})$ through the projection operator is used, why did we not content ourselves with the orthogonal projection in the space $L^2(\hat{K})$ onto the subspace $P_0(\hat{K})$? Let us denote by $\hat{\Pi}_0$ such a projection mapping.

If $k \geq 2$, then the whole argument holds with $\hat{\Pi}_0$ instead of $\hat{\Pi}$. If $k = 1$ however, part (i) of the proof yields the inequality $|\hat{E}(\hat{f}\hat{\Pi}_0\hat{p})| \leq \hat{C}|\hat{f}|_{k,q,\hat{K}}|\hat{p}|_{0,\hat{K}}$, which is perfectly admissible, but then part (ii) of the proof is no longer empty and it is necessary to estimate the quantity $\hat{E}(\hat{f}(\hat{p} - \hat{\Pi}_0\hat{p}))$ for $\hat{p} \in P_1(\hat{K})$. But then it is impossible to find a space $W^{0,p}(\hat{K}) = L^p(\hat{K})$ which would be contained in the space $\mathscr{C}^0(\hat{K})$ with a continuous injection. Thus it is simply to avoid two distinct proofs (one with $\hat{\Pi}$ if $k = 1$, another one with $\hat{\Pi}_0$ if $k \geq 2$) that we have used the single mapping $\hat{\Pi}$.

(v) Why is it necessary to introduce the intermediate space $W^{k-1,p}(\hat{K})$? For all $\hat{p} \in P_k(\hat{K})$, the function $(\hat{p} - \hat{\Pi}\hat{p})$ is also a polynomial of degree $\leqslant k$. Since, on the other hand, the quadrature scheme is exact for polynomials of degree $\leqslant (2k-2)$, the application of the Bramble-Hilbert lemma to the linear form $\hat{f} \to \hat{E}(\hat{f}(\hat{p} - \hat{\Pi}\hat{p}))$ necessitates that the function $\hat{f}$ be taken in a Sobolev space which involves derivatives up to and including the order $(k-1)$, and no more. □

Estimate of the error $\|u - u_h\|_{1,\Omega}$

Combining the previous theorems, we obtain the main result of this section (compare with Theorem 3.2.2).

Theorem 4.1.6. *In addition to (H1), (H2) and (H3), assume that there exists an integer* $k \geqslant 1$ *such that the following relations are satisfied*:

$$\hat{P} = P_k(\hat{K}), \tag{4.1.56}$$

$$H^{k+1}(\hat{K}) \hookrightarrow \mathscr{C}^s(\hat{K}), \tag{4.1.57}$$

where s is the maximal order of partial derivatives occurring in the definition of the set $\hat{\Sigma}$,

$$\forall \hat{\varphi} \in P_{2k-2}(\hat{K}), \quad \hat{E}(\hat{\varphi}) = 0. \tag{4.1.58}$$

Then if the solution $u \in H_0^1(\Omega)$ *of the variational problem corresponding to the data* (4.1.1) *belongs to the space* $H^{k+1}(\Omega)$, *if* $a_{ij} \in W^{k,\infty}(\Omega)$, $1 \leqslant i, j \leqslant n$, *and if* $f \in W^{k,q}(\Omega)$ *for some number* $q \geqslant 2$ *with* $k > (n/q)$, *there exists a constant C independent of h such that*

$$\|u - u_h\|_{1,\Omega} \leqslant Ch^k\Big(|u|_{k+1,\Omega} + \sum_{i,j=1}^{n} \|a_{ij}\|_{k,\infty,\Omega}\|u\|_{k+1,\Omega} + \|f\|_{k,q,\Omega}\Big), \tag{4.1.59}$$

where $u_h \in V_h$ *is the discrete solution.*

Proof. By virtue of the inclusion (4.1.57), we have (Theorem 3.2.1)

$$\|u - \Pi_h u\|_{1,\Omega} \leqslant Ch^k|u|_{k+1,\Omega},$$

where, here and subsequently, C stands for a constant independent of h.

Using (4.1.36), Theorem 4.1.4 and the Cauchy-Schwarz inequality, we obtain for any $w_h \in V_h$,

$$\begin{aligned}|a(\Pi_h u, w_h) - a_h(\Pi_h u, w_h)| &\leq \sum_{K\in\mathcal{T}_h} \sum_{i,j=1}^{n} |E_K(a_{ij}\partial_i(\Pi_h u|_K)\partial_j(w_{h|K}))| \\ &\leq C \sum_{K\in\mathcal{T}_h} h_K^k \sum_{i,j=1}^{n} \|a_{ij}\|_{k,\infty,K}\|\Pi_h u\|_{k,K}|w_h|_{1,K} \\ &\leq Ch^k\Big(\sum_{i,j=1}^{n} \|a_{ij}\|_{k,\infty,\Omega}\Big) \\ &\quad\times\Big(\sum_{K\in\mathcal{T}_h} \|\Pi_h u\|_{k,K}^2\Big)^{1/2} |w_h|_{1,\Omega}.\end{aligned}$$

By Theorem 3.2.1, we have

$$\begin{aligned}\Big(\sum_{K\in\mathcal{T}_h} \|\Pi_h u\|_{k,K}^2\Big)^{1/2} &\leq \|u\|_{k,\Omega} + \Big(\sum_{K\in\mathcal{T}_h} \|u - \Pi_h u\|_{k,K}^2\Big)^{1/2} \\ &\leq \|u\|_{k,\Omega} + Ch|u|_{k+1,\Omega} \leq C\|u\|_{k+1,\Omega},\end{aligned}$$

and thus,

$$\begin{aligned}&\inf_{v_h\in V_h}\Big(\|u - v_h\|_{1,\Omega} + \sup_{w_h\in V_h} \frac{|a(v_h, w_h) - a_h(v_h, w_h)|}{\|w_h\|_{1,\Omega}}\Big) \\ &\quad\leq \|u - \Pi_h u\|_{1,\Omega} + \sup_{w_h\in V_h} \frac{|a(\Pi_h u, w_h) - a_h(\Pi_h u, w_h)|}{\|w_h\|_{1,\Omega}} \\ &\quad\leq Ch^k\Big(|u|_{k+1,\Omega} + \sum_{i,j=1}^{n} \|a_{ij}\|_{k,\infty,\Omega}\|u\|_{k+1,\Omega}\Big).\end{aligned}$$

Likewise, using (4.1.37) and Theorem 4.1.5, we obtain

$$\begin{aligned}|f(w_h) - f_h(w_h)| &\leq \sum_{K\in\mathcal{T}_h} |E_K(fw_h)| \\ &\leq C \sum_{K\in\mathcal{T}_h} h_K^k (\mathrm{meas}(K))^{(1/2)-(1/q)}\|f\|_{k,q,K}\|w_h\|_{1,K} \\ &\leq Ch^k\, \mathrm{meas}(\Omega)^{(1/2)-(1/q)}\|f\|_{k,q,\Omega}\|w_h\|_{1,\Omega},\end{aligned}$$

where, in the last inequality, we have made use of the inequality

$$\sum_K |a_K b_K c_K| \leq \Big(\sum_K |a_K|^\alpha\Big)^{1/\alpha}\Big(\sum_K |b_K|^\beta\Big)^{1/\beta}\Big(\sum_K |c_K|^\gamma\Big)^{1/\gamma}$$

valid for any numbers $\alpha \geq 1$, $\beta \geq 1$, $\gamma \geq 1$ which satisfy $(1/\alpha)+(1/\beta)+(1/\gamma)=1$. Here, $(1/\alpha)=(1/2)-(1/q)$, $\beta = q$, $\gamma = 2$ (this is why the assumption $q \geq 2$ was needed).

Therefore we obtain

$$\sup_{w_h \in V_h} \frac{|f(w_h) - f_h(w_h)|}{\|w_h\|_{1,\Omega}} \leq Ch^k \operatorname{meas}(\Omega)^{(1/2)-(1/q)} \|f\|_{k,q,\Omega}.$$

To complete the proof, it suffices to use the abstract error estimate of Theorem 4.1.1 which we may indeed apply since, by virtue of assumptions (4.1.56) and (4.1.58), the approximate bilinear forms are uniformly V_h-elliptic (Theorem 4.1.2). □

Remark 4.1.8. When $\hat{P} = P_k(\hat{K})$, the condition that the quadrature scheme be exact for the space $P_{2k-2}(\hat{K})$ has a simple interpretation: It means that *all integrals* $\int_K a_{ij}\partial_i u_h \partial_j v_h \, dx$ *are exactly computed when all coefficients* a_{ij} *are constant functions*. To see this, notice that

$$\forall p', p \in P_K, \quad \int_K \partial_i p' \partial_j p \, dx = \int_{\hat{K}} \det B_K (\partial_i p')\hat{}\, (\partial_j p)\hat{}\, d\hat{x},$$

with

$$\det B_K = \text{constant}, \; (\partial_i p')\hat{} \in P_{k-1}(\hat{K}), \; (\partial_j p)\hat{} \in P_{k-1}(\hat{K}). \qquad \square$$

To conclude, let us examine some applications of the last theorem:

If we are using n-simplices of type (1), then we still get $\|u - u_h\|_{1,\Omega} = O(h)$ provided we use a quadrature scheme exact for constant functions, such as that of (4.1.14).

If we use triangles of type (2), then we still get $\|u - u_h\|_{1,\Omega} = O(h^2)$ provided we use a quadrature scheme exact for polynomials of degree ≤ 2, such as that of (4.1.17).

If we use triangles of type (3), it would be necessary to use a quadrature scheme exact for polynomials of degree ≤ 4, in order to preserve the error estimate $\|u - u_h\|_{1,\Omega} = O(h^3)$, etc. . . .

Exercises

4.1.1. (i) Let K be an n-simplex, and let $\lambda_i(x)$, $1 \leq i \leq n+1$, denote the barycentric coordinates of a point x with respect to the vertices of the n-simplex. Show that for any integers $\alpha_i \geq 0$, $1 \leq i \leq n+1$, one has

$$\int_K \lambda_1^{\alpha_1}(x)\lambda_2^{\alpha_2}(x) \ldots \lambda_{n+1}^{\alpha_{n+1}}(x)\, dx =$$

$$= \frac{\alpha_1!\alpha_2! \ldots \alpha_{n+1}! n!}{(\alpha_1 + \alpha_2 + \cdots + \alpha_{n+1} + n)!} \operatorname{meas}(K).$$

(ii) For $n = 2$, let $\hat{\Pi}_2$ denote the $P_2(\hat{K})$-interpolation operator associated with the set $\hat{\Sigma} = \{p(\hat{a}_i),\ 1 \leqslant i \leqslant 3;\ \hat{p}(\hat{a}_{ij}),\ 1 \leqslant i < j \leqslant 3\}$. Using (i), show that the quadrature scheme of (4.1.16) can also be written

$$\int_{\hat{K}} \hat{\varphi}(\hat{x})\,\mathrm{d}\hat{x} \sim \int_{\hat{K}} \hat{\Pi}_2\hat{\varphi}(\hat{x})\,\mathrm{d}\hat{x},$$

and consequently this scheme is exact for the space $P_2(\hat{K})$.

(iii) Show that in dimension 3, the same derivation would result in some strictly negative weights.

(iv) For $n = 2$, show that the set

$$\hat{\Sigma} = \{\hat{p}(\hat{a}_i),\quad 1 \leqslant i \leqslant 3;\quad \hat{p}(\hat{a}_{ij}),\quad 1 \leqslant i < j \leqslant 3;\quad \hat{p}(\hat{a}_{123})\}$$

is $\hat{P}$-unisolvent, where

$$\hat{P} = P_2(\hat{K}) \oplus \mathsf{V}\{\hat{\lambda}_1\hat{\lambda}_2\hat{\lambda}_3\}.$$

Using this fact combined with (i), show that the quadrature scheme of (4.1.18) can also be written

$$\int_{\hat{K}} \hat{\varphi}(\hat{x})\,\mathrm{d}\hat{x} \sim \int_{\hat{K}} \hat{\Pi}_2\hat{\varphi}(\hat{x})\,\mathrm{d}\hat{x},$$

where $\hat{\Pi}$ is the $\hat{P}$-interpolation operator.

(v) Show that the quadrature scheme of (4.1.18) is exact for the space $P_3(\hat{K})$, but not for the space $P_4(\hat{K})$.

4.1.2. Let there be given a quadrature scheme over the reference finite element for which the weights are not necessarily positive. Assume that there exists an integer k' such that the inclusion $\hat{P} \subset P_{k'}(\hat{K})$ holds and that the quadrature scheme is exact for the space $P_{2k'-2}(\hat{K})$.

(i) Show that there exists a constant C independent of $K \in \mathcal{T}_h$ and h such that

$$\forall p_K \in P_K,\quad \sum_{l=1}^{L} \omega_{l,K} \sum_{i,j=1}^{n} (a_{ij}\partial_i p_K \partial_j p_K)(b_{l,K}) \geqslant$$

$$\geqslant (\beta - C \max_{1 \leqslant i,j \leqslant n} \operatorname{osc}(a_{ij};\, K))|p|_{1,K}^2.$$

(ii) Deduce from (i) that the approximate bilinear forms of the form (4.1.21) are uniformly V_h-elliptic for sufficiently small values of the parameter h, when the functions a_{ij}, $1 \leqslant i, j \leqslant n$, are continuous.

4.1.3. The purpose of this exercise is to obtain an abstract error estimate which generalizes that of Theorem 3.2.4 in the abstract setting of Theorem 4.1.1. Let H be a Hilbert space such that $\bar{V} = H$ with a continuous injection. With the same notations as in the text, show that

$$|u - u_h| \leq \sup_{g \in H} \frac{1}{|g|} \inf_{\varphi_h \in V_h} \{M\|u - u_h\| \, \|\varphi_g - \varphi_h\| + \\ + |a(u_h, \varphi_h) - a_h(u_h, \varphi_h)| + |f(\varphi_h) - f_h(\varphi_h)|\},$$

where $|\cdot|$ denotes the norm in H, and for each $g \in H$, the function $\varphi_g \in V$ is the unique solution of the variational problem

$$\forall v \in V, \quad a(v, \varphi_g) = g(v).$$

4.1.4. Let there be given a quadrature scheme over the reference finite element such that

$$\forall \hat{\varphi} \in P_l(\hat{K}), \quad \hat{E}(\hat{\varphi}) = 0$$

for some integer $l \geq 0$, and let $r \in [1, \infty]$ be such that the inclusion $W^{l+1,r}(\hat{K}) \hookrightarrow \mathscr{C}^0(\hat{K})$ holds.

Using the Bramble-Hilbert lemma, show that there exists a constant C independent of $K \in \mathscr{T}_h$ and h such that

$$\forall \varphi \in W^{l+1,r}(K), \quad |E_K(\varphi)| \leq C(\det B_K)^{1-(1/r)} h_K^{l+1} |\varphi|_{l+1,r,K}.$$

4.1.5. The purpose of this problem is to analyze the effect of numerical integration for the homogeneous Neumann problem corresponding to the following data:

$$\begin{cases} V = H^1(\Omega), \\ a(u, v) = \displaystyle\int_\Omega \Big\{ \sum_{i,j=1}^n a_{ij} \partial_i u \partial_j v + auv \Big\} \, dx, \\ f(v) = \displaystyle\int_\Omega fv \, dx, \end{cases}$$

where, in addition to the assumptions made at the beginning of the section, it is assumed that the function a is defined everywhere over the set $\bar{\Omega}$ and that

$$\exists a_0 > 0, \quad \forall x \in \bar{\Omega}, \quad a(x) \geq a_0 > 0.$$

Thus the discrete problem corresponds to the approximate bilinear form

$$a_h(u_h, v_h) = \sum_{K\in\mathcal{T}_h} \sum_{l=1}^{L} \omega_{l,K} \sum_{i,j=1}^{n} (a_{ij}\partial_i u_h \partial_j v_h)(b_{l,K}) +$$
$$+ \sum_{K\in\mathcal{T}_h} \sum_{l=1}^{L} \omega_{l,K}(a u_h v_h)(b_{l,K}).$$

(i) With the same assumptions as in Theorem 4.1.2, show that there exists a constant $\tilde{\alpha} > 0$ such that

$$\forall v_h \in V_h, \quad \tilde{\alpha}\|v_h\|_{1,\Omega}^2 \leq a_h(v_h, v_h).$$

(ii) Assume that, for some integer $k \geq 1$,

$$\hat{P} = P_k(\hat{K}),$$
$$\forall \hat{\varphi} \in P_{2k-2}(\hat{K}), \qquad \hat{E}(\hat{\varphi}) = 0.$$

Show that there exists a constant C independent of $K \in \mathcal{T}_h$ and h such that

$$\forall a \in W^{k,\infty}(K), \quad \forall p \in P_k(K), \quad \forall p' \in P_k(K),$$
$$|E_K(ap'p)| \leq C h_K^k \|a\|_{k,\infty,K} \|p'\|_{k,K} \|p\|_{1,K}.$$

(iii) State and prove the analogue of Theorem 4.1.6 in this case.

4.1.6. The purpose of this problem is to consider the case where the space $\hat{P}$ satisfies the inclusions

$$P_k(\hat{K}) \subset \hat{P} \subset P_{k'}(\hat{K}).$$

In this case the question of V_h-ellipticity is already settled (cf. Theorem 4.1.2).

(i) Show that the analogues of Theorems 4.1.4 and 4.1.5 hold if the quadrature scheme is exact for the space $P_{k+k'-2}(\hat{K})$.

(ii) Deduce that the analogue of Theorem 4.1.6 holds if all the weights are positive, the union $\bigcup_{l=1}^{L} \{\hat{b}_l\}$ contains a $P_{k'-1}(\hat{K})$-unisolvent subset and the quadrature scheme is exact for the space $P_{k+k'-2}(\hat{K})$.

(iii) Deduce from this analysis that triangles of type (3′) may be used in conjunction with the quadrature scheme of (4.1.18). Could the quadrature scheme of (4.1.16) be used?

4.1.7. The purpose of this problem is to consider the case where the space $\hat{P}$ satisfies the inclusions

$$P_k(\hat{K}) \subset \hat{P} \subset Q_k(\hat{K}),$$

i.e., essentially the case of rectangular finite elements.

(i) Let $n = 1$ and $K = [0, 1]$. It is well known that for each integer $k \geq 0$, there exist $(k+1)$ points $b_i \in [0, 1]$ and $(k+1)$ weights $\omega_i > 0$, $1 \leq i \leq k+1$, such that the quadrature scheme

$$\int_{[0,1]} \varphi(x)\,\mathrm{d}x \sim \sum_{i=1}^{k+1} \omega_i \varphi(b_i)$$

is exact for the space $P_{2k+1}([0, 1])$. This particular quadrature formula is known as the *Gauss-Legendre formula.*

Then show that the quadrature scheme

$$\int_{[0,1]^n} \varphi(x)\,\mathrm{d}x \sim \sum_{\substack{i_j=1\\ 1\leq j\leq n}}^{k+1} (\omega_{i_1}\omega_{i_2}\ldots\omega_{i_n})\varphi(b_{i_1}, b_{i_2}, \ldots, b_{i_n})$$

is exact for the space $Q_{2k+1}([0, 1]^n)$.

(ii) Assuming the positivity of the weights, show that the approximate bilinear forms are uniformly V_h-elliptic if the union $\bigcup_{l=1}^{L} \{\hat{b}_l\}$ contains a $Q_k(\hat{K}) \cap P_{nk-1}(\hat{K})$-unisolvent subset.

(iii) Show that the analogues of Theorems 4.1.4 and 4.1.5 hold if the quadrature scheme is exact for the space $Q_{2k-1}(\hat{K})$.

(iv) Deduce that the analogue of Theorem 4.1.6 holds if all the weights are positive, if the union $\bigcup_{l=1}^{L} \{\hat{b}_l\}$ contains a $Q_k(\hat{K}) \cap P_{nk-1}(\hat{K})$-unisolvent subset, and if the quadrature scheme is exact for the space $Q_{2k-1}(\hat{K})$.

As a consequence, and contrary to the case where $\hat{P} = P_k(\hat{K})$ (cf. Remark 4.1.8), it is no longer necessary to exactly compute the integrals

$$\int_K a_{ij}\partial_i u_h \partial_j v_h \,\mathrm{d}x$$

when the coefficients a_{ij} are constant functions.

(v) Show that consequently one may use the Gauss-Legendre formula described in (i).

4.1.8. Let $\bar{\Omega} = [0, I\rho] \times [0, J\rho]$ where I and J are integers and ρ is a strictly positive number, and let $\mathcal{T}_h$ be a triangulation of the set $\bar{\Omega}$ made up of rectangles of type (1) of the form

$$[i\rho, (i+1)\rho] \times [j\rho, (j+1)\rho], \quad 0 \leq i \leq I-1, \quad 0 \leq j \leq J-1.$$

Let U_{ij} denote the unknown (usually denoted u_k) corresponding to the (k-th) node (ih, jh), $1 \leq i \leq I-1$, $1 \leq j \leq J-1$.

In what follows, we only consider nodes $(i\rho, j\rho)$ which are at least two squares away from the boundary of the set Ω, i.e., for which $2 \leq i \leq$ $\leq I-2$, $2 \leq j \leq J-2$.

Finally, we assume that the bilinear form is of the form

$$a(u, v) = \int_\Omega \sum_{l=1}^{n} \partial_l u \partial_l v \, dx,$$

i.e., the corresponding partial differential equation is the Poisson equation $-\Delta u = f$ in Ω.

(i) Show that, in the absence of numerical integration, the expression (usually denoted) $\Sigma_{k=1}^{M} a(w_k, w_m) u_k$ corresponding to the (m-th) node $(i\rho, j\rho)$ is, up to a constant factor, given by the expression

$$8U_{ij} - (U_{i+1,j} + U_{i+1,j+1} + U_{i,j+1} + U_{i-1,j+1} + \\ + U_{i-1,j} + U_{i-1,j-1} + U_{i,j-1} + U_{i+1,j-1}).$$

(ii) Assume that the quadrature scheme over the reference square $\hat{K} = [0, 1]^2$ is

$$\int_{[0,1]^2} \hat{\varphi}(\hat{x}) \, d\hat{x} \sim \frac{1}{4}(\hat{\varphi}(0,0) + \hat{\varphi}(0,1) + \hat{\varphi}(1,1) + \hat{\varphi}(1,0)).$$

Show that this quadrature scheme is exact for the space $Q_1(\hat{K})$. Since the set of nodes is $Q_1(\hat{K})$-unisolvent, the associated approximate bilinear forms are uniformly V_h-elliptic and therefore this scheme preserves the convergence in the norm $\|\cdot\|_{1,\Omega}$ (cf. Exercise 4.1.7). Show that the corresponding equality (usually denoted) $\Sigma_{k=1}^{M} a_h(w_k, w_m) u_k = f_h(w_m)$ is given by

$$4U_{ij} - (U_{i+1,j} + U_{i,j+1} + U_{i-1,j} + U_{i,j-1}) = \rho^2 f(i\rho, j\rho),$$

which is exactly *the standard five-point difference approximation to the equation*

$$-\Delta u = f.$$

(iii) Assume that the quadrature scheme over the reference square is

$$\int_{[0,1]^2} \hat{\varphi}(\hat{x}) \, d\hat{x} \sim \hat{\varphi}\left(\frac{1}{2}, \frac{1}{2}\right).$$

Show that this quadrature scheme is exact for the space $Q_1(\hat{K})$.

Show that the associated approximate bilinear forms are not uniformly V_h-elliptic, however.

Show that the expression (usually denoted) $\Sigma_{k=1}^{M} a_h(w_h, w_m)u_k$ is, up to a constant factor, given by the expression

$$4U_{ij} - (U_{i+1,j+1} + U_{i-1,j+1} + U_{i-1,j-1} + U_{i+1,j-1}).$$

It is interesting to notice that the predictably poor performance of such a method is confirmed by the geometrical structure of the above finite difference scheme, which is subdivided in two distinct schemes!

4.2. A nonconforming method

Nonconforming methods for second-order problems.
Description of the resulting discrete problem

Let us assume for definiteness that we are solving a second-order boundary value problem corresponding to the following data:

$$\begin{cases} V = H_0^1(\Omega), \\ a(u, v) = \displaystyle\int_\Omega \sum_{i,j=1}^{n} a_{ij}\partial_i u \partial_j v \,\mathrm{d}x, \\ f(v) = \displaystyle\int_\Omega fv \,\mathrm{d}x. \end{cases} \tag{4.2.1}$$

At this essentially descriptive stage, the only assumptions which we need to record are that

$$a_{ij} \in L^\infty(\Omega), \quad 1 \leq i,j \leq n, \quad f \in L^2(\Omega), \tag{4.2.2}$$

and that the set $\bar{\Omega}$ is *polygonal*. Just as in the previous section, this last assumption is made so as to insure that the set $\bar{\Omega}$ can be exactly covered with triangulations. Given such a triangulation $\bar{\Omega} = \bigcup_{K\in\mathcal{T}_h} K$, we construct *a finite element space X_h whose generic finite element is not of class $\mathscr{C}^0$*. Then the space X_h will not be contained in the space $H^1(\Omega)$, as we show in the next theorem, which is the converse of Theorem 2.1.1.

Theorem 4.2.1. *Assume that the inclusions $P_K \subset \mathscr{C}^0(K)$ for all $K \in \mathcal{T}_h$*

and $X_h \subset H^1(\Omega)$ *hold. Then the inclusion*

$$X_h \subset \mathscr{C}^0(\bar{\Omega})$$

holds.

Proof. Let us assume that the conclusion is false. Then there exists a function $v \in X_h$, there exist two adjacent finite elements K_1 and K_2, and there exists a non empty open set $\mathcal{O} \subset K_1 \cup K_2$ such that (for example)

$$(v_{|K_1} - v_{|K_2}) > 0 \quad \text{along} \quad K' \cap \mathcal{O}, \tag{4.2.3}$$

where K' is the face common to K_1 and K_2. Let then φ be a (non zero) positive function in the space $\mathscr{D}(\mathcal{O}) \subset \mathscr{D}(\Omega)$. Using Green's formula (1.2.4), we have (with standard notations)

$$\begin{aligned}
\int_\Omega \partial_i v\varphi \, \mathrm{d}x &= \sum_{\lambda=1,2} \int_{K_\lambda} \partial_i v\varphi \, \mathrm{d}x \\
&= -\sum_{\lambda=1,2} \int_{K_\lambda} v\partial_i\varphi \, \mathrm{d}x + \sum_{\lambda=1,2} \int_{\partial K_\lambda} v|_{K_\lambda} \varphi \nu_{K_\lambda} \, \mathrm{d}\gamma \\
&= -\int_\Omega v\partial_i\varphi \, \mathrm{d}x + \int_{K'} (v_{|K_1} - v_{|K_2})\varphi\nu_{K_1} \, \mathrm{d}\gamma,
\end{aligned}$$

and thus we reach a contradiction since the integral along K' should be strictly positive by (4.2.3). □

For the time being, we shall simply assume that the inclusions

$$\forall K \in \mathscr{T}_h, \quad P_K \subset H^1(K), \tag{4.2.4}$$

hold, so that, in particular, the inclusion

$$X_h \subset L^2(\Omega) \tag{4.2.5}$$

holds. Then one defines a subspace X_{0h} of X_h which takes as well as possible into account the boundary condition $v = 0$ along the boundary Γ of Ω. For example, if the generic finite element is a Lagrange element, all degrees of freedom are set equal to zero at the boundary nodes. But, again because the finite element is not of class $\mathscr{C}^0$ (cf. Remark 2.3.10), *the functions in the space* X_{0h} *will in general vanish only at the boundary nodes.*

In order to define a discrete problem over the space $V_h = X_{0h}$, we observe that, if the linear form f is still defined over the space V_h by virtue of the inclusion (4.2.5), this is not the case of the bilinear form $a(\cdot, \cdot)$. To obviate this difficulty, we *define*, in view of (4.2.1) and (4.2.4), *the approximate bilinear form*

$$a_h(u_h, v_h) = \sum_{K \in \mathcal{T}_h} \int_K \sum_{i,j=1}^{n} a_{ij} \partial_i u_h \partial_j v_h \, dx, \tag{4.2.6}$$

and *the discrete problem consists in finding a functio* $_h \in V_h$ *such that*

$$\forall v_h \in V_h, \quad a_h(u_h, v_h) = f(v_h). \tag{4.2.7}$$

We shall say that such a process of constructing a finite element approximation of a second-order boundary value problem is a *nonconforming finite element method.* By extension, any generic finite element which is used in such method is often called a *nonconforming finite element.*

Abstract error estimate. The second Strang lemma

In view of our subsequent analysis, we need, of course, to equip the space V_h with a norm. In analogy with the norm $|\cdot|_{1,\Omega}$ of the space $V = H_0^1(\Omega)$, a natural candidate is the mapping

$$v_h \to \|v_h\|_h = \Big(\sum_{K \in \mathcal{T}_h} |v_h|_{1,K}^2\Big)^{1/2}, \tag{4.2.8}$$

which is *a priori only a semi-norm over the space* V_h. Thus, given a specific nonconforming finite element, the first task is to check that the mapping of (4.2.8) is indeed a norm on the space V_h. Once this is done, we shall be interested in showing that, for a family of spaces V_h, the approximate bilinear forms of (4.2.6) are *uniformly* V_h*-elliptic* in the sense that

$$\exists \tilde{\alpha} > 0, \quad \forall V_h, \forall v_h \in V_h, \quad \tilde{\alpha}\|v_h\|_h^2 \le a_h(v_h, v_h). \tag{4.2.9}$$

This is the case if the ellipticity condition (cf. (4.1.2)) is satisfied.

Apart from implying the existence and uniqueness of the solution of the discrete problem, this condition is essential in order to obtain the abstract error estimate of Theorem 4.2.2 below.

From now on, we shall consider that the domain of definition of both the approximate bilinear form of (4.2.6) and the semi-norm of (4.2.8) is the space $V_h + V$. This being the case, notice that

$$\forall v \in V, \quad a_h(v, v) = a(v, v) \quad \text{and} \quad \|v\|_h = |v|_{1,\Omega}. \tag{4.2.10}$$

Also, the first assumptions (4.2.2) imply that there exists a constant $\tilde{M}$ independent of the space V_h such that

$$\forall u, v \in (V_h + V), \quad |a_h(u, v)| \leq \tilde{M} \|u\|_h \|v\|_h. \tag{4.2.11}$$

Theorem 4.2.2 (*second Strang lemma*). *Consider a family of discrete problems for which the associated approximate bilinear forms are uniformly V_h-elliptic.*

Then there exists a constant C independent of the subspace V_h such that

$$\|u - u_h\|_h \leq C\left(\inf_{v_h \in V_h} \|u - v_h\|_h + \sup_{w_h \in V_h} \frac{|a_h(u, w_h) - f(w_h)|}{\|w_h\|_h} \right). \tag{4.2.12}$$

Proof. Let v_h be an arbitrary element in the space V_h. Then in view of the uniform V_h-ellipticity and continuity of the bilinear forms a_h (cf. (4.2.9) and (4.2.11)) and of the definition (4.2.7) of the discrete problem, we may write

$$\begin{aligned}\tilde{\alpha}\|u_h - v_h\|_h^2 &\leq a_h(u_h - v_h, u_h - v_h)\\ &= a_h(u - v_h, u_h - v_h) + \{f(u_h - v_h) - a_h(u, u_h - v_h)\},\end{aligned}$$

from which we deduce

$$\begin{aligned}\tilde{\alpha}\|u_h - v_h\|_h &\leq \tilde{M}\|u - v_h\|_h + \frac{|f(u_h - v_h) - a_h(u, u_h - v_h)|}{\|u_h - v_h\|_h}\\ &\leq \tilde{M}\|u - v_h\|_h + \sup_{w_h \in V_h} \frac{|f(w_h) - a_h(u, w_h)|}{\|w_h\|_h}.\end{aligned}$$

Then inequality (4.2.12) follows from the above inequality and the triangular inequality

$$\|u - u_h\|_h \leq \|u - v_h\|_h + \|u_h - v_h\|_h. \qquad \square$$

Remark 4.2.1. The error estimate (4.2.12) indeed generalizes the error estimate which was established in Céa's lemma (Theorem 2.4.1) for conforming methods, since the difference $f(w_h) - a_h(u, w_h)$ is identically zero for all $w_h \in V_h$ when the space V_h is contained in the space V. $\square$

An example of a nonconforming finite element: Wilson's brick

Let us now describe a specific example of a nonconforming finite element known as *Wilson's brick*, which is used in particular in the approximation of problems of three-dimensional and two-dimensional elasticity posed over rectangular domains. We shall confine ourselves to the three-dimensional case, leaving the other case as a problem (Exercise 4.2.1).

Wilson's brick is an example of a *rectangular* finite element in $\mathbf{R}^3$, i.e., the set K is a 3-rectangle, whose vertices will be denoted a_i, $1 \leq i \leq 8$ (Fig. 4.2.1).

The space P_K is the space $P_2(K)$ to which are added linear combinations of the function $(x_1x_2x_3)$. Equivalently, we can think of the space P_K as being the space $Q_1(K)$ to which have been added linear combinations of the three functions x_j^2, $1 \leq j \leq 3$. We shall therefore record this definition by writing

$$P_K = P_2(K) \oplus \mathsf{V}\{x_1x_2x_3\} = Q_1(K) \oplus \mathsf{V}\{x_j^2, 1 \leq j \leq 3\}. \tag{4.2.13}$$

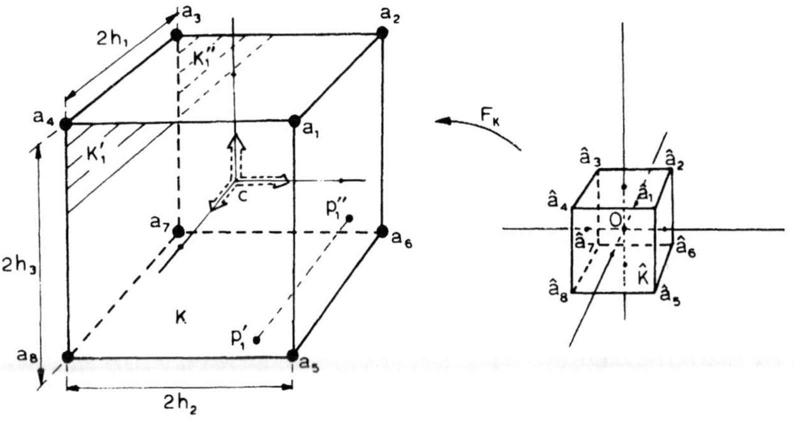

Wilson's brick; $n = 3$
$P_K = Q_1(K) \oplus \mathsf{V}\{x_j^2, 1 \leq j \leq 3\}$; $\dim(P_K) = 11$;
$\Sigma_K = \left\{ p(a_i), 1 \leq i \leq 8; \dfrac{h_j^2}{h_1h_2h_3} \int_K \partial_{jj} p \, dx, 1 \leq j \leq 3 \right\}$

Fig. 4.2.1

Notice that the inclusions

$$P_2(K) \subset P_K, \quad Q_1(K) \subset P_K \tag{4.2.14}$$

hold and that

$$\dim(P_K) = 11. \tag{4.2.15}$$

It is easily seen that *the values* $p(a_i)$, $1 \leq i \leq 8$, *at the vertices, together with the values of the* (constant) *second derivatives* $\partial_{jj}p$, $1 \leq j \leq 3$, *form a* P_K*-unisolvent set*. To see this, it suffices to check the validity of the following identity: For all functions $\hat{p} \in P_{\hat{K}}$, with $\hat{K} = [-1, +1]^3$, one has

$$\begin{aligned}\hat{p} = &\tfrac{1}{8}(1+x_1)(1+x_2)\{(1+x_3)\hat{p}(\hat{a}_1) + (1-x_3)\hat{p}(\hat{a}_5)\}\\ &+\tfrac{1}{8}(1-x_1)(1+x_2)\{(1+x_3)\hat{p}(\hat{a}_2) + (1-x_3)\hat{p}(\hat{a}_6)\}\\ &+\tfrac{1}{8}(1-x_1)(1-x_2)\{(1+x_3)\hat{p}(\hat{a}_3) + (1-x_3)\hat{p}(\hat{a}_7)\}\\ &+\tfrac{1}{8}(1+x_1)(1-x_2)\{(1+x_3)\hat{p}(\hat{a}_4) + (1-x_3)\hat{p}(\hat{a}_8)\}\\ &+\tfrac{1}{2}(x_1^2-1)\partial_{11}\hat{p} + \tfrac{1}{2}(x_2^2-1)\partial_{22}\hat{p} + \tfrac{1}{2}(x_3^2-1)\partial_{33}\hat{p}.\end{aligned} \tag{4.2.16}$$

Therefore if we denote by $c = \frac{1}{8}\sum_{i=1}^{8} a_i$ the center of the finite element K, one is naturally tempted to define the following set of degrees of freedom:

$$\Xi_K = \{p(a_i),\ 1 \leq i \leq 8;\ \partial_{jj}p(c),\ 1 \leq j \leq 3\}, \tag{4.2.17}$$

whose degrees of freedom are all in a familiar form. Of course, nothing obliges us to attach the last three degrees of freedom to the particular point c (except perhaps an aesthetical reason of symmetry), since the second derivatives $\partial_{jj}p$, $1 \leq j \leq 3$, are constant for any function $p \in P_K$.

Keeping this last property in mind, we may also choose for degrees of freedom the averages $\int_K \partial_{jj}p \, dx$, $1 \leq j \leq 3$, and we shall indeed show that this choice is more appropriate. For the time being, we observe that such degrees of freedom are of a new type, although they are still linear forms over the space $\mathscr{C}^2(K)$ as indeed they should be, to comply with the general definition given in Section 2.3.

Notice that since any function $p \in P_K$ satisfies

$$\partial_{jj}p(c) = \frac{1}{8h_1h_2h_3}\int_K \partial_{jj}p \, dx, \quad 1 \leq j \leq 3, \tag{4.2.18}$$

where $2h_j$, $1 \leq j \leq 3$, denote the lengths of the sides as indicated in Fig. 4.2.1, the two types of degrees of freedom are interchangeable over the space P_K. However, relations (4.2.18) do not hold in general for arbitrary functions in the space $\mathscr{C}^2(K)$, and this is the basic reason why we obtain in this fashion two *different* finite elements (cf. Remark 4.2.2 below; also, this is an instance of a phenomenon that was mentioned in Remark 2.3.3).

Let us then equip Wilson's brick with degrees of freedom of the form (4.2.18). Our next objective is to extend the definition of affine-equivalence so that Wilson's bricks can be imbedded in an affine family, the reference finite element being in this case the hypercube $\hat{K} = [-1, +1]^3$. To do this, it suffices, according to Remark 2.3.5, to write the degrees of freedom in such a way that if we have the identity

$$\forall \hat{p} \in P_{\hat{K}}, \quad \hat{p} = \sum_{i=1}^{8} \hat{p}(\hat{a}_i)\hat{p}_i + \sum_{j=1}^{3} \hat{\phi}_j(\hat{p})\hat{q}_j, \tag{4.2.19}$$

then we also have the identity

$$\forall p \in P_K, \quad p = \sum_{i=1}^{8} p(a_i)p_i + \sum_{j=1}^{3} \phi_j(p)q_j, \tag{4.2.20}$$

where the basis functions $\hat{p}_i$ and p_i, resp. $\hat{q}_j$ and q_j, are in the usual correspondence (2.3.18), and $\hat{\phi}_j$ and ϕ_j, $1 \leq j \leq 3$, denote the degrees of freedom of the form $\int_K \partial_{jj}p \, \mathrm{d}x$, attached to the sets $\hat{K}$ and K, respectively. Using (4.2.16), we easily deduce that any function p in the space P_K satisfies the following identity, where c_i, $1 \leq i \leq 3$, denote the coordinates of the point c:

$$\begin{aligned} p = {} & \frac{1}{8}\left(1 + \frac{(x_1 - c_1)}{h_1}\right)\left(1 + \frac{(x_2 - c_2)}{h_2}\right)\left\{\left(1 + \frac{(x_3 - c_3)}{h_3}\right)p(a_1)\right. \\ & \left. + \left(1 - \frac{(x_3 - c_3)}{h_3}\right)p(a_5)\right\} + \cdots \\ & + \sum_{j=1}^{3} \frac{1}{16}\left\{\left(\frac{x_j - c_j}{h_j}\right)^2 - 1\right\}\frac{h_j^2}{h_1h_2h_3}\int_K \partial_{jj}p \, \mathrm{d}x. \end{aligned} \tag{4.2.21}$$

Upon comparing (4.2.20) and (4.2.21) we find that the proper choices for ϕ_j and q_j are:

$$\begin{aligned} \phi_j(p) &= \frac{h_j^2}{h_1h_2h_3}\int_K \partial_{jj}p \, \mathrm{d}x, \\ q_j &= \frac{1}{16}\left(\left(\frac{x_j - c_j}{h_j}\right)^2 - 1\right), \quad 1 \leq j \leq 3. \end{aligned} \tag{4.2.22}$$

These choices insure that the following relations hold:

$$\begin{aligned} &p_i(a_k) = \delta_{ik}, \quad 1 \leqslant i, k \leqslant 8, \\ &q_j(a_i) = 0, \quad 1 \leqslant i \leqslant 8, \quad 1 \leqslant j \leqslant 3, \\ &\phi_j(p_i) = 0, \quad 1 \leqslant i \leqslant 8, \quad 1 \leqslant j \leqslant 3, \\ &\phi_l(q_j) = \delta_{lj}, \quad 1 \leqslant j, l \leqslant 3. \end{aligned} \tag{4.2.23}$$

Consequently, *we shall henceforth consider that the set of degrees of freedom of Wilson's brick is*

$$\Sigma_K = \left\{ p(a_i), \quad 1 \leqslant i \leqslant 8; \quad \frac{h_j^2}{h_1 h_2 h_3} \int_K \partial_{jj} p \, \mathrm{d}x, \quad 1 \leqslant j \leqslant 3 \right\}. \tag{4.2.24}$$

Notice that we could drop the multiplicative factors $h_j^2/h_1h_2h_3$ in the last degrees of freedom without changing the definition of the finite element.

Following definition (2.3.6), the associated operator Π_K is such that, for any sufficiently smooth function $v: K \to \mathbf{R}$, the function $\Pi_K v$ belongs to the space P_K and is uniquely determined by the conditions

$$\begin{aligned} &\Pi_K v(a_i) = v(a_i), \quad 1 \leqslant i \leqslant 8, \\ &\text{and} \quad \phi_j(\Pi_K v) = \phi_j(v), \ 1 \leqslant j \leqslant 3. \end{aligned} \tag{4.2.25}$$

Notice that the last three conditions can also be written as

$$\int_K \partial_{jj}(\Pi_K v) \, \mathrm{d}x = \int_K \partial_{jj} v \, \mathrm{d}x, \quad 1 \leqslant j \leqslant 3. \tag{4.2.26}$$

By construction, the P_K-interpolation operator satisfies

$$(\Pi_K v)\hat{} = \Pi_{\hat{K}} \hat{v} \tag{4.2.27}$$

for functions v and $\hat{v}$ in the usual correspondence. Also, by virtue of the first relation (4.2.13), we have

$$\forall \hat{p} \in P_2(\hat{K}), \quad \Pi_{\hat{K}} \hat{p} = \hat{p}. \tag{4.2.28}$$

Remark 4.2.2. According to definition (2.3.9), the finite elements (K, P_K, Ξ_K) and (K, P_K, Σ_K) (cf. (4.2.17) and (4.2.24)) are *not* identical since the associated interpolation operators do not coincide over the space $\mathscr{C}^2(K)$ (ignoring momentarily that the domain of the interpolation operator corresponding to the set Σ_K is wider, as we next indicate). □

We are now in a position to explain the definite advantage of choosing the forms ϕ_i as degrees of freedom, rather than the point values $\partial_{jj}p(c)$. On the one hand the basic properties (4.2.27) and (4.2.28) of the interpolation operator are unaltered, but on the other hand, *the interpolation operator Π_K has a wider domain*: Whereas in the first case, one is led to assume that the function $v: K \to \mathbf{R}$ is twice differentiable over K in order to define its P_K-interpolant, in the second case the P_K-interpolant is well-defined for functions "only" in the space $H^2(K)$ (which is contained in the space $\mathscr{C}^0(K)$ for $n = 3$). This property will later avoid unnecessary restrictions on the smoothness of the solution u of our original problem (cf. Theorem 4.2.6).

Although the larger Sobolev space over which the P_K-interpolant is defined is the space $W^{2,p}(K)$ for $p > \frac{3}{2}$, we shall consider for simplicity that

$$\operatorname{dom} \Pi_K = H^2(K). \tag{4.2.29}$$

In the next theorem, we shall estimate the interpolation errors $|v - \Pi_K v|_{m,K}$. The notations h_K and ρ_K represent the usual geometrical parameters (cf. (3.1.40)).

Theorem 4.2.3. *There exists a constant C such that, for all Wilson's bricks,*

$$\forall v \in H^l(\Omega), \quad |v - \Pi_K v|_{m,K} \leq C \frac{h_K^l}{\rho_K^m} |v|_{l,K}, \quad 0 \leq m \leq l, \quad l = 2, 3. \tag{4.2.30}$$

Proof. Using an argument similar to that used in the proof of Theorem 3.1.5, it can be checked that the mapping

$$\Pi_{\hat{K}}: H^l(\hat{K}) \subset H^2(\hat{K}) = \operatorname{dom} \Pi_{\hat{K}} \to H^m(\hat{K})$$

is continuous for $0 \leq m \leq l$, $l = 2$ or 3. Combining this fact with relations (4.2.27) and (4.2.28), it only remains to apply Theorem 3.1.4. □

Let us assume that the set $\bar{\Omega}$ is rectangular so that it may be covered by triangulations $\mathscr{T}_h$ composed of 3-rectangles.

We then let X_h denote the finite element space whose functions v_h have the following properties: (i) For each $K \in \mathscr{T}_h$, the restrictions $v_{h|K}$ belong to the space P_K defined in (4.2.13). (ii) Each function $v_h \in X_h$ is defined by its values at all the vertices and by the averages $\int_K \partial_{jj} v_h|_K \, dx$, $1 \leq j \leq 3$, $K \in \mathcal{T}_h$.

Since the basis functions q_j given in (4.2.22) do not vanish on the boundary of Wilson's brick, *this element is not of class* $\mathscr{C}^0$ *and the space* X_h *is not contained in the space* $H^1(\Omega)$, by Theorem 4.2.1. Continuity is however guaranteed at the vertices of the triangulations, since the functions q_j vanish at all nodes of Wilson's brick (cf. (4.2.23)).

Finally, we let $V_h = X_{0h}$, where X_{0h} denotes the space of all functions $v_h \in X_h$ which vanish at the boundary nodes. For the same reasons as before, the functions in the space X_{0h} do not vanish along the boundary Γ, but they vanish at the boundary nodes.

According to the analysis made at the beginning of this section, we need first to verify that the mapping $\|\cdot\|_h$ defined in (4.2.8) is indeed a norm over the space V_h.

Theorem 4.2.4. *The mapping*

$$v_h \to \|v_h\|_h = \Big(\sum_{K \in \mathcal{T}_h} |v_h|_{1,K}^2\Big)^{1/2} \tag{4.2.31}$$

is a norm over the space V_h.

Proof. Let v_h be a function in the space V_h which satisfies

$$\|v_h\|_h = \Big(\sum_{K \in \mathcal{T}_h} |v_{h|K}|_{1,K}^2\Big)^{1/2} = 0.$$

Then each polynomial $v_{h|K}$ is a constant so that one has $\partial_{jj}(v_{h|K}) = 0$, $1 \leq j \leq 3$, $K \in \mathcal{T}_h$, on the one hand. On the other, the function $v_h : \bar{\Omega} \to \mathbf{R}$ is a single constant since it is continuous at all the vertices and thus, it is identically zero since it vanishes at the boundary nodes. □

In order to simplify the exposition, *we shall henceforth assume that the bilinear form of* (4.2.1) *is*

$$a(u, v) = \int_\Omega \sum_{i=1}^{3} \partial_i u \partial_i v \, dx, \tag{4.2.32}$$

i.e., the corresponding boundary value problem is a homogeneous Dirichlet problem for the operator $-\Delta$. In this particular case, the uniform V_h-ellipticity of the approximate bilinear forms is a consequence of the identity

$$\forall v_h \in V_h, \quad \|v_h\|_h^2 = a_h(v_h, v_h). \tag{4.2.33}$$

This being the case, we may apply the abstract error estimate of Theorem 4.2.2. The first term, $\inf_{v_h \in V_h} \|u - v_h\|_h$, is easily taken care of: Assuming that we consider a family of discrete problems associated with a regular family of triangulations, and assuming that the solution u is in the space $H^2(\Omega)$, we deduce from Theorem 4.2.3 that

$$\inf_{v_h \in V_h} \|u - v_h\|_h \leq \Big(\sum_{K \in \mathcal{T}_h} |u - \Pi_K u|_{1,K}^2 \Big)^{1/2} \leq Ch|u|_{2,\Omega}. \tag{4.2.34}$$

Notice that the derivation of this interpolation error estimate uses in an essential manner the familiar implication (cf. (2.3.38))

$$v \in \operatorname{dom} \Pi_h = H^2(\Omega) \quad \text{and} \quad v_{|\Gamma} = 0 \Rightarrow \Pi_h v \in X_{0h},$$

where Π_h is the X_h-interpolation operator.

Remark 4.2.3. Of course, we could assume that $u \in H^3(\Omega)$, thus getting an $O(h^2)$ estimate instead of (4.2.34). However the eventual gain is nil because the other term in the right-hand side of inequality (4.2.12) is of order h, whatever the additional smoothness of the solution may be. Besides, we recall that the assumption $u \in H^2(\Omega)$ is realistic: One does not have a smoother solution in general on convex polygonal domains. □

Consistency error estimate. The bilinear lemma

Thus it remains to evaluate the other term, $\sup_{w_h \in V_h} |a_h(u, w_h) - f(w_h)|/\|w_h\|_h$, appearing in inequality (4.2.12) and this will be achieved through a careful analysis of the difference

$$D_h(u, w_h) = a_h(u, w_h) - f(w_h), \quad w_h \in V_h \tag{4.2.35}$$

(the consideration of the simplified bilinear form of (4.2.32) will allow for shorter computations in this process).

Since $-\Delta u = f$, we can write for any function $w_h \in V_h$,

$$\begin{aligned} D_h(u, w_h) &= \sum_{K \in \mathcal{T}_h} \int_K \sum_{i=1}^{3} \partial_i u \partial_i w_h \, dx - \int_\Omega f w_h \, dx \\ &= \sum_{K \in \mathcal{T}_h} \int_K \Big\{ \sum_{i=1}^{3} \partial_i u \partial_i w_h \, dx + \Delta u w_h \Big\} \, dx, \end{aligned} \tag{4.2.36}$$

i.e., we have obtained *one* decomposition of the form

$$D_h(u, w_h) = \sum_{K \in \mathcal{T}_h} D_K(u_{|K}, w_{h|K}), \tag{4.2.37}$$

where, for each $K \in \mathcal{T}_h$, the mapping $D_K(.\,,.)$ appears as a bilinear form over the space $H^2(K) \times P_K$. Ignoring for the time being that such a decomposition is not unique (we shall return to this crucial point later), let us assume that, for one decomposition of the form (4.2.37), we can show that there exists a constant C independent of $K \in \mathcal{T}_h$ and h such that

$$\forall v \in H^2(K),\quad \forall p \in P_K,\quad |D_K(v, p)| \leq C h_K |v|_{2,K} |p|_{1,K}. \tag{4.2.38}$$

Then an application of Cauchy–Schwarz inequality yields

$$|D_h(u, w_h)| \leq C h |u|_{2,\Omega} \|w_h\|_h, \tag{4.2.39}$$

and therefore we obtain

$$\sup_{w_h \in V_h} \frac{|a_h(u, w_h) - f(w_h)|}{\|w_h\|_h} \leq C h |u|_{2,\Omega}, \tag{4.2.40}$$

i.e., *an estimate similar to that of* (4.2.34).

Remark 4.2.4. The term

$$\sup_{w_h \in V_h} \frac{|a_h(u, w_h) - f(w_h)|}{\|w_h\|_h}$$

is a *consistency error* term due to the "non conformity" of the method. Consequently, a sufficient condition for convergence is the *consistency condition*:

$$\lim_{h \to 0} \sup_{w_h \in V_h} \frac{|a_h(u, w_h) - f(w_h)|}{\|w_h\|_h} = 0. \qquad \square$$

For proving estimates such as (4.2.38), the following result turns out to be useful. It plays with respect to bilinear forms the role played by the Bramble–Hilbert lemma (Theorem 4.1.3) with respect to linear forms. For this reason, we shall at times refer to this result as the "*bilinear lemma*".

Theorem 4.2.5. *Let Ω be an open subset of $\mathbf{R}^n$ with a Lipschitz-continuous boundary. Let b be a continuous bilinear form over the space*

$W^{k+1,p}(\Omega)\times W$, where the space W satisfies the inclusions

$$P_l(\Omega)\subset W\subset W^{l+1,q}(\Omega), \tag{4.2.41}$$

and is equipped with the norm $\|\cdot\|_{l+1,q,\Omega}$. We assume that

$$\forall p\in P_k(\Omega),\quad \forall w\in W,\quad b(p,w)=0, \tag{4.2.42}$$

$$\forall v\in W^{k+1,p}(\Omega),\quad \forall q\in P_l(\Omega),\quad b(v,q)=0. \tag{4.2.43}$$

Then there exists a constant $C(\Omega)$ such that

$$\forall v\in W^{k+1,p}(\Omega),\quad \forall w\in W,\quad |b(v,w)|\leqslant \leqslant C(\Omega)\|b\|\,|v|_{k+1,p,\Omega}|w|_{l+1,q,\Omega}, \tag{4.2.44}$$

where $\|b\|$ is the norm of the bilinear form b in the space

$$\mathcal{L}_2(W^{k+1,p}(\Omega)\times W;\mathbf{R}).$$

Proof. Given a function $w\in W$, the linear form $b(.\,,w)\colon v\in W^{k+1,p}(\Omega)\to b(v,w)$ is continuous and it vanishes over the space $P_k(\Omega)$, by (4.2.42). Thus, by the Bramble–Hilbert lemma, there exists a constant $C_1(\Omega)$ such that

$$\forall v\in W^{k+1,p}(\Omega),\quad |b(v,w)|\leqslant C_1(\Omega)\|b(.\,,w)\|^\star_{k+1,p,\Omega}|v|_{k+1,p,\Omega}. \tag{4.2.45}$$

Using (4.2.43), we may write $b(v,w)=b(v,w+q)$ for all $q\in P_l(\Omega)$ so that we get

$$|b(v,w)|=|b(v,w+q)|\leqslant\|b\|\,\|v\|_{k+1,p,\Omega}\|w+q\|_{l+1,q,\Omega}.$$

Therefore,

$$\forall v\in W^{k+1,p}(\Omega),\quad \forall w\in W,$$

$$|b(v,w)|\leqslant\|b\|\,\|v\|_{k+1,p,\Omega}\inf_{q\in P_l(\Omega)}\|w+q\|_{l+1,q,\Omega}$$

$$\leqslant C_2(\Omega)\|b\|\,\|v\|_{k+1,p,\Omega}|w|_{l+1,q,\Omega},$$

as an application of Theorem 3.1.1 shows. Consequently,

$$\|b(.\,,w)\|^\star_{k+1,p,\Omega}=\sup_{v\in W^{k+1,p}(\Omega)}\frac{|b(v,w)|}{\|v\|_{k+1,p,\Omega}}\leqslant C_2(\Omega)\|b\|\,|w|_{l+1,q,\Omega}, \tag{4.2.46}$$

and inequality (4.2.44) follows by combining inequalities (4.2.45) and (4.2.46). □

Estimate of the error $(\sum_{K\in\mathcal{T}_h}|u-u_h|_{1,K}^2)^{1/2}$

We now prove our main result.

Theorem 4.2.6. *Assume that the solution u is in the space* $H^2(\Omega)$. *Then for any regular family of triangulations there exists a constant C independent of h such that*

$$\|u-u_h\|_h = \Big(\sum_{K\in\mathcal{T}_h}|u-u_h|_{1,K}^2\Big)^{1/2} \leq Ch|u|_{2,\Omega}. \tag{4.2.47}$$

Proof. The central idea of the proof is to apply the bilinear lemma to each term $D_K(u, w_h)$ occuring in a decomposition of the expression $D_h(u, w_h)$ of the form (4.2.37). Some care has to be exercised, however: From (4.2.36), an obvious choice for the bilinear forms D_K is

$$v\in H^2(K),\quad p\in P_K\to \int_K\Big\{\sum_{i=1}^{3}\partial_i v\partial_i p+\Delta vp\Big\}\,\mathrm{d}x = \int_{\partial K}\partial_{\nu_K}vp\,\mathrm{d}\gamma,$$

where ν_K denotes the outer normal along the boundary ∂K of the element K. However, there are not "enough" polynomial invariances at our disposal in such bilinear forms D_K in order to eventually obtain estimates of the form (4.2.38) (the reader should check this statement). Fortunately, there are other choices for a decomposition of the form (4.2.37) which will yield the right estimates. The key idea is *to obtain the desired additional "local" polynomial invariances from a "global" polynomial invariance*, as we now show.

Let Y_h denote the finite element space whose generic finite element is the rectangle of type (1). In other words:

(i) For each $K\in\mathcal{T}_h$, the restrictions $v_h|_K$ span the space $Q_1(K)$.

(ii) Each function $v_h\in Y_h$ is defined by its values at all the vertices of the triangulation. Then we let $W_h = Y_{0h}$ denote the space of all functions $v_h\in Y_h$ which vanish at the boundary nodes. Therefore the *inclusion*

$$W_h\subset \mathscr{C}^0(\bar{\Omega})\cap H_0^1(\Omega)$$

holds, and consequently (cf. Remark 4.2.1), we have

$$\forall v\in H^2(\Omega),\quad \forall w_h\in W_h,\quad D_h(v, w_h)=0, \tag{4.2.48}$$

where it is henceforth understood that the function $D_h\colon H^2(\Omega)\times X_h$ is given by the second expression of (4.2.36), i.e.,

$$D_h(v, w_h) = \sum_{K\in\mathcal{T}_h}\int_K\Big\{\sum_{i=1}^{3}\partial_i v\partial_i w_h+\Delta v w_h\Big\}\,\mathrm{d}x. \tag{4.2.49}$$

Notice that the second inclusion of (4.2.14) implies that the inclusions

$$Y_h \subset X_h \quad \text{and} \quad Y_{0h} = W_h \subset X_{0h} = V_h \tag{4.2.50}$$

hold.

For any function $w_h \in X_h$, let $\Lambda_h w_h$ denote the unique function in the space Y_h which takes the same values as w_h at all the vertices of the triangulation. Notice that, for each $K \in \mathcal{T}_h$, $\Lambda_h w_{h|K} = \Lambda_K(w_{h|K})$, where Λ_K denotes the corresponding $Q_1(K)$-interpolation operator, and that the function $\Lambda_h w_h$ belongs to the space $W_h = Y_{0h}$ if the function w_h belongs to the space $V_h = X_{0h}$. Using relations (4.2.49), we deduce that

$$\forall v \in H^2(\Omega), \quad \forall w_h \in V_h, \quad D_h(v, w_h) = D_h(v, w_h - \Lambda_h w_h), \tag{4.2.51}$$

so that another possible decomposition of the difference $D_h(.,.)$ of (4.2.49) consists in writing

$$\forall v \in H^2(\Omega), \quad \forall w_h \in V_h, \quad D_h(v, w_h) = \sum_{K \in \mathcal{T}_h} D_K(v, w_h),$$

where the bilinear forms $D_K(.,.)$ are now given by

$$\forall v \in H^2(K), \quad \forall p \in P_K, \quad D_K(v, p) = \int_{\partial K} \partial_{\nu_K} v (p - \Lambda_K p)\, \mathrm{d}\gamma. \tag{4.2.52}$$

We observe that, by definition of the operator Λ_K, we have

$$\forall v \in H^2(K), \quad \forall p \in Q_1(K), \quad D_K(v, p) = 0, \tag{4.2.53}$$

and thus we get a *first polynomial invariance*.

To obtain the other polynomial invariance, assume that the function v belongs to the space $P_1(K)$. Then the expression $D_K(v, p)$ is a sum of three terms, each of which is, up to a constant multiplicative factor, the difference between integrals of the expression $(p - \Lambda_K p)$ over opposite faces. Consider one such term, say (with the notations of Fig. 4.2.1):

$$\delta_1 = \int_{K_1'} (p - \Lambda_K p)\, \mathrm{d}x_2\, \mathrm{d}x_3 - \int_{K_1''} (p - \Lambda_K p)\, \mathrm{d}x_2\, \mathrm{d}x_3. \tag{4.2.54}$$

Using the properties of the interpolation operator Λ_K, the identity (4.2.21), and the equations $\partial_{jj}(\Lambda_h p) = 0$, $1 \leqslant j \leqslant 3$, we deduce that

$$p - \Lambda_K p = \sum_{j=1}^{3} \frac{1}{16} \left(\left(\frac{x_j - c_j}{h_j} \right)^2 - 1 \right) \frac{h_j^2}{h_1 h_2 h_3} \int_K \partial_{jj} p\, \mathrm{d}x. \tag{4.2.55}$$

Since the function $(((x_1 - c_1)/h_1)^2 - 1)$ vanishes along the faces K_1' and

K_1'', and since the functions $(((x_j - c_j)/h_j)^2 - 1)$, $j = 2, 3$, take on the same values at the points P_1' and P_1'' (cf. Fig. 4.2.1), we conclude that $\delta_1 = 0$. Likewise, the other similar terms vanish. Consequently, we obtain a *second polynomial invariance*:

$$\forall v \in P_1(K), \quad \forall p \in P_K, \quad D_K(v, p) = 0. \tag{4.2.56}$$

Each expression $D_K(v, p)$ found in (4.2.52) is of the form

$$D_K(v, p) = \sum_{j=1}^{3} \Delta_{j,K}(v, p), \tag{4.2.57}$$

where

$$\Delta_{1,K}(v, p) = \int_{K_1'} \partial_1 v(p - \Lambda_K p)\, dx_2\, dx_3 - \\ - \int_{K_1''} \partial_1 v(p - \Lambda_K p)\, dx_2\, dx_3, \tag{4.2.58}$$

and the expressions $\Delta_{2,K}(v, p)$ and $\Delta_{3,K}(v, p)$ are analogously defined.

Using the standard correspondences $\hat{v} \to v$ between the functions $\hat{v}: \hat{K} \to \mathbf{R}$ and $v: K \to \mathbf{R}$, we obtain

$$\Delta_{1,K}(v, p) = \frac{h_2 h_3}{h_1} \Delta_{1,\hat{K}}(\hat{v}, \hat{p}). \tag{4.2.59}$$

The previous analysis implies that, for each $j \in \{1, 2, 3\}$,

$$\begin{cases} \forall \hat{v} \in H^2(\hat{K}), & \forall \hat{p} \in P_0(\hat{K}), \quad \Delta_{j,\hat{K}}(\hat{v}, \hat{p}) = 0, \\ \forall \hat{v} \in P_1(\hat{K}), & \forall \hat{p} \in P_{\hat{K}}, \quad \Delta_{j,\hat{K}}(\hat{v}, \hat{p}) = 0, \end{cases} \tag{4.2.60}$$

so that, by the bilinear lemma, there exists a constant $\hat{C}$ such that

$$\forall \hat{v} \in H^2(\hat{K}), \quad \forall \hat{p} \in P_{\hat{K}}, \quad |\Delta_{j,\hat{K}}(\hat{v}, \hat{p})| \leq \hat{C} |\hat{v}|_{2,\hat{K}} |\hat{p}|_{1,\hat{K}}. \tag{4.2.61}$$

Using Theorem 3.1.2 and the regularity assumption, there exist constants C such that

$$|\hat{v}|_{2,\hat{K}} \leq C \|B_K\|^2 |\det(B_K)|^{-1/2} |v|_{2,K} \leq C h_K^{1/2} |v|_{2,K}, \tag{4.2.62}$$

$$|\hat{p}|_{1,\hat{K}} \leq C \|B_K\| \, |\det(B_K)|^{-1/2} |p|_{1,K} \leq C h_K^{-1/2} |p|_{1,K}, \tag{4.2.63}$$

so that, upon combining (4.2.57), (4.2.59), (4.2.61), (4.2.62), (4.2.63), we find that there exists a constant C such that

$$\forall K \in \mathcal{T}_h, \quad \forall v \in H^2(K), \quad \forall p \in P_K, \\ |D_K(v, p)| \leq C h_K |v|_{2,K} |p|_{1,K}.$$

This last inequality is of the form (4.2.38) and therefore the proof is complete. □

Remark 4.2.5. Loosely speaking, one may think of the space W_h introduced in the above proof as representing the "conforming" part of the otherwise "nonconforming" space V_h. □

Remark 4.2.6. Adding up equations (4.2.56), we find that

$$\forall p \in P_1(\bar{\Omega}), \quad \forall w_h \in V_h, \quad D_h(p, w_h) = 0.$$

In particular if we restrict ourselves to a basis function $w_i \in V$, whose support is a *patch* $\mathcal{P}_i$, i.e., a union of finite elements $K \in \mathcal{T}_h$, we find that

$$\forall p \in P_1(\mathcal{P}_i), \quad D_h(p, w_i) = 0.$$

This is an instance of *Irons patch test*, which B. Irons was the first to (empirically) recognize as a condition for getting convergence of a nonconforming finite element method. For further details about the patch test, see STRANG & FIX (1973, Section 4.2). □

Exercises

4.2.1. Describe the analog of Wilson's brick in dimension 2, which is known as *Wilson's rectangle* (there should be six degrees of freedom). For the application of this element to the system of plane elasticity, see LESAINT (1976).

4.2.2. Extend the analysis carried out in the text to the case of more general bilinear forms such as $a(u, v) = \int_\Omega \{\sum_{i,j=1}^n a_{ij}\partial_i u \partial_j v + auv\}\,dx$.

4.2.3. (i) Let H be a Hilbert space such that $\bar{V} = H$, $V \hookrightarrow H$, and $V_h \subset H$ for all h, and let, for all $u, v \in V_h + V$,

$$D_h(u, v) = a_h(u, v) - f(v).$$

Finally, assume that the bilinear form is symmetric.

Show that the estimate of the Aubin–Nitsche lemma (Theorem 3.2.4) is replaced in the present situation by

$$|u - u_h| \leq \sup_{g \in H} \frac{1}{|g|} \inf_{\varphi_h \in V_h} \{\tilde{M}\|u - u_h\|\,\|\varphi_g - \varphi_h\| + \\ + |D_h(u, \varphi_g - \varphi_h)| + |D_h(\varphi_g, u - u_h)|\},$$

where $|\cdot|$ denotes the norm in H, and for each $g \in H$, $\varphi_g \in V$ denotes the unique solution of the variational problem

$$\forall v \in V, \quad a(v, \varphi_g) = g(v).$$

This abstract error estimate is found in NITSCHE (1974) and LASCAUX & LESAINT (1975).

(ii) Using part (i), show that, if the solution u is in the space $H^2(\Omega)$, one has (LESAINT (1976))

$$|u - u_h|_{0,\Omega} \leq Ch^2 |u|_{2,\Omega}.$$

It is worth pointing out that, by contrast with (4.2.60), the "full" available polynomial invariances are used in the derivation of the above error estimate in the norm $|\cdot|_{0,\Omega}$.

4.3. Isoparametric finite elements

Isoparametric families of finite elements

Our first task consists in extending the notions of affine-equivalence and affine families which we discussed in Section 2.3. There, we saw how to generate finite elements through affine maps, a construction that will be generalized in Theorem 4.3.1 below. For simplicity we shall restrict ourselves in this section to Lagrange finite elements, leaving the case of Hermite finite elements as a problem (Exercise 4.3.1).

Theorem 4.3.1. *Let $(\hat{K}, \hat{P}, \hat{\Sigma})$ be a Lagrange finite element in $\mathbf{R}^n$ with $\hat{\Sigma} = \{\hat{p}(\hat{a}_i),\ 1 \leq i \leq N\}$ and let there be given a one-to-one mapping $F: \hat{x} \in \hat{K} \to (F_j(\hat{x}))_{j=1}^n \in \mathbf{R}^n$ such that*

$$F_j \in \hat{P}, \quad 1 \leq j \leq n. \tag{4.3.1}$$

Then if we let

$$\begin{cases} K = F(\hat{K}), \\ P = \{p : K \to \mathbf{R};\ \ p = \hat{p} \cdot F^{-1},\ \ \hat{p} \in \hat{P}\}, \\ \Sigma = \{p(F(\hat{a}_i));\ \ 1 \leq i \leq N\}, \end{cases} \tag{4.3.2}$$

the set Σ is P-unisolvent. Consequently, if K is a closed subset of $\mathbf{R}^n$ with a non-empty interior, the triple (K, P, Σ) is a Lagrange finite element.

Proof. Let us establish the bijections:

$$\hat{x} \in \hat{K} \rightarrow x = F(\hat{x}) \in K,$$

$$\hat{p} \in \hat{P} \rightarrow p = \hat{p} \cdot F^{-1} \in P.$$

If $\hat{p}_i$, $1 \leq i \leq N$, denote the basis functions of the finite element $(\hat{K}, \hat{P}, \hat{\Sigma})$, we have for all $p \in P$ and all $x \in K$,

$$p(x) = \hat{p}(\hat{x}) = \sum_{i=1}^{N} \hat{p}(\hat{a}_i)\hat{p}_i(\hat{x}) = \sum_{i=1}^{N} p(a_i)p_i(x),$$

i.e.,

$$\forall p \in P, \quad p = \sum_{i=1}^{N} p(a_i)p_i.$$

The functions p_i, $1 \leq i \leq N$, are linearly independent since $\Sigma_{i=1}^{N} \lambda_i p_i = 0$ implies $\Sigma_{i=1}^{N} \lambda_i \hat{p}_i = 0$ and therefore $\lambda_i = 0$, $1 \leq i \leq N$. In other words, we have shown that the set Σ is P-unisolvent, which completes the proof. □

We shall henceforth use the following notation: To indicate that a mapping $F: \hat{x} \in \hat{K} \rightarrow F(\hat{x}) = (F_j(\hat{x}))_{j=1}^{n} \in \mathbf{R}^n$ satisfies relations (4.3.1), we shall write:

$$F \in (\hat{P})^n \Leftrightarrow F_j \in \hat{P}, \quad 1 \leq j \leq n.$$

Notice that the construction of Theorem 4.3.1 is indeed a generalization of the construction which led to affine-equivalent finite elements, because the inclusion $P_1(\hat{K}) \subset \hat{P}$ is satisfied by all the finite elements hitherto considered.

With Theorem 4.3.1 in mind, we proceed to give several definitions: First, any finite element (K, P, Σ) constructed from another finite element $(\hat{K}, \hat{P}, \hat{\Sigma})$ through the process given in this theorem will be called an *isoparametric finite element*, and the finite element (K, P, Σ) will be said to be *isoparametrically equivalent* to the finite element $(\hat{K}, \hat{P}, \hat{\Sigma})$. Observe that this is *not* a symmetric relation in general, by contrast with the definition of affine-equivalence.

Next, we shall say that the family of finite elements (K, P_K, Σ_K) is an *isoparametric family* if all its elements are isoparametrically equivalent to a single finite element $(\hat{K}, \hat{P}, \hat{\Sigma})$, called the *reference finite element* of the family. In other words, for each K, there exists an *isoparametric mapping* $F_K: \hat{K} \to \mathbf{R}^n$, i.e., which satisfies the relations

$$F_K \in (\hat{P})^n, \quad \textit{and} \quad F_K \textit{ is one-to-one,} \tag{4.3.3}$$

such that

$$\begin{cases} K = F_K(\hat{K}), \\ P_K = \{p : K \to \mathbf{R}; \quad p = \hat{p} \cdot F_K^{-1};\ \hat{p} \in \hat{P}\}, \\ \Sigma_K = \{p(F_K(\hat{a}_i)); \quad 1 \leq i \leq N\}. \end{cases} \tag{4.3.4}$$

As exemplified by the special case of affine-equivalent finite elements, one may consider families of isoparametric finite elements for which the associated mappings F_K belong to some space $(\hat{Q})^n$, where $\hat{Q}$ is a strict subspace of the space $\hat{P}$. Such finite elements are sometimes called *subparametric finite elements*. For examples, see in particular Fig. 4.3.4.

Remark 4.3.1. The prefix "iso" in the adjective "isoparametric" refers to the fact that it is precisely the space $\hat{P}$ of the finite element $(\hat{K}, \hat{P}, \hat{\Sigma})$ which is used in the definition of the mapping F_K (and, consequently, in the definitions of the set K, the space P_K, and the set Σ_K, which use in turn the mapping F_K). □

It is worth pointing out that, by contrast with affine-equivalent finite elements, *the space P_K defined* in (4.3.4) *generally contains functions which are not polynomials, even when the space $\hat{P}$ consists of polynomials only* (see Exercise 4.3.3). However *this complication is ignored in practical computation*, inasmuch as all the computations are performed on the set $\hat{K}$, not on the set K. All that is needed is the knowledge of the mapping F_K, as we shall see in the next section.

In practice, an isoparametric finite element is not directly determined by a mapping F but, rather, by the data of N distinct points a_i, $1 \leq i \leq N$, which in turn uniquely determine a mapping F satisfying

$$F \in (\hat{P})^n \quad \text{and} \quad F(\hat{a}_i) = a_i, \quad 1 \leq i \leq N. \tag{4.3.5}$$

Such a mapping is given by

$$F: \hat{x} \in \hat{K} \to F(\hat{x}) = \sum_{i=1}^{N} \hat{p}_i(\hat{x}) a_i, \tag{4.3.6}$$

as it is readily verified, and it is uniquely defined since for each

$j \in \{1, 2, \ldots, n\}$, we must have, with $a_i = (a_{ji})_{j=1}^n$,

$$F_j \in \hat{P} \quad \text{and} \quad F_j(\hat{a}_i) = a_{ji}, \quad 1 \leq i \leq N,$$

and the set $\hat{\Sigma}$ is $\hat{P}$-unisolvent. However, *in the absence of additional assumptions, nothing guarantees that the mapping* $F: \hat{K} \to F(\hat{K})$ *is invertible*, and indeed this property will require a verification for each example.

Notice that the points a_i are the *nodes* of the finite element (K, P, Σ).

The main interest of isoparametric finite elements is that *the freedom in the choice of the points* a_i *yields more general geometric shapes of sets* K than the polygonal shapes considered up to now. As we shall show in the next section, this property is crucial for getting a good approximation of curved boundaries.

Examples of isoparametric finite elements

Let us next examine several instances of commonly used isoparametric finite elements. For brevity, we shall give a detailed discussion only for our first example, the *isoparametric n-simplex of type* (2), i.e., for which the finite element $(\hat{K}, \hat{P}, \hat{\Sigma})$ is the n-simplex of type (2). Such an isoparametric finite element is determined by the data of $(n+1)$ *vertices* a_i, $1 \leq i \leq n+1$, and $n(n+1)/2$ points which we shall denote by a_{ij}, $1 \leq i < j \leq n+1$. Then (cf. (4.3.5)) there exists a unique mapping F such that

$$\begin{cases} F \in (P_2(\hat{K}))^n, \\ F(\hat{a}_i) = a_i, \quad 1 \leq i \leq n+1, \\ F(\hat{a}_{ij}) = a_{ij}, \quad 1 \leq i < j \leq n+1. \end{cases}$$

This mapping is given by (cf. (2.2.9) and (4.3.6))

$$F: \hat{x} \in \hat{K} \to F(\hat{x}) = \sum_i \lambda_i(\hat{x})(2\lambda_i(\hat{x}) - 1)a_i + \sum_{i<j} 4\lambda_i(\hat{x})\lambda_j(\hat{x})a_{ij}. \tag{4.3.7}$$

Observe that *if it so happened that the points* a_{ij} *were exactly the mid-points* $(a_i + a_j)/2$, *then*, by virtue of the uniqueness of the mapping F, *the mapping* F *would "degenerate" and become affine.*

These considerations are illustrated in Fig. 4.3.1 for $n = 2$, i.e., in the case of the *isoparametric triangle of type* (2).

It is only later (Theorem 4.3.3) that we shall give sufficient conditions

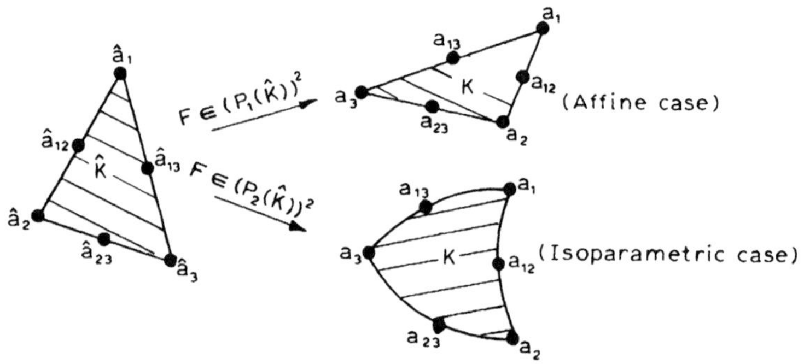

Isoparametric triangle of type (2)
Fig. 4.3.1

which guarantee the invertibility of the mapping F of (4.3.7), but at least we can already indicate that these conditions proceed from a natural idea: When $n = 2$ (cf. Fig. 4.3.1), let us assume that the three vertices a_i, $1 \leqslant i \leqslant 3$, are the vertices of a nondegenerate triangle $\tilde{K}$. Then the mapping $F: \hat{K} \to K$ is invertible if the points a_{ij} are not "too far" from the actual mid-points $(a_i + a_j)/2$ of the triangle $\tilde{K}$ (for a counter-example, see Exercise 4.3.4).

The *boundary* of the set $K = F(\hat{K})$ is composed of *faces*, i.e., the images $F(\hat{K}')$ of the faces $\hat{K}'$ of the n-simplex $\hat{K}$. Since each basis function $\hat{\varphi}$ of the n-simplex $\hat{K}$ of type (2) vanishes along any face of $\hat{K}$ which does not contain the node associated with $\hat{\varphi}$ (cf. Remark 2.3.10), we conclude that each face of the isoparametric n-simplex of type (2) is solely determined by the nodes through which it passes (see also Exercise 4.3.4). This property, which is true of all isoparametric finite elements considered in the sequel (as the reader may check) allows the construction of triangulations made up of isoparametric finite elements (cf. Section 4.4).

We can similarly consider the *isoparametric n-simplex of type* (3) (cf. Fig. 4.3.2 for $n = 2$), for which the mapping F is given by (cf. (2.2.10)):

$$F: \hat{x} \in \hat{K} \to F(\hat{x}) = \sum_i \frac{\lambda_i(\hat{x})(3\lambda_i(\hat{x}) - 1)(3\lambda_i(\hat{x}) - 2)}{2} a_i$$
$$+ \sum_{i \neq j} \frac{9\lambda_i(\hat{x})\lambda_j(\hat{x})(3\lambda_i(\hat{x}) - 1)}{2} a_{iij}$$
$$+ \sum_{i<j<k} 27\lambda_i(\hat{x})\lambda_j(\hat{x})\lambda_k(\hat{x}) a_{ijk}. \tag{4.3.8}$$

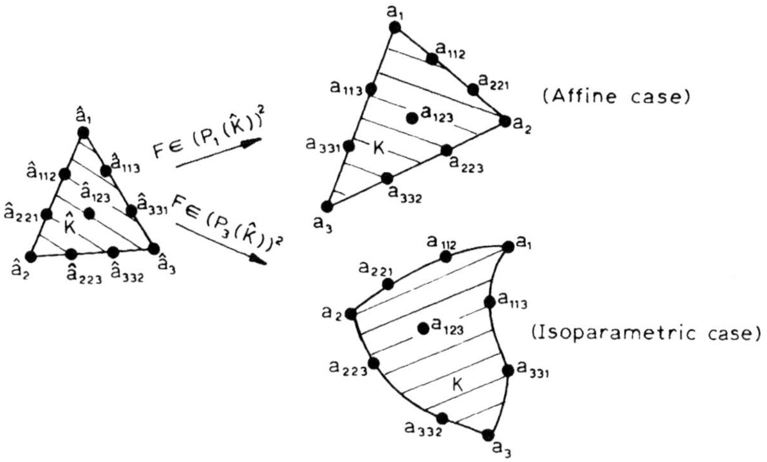

Isoparametric triangle of type (3).
Fig. 4.3.2

Observe in this case that, for $n = 2$, even if *the point a_{123} plays no role in the definition of the boundary of the set K*, the space P_K still depends on its position. We leave it to the reader to similarly define the *isoparametric n-simplex of type* (3′), and the *isoparametric n-simplex of type* (k) for any integer $k \geq 1$. All these isoparametric finite elements are instances of *simplicial* (or *triangular* if $n = 2$, or *tetrahedral* if $n = 3$) *isoparametric finite elements* in the sense that they are isoparametrically equivalent to a finite element for which the set $\hat{K}$ is an n-simplex.

We next describe some examples of *quadrilateral finite elements*, in the sense that they are isoparametrically equivalent to a finite element for which the set $\hat{K}$ is an n-rectangle, for example the unit hypercube $\hat{K} = [0, 1]^n$. In this fashion we obtain the *quadrilateral of type* (1) (cf. Fig. 4.3.3 for $n = 2$).

For $n = 2$, this is an example of a true isoparametric finite element whose sides are not curved! This is so because the functions in the space $Q_1([0, 1]^2)$ are affine in the direction of each coordinate axis. However, this is special to dimension 2. If $n = 3$ for instance, the faces of the set K are portions of hyperbolic paraboloids and are therefore generally curved.

Another example of a quadrilateral finite element is the *quadrilateral of type* (2). In Fig. 4.3.4, we have indicated various subparametric cases of interest for this element, when $n = 2$.

Given a finite element (K, P, Σ) isoparametrically equivalent to a finite

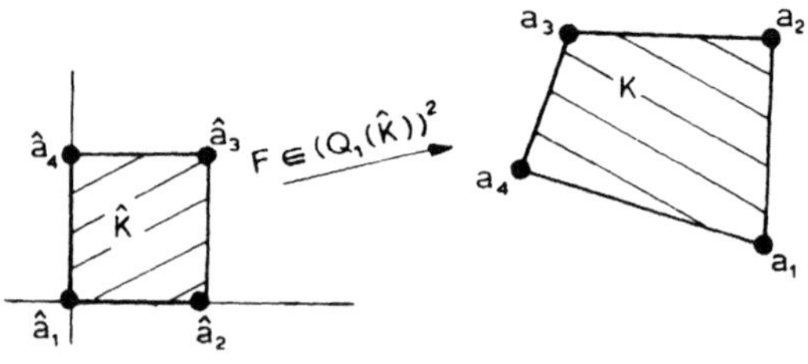

Quadrilateral of type (1) for $n = 2$
Fig. 4.3.3

element $(\hat{K}, \hat{P}, \hat{\Sigma})$ through a mapping F, we shall use the usual correspondences

$$\hat{x} \in \hat{K} \to x = F(\hat{x}) \in K, \tag{4.3.9}$$

$$\hat{p} \in \hat{P} \to p = \hat{p} \cdot F^{-1} \in P, \tag{4.3.10}$$

between the points in the sets $\hat{K}$ and K, and between the functions in the spaces $\hat{P}$ and P, respectively. We shall extend the correspondence (4.3.10) to functions defined over the sets $\hat{K}$ and K by letting

$$(\hat{v}: \hat{K} \to \mathbf{R}) \to (v = \hat{v} \cdot F^{-1}: K \to \mathbf{R}). \tag{4.3.11}$$

Then it is an easy matter to see that the associated $\hat{P}$-interpolation and P-interpolation operators $\hat{\Pi}$ and Π are such that

$$\forall \hat{v} \in \operatorname{dom} \hat{\Pi} = \mathscr{C}^0(\hat{K}), \quad (\Pi v)^{\hat{}} = \hat{\Pi}\hat{v}, \tag{4.3.12}$$

provided $\hat{v} \in \operatorname{dom} \hat{\Pi} \Rightarrow v = \hat{v} \cdot F^{-1} \in \operatorname{dom} \hat{\Pi} = \mathscr{C}^0(K)$ (this condition excludes situations where the mapping F^{-1} would not be continuous).

Estimates of the interpolation errors $|v - \Pi_K v|_{m,q,K}$

The remainder of this section will be devoted to the derivation of an interpolation theory for isoparametric finite elements, i.e., we shall estimate the *interpolation errors* $|v - \Pi_K v|_{m,q,K}$ for finite elements (K, P, Σ) isoparametrically equivalent to a reference finite element $(\hat{K}, \hat{P}, \hat{\Sigma})$. This analysis is carried out in three stages, which parallel those used for affine-equivalent finite elements:

(i) Assuming the $\hat{P}$-interpolation operator $\hat{\Pi}$ leaves the space $P_k(\hat{K})$ invariant, an argument similar to that used in Theorem 3.1.4 yields

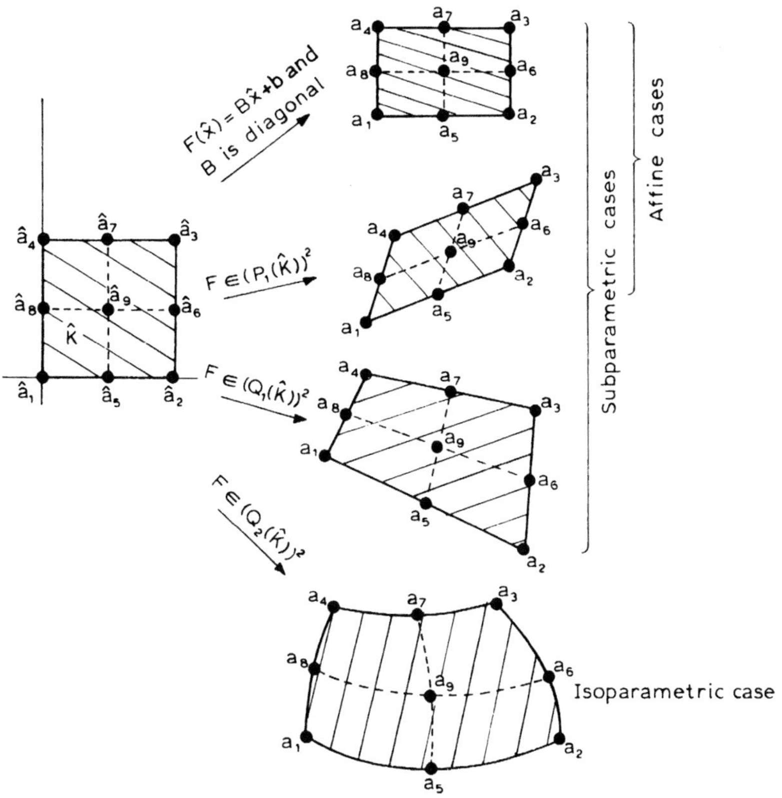

Quadrilateral of type (2) for $n = 2$
Fig. 4.3.4

inequalities of the form

$$|\hat{v} - \hat{\Pi}\hat{v}|_{m,q,\hat{K}} \leq C(\hat{K}, \hat{P}, \hat{\Sigma})|\hat{v}|_{k+1,p,\hat{K}}. \tag{4.3.13}$$

Thus this step is the same as before.

(ii) We then examine how the semi-norms occuring in (4.3.13) are transformed from $\hat{K}$ to K and vice versa. Recall that for affine families, we found inequalities of the form (cf. Theorem 3.1.2):

$$|v|_{m,p,K} \leq C\|B^{-1}\|^m |\det(B)|^{1/p} |\hat{v}|_{m,p,\hat{K}},$$

with $F: \hat{x} \in \hat{K} \to F(\hat{x}) = B\hat{x} + b$. In the present case, we shall find for example (Theorem 4.3.2) that the semi-norms $|v|_{m,p,K}$ are bounded above not only in terms of the semi-norm $|\hat{v}|_{m,p,\hat{K}}$, but instead in terms of all the semi-norms $|\hat{v}|_{l,p,\hat{K}}$, $1 \leq l \leq m$.

(iii) Just as the quantities $\|B\|$, $|\det(B)|$, etc..., which appeared in the affine case were eventually expressed in terms of the geometrical parameters meas(K), h and ρ_K (cf. Theorem 3.1.3), we shall subsequently turn (Theorem 4.3.3) to the problem of estimating analogous quantities (found in step (ii)) in terms of simple geometrical parameters attached to the finite element K.

Thus there are essentially two new steps ((ii) and (iii)) to develop, and rather than giving the general theory (for which the reader is referred to CIARLET & RAVIART (1972b)), we shall concentrate on one example, the *isoparametric n-simplex of type* (2).

We shall use the following notations:

$$\begin{cases} J_F(\hat{x}) = \text{Jacobian of } F \text{ at } \hat{x} = \det(\partial_j F_i(\hat{x})), \\ J_{F^{-1}}(x) = \text{Jacobian of } F^{-1} \text{ at } x = (J_F(\hat{x}))^{-1}, \end{cases} \tag{4.3.14}$$

$$\begin{cases} |F|_{l,\infty,\hat{K}} = \sup\limits_{\hat{x}\in\hat{K}} \|D^l F(\hat{x})\|_{\mathcal{L}_l(\mathbf{R}^n;\mathbf{R}^n)}, \\ |F^{-1}|_{l,\infty,K} = \sup\limits_{x\in K} \|D^l F^{-1}(x)\|_{\mathcal{L}_l(\mathbf{R}^n;\mathbf{R}^n)}, \end{cases} \tag{4.3.15}$$

whenever $F: \hat{K} \to K = F(\hat{K})$ is a sufficiently smooth mapping defined on any subset $\hat{K}$ of $\mathbf{R}^n$ with a sufficiently smooth inverse $F^{-1}: K \to \hat{K}$. Notice that when the mapping F is of the form $F: \hat{x} \to F(\hat{x}) = B\hat{x} + b$, then

$$J_F = \det(B), \quad J_{F^{-1}} = \det(B^{-1}), \; |F|_{1,\infty,\hat{K}} = \|B\|, \; \|F^{-1}\|_{1,\infty,K} = \|B^{-1}\|.$$

Since we are considering n-simplices of type (2), we need apply inequality (4.3.13) with the values $m = 0, 1, 2, 3$ and $k + 1 = 3$ only and thus we shall restrict ourselves to the semi-norms $|\cdot|_{l,p,\Omega}$ with $0 \leq l \leq 3$ in the next theorem. Notice that the following result is valid for general mappings F, i.e., it is irrelevant that the mapping F be in the space $(\hat{P})^n$ for some finite element $(\hat{K}, \hat{P}, \hat{\Sigma})$.

Theorem 4.3.2. *Let Ω and $\hat{\Omega}$ be two bounded open subsets of $\mathbf{R}^n$ such that $\Omega = F(\hat{\Omega})$, where F is a sufficiently smooth one-to-one mapping with a sufficiently smooth inverse $F^{-1}: \Omega \to \hat{\Omega}$.*

Then if a function $\hat{v}: \hat{\Omega} \to \mathbf{R}$ belongs to the space $W^{l,p}(\hat{\Omega})$ for some *integer $l \geqslant 0$ and some number $p \in [1, \infty]$, the function $v = \hat{v} \cdot F^{-1}: \Omega \to \mathbf{R}$ belongs to the space $W^{l,p}(\Omega)$ and, in addition, there exist constants C such that*

$$\forall \hat{v} \in L^p(\hat{\Omega}), \quad |v|_{0,p,\Omega} \leqslant |J_F|_{0,\infty,\hat{\Omega}}^{1/p} |\hat{v}|_{0,p,\hat{\Omega}}, \tag{4.3.16}$$

$$\forall \hat{v} \in W^{1,p}(\hat{\Omega}), \quad |v|_{1,p,\Omega} \leqslant C |J_F|_{0,\infty,\hat{\Omega}}^{1/p} |F^{-1}|_{1,\infty,\Omega} |\hat{v}|_{1,p,\hat{\Omega}}, \tag{4.3.17}$$

$$\begin{aligned} &\forall \hat{v} \in W^{2,p}(\hat{\Omega}), \quad |v|_{2,p,\Omega} \leqslant \\ &\quad \leqslant C |J_F|_{0,\infty,\hat{\Omega}}^{1/p} (|F^{-1}|_{1,\infty,\Omega}^2 |\hat{v}|_{2,p,\hat{\Omega}} + |F^{-1}|_{2,\infty,\Omega} |\hat{v}|_{1,p,\hat{\Omega}}), \end{aligned} \tag{4.3.18}$$

$$\begin{aligned} &\forall \hat{v} \in W^{3,p}(\hat{\Omega}), \quad |v|_{3,p,\Omega} \leqslant \\ &\quad \leqslant C |J_F|_{0,\infty,\hat{\Omega}}^{1/p} (|F^{-1}|_{1,\infty,\Omega}^3 |\hat{v}|_{3,p,\hat{\Omega}} + |F^{-1}|_{1,\infty,\Omega} |F^{-1}|_{2,\infty,\Omega} |\hat{v}|_{2,p,\hat{\Omega}} \\ &\qquad + |F^{-1}|_{3,\infty,\Omega} |\hat{v}|_{1,p,\hat{\Omega}}). \end{aligned} \tag{4.3.19}$$

Proof. As in Theorem 3.1.2, it suffices for $p < \infty$ to prove inequalities (4.3.16) through (4.3.19) for smooth functions (the case $p = \infty$ is left to the reader).

Using the formula for change of variables in multiple integrals, we obtain

$$|v|_{0,p,\Omega}^p = \int_\Omega |v(x)|^p \, dx = \int_\Omega |\hat{v}(F^{-1}(x))|^p \, dx = \int_{\hat{\Omega}} |J_F(\hat{x})| \, |\hat{v}(\hat{x})|^p \, d\hat{x},$$

for which we deduce inequality (4.3.16).

Since $v = \hat{v} \cdot F^{-1}$, we infer that

$$\forall x = F(\hat{x}), \quad Dv(x) = D\hat{v}(\hat{x}) \cdot DF^{-1}(x),$$

and thus,

$$\forall x \in \Omega, \quad \|Dv(x)\| \leqslant |F^{-1}|_{1,\infty,\Omega} \|D\hat{v}(F^{-1}(x))\|.$$

Consequently,

$$\begin{aligned} \int_\Omega \|Dv(x)\|^p \, dx &\leqslant |F^{-1}|_{1,\infty,\Omega}^p \int_\Omega \|D\hat{v}(F^{-1}(x))\|^p \, dx \\ &= |F^{-1}|_{1,\infty,\Omega}^p \int_{\hat{\Omega}} |J_F(\hat{x})| \, \|D\hat{v}(\hat{x})\|^p \, d\hat{x} \\ &\leqslant |F^{-1}|_{1,\infty,\Omega}^p |J_F|_{0,\infty,\hat{\Omega}} \int_{\hat{\Omega}} \|D\hat{v}(\hat{x})\|^p \, d\hat{x}, \end{aligned}$$

from which we deduce inequality (4.3.17), using the equivalence between the semi-norms (cf. (3.1.22) and (3.1.25))

$$v \to |v|_{m,p,\Omega} \quad \text{and} \quad v \to \left(\int_\Omega \|D^m v(x)\|^p \, dx \right)^{1/p}.$$

Likewise, we have for all $x \in \Omega$, $\xi_1 \in \mathbf{R}^n$, $\xi_2 \in \mathbf{R}^n$,

$$\begin{aligned} D^2 v(x)(\xi_1, \xi_2) = {} & D\hat{v}(\hat{x})(D^2 F^{-1}(x)(\xi_1, \xi_2)) + \\ & + D^2 \hat{v}(\hat{x})(DF^{-1}(x)\xi_1, DF^{-1}(x)\xi_2), \end{aligned}$$

so that we obtain, for all $x = F(\hat{x}) \in \Omega$,

$$\begin{aligned} \|D^2 v(x)\| &= \sup_{\left\{ \substack{\|\xi_l\| \leq 1 \\ l=1,2} \right.} |D^2 v(x)(\xi_1, \xi_2)| \leq \\ &\leq |F^{-1}|_{2,\infty,\Omega} \|D\hat{v}(\hat{x})\| + |F^{-1}|_{1,\infty,\Omega}^2 \|D^2 \hat{v}(\hat{x})\|. \end{aligned}$$

Therefore,

$$\begin{aligned} \left(\int_\Omega \|D^2 v(x)\|^p \, dx \right)^{1/p} \leq {} & |F^{-1}|_{2,\infty,\Omega} \left(\int_\Omega \|D\hat{v}(F^{-1}(x))\|^p \, dx \right)^{1/p} + \\ & + |F^{-1}|_{1,\infty,\Omega}^2 \left(\int_\Omega \|D^2 \hat{v}(F^{-1}(x))\|^p \, dx \right)^{1/p}. \end{aligned}$$

Arguing as before, we find that, for any integer l,

$$\begin{aligned} \left(\int_\Omega \|D^l \hat{v}(F^{-1}(x))\|^p \, dx \right)^{1/p} &= \left(\int_{\hat{\Omega}} |J_F(\hat{x})| \, \|D^l \hat{v}(\hat{x})\|^p \, d\hat{x} \right)^{1/p} \\ &\leq |J_F|_{0,\infty,\hat{\Omega}}^{1/p} \left(\int_{\hat{\Omega}} \|D^l \hat{v}(\hat{x})\|^p \, d\hat{x} \right)^{1/p}, \end{aligned}$$

and thus inequality (4.3.18) is proved.

Inequality (4.3.19) is proved analogously by using the following inequality:

$$\begin{aligned} \|D^3 v(x)\| \leq {} & |F^{-1}|_{3,\infty,\Omega} \|D\hat{v}(\hat{x})\| + 3 |F^{-1}|_{1,\infty,\Omega} |F^{-1}|_{2,\infty,\Omega} \|D^2 \hat{v}(\hat{x})\| \\ & + |F^{-1}|_{1,\infty,\Omega}^3 \|D^3 \hat{v}(\hat{x})\| \end{aligned}$$

which the reader may easily establish for all $x = F(\hat{x}) \in \Omega$. □

To apply the previous theorem, we must next obtain estimates of the following quantities:

$$|J_F|_{0,\infty,\hat{K}};\ |J_{F^{-1}}|_{0,\infty,K};\ |F|_{l,\infty,\hat{K}},\ l = 1, 2, 3;$$

$$|F^{-1}|_{l,\infty,K},\ l = 1, 2,$$

for an isoparametric n-simplex of type (2). To do this, the key idea is the following: Since the affine case is a special case of the isoparametric case, we may expect the same type of error bounds, provided the mapping F is not "too far" from the unique affine mapping $\tilde{F}$ which satisfies

$$\tilde{F}(\hat{a}_i) = a_i, \quad 1 \leq i \leq n+1. \tag{4.3.20}$$

Therefore we are naturally led to introduce the n-simplex

$$\tilde{K} = \tilde{F}(\hat{K}) \tag{4.3.21}$$

and the points

$$\tilde{a}_{ij} = \tilde{F}(\hat{a}_{ij}), \quad 1 \leq i < j \leq n+1. \tag{4.3.22}$$

Looking at Fig. 4.3.5 (where we have represented the case $n = 2$), we expect the vectors $(a_{ij} - \tilde{a}_{ij})$ to serve as a good measure of the discrepancy between the mappings F and $\tilde{F}$: Indeed, if we let $\hat{p}_{ij}$ denote the basis functions of the n-simplex of type (2) attached to the node $\hat{a}_{ij}$, we have

$$F = \tilde{F} + \sum_{i<j} \hat{p}_{ij}(a_{ij} - \tilde{a}_{ij}). \tag{4.3.23}$$

To see this, it suffices to verify that the mapping

$$G = \tilde{F} + \sum_{i<j} \hat{p}_{ij}(a_{ij} - \tilde{a}_{ij})$$

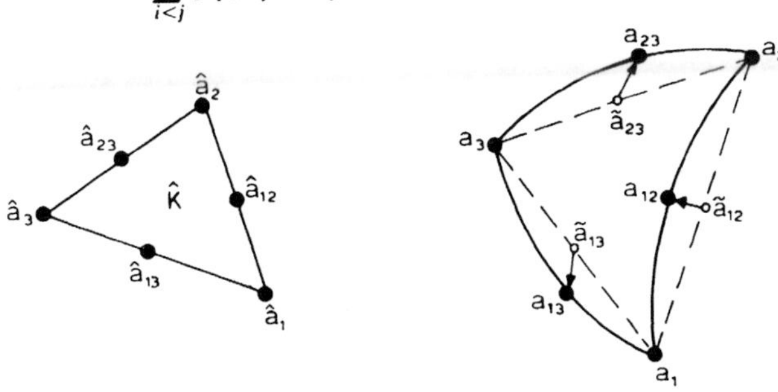

Fig. 4.3.5

satisfies the relations

$$\begin{cases} G \in (P_2(\hat{K}))^n, \\ G(\hat{a}_i) = a_i, \quad 1 \leqslant i \leqslant n+1, \\ G(\hat{a}_{ij}) = a_{ij}, \quad 1 \leqslant i < j \leqslant n+1, \end{cases}$$

which precisely characterize the unique isoparametric mapping F.

Let there be given an isoparametric family of n-simplices K of type (2), each of which is determined by the data of vertices $a_{i,K}$, $1 \leqslant i \leqslant n+1$, and points $a_{ij,K}$, $1 \leqslant i < j \leqslant n+1$. Then in view of (4.3.20) and (4.3.21), we let $\tilde{F}_K$ denote for each K the unique affine mapping which satisfies $\tilde{F}_K(\hat{a}_i) = a_{i,K}$, $1 \leqslant i \leqslant n+1$, and we define the n-simplex $\tilde{K} = \tilde{F}_K(\hat{K})$. Finally, we let, for each $\tilde{K}$,

$$h_K = \operatorname{diam}(\tilde{K}), \tag{4.3.24}$$

$$\rho_K = \text{diameter of the sphere inscribed in } \tilde{K}. \tag{4.3.25}$$

We shall say that *an isoparametric family of n-simplices K of type* (2) *is regular* if the following three conditions are satisfied:

(i) There exists a constant σ such that

$$\forall K, \quad \frac{h_K}{\rho_K} \leqslant \sigma. \tag{4.3.26}$$

(ii) The quantities h_K approach zero.
(iii) We have

$$\|a_{ij,K} - \tilde{a}_{ij,K}\| = \mathrm{O}(h_K^2), \quad 1 \leqslant i < j \leqslant n+1, \tag{4.3.27}$$

where, for each K, we let $\tilde{a}_{ij,K} = \tilde{F}_K(\hat{a}_{ij})$.

Remark 4.3.2. In the special case of an affine family, condition (4.3.27) is automatically satisfied ($a_{ij,K} = \tilde{a}_{ij,K}$), so that the above definition contains the definition of a regular affine family which was given in Section 3.1. □

Although it is clear that condition (4.3.27) does insure that the mappings F_K and $\tilde{F}_K$ do not differ too much, the reason the vectors $(a_{ij,K} - \tilde{a}_{ij,K})$ have to be precisely of order $\mathrm{O}(h_K^2)$ may seem arbitrary at this stage. As we shall show later (cf. Theorem 4.3.4), the basic justification of this assumption is that it yields the same interpolation error estimates as in the affine case.

To begin with, we show that this assumption allows us to obtain upper bounds of the various quantities found in the inequalities of Theorem 4.3.2.

We are also able to show in this particular case that the mappings F_K are invertible (the invertibility of the mapping F_K is part of the definition of an isoparametric family).

Theorem 4.3.3. *Let there be given a regular isoparametric family of n-simplices of type* (2). *Then, provided h_K is small enough, the mappings $F_K: \hat{K} \to K = F_K(\hat{K})$ are one-to-one, their Jacobians J_{F_K} do not vanish, and there exist constants C such that*

$$|F_K|_{1,\infty,\hat{K}} \leq Ch_K, \quad |F_K|_{2,\infty,\hat{K}} \leq Ch_K^2, \quad |F_K|_{3,\infty,\hat{K}} = 0, \tag{4.3.28}$$

$$|F_K^{-1}|_{1,\infty,K} \leq \frac{C}{h_K}, \quad |F_K^{-1}|_{2,\infty,K} \leq \frac{C}{h_K}, \tag{4.3.29}$$

$$|J_{F_K}|_{0,\infty,\hat{K}} \leq C\,\text{meas}(\tilde{K}), \quad |J_{F_K^{-1}}|_{0,\infty,K} \leq \frac{C}{\text{meas}(\tilde{K})}. \tag{4.3.30}$$

Proof. For notational convenience, we shall drop the index K throughout the proof. Using the decomposition (4.3.23) of the mapping F, we deduce that, for all $\hat{x} \in \hat{K}$,

$$DF(\hat{x}) = D\tilde{F}(\hat{x}) + E(\hat{x}) = B + E(\hat{x}), \tag{4.3.31}$$

where

$$E(\hat{x}) = \sum_{i<j} (a_{ij} - \tilde{a}_{ij}) D\hat{p}_{ij}(\hat{x}).$$

Therefore, by virtue of assumption (4.3.27), and since the basis functions $\hat{p}_{ij}$ are independent of K, we find that

$$\sup_{\hat{x}\in\hat{K}} \|E(\hat{x})\| \leq Ch^2 \tag{4.3.32}$$

(as usual the same letter C stands for various constants). Thus we have

$$|F|_{1,\infty,\hat{K}} = \sup_{\hat{x}\in\hat{K}} \|DF(\hat{x})\| \leq \|B\| + \sup_{\hat{x}\in\hat{K}} \|E(\hat{x})\| \leq Ch,$$

since $\|B\| \leq Ch$ (cf. Theorem 3.1.3). Likewise, we have

$$D^2F(\hat{x}) = DE(\hat{x}),$$

since $D^2\tilde{F} = 0$, and, arguing as before, we find that

$$\sup_{\hat{x}\in\hat{K}} \|DE(\hat{x})\| \leq Ch^2,$$

so that

$$|F|_{2,\infty,\hat{K}} = \sup_{\hat{x}\in\hat{K}} \|D^2F(\hat{x})\| \leq Ch^2.$$

Hence all relations (4.3.28) are proved, the last one being obvious since $F \in (P_2(\hat{K}))^n$.

Considered as a function of its column vectors $\partial_j F(\hat{x})$, $1 \leq j \leq n$, the determinant $J_F(\hat{x}) = \det(DF(\hat{x}))$ is a continuous multilinear mapping and therefore there exists a constant $C = C(n)$ such that

$$\forall \hat{x} \in \hat{K}, \quad J_F(\hat{x}) \leq C \sum_{j=1}^{n} \|\partial_j F(\hat{x})\|.$$

Since the inequality $|F|_{1,\infty,\hat{K}} \leq Ch$ proved above implies the similar inequalities $\sup_{\hat{x}\in\hat{K}} \|\partial_j F(\hat{x})\| \leq Ch$, $1 \leq j \leq n$, we deduce that

$$|J_F|_{0,\infty,\hat{K}} = \sup_{\hat{x}\in\hat{K}} |J_F(\hat{x})| \leq Ch^n \leq C \text{ meas } \tilde{K},$$

and the first inequality of (4.3.30) is proved.

Because of assumption (4.3.26), the matrices B are all invertible so that we can write (4.3.31) in the form

$$DF(\hat{x}) = B(I + B^{-1}E(\hat{x})).$$

Using inequality $\|B^{-1}\| \leq C/h$ (cf. Theorem 3.1.3 and assumptions (4.3.26)) and inequality (4.3.32), we deduce that $\sup_{\hat{x}\in\hat{K}} \|B^{-1}E(\hat{x})\| \leq Ch$. Let then γ be a fixed number in the interval $]0, 1[$. There exists $h_0 > 0$ such that

$$\forall h \leq h_0, \quad \sup_{\hat{x}\in\hat{K}} \|B^{-1}E(\hat{x})\| \leq \gamma,$$

so that, for $h \leq h_0$, the operator $(I + B^{-1}E(\hat{x}))$ is invertible for each $\hat{x} \in \hat{K}$, and

$$\sup_{\hat{x}\in\hat{K}} \|(I + B^{-1}E(\hat{x}))^{-1}\| \leq \frac{1}{1-\gamma}. \tag{4.3.33}$$

This shows that the derivative $DF(\hat{x})$ is invertible for all $\hat{x} \in \hat{K}$, with

$$(DF(\hat{x}))^{-1} = (I + B^{-1}E(\hat{x}))^{-1}B^{-1}. \tag{4.3.34}$$

We next prove that the mapping $F: \hat{K} \to K$ *is invertible* (by the implicit function theorem, we can only deduce that the mapping F is invertible *locally*, i.e., in a sufficiently small neighborhood of each point of $\hat{K}$; this is why the global invertibility requires an additional analysis; for a more general approach, see Exercise 4.3.6). Let $\hat{x}$, $\hat{y} \in \hat{K}$ be such that $F(\hat{x}) = F(\hat{y})$. Since the set $\hat{K}$ is convex, we may apply Taylor formula:

$$F(\hat{y}) = F(\hat{x}) + DF(\hat{x})(\hat{y} - \hat{x}) + \frac{A}{2}(\hat{y} - \hat{x})^2,$$

where $A \in \mathscr{L}_2(\mathbf{R}^n; \mathbf{R}^n)$ is the constant second derivative of the mapping F. We deduce that

$$DF(\hat{x})(\hat{y} - \hat{x}) = -\frac{A}{2}(\hat{y} - \hat{x})^2 = -\frac{A}{2}(\hat{x} - \hat{y})^2 = DF(\hat{y})(\hat{x} - \hat{y}),$$

and consequently,

$$(DF(\hat{x}) + DF(\hat{y}))(\hat{y} - \hat{x}) = 0.$$

Each component F_i of the mapping F is in the space $P_2(\hat{K})$, so that we have $((\hat{x} + \hat{y})/2 \in \hat{K})$:

$$\partial_j F_i \in P_1(\hat{K}) \Rightarrow \partial_j F_i(\hat{x}) + \partial_j F_i(\hat{y}) = 2\partial_j F_i\left(\frac{\hat{x} + \hat{y}}{2}\right), \quad 1 \leq i, j \leq n,$$

i.e.,

$$0 = (DF(\hat{x}) + DF(\hat{y}))(\hat{y} - \hat{x}) = 2DF\left(\frac{\hat{x} + \hat{y}}{2}\right)(\hat{y} - \hat{x}).$$

Since the derivative $DF((\hat{x} + \hat{y})/2)$ is an invertible mapping in $\mathscr{L}(\mathbf{R}^n)$, we conclude that $\hat{x} = \hat{y}$.

We can write

$$\forall \hat{x} \in \hat{K}, \ (DF(\hat{x}))^{-1} = DF^{-1}(x),$$

and thus, by relations (4.3.33) and (4.3.34),

$$|F^{-1}|_{1,\infty,K} = \sup_{x \in K} \|DF^{-1}(x)\| \leq \frac{C}{h},$$

which proves the first inequality of (4.3.29).

Given functions $F: \mathbf{R}^n \to \mathbf{R}^n$ and $G: \mathbf{R}^n \to \mathbf{R}^n$, the function $H = G \cdot F: \mathbf{R}^n \to \mathbf{R}^n$ is such that, for all vectors $\xi_1, \xi_2 \in \mathbf{R}^n$,

$$D^2H(\hat{x})(\xi_1, \xi_2) = DG(x)(D^2F(\hat{x})(\xi_1, \xi_2)) +$$
$$+ D^2G(x)(DF(\hat{x})\xi_1, DF(\hat{x})\xi_2).$$

If we apply this formula with $G = F^{-1}$, so that $H = I$, we obtain, for all $x = F(\hat{x}) \in K$,

$$D^2F^{-1}(x)(DF(\hat{x})\xi_1, DF(\hat{x})\xi_2) = -DF^{-1}(x)(D^2F(\hat{x})(\xi_1, \xi_2)).$$

Since for each $x = F(\hat{x}) \in K$, the mapping $DF(\hat{x}): \mathbf{R}^n \to \mathbf{R}^n$ is invertible, we deduce that for all vectors $\eta_1, \eta_2 \in \mathbf{R}^n$,

$$D^2F^{-1}(x)(\eta_1, \eta_2) = -DF^{-1}(x)(D^2F(\hat{x})(DF^{-1}(x)\eta_1, DF^{-1}(x)\eta_2)),$$

and thus,

$$\|D^2F^{-1}(x)\| = \sup_{\substack{\|\eta_l\| \le 1 \\ l=1,2}} \|D^2F^{-1}(x)(\eta_1, \eta_2)\| \le \|D^2F(\hat{x})\| \, \|DF^{-1}(x)\|^3,$$

so that, using the second inequality of (4.3.28) and the first inequality (4.3.29),

$$|F^{-1}|_{2,\infty,K} = \sup_{x \in K} \|D^2F^{-1}(x)\| \le |F|_{2,\infty,\hat{K}} |F^{-1}|^3_{1,\infty,K} \le \frac{C}{h},$$

and the second inequality of (4.3.29) is proved.

Using (4.3.34), we can write

$$\forall \hat{x} \in \hat{K}, \quad B = DF(\hat{x})(I + B^{-1}E(\hat{x}))^{-1},$$

and thus, by (4.3.33),

$$\forall \hat{x} \in \hat{K}, \; |\det(B)| = |J_F(\hat{x})| |\det(I + B^{-1}E(\hat{x}))^{-1}| \le \frac{|J_F(\hat{x})|}{(1-\gamma)^n}.$$

Therefore, we deduce that

$$\frac{1}{|J_{F^{-1}}|_{0,\infty,K}} = \frac{1}{\sup_{x \in K} |J_{F^{-1}}(x)|} = \inf_{\hat{x} \in \hat{K}} J_F(\hat{x}) \ge$$
$$\ge (1-\gamma)^n |\det(B)| \ge C \operatorname{meas}(\tilde{K}),$$

and the second inequality of (4.3.30) is proved, which completes the proof. □

Combining Theorems 4.3.2 and 4.3.3, we are in a position to prove our main result (compare with Theorems 3.1.6).

Theorem 4.3.4. *Let there be given a regular isoparametric family of n-simplices K of type* (2) *and let there be given an integer $m \geq 0$ and two numbers $p, q \in [1, \infty]$ such that the following inclusions hold*:

$$W^{3,p}(\hat{K}) \hookrightarrow \mathscr{C}^0(\hat{K}), \tag{4.3.35}$$

$$W^{3,p}(\hat{K}) \hookrightarrow W^{m,q}(\hat{K}), \tag{4.3.36}$$

where $\hat{K}$ is the reference n-simplex of type (2) *of the family.*

Then provided the diameters h_K are small enough, there exists a constant C such that, for all finite elements in the family, and all functions $v \in W^{3,p}(K)$,

$$\|v - \Pi_K v\|_{m,q,K} \leq C(\operatorname{meas}(\tilde{K}))^{1/q-1/p} h_K^{3-m}(|v|_{2,p,K} + |v|_{3,p,K}). \tag{4.3.37}$$

Proof. The inclusion (4.3.35) guarantees the existence of the interpolation operators $\hat{\Pi}$ and Π_K, which satisfy the relation (4.3.12). Combining the inequalities of Theorems 4.3.2 and 4.3.3, we obtain, if $m = 0$, 1 or 2, respectively,

$$\begin{aligned}
|v - \Pi_K v|_{0,q,K} &\leq |J_{F_K}|_{0,\infty,\hat{K}}^{1/q} |\hat{v} - \hat{\Pi}\hat{v}|_{0,q,\hat{K}} \\
&\leq C(\operatorname{meas}(\tilde{K}))^{1/q} |\hat{v} - \hat{\Pi}\hat{v}|_{0,q,\hat{K}}, \\
|v - \Pi_K v|_{1,q,K} &\leq C |J_{F_K}|_{0,\infty,\hat{K}}^{1/q} |F_K^{-1}|_{1,\infty,K} |\hat{v} - \hat{\Pi}\hat{v}|_{1,q,\hat{K}} \\
&\leq C(\operatorname{meas}(\tilde{K}))^{1/q} \left(\frac{1}{h_K} |\hat{v} - \hat{\Pi}\hat{v}|_{1,q,\hat{K}}\right), \\
|v - \Pi_K v|_{2,q,K} &\leq C |J_{F_K}|_{0,\infty,\hat{K}}^{1/q} (|F_K^{-1}|_{1,\infty,K}^2 |\hat{v} - \hat{\Pi}\hat{v}|_{2,q,\hat{K}} \\
&\quad + |F_K^{-1}|_{2,\infty,K} |\hat{v} - \hat{\Pi}\hat{v}|_{1,q,\hat{K}}) \\
&\leq C(\operatorname{meas}(\tilde{K}))^{1/q} \left(\frac{1}{h_K^2}\left(|\hat{v} - \hat{\Pi}\hat{v}|_{2,q,\hat{K}}\right.\right. \\
&\quad \left.\left. + \frac{1}{h_K} |\hat{v} - \hat{\Pi}\hat{v}|_{1,q,\hat{K}}\right)\right).
\end{aligned}$$

By virtue of the inclusions (4.3.35) and (4.3.36), we may argue as in Theorem 3.1.4 and infer that there exists a constant C depending only on the set $\hat{K}$ such that for all $\hat{v} \in W^{3,p}(\hat{K})$,

$$|\hat{v} - \hat{\Pi}\hat{v}|_{l,q,\hat{K}} \leq C|\hat{v}|_{3,p,\hat{K}}, \quad l \leq m.$$

Upon combining the above inequalities, we obtain

$$|v - \Pi_K v|_{m,q,K} \leq C(\text{meas}(\tilde{K}))^{1/q} \frac{1}{h_K^m} |\hat{v}|_{3,p,\hat{K}},$$

and another application of Theorems 4.3.2 and 4.3.3 yields:

$$\begin{aligned} |\hat{v}|_{3,p,\hat{K}} &\leq C|J_{F_K^{-1}}|_{0,\infty,K}^{1/p}(|F_K|_{1,\infty,\hat{K}}^3|v|_{3,p,K} + \\ &\quad + |F_K|_{1,\infty,\hat{K}}|F_K|_{2,\infty,\hat{K}}|v|_{2,p,K} + |F_K|_{3,\infty,\hat{K}}|v|_{1,p,K}) \\ &\leq C(\text{meas}(\tilde{K}))^{-1/p} h_K^3(|v|_{2,p,K} + |v|_{3,p,K}). \end{aligned}$$

Thus inequality (4.3.37) is proved for the values $m = 0, 1$ and 2. The case $m = 3$ is left as a problem (Exercise 4.3.7). □

It is interesting to compare the estimates of the above theorem with the analogous estimates obtained for a regular *affine* family of n-simplices of type (2) (cf. Theorem 3.1.6):

$$\|v - \Pi_K v\|_{m,q,K} \leq C(\text{meas}(\tilde{K}))^{1/q-1/p} h_K^{3-m} |v|_{3,p,K}.$$

We conclude that the two estimates coincide except for the additional semi-norm $|v|_{2,p,K}$ (which appears when one differentiates a function composed with other than an affine function; cf. the end of the proof of Theorem 4.3.2). Also, the present estimates have been established under the additional assumption that the diameters h_K are sufficiently small, basically to insure the invertibility of the derivatives $DF_K(\hat{x})$, $\hat{x} \in \hat{K}$ (cf. the proof of Theorem 4.3.3).

Remark 4.3.3. (i) Just as in the case of affine families (cf. Remark 3.1.3), the parameter meas$(\tilde{K})$ can be replaced by h_K^n in inequality (4.3.37), since it satisfies (cf. (4.3.26)) the inequalities

$$\sigma_n \sigma^{-n} h_K^n \leq \text{meas}(\tilde{K}) \leq \sigma_n h_K^n,$$

where σ_n denotes the dx-measure of the unit sphere in $\mathbf{R}^n$.

(ii) If necessary, the expression $(|v|_{2,p,K} + |v|_{3,p,K})$ appearing in the right-hand side of inequality (4.3.37) can be of course replaced by the expression $(|v|_{2,p,K}^p + |v|_{3,p,K}^p)^{1/p}$. □

Similar analyses can be carried out for other types of simplicial finite elements, such as the isoparametric n-simplex of type (3) (cf. Exercise 4.3.8). For the general theory, which also applies to isoparametric Hermite finite elements, see CIARLET & RAVIART (1972b).

If we turn to quadrilateral finite elements, the situation is different. Of course, we could again consider this case as a perturbation of the affine case. But, as exemplified by Fig. 4.3.4, this would reduce the possible shapes to "nearly parallelograms". Hopefully, a new approach can be developed whereby the admissible shapes correspond to mappings F_K which are perturbations of mappings $\tilde{F}_K$ in the space $(Q_1(\hat{K}))^n$, instead of the space $(P_1(\hat{K}))^n$. Accordingly, a new theory has to be developed, in particular for the quadrilateral of type (1), as indicated in Exercise 4.3.9.

Exercises

4.3.1. Let $(\hat{K}, \hat{P}, \hat{\Sigma})$ be a Hermite finite element where the order of directional derivatives occuring in the definition is one, i.e., the set $\hat{\Sigma}$ is of the form

$$\hat{\Sigma} = \{\hat{\varphi}_i^0,\ 1 \leqslant i \leqslant N_0;\ \hat{\varphi}_{ik}^1,\ 1 \leqslant k \leqslant d_i,\ 1 \leqslant i \leqslant N_1\},$$

with degrees of freedom of the following form:

$$\hat{\varphi}_i^0\colon \hat{p} \to \hat{p}(\hat{a}_i^0),\ \hat{\varphi}_{ik}^1\colon \hat{p} \to D\hat{p}(\hat{a}_i^1)\hat{\xi}_{ik}^1.$$

(i) Let $F\colon \hat{K} \to \mathbf{R}^n$ be a differentiable one-to-one mapping. Let $a_i^0 = F(\hat{a}_i^0)$, $1 \leqslant i \leqslant N_0$; $a_i^1 = F(\hat{a}_i^1)$ and $\xi_{ik}^1 = DF(\hat{a}_i^1)\hat{\xi}_{ik}^1$, $1 \leqslant k \leqslant d_i$, $1 \leqslant i \leqslant N_1$. Then show that the triple (K, P, Σ) is a Hermite finite element, where

$$\begin{cases} K = F(\hat{K}),\ P = \{p\colon K \to \mathbf{R};\ p = \hat{p} \cdot F^{-1},\ \hat{p} \in \hat{P}\}, \\ \Sigma = \{\varphi_i^0,\ 1 \leqslant i \leqslant N_0;\ \varphi_{ik}^1,\ 1 \leqslant k \leqslant d_i,\ 1 \leqslant i \leqslant N_1\}, \\ \varphi_i^0\colon p \to p(a_i^0),\ \varphi_{ik}^1\colon p \to Dp(a_i^1)\xi_{ik}^1, \end{cases}$$

and show that $(\Pi v)\hat{} = \hat{\Pi}\hat{v}$.

(ii) If the mapping F belongs to the space $(\hat{P})^n$, one obtains in this fashion an *isoparametric Hermite finite element*. In this case, write the mapping F in terms of the basis functions of the finite element $(\hat{K}, \hat{P}, \hat{\Sigma})$. Deduce that, in practice, the isoparametric finite element is completely determined by the data of the points a_i^0 and of the vectors ξ_{ik}^1. Why are the points a_i^1 not arbitrary? Show that this is again a generalization of affine equivalence.

(iii) Using (ii), construct the *isoparametric Hermite triangle of type* (3) which is thus defined by the data of three "vertices" a_i, $1 \leq i \leq 3$, two directions at each point a_i, $1 \leq i \leq 3$, and a point a_{123}.

(iv) Examine whether the construction of (i) and (ii) could be extended to Hermite finite elements in the definition of which higher order directional derivatives are used.

For a reference about questions (i), (ii) and (iii), see CIARLET & RAVIART (1972b).

4.3.2. Let (K, P, Σ) be an isoparametric finite element derived from a finite element $(\hat{K}, \hat{P}, \hat{\Sigma})$ by the construction of Theorem 4.3.1. Show that if the space $\hat{P}$ contains constant functions, then the space P always contains polynomials of degree 1 in the variables $x_1, x_2, \ldots, x_n$. Isn't there a paradox?

4.3.3. Give a description of the space P_K corresponding to the isoparametric triangle of type (2). In particular show that one has $P_K \neq P_2(K)$ in general (although the inclusion $P_1(K) \subset P_K$ holds; cf. Exercise 4.3.2).

4.3.4. (i) Let a_i and a_j, $i \neq j$, be two "vertices" of an isoparametric triangle of type (2). Show that the curved "side" joining these two points is an arc of parabola uniquely determined by the following conditions: It passes through the points a_i, a_j, a_{ij} and its asymptotic direction is parallel to the vector $a_{ij} - (a_i + a_j)/2$.

(ii) Use the result of (i) to show that the mapping F corresponding to the following data is not invertible:

$$a_1 = (0, 0),\ a_2 = (2, 0),\ a_3 = (0, 2),\ a_{12} = (1, 0),\ a_{13} = (1, 1),\ a_{23} = (0, 1).$$

4.3.5. Given a regular isoparametric family of n-simplices of type (2), do we have $\operatorname{diam} K = \operatorname{diam} \tilde{K}$ for $h_K = \operatorname{diam} \tilde{K}$ sufficiently small?

4.3.6. In Theorem 4.3.3, it was shown that the isoparametric mappings F_K are one-to-one (for h_K small enough) by an argument special to isoparametric n-simplices of type (2). Give a more general proof, which would apply to other isoparametric finite elements.

4.3.7. Complete the proof of inequalities (4.3.37) by considering the case $m = 3$.

4.3.8. (i) Carry out an analysis similar to the one given in the text for the isoparametric n-simplex of type (3) (cf. Fig. 4.3.2 for $n = 2$). Introducing the unique affine mapping $\tilde{F}_K$ which satisfies $\tilde{F}_K(\hat{a}_i) = a_i$, $1 \leq i \leq n + 1$, show that one obtains interpolation error estimates of the form

$$\|v - \Pi_K v\|_{m,q,K} \leq C(\text{meas}(\tilde{K}))^{1/q-1/p} h_K^{4-m} \|v\|_{4,p,K},$$

i.e., as in the affine case, provided we consider a *regular* isoparametric family for which condition (iii) of (4.3.27) is replaced by the following:

$$(*)\ \|a_{iij,K} - \tilde{a}_{iij,K}\| = O(h_K^3), \quad 1 \leq i, j \leq n+1,\ i \neq j,$$

$$(**)\ \|a_{ijk,K} - \tilde{a}_{ijk,K}\| = O(h_K^3), \quad 1 \leq i < j < k \leq n+1,$$

where $\tilde{a}_{iij,K} = \tilde{F}_K(\hat{a}_{iij})$ and $\tilde{a}_{ijk,K} = \tilde{F}(\hat{a}_{ijk})$.

(ii) It is clear however that if the points $a_{iij,K}$ are taken from an actual boundary (as they would be in practice), the above condition (*) cannot be satisfied since one has instead in this situation $\|a_{iij,K} - \tilde{a}_{iij,K}\| = O(h_K^2)$. There is nevertheless one case where this difficulty can be circumvented: Assume that $n = 2$ and that (cf. Fig. 4.3.6 where the indices K have been dropped for convenience)

$$a_{331,K} = \tilde{a}_{331,K}, \quad a_{113,K} = \tilde{a}_{113,K}, \quad a_{332,K} = \tilde{a}_{332,K}, \quad a_{223,K} = \tilde{a}_{223,K}.$$

Then show that the estimates of (i) hold with assumptions (*) and (**)

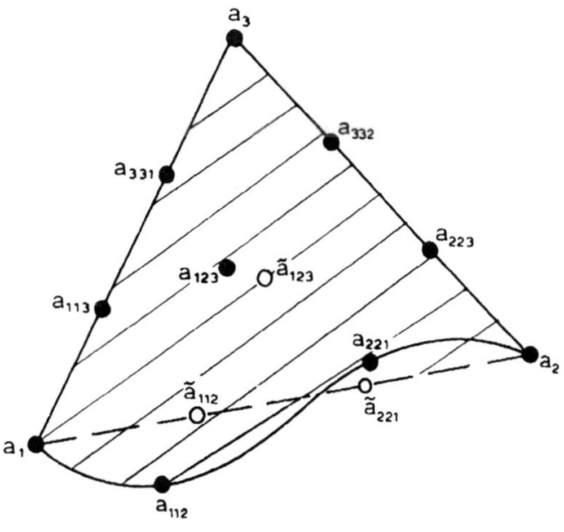

Fig. 4.3.6

replaced by the following:

$$\|a_{112,K}-\tilde{a}_{112,K}\| = O(h_K^2), \quad \|a_{221,K}-\tilde{a}_{221,K}\| = O(h_K^2),$$

$$\|(a_{112,K}-\tilde{a}_{112,K})-(a_{221,K}-\tilde{a}_{221,K})\| = O(h_K^3),$$

$$\|4(a_{123,K}-\tilde{a}_{123,K})-(a_{112,K}-\tilde{a}_{112,K})-(a_{221,K}-\tilde{a}_{221,K})\| = O(h_K^3).$$

Show that the above assumptions are now realistic in the sense that $a_{112,K}$ and $a_{221,K}$ can be actually chosen along a smooth boundary so as to fulfill the above conditions.

For a reference for this problem, see CIARLET & RAVIART (1972b).

4.3.9. (i) Let

$$[v]_{m,p,\Omega} = \left(\int_\Omega \sum_{i=1}^n |D^m v(x)(e_i)^m|^p \, dx\right)^{1/p}.$$

Then (cf. Exercise 3.1.1) the semi-norm $[.]_{k,p,\Omega}$ is a norm over the quotient space $W^{k+1,p}(\Omega)/Q_k(\Omega)$, equivalent to the quotient norm. Let there be given two Sobolev spaces $W^{k+1,p}(\Omega)$ and $W^{m,q}(\Omega)$ with $W^{k+1,p}(\Omega) \subset W^{m,p}(\Omega)$ and let $\Pi \in \mathscr{L}(W^{k+1,p}(\Omega);\ W^{m,q}(\Omega))$ be a mapping which satisfies

$$\forall q \in Q_k(\Omega), \quad \Pi q = q.$$

Show that there exists a constant $C(\Omega, \Pi)$ such that

$$\forall v \in W^{k+1,p}(\Omega), \quad |v - \Pi v|_{m,q,\Omega} \leq C(\Omega, \Pi)[v]_{k+1,p,\Omega}.$$

(ii) For each integer $l \geq 1$, let

$$[\![F]\!]_{l,\infty,\hat{K}} = \max_{1\leq i\leq n} \sup_{\hat{x}\in\hat{K}} \|D^l F(\hat{x})(e_i)^l\|.$$

With the same assumptions as in Theorem 4.3.2, show that

$$\forall \hat{v} \in W^{2,p}(\hat{\Omega}), \quad [\hat{v}]_{2,p,\hat{\Omega}} \leq C|J_{F^{-1}}|_{0,\infty,\Omega}^{1/p}([\![F]\!]_{1,\infty,\hat{\Omega}}^2 |v|_{2,p,\Omega} + \\ + [\![F]\!]_{2,\infty,\hat{\Omega}} |v|_{1,p,\Omega}),$$

$$\forall \hat{v} \in W^{3,p}(\hat{\Omega}), \quad [\hat{v}]_{3,p,\hat{\Omega}} \leq C|J_{F^{-1}}|_{0,\infty,\Omega}^{1/p}([\![F]\!]_{1,\infty,\hat{\Omega}}^3 |v|_{3,p,\Omega} + \\ + [\![F]\!]_{1,\infty,\hat{\Omega}} [\![F]\!]_{2,\infty,\hat{\Omega}} |v|_{2,p,\Omega} \\ + [\![F]\!]_{3,\infty,\hat{\Omega}} |v|_{1,p,\Omega}).$$

(iii) Consider an isoparametric family of quadrilaterals K of type (1) for $n = 2$ (cf. Fig. 4.3.3). We let for each K,

$$h_K = \operatorname{diam}(K),$$
$$h'_K = \text{smallest length of the sides of } K,$$
$$\gamma_K = \max\{|\cos\{(a_{i+1} - a_i) \cdot (a_{i-1} - a_i)\}|, \quad 1 \leqslant i \leqslant 4 (\operatorname{mod} 4)\}.$$

Then such a family is said to be *regular* if all the sets K are convex, if there exist constants σ' and γ such that

$$(*) \quad \forall K, \frac{h_K}{h'_K} \leqslant \sigma' \quad \text{and} \quad \gamma_K \leqslant \gamma < 1,$$

and if the quantity h_K approaches zero. Show that condition (*) implies the usual condition that the ratios h_K/ρ_K be bounded (the converse is clearly false).

Show that, given such a family, the mappings $F_K: \hat{K} = [0, 1]^2 \to K$ are one-to-one and that the following estimates hold:

$$[\![F_K]\!]_{1,\infty,\hat{K}} \leqslant Ch_K, \quad [\![F_K]\!]_{2,\infty,\hat{K}} = 0,$$
$$|F_K^{-1}|_{1,\infty,K} \leqslant \frac{C}{h_K},$$
$$|J_{F_K}|_{0,\infty,\hat{K}} \leqslant Ch_K^2, \quad |J_{F_K^{-1}}|_{0,\infty,K} \leqslant \frac{C}{h_K^2}.$$

Using the above results, derive the following interpolation error estimates (under the assumptions $W^{2,p}(\hat{K}) \hookrightarrow \mathscr{C}^0(\hat{K})$ and $W^{2,p}(\hat{K}) \hookrightarrow W^{m,q}(\hat{K})$): For all $v \in W^{2,p}(K)$,

$$\|v - \Pi_K v\|_{m,q,K} \leqslant C(h_K^2)^{1/q - 1/p} h_K^{2-m} \|v\|_{2,p,K}, \quad m = 0, 1.$$

(iv) Consider an isoparametric family of quadrilaterals K of type (2) for $n = 2$.

For each $K = F_K(\hat{K})$, where the mapping $F_K \in (Q_2(\hat{K}))^2$ is uniquely determined by the data of nine points $a_{i,K}$, $1 \leqslant i \leqslant 9$, (cf. Fig. 4.3.4), we let $\tilde{F}_K$ denote the mapping uniquely determined by the conditions

$$\tilde{F}_K \in (Q_1(\hat{K}))^2, \quad \tilde{F}_K(\hat{a}_i) = a_{i,K}, \quad 1 \leqslant i \leqslant 4.$$

Then we say that the family is *regular* if the family of quadrilaterals $\tilde{K} = \tilde{F}_K(\hat{K})$ is regular in the sense of (iii) and if one has (compare with (4.3.27)):

$$\|a_{i,K} - \tilde{a}_{i,K}\| = O(h_K^2), \quad 5 \leqslant i \leqslant 9,$$

where $\tilde{a}_{i,K} = \tilde{F}_K(\hat{a}_i)$, $5 \leqslant i \leqslant 9$.

Given such a family, show that the mappings $F_K\colon \hat{K} = [0,1]^2 \to K$ are one-to-one for h_K small enough and that the following estimates hold:

$$[\![F_K]\!]_{1,\infty,\hat{K}} \leq Ch_K, \quad [\![F_K]\!]_{2,\infty,\hat{K}} \leq Ch_K^2, \quad [\![F_K]\!]_{3,\infty,\hat{K}} = 0,$$

$$|F_K^{-1}|_{1,\infty,K} \leq \frac{C}{h_K}, \quad |F_K^{-1}|_{2,\infty,K} \leq \frac{C}{h_K},$$

$$|J_{F_K}|_{0,\infty,\hat{K}} \leq Ch_K^2, \quad |J_{F_K^{-1}}|_{0,\infty,K} \leq \frac{C}{h_K^2}.$$

Using the above results, derive the following interpolation error estimates (under the assumptions $W^{3,p}(\hat{K}) \hookrightarrow \mathscr{C}^0(\hat{K})$ and $W^{3,p}(\hat{K}) \hookrightarrow W^{m,q}(\hat{K})$): For all $v \in W^{3,p}(K)$,

$$\|v - \Pi_K v\|_{m,q,K} \leq C(h_K^2)^{1/q-1/p} h_K^{3-m} \|v\|_{3,p,K}, \quad m = 0, 1, 2.$$

A general theory for isoparametric quadrilateral finite elements is given in CIARLET & RAVIART (1972b). Significant improvements have recently been obtained by JAMET (1976b).

4.4 Application to second-order problems over curved domains

Approximation of a curved boundary with isoparametric finite elements

As in Section 4.1, we consider the homogeneous second-order Dirichlet problem which corresponds to the following data:

$$\begin{cases} V = H_0^1(\Omega), \\ a(u, v) = \displaystyle\int_\Omega \sum_{i,j=1}^n a_{ij}\partial_i u \partial_j v \,\mathrm{d}x, \\ f(v) = \displaystyle\int_\Omega fv \,\mathrm{d}x, \end{cases} \tag{4.4.1}$$

where Ω is a bounded open subset of $\mathbf{R}^n$ with a *curved* boundary Γ (the main novelty) and the functions $a_{ij} \in L^\infty(\Omega)$ and $f \in L^2(\Omega)$ are *everywhere* defined over the set $\bar{\Omega}$. We shall assume that the ellipticity condition holds, i.e.,

$$\exists \beta > 0, \quad \forall x \in \bar{\Omega}, \quad \forall \xi_i, \quad 1 \leq i \leq n, \quad \sum_{i,j=1}^n a_{ij}(x)\xi_i\xi_j \geq \beta \sum_{i=1}^n \xi_i^2. \tag{4.4.2}$$

The elaboration of the discrete problem comprises three steps:

(i) Construction of a *triangulation* of the set $\bar{\Omega}$ using isoparametric finite elements.

(ii) Definition of a discrete problem *without* numerical integration.

(iii) Definition of a discrete problem *with* numerical integration.

We begin by constructing a set $\bar{\Omega}_h$ as a finite union $\bar{\Omega}_h = \bigcup_{K \in \mathcal{T}_h} K$ of isoparametric finite elements (K, P_K, Σ_K), $K \in \mathcal{T}_h$, which we shall assume to be of *Lagrange type*. Following the description given in the previous section, each finite element (K, P_K, Σ_K) is obtained from a reference finite element $(\hat{K}, \hat{P}, \hat{\Sigma})$ through an isoparametric mapping $F_K \in (\hat{P})^n$ which is uniquely determined by the data of the nodes of the finite element K. *These nodes will always be assumed to belong to the set* $\bar{\Omega}$.

In addition, we shall restrict ourselves to finite elements which possess the following property (cf. Remark 2.3.10):

$$\text{\textit{Each basis function } } \hat{p} \text{\textit{ of the reference finite element }} (\hat{K}, \hat{P}, \hat{\Sigma}) \text{\textit{ vanishes along any face of the set }} \hat{K} \text{\textit{ which does not contain the node associated with }} \hat{p}. \tag{4.4.3}$$

As shown by the examples given in the preceding section, this is not a restrictive assumption.

Of course, we shall take advantage of the isoparametric mappings F_K for getting a good approximation of the boundary Γ: By an appropriate choice of nodes along Γ, we construct finite elements with (at least) one curved face which should be very close to Γ, at any rate closer than a straight face would be. Let us assume for definiteness that we are using *simplicial* finite elements. We may then distinguish two cases, depending upon whether the mapping F_K is affine, i.e., $F_K \in (P_1(\hat{K}))^n$, or the mapping F_K is "truly" isoparametric, i.e., $F_K \in (\hat{P})^n$ but $F_K \notin (\hat{P}_1(\hat{K}))^n$. The latter case will in particular apply to "boundary" finite elements, while the former will rather apply to "interior" finite elements. These considerations are illustrated in Fig. 4.4.1, where we consider the case of triangles of type (2).

For computational simplicity, it is clear that we shall try to keep to a minimum the number of curved faces, and this is why, in general, only the "boundary" finite elements will have one curved face. However, all the subsequent analysis applies equally well to all possible cases, including those in which *all* finite elements $K \in \mathcal{T}_h$ are of the isoparametric type.

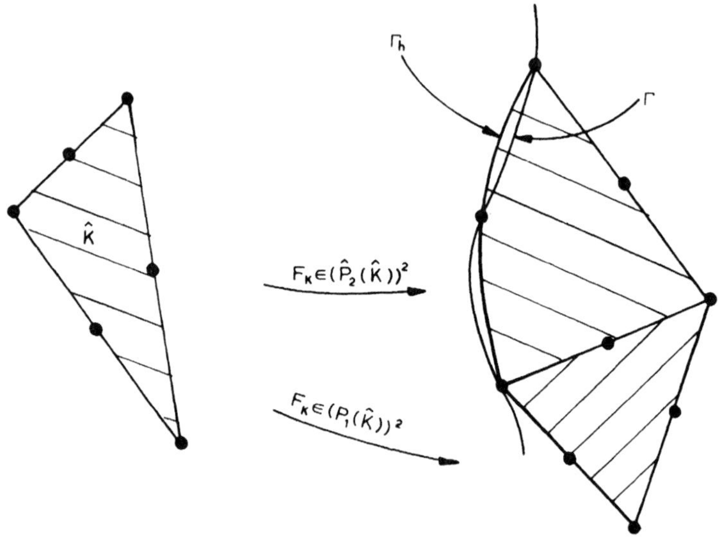

Fig. 4.4.1

In this last case, we must check that the intersection of "adjacent" finite elements is indeed a face for both of them. In other words, there should be no holes and no overlaps. This is true because the finite elements which satisfy (4.4.3) are such that any one of their faces is solely determined by the nodes which are on it (of course, the nodes which define a common face are assumed to be the same for two adjacent finite elements). As an example, we have represented in Fig. 4.4.2 three isoparametric tetrahedra of type (2) "just before assembly": Then the face K' is completely determined by the data of the nodes a_1, a_2, a_3, a_{12}, a_{23}, a_{13}, and the arc $\mathscr{A}$ is completely defined by the data of the nodes a_1, a_2 and a_{12}.

Returning to the general case, we shall assume that *all the nodes which are used in the definition of the faces which approximate the boundary Γ are also the nodes which are on Γ*. Thus, the situation indicated in Fig. 4.4.4 (a) (cf. Exercise 4.4.4) is excluded.

Because each face K' of an isoparametric finite element is necessarily of the form $K' = F_K(\hat{K}')$ with $F_K \in (\hat{P})^n$ and $\hat{K}'$ a face of $\hat{K}$, it is clear that the boundary Γ_h of the set $\bar{\Omega}_h = \bigcup_{K \in \mathscr{T}_h} K$ does not coincide in general with the boundary Γ of the set Ω. Nevertheless, we shall call $\mathscr{T}_h$

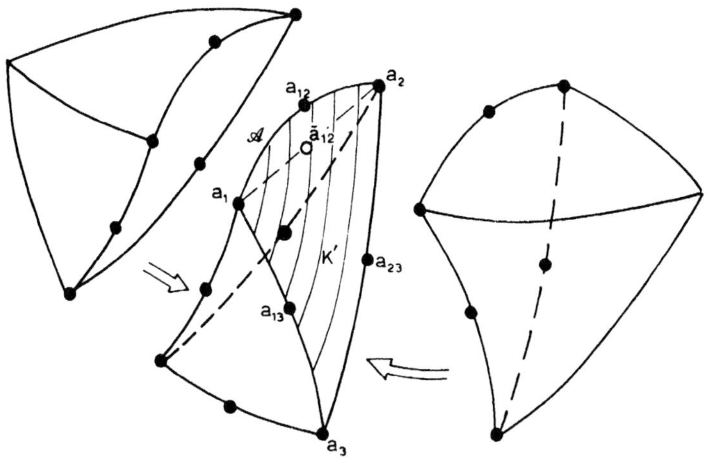

Fig. 4.4.2

a *triangulation of the set* $\bar{\Omega}$ (even though it should more appropriately be called a triangulation of the set $\bar{\Omega}_h$).

Then we let X_h denote the *finite element space* whose functions $v_h : \bar{\Omega}_h \to \mathbf{R}$ are defined as follows:

(i) For each $K \in \mathcal{T}_h$, the restrictions $v_{h|K}$ span the space

$$P_K = \{p : K \to \mathbf{R};\, p = \hat{p} \cdot F_K^{-1},\, \hat{p} \in \hat{P}\}.$$

(ii) Over each $K \in \mathcal{T}_h$, the restrictions $v_{h|K}$ are defined by their values at the nodes of the finite element K.

If the functions of the space $\hat{P}$ are smooth enough, such a space X_h is contained in the space $\mathscr{C}^0(\bar{\Omega}_h)$ (this is again implied by property (4.4.3)), and consequently the inclusion $X_h \subset H^1(\Omega_h)$ holds by Theorem 2.1.1.

We let X_{0h} denote the subspace of X_h whose functions vanish at the *boundary nodes*, i.e., those nodes which are on the boundary Γ. We recall that, by construction, these nodes coincide with those which are on the boundary Γ_h. Therefore another application of property (4.4.3) shows that *the functions in the space* X_{0h} *vanish along the boundary* Γ_h, and thus the inclusion

$$V_h = X_{0h} \subset H_0^1(\Omega_h) \tag{4.4.4}$$

holds (Ω_h denotes the interior of the set $\bar{\Omega}_h$).

Since we expect that the boundaries Γ_h and Γ are close, we shall henceforth assume that *there exists a bounded open set* $\tilde{\Omega}$ *such that*

$$\Omega \subset \tilde{\Omega} \quad \text{and} \quad \Omega_h \subset \tilde{\Omega} \tag{4.4.5}$$

for all the triangulations $\mathcal{T}_h$ *which we shall consider.*

Then the most straightforward definition of a *discrete problem* associated with the space V_h consists in finding a function $\tilde{u}_h \in V_h$ such that

$$\forall v_h \in V_h, \quad \int_{\Omega_h} \sum_{i,j=1}^{n} \tilde{a}_{ij} \partial_i \tilde{u}_h \partial_j v_h \, \mathrm{d}x = \int_{\Omega_h} \tilde{f} v_h \, \mathrm{d}x, \tag{4.4.6}$$

where the functions $\tilde{a}_{ij}$ and $\tilde{f}$ are some *extensions* of the functions a_{ij} and f to the set $\tilde{\Omega}$.

Taking into account isoparametric numerical integration. Description of the resulting discrete problem

In spite of the simplicity and of the natural character of this definition, several questions immediately arise: How should one choose between all possible extensions? How should one construct such extensions in practice? What is the dependence of the discrete solution $\tilde{u}_h$ upon these extensions? Surprisingly, it turns out that these ambiguities will be circumvented by taking into account the effect of *isoparametric numerical integration*:

Just as in Section 4.1, we assume that we have at our disposal a quadrature scheme over the set $\hat{K}$:

$$\int_{\hat{K}} \hat{\varphi}(\hat{x}) \, \mathrm{d}\hat{x} \sim \sum_{l=1}^{L} \hat{\omega}_l \hat{\varphi}(\hat{b}_l), \quad \text{with} \quad \hat{\omega}_l \in \mathbf{R}, \hat{b}_l \in \hat{K}, \ 1 \le l \le L. \tag{4.4.7}$$

Given two functions $\hat{\varphi}: \hat{K} \to \mathbf{R}$ and $\varphi: K = F_K(\hat{K}) \to \mathbf{R}$ in the usual correspondence (i.e., $\varphi = \hat{\varphi} \cdot F_K^{-1}$), we have

$$\int_K \varphi(x) \, \mathrm{d}x = \int_{\hat{K}} \hat{\varphi}(\hat{x}) J_{F_K}(\hat{x}) \, \mathrm{d}\hat{x},$$

where the Jacobian J_{F_K} of the mapping F_K may be assumed without loss of generality to be strictly positive over the set $\hat{K}$. Therefore, *the quadrature scheme* (4.4.7) *over the reference element* $\hat{K}$ *automatically induces a quadrature scheme over the finite element* K (compare with

(4.1.9) and (4.1.10)), namely

$$\int_K \varphi(x)\,\mathrm{d}x \sim \sum_{l=1}^{L} \omega_{l,K}\varphi(b_{l,K}), \tag{4.4.8}$$

with *weights* $\omega_{l,K}$ and *nodes* $b_{l,K}$ defined by

$$\omega_{l,K} = \hat{\omega}_l J_{F_K}(\hat{b}_l) \quad \text{and} \quad b_{l,K} = F_K(\hat{b}_l), \quad 1 \leq l \leq L. \tag{4.4.9}$$

Accordingly, we define the *quadrature error functionals*

$$E_K(\varphi) = \int_K \varphi(x)\,\mathrm{d}x - \sum_{l=1}^{L} \omega_{l,K}\varphi(b_{l,K}), \tag{4.4.10}$$

$$\hat{E}(\hat{\varphi}) = \int_{\hat{K}} \hat{\varphi}(\hat{x})\,\mathrm{d}\hat{x} - \sum_{l=1}^{L} \hat{\omega}_l\hat{\varphi}(\hat{b}_l), \tag{4.4.11}$$

which are related through the equation

$$E_K(\varphi) = \hat{E}(\hat{\varphi} J_{F_K}). \tag{4.4.12}$$

Let us now examine how isoparametric numerical integration affects the definition of the discrete problem (4.4.6). Assuming that the extensions $\tilde{a}_{ij}$ and $\tilde{f}$ are defined everywhere over the set $\bar{\tilde{\Omega}}$, we have to find a discrete solution $u_h \in V_h$ which satisfies (compare with (4.1.24) and (4.1.25)):

$$\forall v_h \in V_h, \sum_{K\in\mathcal{T}_h} \sum_{l=1}^{L} \omega_{l,K} \sum_{i,j=1}^{n} (\tilde{a}_{ij}\partial_i u_h \partial_j v_h)(b_{l,K}) =$$

$$= \sum_{K\in\mathcal{T}_h} \sum_{l=1}^{L} \omega_{l,K}(\tilde{f} v_h)(b_{l,K}). \tag{4.4.13}$$

Then it is clear that *the extensions $\tilde{a}_{ij}$ and $\tilde{f}$ are not needed in the definition of the above discrete problem if all the quadrature nodes $b_{l,K}$, $1 \leq l \leq L$, $K \in \mathcal{T}_h$, belong to the set $\bar{\Omega}$*. To show that this is indeed a common circumstance, let us consider one typical example. Let $n = 2$ and assume that we are using isoparametric triangles of type (2) and that *each node of the quadrature scheme over the set $\hat{K}$ either coincides with a node of the triangle $\hat{K}$ of type (2) or is in the interior $\mathring{\hat{K}}$ of the set $\hat{K}$*. As shown by the examples given in Section 4.1 (cf. Figs. 4.1.1, 4.1.2 and 4.1.3), this is a realistic situation.

To prove our assertion, we need of course consider only the case of a "boundary" finite element and, at this point, it becomes necessary to indicate *how the boundary nodes are actually chosen*. With the notations of Fig. 4.4.3, the point $a_{12,K}$ is chosen at the intersection between the boundary Γ and the line perpendicular to the segment $[a_{1,K}, a_{2,K}]$ which passes through the point $\tilde{a}_{12,K} = (a_{1,K} + a_{2,K})/2$.

This choice has three important consequences:

First, if the boundary Γ is smooth enough, we automatically have

$$\|a_{12,K} - \tilde{a}_{12,K}\| = \mathrm{O}(h_K^2), \tag{4.4.14}$$

where h_K is the diameter of the triangle with vertices $a_{i,K}$, $1 \leq i \leq 3$. This estimate will insure that a family made up of such isoparametric triangles of type (2) is regular in the sense understood in Section 4.3. We shall use this property in Theorem 4.4.3.

Secondly, *the image $b_K = F_K(\hat{b})$ of any point $\hat{b} \in \mathring{\hat{K}}$ belongs to the set $\bar{\Omega} \cap K$ provided h_K is small enough.* Intuitively, this seems reasonable from a geometrical point of view, and we leave the complete proof to the reader (Exercise 4.4.1).

Thirdly, it is clear that there exists a bounded open set $\tilde{\Omega}$ such that the inclusions (4.4.5) hold.

Remark 4.4.1. The above construction can be easily extended to an open set with a *piecewise smooth boundary*, i.e., a Lipschitz-continuous boundary which is composed of a finite number of smooth arcs, provided each intersection of adjacent arcs is a "vertex" of at least one isoparametric triangle of type (2). □

Remark 4.4.2. When $n = 3$, a node such as a_{12} (cf. Fig. 4.4.2) may be chosen in such a way that the distance between the points $\tilde{a}_{12}$ and a_{12} is equal to the distance between the point $\tilde{a}_{12}$ and the boundary Γ. □

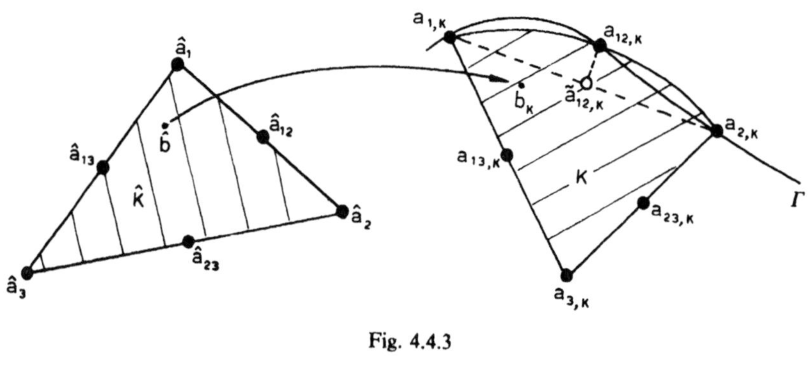

Fig. 4.4.3

Returning to the general case, we are therefore justified in assuming from now on that *the relations*

$$\forall K \in \mathcal{T}_h, \quad b_{l,K} = F_K(\hat{b}_l) \in \bar{\Omega}, \quad 1 \leq l \leq L, \tag{4.4.15}$$

hold for all the triangulations $\mathcal{T}_h$ *to be considered.*

This being the case, the *discrete problem* (4.4.13) consists in finding a *discrete solution* $u_h \in V_h$ such that

$$\forall v_h \in V_h, \quad a_h(u_h, v_h) = f_h(v_h), \tag{4.4.16}$$

where, for all functions $u_h, v_h \in V_h$, the *approximate bilinear form* $a_h(.,.)$ and the *approximate linear form* $f_h(.)$ are given by

$$a_h(u_h, v_h) = \sum_{K \in \mathcal{T}_h} \sum_{l=1}^{L} \omega_{l,K} \sum_{i,j=1}^{n} (a_{ij}\partial_i u_h \partial_j v_h)(b_{l,K}), \tag{4.4.17}$$

and

$$f_h(v_h) = \sum_{K \in \mathcal{T}_h} \sum_{l=1} \omega_{l,K}(f v_h)(b_{l,K}). \tag{4.4.18}$$

In other words, *thanks to the effect of numerical integration, the discrete problem can be defined without any reference to possible extensions of the functions* a_{ij} *and* f, and this observation is of course of great practical value (by contrast, extensions explicitly appear in the final error estimate; cf. Theorem 4.4.6).

Remark 4.4.3. Conceivably, *several* quadrature schemes over the reference finite element may be used, depending upon the finite elements $K \in \mathcal{T}_h$. In particular, one would naturally expect that more sophisticated schemes are necessary for dealing with the "truly" isoparametric finite elements. Since our final result (Theorem 4.4.6) shows however that this is not the case, we shall deliberately ignore this possibility (which would require straightforward notational modifications in the writing of (4.4.17) and (4.4.18)). □

Abstract error estimate

Given a family of discrete problems of the form (4.4.16), we shall say that the approximate bilinear forms $a_h(.,.)$ of (4.4.17) are *uniformly*

V_h-elliptic if

$$\exists \tilde{\alpha} > 0, \quad \forall v_h \in V_h, \quad \tilde{\alpha}\|v_h\|^2_{1,\Omega_h} \leq a_h(v_h, v_h), \tag{4.4.19}$$

where the constant $\tilde{\alpha}$ is independent of the subspace V_h.

As usual, we first prove an *abstract error estimate.* The reader should not be surprised by the arbitrariness at this stage in the definition of the functions $\tilde{u}$ and $\tilde{a}_{ij}$ which appear in the next theorem: When this error estimate is actually applied, these will be taken as *extensions* of the functions u and a_{ij} (cf. Theorem 4.4.6).

Theorem 4.4.1. *Given an open set $\tilde{\Omega}$ which contains all the sets Ω_h, and given functions $\tilde{a}_{ij} \in L^\infty(\tilde{\Omega})$, we let*

$$\forall v, w \in H^1(\Omega_h), \quad \tilde{a}_h(v, w) = \int_{\Omega_h} \sum_{i,j=1}^n \tilde{a}_{ij}\partial_i v \partial_j w \, dx. \tag{4.4.20}$$

Then if we consider a family of discrete problems of the form (4.4.16), *for which the associated approximate bilinear forms are uniformly V_h-elliptic, there exists a constant C independent of the space V_h such that*

$$\begin{aligned}\|\tilde{u} - u_h\|_{1,\Omega_h} \leq C\Bigg(&\inf_{v_h \in V_h} \|\tilde{u} - v_h\|_{1,\Omega_h} \\ &+ \sup_{w_h \in V_h} \frac{|\tilde{a}_h(v_h, w_h) - a_h(v_h, w_h)|}{\|w_h\|_{1,\Omega_h}} \\ &+ \sup_{w_h \in V_h} \frac{|\tilde{a}_h(\tilde{u}, w_h) - f_h(w_h)|}{\|w_h\|_{1,\Omega_h}}\Bigg),\end{aligned} \tag{4.4.21}$$

where $\tilde{u}$ is any function in the space $H^1(\tilde{\Omega})$, and u_h denotes the solutions of the discrete problems (4.4.16).

Proof. The assumption of uniform V_h-ellipticity insures in particular that each discrete problem has a unique solution u_h. Also, there exists a constant $\tilde{M}$ independent of h such that

$$\forall v, w \in H^1(\Omega_h), \quad |\tilde{a}_h(v, w)| \leq \tilde{M}\|v\|_{1,\Omega_h}\|w\|_{1,\Omega_h}. \tag{4.4.22}$$

Let then v_h denote an arbitrary element in the space V_h. We have

$$\begin{aligned}\tilde{\alpha}\|u_h - v_h\|^2_{1,\Omega_h} &\leq a_h(u_h - v_h, u_h - v_h) \\ &= \tilde{a}_h(\tilde{u} - v_h, u_h - v_h) + \{\tilde{a}_h(v_h, u_h - v_h) \\ &\quad - a_h(v_h, u_h - v_h) + \{f_h(u_h - v_h) - \tilde{a}_h(\tilde{u}, u_h - v_h)\},\end{aligned}$$

so that, using (4.4.22),

$$\begin{aligned}\tilde{\alpha}\|u_h - v_h\|_{1,\Omega_h} &\leq \tilde{M}\|\tilde{u} - v_h\|_{1,\Omega_h} + \frac{|\tilde{a}_h(v_h, u_h - v_h) - a_h(v_h, u_h - v_h)|}{\|u_h - v_h\|_{1,\Omega_h}} \\ &\quad + \frac{|\tilde{a}_h(\tilde{u}, u_h - v_h) - f_h(u_h - v_h)|}{\|u_h - v_h\|_{1,\Omega_h}} \\ &\leq \tilde{M}\|\tilde{u} - v_h\|_{1,\Omega_h} + \sup_{w_h \in V_h} \frac{|\tilde{a}_h(v_h, w_h) - a_h(v_h, w_h)|}{\|w_h\|_{1,\Omega_h}} \\ &\quad + \sup_{w_h \in V_h} \frac{|\tilde{a}_h(\tilde{u}, w_h) - f_h(w_h)|}{\|w_h\|_{1,\Omega_h}}.\end{aligned}$$

Combining the above inequality with the triangular inequality

$$\|\tilde{u} - u_h\|_{1,\Omega_h} \leq \|\tilde{u} - v_h\|_{1,\Omega_h} + \|u_h - v_h\|_{1,\Omega_h},$$

and taking the infimum with respect to $v_h \in V_h$, we obtain inequality (4.4.21). □

Accordingly, the remainder of this section will be devoted to giving sufficient conditions which insure the uniform V_h-ellipticity of the approximate bilinear forms (Theorem 4.4.2) and to estimating the various terms which appear in the right-hand side of inequality (4.4.21). To keep the development within reasonable limits, *we shall however restrict ourselves to finite element spaces made up of isoparametric n-simplices of type* (2).

Finally, we shall make the following assumption:

(H1) *The associated family of triangulations $\mathcal{T}_h$ is regular* in the sense that the family (K), $K \in \bigcup_h \mathcal{T}_h$, is a regular isoparametric family of n-simplices K of type (2) (in the sense understood in Section 4.3; cf. (4.3.26) and (4.3.27)). It is crucial to notice that, in particular, *condition* (4.3.27) *is perfectly compatible with the construction of boundary finite elements* (cf. (4.4.14) and Remark 4.4.2).

Sufficient conditions for uniform V_h-ellipticity

Let us first examine the question of uniform V_h-ellipticity.

Theorem 4.4.2. *Let (V_h) be a family of finite element spaces made up of isoparametric n-simplices of type* (2), *and let there be given a quadrature* *scheme*

$$\int_{\hat{K}} \hat{\varphi}(\hat{x})\,\mathrm{d}\hat{x} \sim \sum_{l=1}^{L} \hat{\omega}_l \hat{\varphi}(\hat{b}_l) \quad with \quad \hat{\omega}_l > 0, \quad 1 \leq l \leq L,$$

such that the union $\bigcup_{l=1}^{L} \{\hat{b}_l\}$ *contains a* $P_1(\hat{K})$*-unisolvent subset and/or the quadrature scheme is exact for the space* $P_2(\hat{K})$.

Then, if hypothesis (H1) *holds, the associated approximate bilinear forms are uniformly* V_h*-elliptic, i.e.,*

$$\exists \tilde{\alpha} > 0, \quad \forall V_h, \quad \forall v_h \in V_h, \quad \tilde{\alpha}\|v_h\|_{1,\Omega_h}^2 \leq a_h(v_h, v_h). \tag{4.4.23}$$

Proof. (i) Arguing as in part (i) of the proof of Theorem 4.1.2, we find that there exists a constant $\hat{C} > 0$ such that

$$\forall \hat{p} \in \hat{P} = P_2(\hat{K}), \quad \hat{C}|\hat{p}|_{1,\hat{K}}^2 \leq \sum_{l=1}^{L} \hat{\omega}_l \sum_{i=1}^{n} (\partial_i \hat{p}(\hat{b}_l))^2. \tag{4.4.24}$$

(ii) Given a finite element $K \in \mathscr{T}_h$ and a function $v_h \in V_h$, let $p = v_{h|K}$. With the ellipticity condition (4.4.2), we obtain

$$\sum_{l=1}^{L} \omega_{l,K} \sum_{i,j=1}^{n} (a_{ij}\partial_i v_h \partial_j v_h)(b_{l,K}) \geq \beta \sum_{l=1}^{L} \omega_{l,K} \sum_{i=1}^{n} (\partial_i p(b_{l,K}))^2. \tag{4.4.25}$$

We recognize in the expression $\Sigma_{i=1}^{n} (\partial_i p(b_{l,K}))^2$ the square of the Euclidean norm $\|.\|$ of the vector $Dp(b_{l,K})$. For all points $x = F_K(\hat{x})$, $\hat{x} \in \hat{K}$, we have (by Theorem 4.3.3, F_K is invertible for h small enough)

$$Dp(x) = D\hat{p}(\hat{x}) DF_K^{-1}(x),$$

where $Dp(x)$ and $D\hat{p}(\hat{x})$ may be identified with the row vectors $(\partial_1 p(x), \ldots, \partial_n p(x))$ and $(\partial_1 \hat{p}(\hat{x}), \ldots, \partial_n \hat{p}(\hat{x}))$ respectively, and where $DF_K^{-1}(x)$ may be identified with the Jacobian matrix of the mapping F_K^{-1} at x. Using the inequality $\xi A A^T \xi^T \geq (1/\|A^{-1}\|^2)\xi\xi^T$ valid for any invertible matrix A and any row vector ξ (the subscript T denotes transposition), we obtain

$$\forall x = F_K(\hat{x}) \in \hat{K}, \quad \sum_{i=1}^{n} (\partial_i p(x))^2 = Dp(x)Dp(x)^T \geq$$

$$\geq \frac{1}{\|DF_K(\hat{x})\|^2} \sum_{i=1}^{n} (\partial_i \hat{p}(\hat{x}))^2. \tag{4.4.26}$$

Since $\omega_{l,K} = \hat{\omega}_l J_{F_K}(\hat{b})$ (cf. (4.4.9)) and since the weights $\hat{\omega}_l$ are positive, we deduce from (4.4.26) and (4.4.24):

$$\sum_{l=1}^{L} \omega_{l,K} \sum_{i=1}^{n} (\partial_i p(b_{l,K}))^2$$
$$\geqslant \inf_{\hat{x}\in\hat{K}} J_{F_K}(\hat{x}) \inf_{\hat{x}\in\hat{K}} \left(\frac{1}{\|DF_K(\hat{x})\|^2}\right) \sum_{l=1}^{L} \hat{\omega}_l \sum_{i=1}^{n} (\partial_i \hat{p}(\hat{b}_l))^2$$
$$\geqslant \hat{C} \frac{1}{|J_{F_K^{-1}}|_{0,\infty,K} |F_K|^2_{1,\infty,\hat{K}}} |\hat{p}|^2_{1,\hat{K}}, \tag{4.4.27}$$

where, here and subsequently, we use the notations introduced in (4.3.15). Using Theorem 4.3.2, we know that there exists a constant C such that

$$\forall p = \hat{p} \cdot F_K^{-1}, \quad \hat{p} \in \hat{P}, \; |\hat{p}|_{1,\hat{K}} \geqslant C \frac{1}{|J_{F_K}|^{1/2}_{0,\infty,\hat{K}} |F_K^{-1}|_{1,\infty,K}} |p|_{1,K}. \tag{4.4.28}$$

Hence, upon combining inequalities (4.4.25), (4.4.27) and (4.4.28), we obtain

$$\sum_{l=1}^{L} \omega_{l,K} \sum_{i,j=1}^{n} (a_{ij}\partial_i v_h \partial_j v_h)(b_{l,K}) \geqslant$$
$$\geqslant \beta \hat{C} C^2 \frac{|v_h|^2_{1,K}}{|J_{F_K^{-1}}|_{0,\infty,K} |J_{F_K}|_{0,\infty,\hat{K}} (|F_K|_{1,\infty,\hat{K}} |F_K^{-1}|_{1,\infty,K})^2}. \tag{4.4.29}$$

If we next make use of the estimates established in Theorem 4.3.3 (which we may apply in view of the assumption of regularity), we find that the denominators appearing in the right-hand side of inequality (4.4.29) are uniformly bounded for all $K \in \mathcal{T}_h$, $v_h \in V_h$ and all V_h. Therefore we have shown that

$$\exists \tilde{\alpha}' > 0, \quad \forall v_h \in V_h, \quad \forall K \in \mathcal{T}_h, \quad \forall h,$$
$$\sum_{l=1}^{L} \omega_{l,K} \sum_{i,j=1}^{n} (a_{ij}\partial_i v_h \partial_j v_h)(b_{l,K}) \geqslant \tilde{\alpha}' |v_h|^2_{1,K}. \tag{4.4.30}$$

(iii) With inequality (4.4.30), we obtain

$$\forall v_h \in V_h, \quad a_h(v_h, v_h) = \sum_{K\in\mathcal{T}_h} \sum_{l=1}^{L} \omega_{l,K} \sum_{i,j=1}^{n} (a_{ij}\partial_i v_h \partial_j v_h)(b_{l,K})$$
$$\geqslant \tilde{\alpha}' \sum_{K\in\mathcal{T}_h} |v_h|^2_{1,K} = \tilde{\alpha}' |v_h|^2_{1,\Omega_h}. \tag{4.4.31}$$

Since all the sets Ω_h are contained in a bounded open set $\tilde{\Omega}$ (cf. (4.4.5)), there exists a constant C independent of h such that

$$\forall v \in H_0^1(\Omega_h), \quad \|v\|_{1,\Omega_h} \leq C|v|_{1,\Omega_h}. \tag{4.4.32}$$

To see this, it suffices, for each function $v \in H_0^1(\Omega_h)$, to apply the Poincaré–Friedrichs inequality over the set $\tilde{\Omega}$ to the function $\tilde{v} \in H_0^1(\tilde{\Omega})$ which equals v on Ω_h and vanishes on $\tilde{\Omega} - \Omega_h$.

Inequality (4.4.23) is then a consequence of inequalities (4.4.31) and (4.4.32) and the proof is complete. □

Interpolation error and consistency error estimates

In what follows, we shall consider the *X_h-interpolation operator Π_h*, whose definition is the natural extension of the definition given in Section 2.3 in the case of straight finite elements: Given a function $v \in \operatorname{dom} \Pi_h = \mathscr{C}^0(\bar{\Omega}_h)$, the *$X_h$-interpolant $\Pi_h v$* is the unique function which satisfies

$$\begin{cases} \Pi_h v \in X_h, \\ \forall K \in \mathscr{T}_h, \quad \Pi_h v(a_{i,K}) = v(a_{i,K}), \quad 1 \leq i \leq n+1, \\ \Pi_h v(a_{ij,K}) = v(a_{ij,K}), \quad 1 \leq i < j \leq n+1, \end{cases} \tag{4.4.33}$$

so that it is clear that the relations

$$\forall K \in \mathscr{T}_h, \quad \Pi_h v|_K = \Pi_K v \tag{4.4.34}$$

hold.

We now estimate the difference $(v - \Pi_h v)$ in various norms. In particular, these estimates will subsequently allow us to obtain (for a specific choice of function $\tilde{u}$) an estimate of the term $\inf_{v_h \in V_h} \|\tilde{u} - v_h\|_{1,\Omega_h}$ which appears in inequality (4.4.21). As usual, the same letter C stands for various constants independent of h and of the various functions involved.

Theorem 4.4.3. *Let (X_h) be a family of finite element spaces made up of isoparametric n-simplices of type* (2), *and assume that $n \leq 5$.*

Then if hypothesis (H1) *holds, there exists a constant C independent of h such that*

$$\forall v \in H^3(\tilde{\Omega}), \quad \|v - \Pi_h v\|_{m,\Omega_h} \leq C h^{3-m} \|v\|_{3,\Omega_h}, \quad m = 0, 1, \tag{4.4.35}$$

$$\left(\sum_{K \in \mathscr{T}_h} \|v - \Pi_K v\|_{m,K}^2 \right)^{1/2} \leq C h^{3-m} \|v\|_{3,\Omega_h}, \quad m = 2, 3, \tag{4.4.36}$$

where

$$h = \max_{K \in \mathcal{T}_h} h_K, \tag{4.4.37}$$

and $\tilde{\Omega}$ is any open set such that the inclusions (4.4.5) *hold. We also have the implication*

$$v \in H^3(\tilde{\Omega}) \quad and \quad v = 0 \quad on \quad \Gamma \Rightarrow \Pi_h v \in X_{0h}. \tag{4.4.38}$$

Proof. Since $n \leqslant 5$, the inclusion $H^3(\hat{K}) \hookrightarrow \mathscr{C}^0(\hat{K})$ holds and thus we may apply Theorem 4.3.4: For all functions $v \in H^3(K)$, we have

$$\|v - \Pi_K v\|_{m,K} \leqslant C h_K^{3-m}(|v|_{2,K} + |v|_{3,K}) \leqslant C h_K^{3-m} \|v\|_{3,K}, \quad 0 \leqslant m \leqslant 3,$$

and inequalities (4.4.35) and (4.4.36) follow from the above inequalities and relations (4.4.34).

If a function vanishes on Γ, its X_h-interpolant vanishes at all the nodes situated on Γ_h (by construction) and therefore it vanishes on the boundary Γ_h of the set $\bar{\Omega}_h = \bigcup_{K \in \mathcal{T}_h} K$. Thus implication (4.4.38) is proved. □

Just as in Section 4.1, the *consistency errors*

$$\sup_{w_h \in V_h} \frac{|\tilde{a}_h(v_h, w_h) - a_h(v_h, w_h)|}{\|w_h\|_{1,\Omega_h}} \quad \text{and} \quad \sup_{w_h \in V_h} \frac{|\tilde{a}_h(\tilde{u}, w_h) - f_h(w_h)|}{\|w_h\|_{1,\Omega_h}}$$

(cf. inequality (4.4.21)) will be estimated as a consequence of careful analyses of similar "local" terms. These are the object of the next two theorems (compare with Theorems 4.1.4 and 4.1.5). The quadrature error functionals $E_K(.)$ and $\hat{E}(.)$ have been defined in (4.4.10) and (4.4.11).

Theorem 4.4.4. *Let there be given a regular isoparametric family of n-simplices K of type* (2), *and let the quadrature scheme over the reference finite element be exact for the space $P_2(\hat{K})$, i.e.,*

$$\forall \hat{\varphi} \in P_2(\hat{K}), \quad \hat{E}(\hat{\varphi}) = 0. \tag{4.4.39}$$

Then there exists a constant C independent of K such that

$$\forall a \in W^{2,\infty}(K), \quad \forall p \in P_K, \quad \forall p' \in P_K, \tag{4.4.40}$$
$$|E_K(a \partial_i p' \partial_j p)| \leqslant C h_K^2 \|a\|_{2,\infty,K} \|p'\|_{2,K} |p|_{1,K}.$$

Proof. For notational convenience, the indices K will be dropped throughout the proof.

(i) To begin with, we shall record some consequences of Theorem 4.3.3. First, inequalities (4.3.28) imply that

$$|\partial_i F_k|_{0,\infty,\hat{K}} \leq Ch, \quad 1 \leq i, k \leq n, \tag{4.4.41}$$

$$|\partial_{ij} F_k|_{0,\infty,\hat{K}} \leq Ch^2, \quad 1 \leq i, j, k \leq n, \tag{4.4.42}$$

and inequalities (4.3.30) imply that

$$|J_F|_{0,\infty,\hat{K}} \leq Ch^n, \quad |J_{F^{-1}}|_{0,\infty,K} = |(J_F)^{-1}|_{0,\infty,\hat{K}} \leq \frac{C}{h^n}. \tag{4.4.43}$$

Next, we show that

$$|\partial_i J_F|_{0,\infty,\hat{K}} \leq Ch^{n+1}, \quad 1 \leq i \leq n, \quad |\partial_{ij} J_F|_{0,\infty,\hat{K}} \leq Ch^{n+2}, \quad 1 \leq i, j \leq n, \tag{4.4.44}$$

Let us denote by $\partial_i F(\hat{x})$ and $\partial_{ij} F(\hat{x})$ the column vectors with components $\partial_i F_k(\hat{x})$, $1 \leq k \leq n$, and $\partial_{ij} F_k(\hat{x})$, $1 \leq k \leq n$, respectively. Then to prove the first inequalities of (4.4.44), we observe that, for any $\hat{x} \in \hat{K}$, we have

$$\partial_i J_F(\hat{x}) = \sum_{j=1}^{n} \det(\partial_1 F(\hat{x}), \ldots, \partial_{j-1} F(\hat{x}), \partial_{ij} F(\hat{x}), \partial_{j+1} F(\hat{x}), \ldots, \partial_n F(\hat{x})),$$

and it suffices to use inequalities (4.4.41) and (4.4.42). The second inequalities of (4.4.44) are proved in a similar fashion (since $F \in (P_2(\hat{K}))^n$, the partial derivatives $\partial_{ijk} F$ are identically zero).

(ii) The expression to be estimated can be written as

$$E_K(a \partial_i p' \partial_j p) = \hat{E}(\hat{a}(\partial_i p')\hat{}\,(\partial_j p)\hat{}\, J_F). \tag{4.4.45}$$

Then it is clear that, by contrast with the affine case, *the functions* $(\partial_i p')\hat{}$ *and* $(\partial_j p)\hat{}$ *no longer belong to the space* $P_1(\hat{K})$ in general. This is why our first task is to determine the nature of these functions: Denoting by e_j the j-th basis vector of $\mathbf{R}^n$, we have

$$(\partial_j p)\hat{}\,(\hat{x}) = \partial_j p(x) = Dp(x) e_j = D\hat{p}(\hat{x}) DF^{-1}(x) e_j = D\hat{p}(\hat{x})(DF(\hat{x}))^{-1} e_j.$$

Expressing the fact that the vector $f_j = (DF(\hat{x}))^{-1} e_j$ is the solution of the linear system $DF(\hat{x}) f_j = e_j$, we find that

$$(\partial_j p)\hat{}\,(\hat{x}) = (J_F(\hat{x}))^{-1} \sum_{k=1}^{n} \partial_k \hat{p}(\hat{x}) \times \det(\partial_1 F(\hat{x}), \ldots, \partial_{k-1} F(\hat{x}), e_j, \partial_{k+1} F(\hat{x}), \ldots, \partial_n F(\hat{x})). \tag{4.4.46}$$

Consequently, the expression $(\partial_j p)\hat{}(\hat{x})J_F(\hat{x})$ is a finite sum of terms of the form $\pm\,\partial_k\hat{p}(\hat{x})\,\Pi_{l\neq k}\,\partial_l F_{j_l}(\hat{x})$, and likewise the quantity $(\partial_i p')\hat{}(\hat{x})$ is a finite sum of terms of the form $\pm(J_F(\hat{x}))^{-1}\partial_r(p')\hat{}(\hat{x})\,\Pi_{s\neq r}\,\partial_s F_{j_s}(\hat{x})$. Using (4.4.45), we have obtained a sum of the form

$$E_K(a\partial_i p'\partial_j p) = \sum_{\substack{\{k;j_l,l\neq k\\ r;j_s,s\neq r}}{}' \pm \hat{E}\Big(J_F^{-1}\hat{a}\prod_{s\neq r}\partial_s F_{j_s}\prod_{l\neq k}\partial_l F_{j_l}\partial_r\hat{p}'\partial_k\hat{p}\Big), \tag{4.4.47}$$

where, by the symbol Σ', we simply mean that the indices j_l and j_s do not take all possible values $1, 2, \ldots, n$.

(iii) We shall now take crucial advantage of the fact that the functions $(\partial_i p')\hat{}$ and $(\partial_j p)\hat{}$ can be expressed in terms of the functions $\partial_k\hat{p}$, $1\leq k\leq n$, which *do* belong to the space $P_1(\hat{K})$: Consider one of the terms occuring in the sum (4.4.47). It can be written as

$$\hat{E}\Big(J_F^{-1}\hat{a}\prod_{s\neq r}\partial_s F_{j_s}\prod_{l\neq k}\partial_l F_{j_l}\partial_r\hat{p}'\partial_k\hat{p}\Big) = \hat{E}(\hat{b}\hat{v}\hat{w}), \tag{4.4.48}$$

with

$$\begin{cases} \hat{b} = J_F^{-1}\hat{a}\prod_{s\neq r}\partial_s F_{j_s}\prod_{l\neq k}\partial_l F_{j_l} \in W^{2,\infty}(\hat{K}), \\ \hat{v} = \partial_r\hat{p}' \in P_1(\hat{K}), \\ \hat{w} = \partial_k\hat{p} \in P_1(\hat{K}), \end{cases} \tag{4.4.49}$$

and consequently, we may apply inequality (4.1.47) with the value $k=2$. We find in this fashion that

$$\begin{aligned} |\hat{E}(\hat{b}\hat{v}\hat{w})| &\leq C(|\hat{b}|_{2,\infty,\hat{K}}|\hat{v}|_{0,\hat{K}} + |\hat{b}|_{1,\infty,\hat{K}}|\hat{v}|_{1,\hat{K}})|\hat{w}|_{0,\hat{K}} \\ &\leq C(|\hat{b}|_{2,\infty,\hat{K}}|\hat{p}'|_{1,\hat{K}} + |\hat{b}|_{1,\infty,\hat{K}}|\hat{p}'|_{2,\hat{K}})|\hat{p}|_{1,\hat{K}}, \end{aligned} \tag{4.4.50}$$

and it remains to express the various semi-norms occuring in the above inequality in terms of appropriate norms over the set K. Using Theorems 4.3.2 and 4.3.3, we obtain:

$$\begin{cases} |\hat{p}|_{l,\hat{K}} \leq Ch^{-n/2}h^l|p|_{l,K}, & l=0,1, \\ |\hat{p}'|_{l,\hat{K}} \leq Ch^{-n/2}h^l\|p'\|_{l,K}, & l=1,2. \end{cases} \tag{4.4.51}$$

Next, we have (cf. (4.1.42))

$$\begin{aligned}|\hat{b}|_{1,\infty,\hat{K}} &= \Big|J_F^{-1}\hat{a}\prod_{s\neq r}\partial_s F_{j_s}\prod_{l\neq k}\partial_l F_{j_l}\Big|_{1,\infty,\hat{K}} \\ &\leq C\Big(|J_F^{-1}|_{0,\infty,\hat{K}}|\hat{a}|_{1,\infty,\hat{K}}\Big|\prod_{s\neq r}\partial_s F_{j_s}\prod_{l\neq k}\partial_l F_{j_l}\Big|_{0,\infty,\hat{K}}+ \\ &\quad +|J_F^{-1}|_{0,\infty,\hat{K}}|\hat{a}|_{0,\infty,\hat{K}}\Big|\prod_{s\neq r}\partial_s F_{j_s}\prod_{l\neq k}\partial_l F_{j_l}\Big|_{1,\infty,\hat{K}}+ \\ &\quad +|J_F^{-1}|_{1,\infty,\hat{K}}|\hat{a}|_{0,\infty,\hat{K}}\Big|\prod_{s\neq r}\partial_s F_{j_s}\prod_{l\neq k}\partial_l F_{j_l}\Big|_{0,\infty,\hat{K}}\Big),\end{aligned} \tag{4.4.52}$$

and we could likewise write an analogous inequality for the semi-norm $|\hat{b}|_{2,\infty,\hat{K}}$. Using inequalities (4.4.41) and (4.4.42), we obtain

$$\Big|\prod_{s\neq r}\partial_s F_{j_s}\prod_{l\neq k}\partial_l F_{j_l}\Big|_{\lambda,\infty,\hat{K}} \leq Ch^{2n-2+\lambda}, \quad \lambda = 0, 1, 2, \tag{4.4.53}$$

and, using inequalities (4.4.43) and (4.4.44), we obtain

$$|J_F^{-1}|_{\mu,\infty,\hat{K}} \leq Ch^{\mu-n}, \quad \mu = 0, 1, 2, \tag{4.4.54}$$

so that, upon combining inequalities (4.4.52), (4.4.53), (4.4.54) with the inequalities (cf. Theorems 4.3.2 and 4.3.3)

$$|\hat{a}|_{\nu,\infty,\hat{K}} \leq Ch^{\nu}\|a\|_{\nu,\infty,K}, \quad \nu = 0, 1, 2, \tag{4.4.55}$$

we eventually find that

$$|\hat{b}|_{1,\infty,\hat{K}} \leq Ch^{n-1}\|a\|_{1,\infty,K}. \tag{4.4.56}$$

By a similar analysis, we would find that

$$|\hat{b}|_{2,\infty,\hat{K}} \leq Ch^{n}\|a\|_{2,\infty,K}. \tag{4.4.57}$$

Then the conjunction of inequalities (4.4.50), (4.4.51), (4.4.56) and (4.4.57) with equation (4.4.48) shows that

$$\begin{aligned}&\Big|\hat{E}\Big(J_F^{-1}\hat{a}\prod_{s\neq r}\partial_s F_{j_s}\prod_{l\neq k}\partial_l F_{j_l}\partial_r\hat{p}'\partial_k\hat{p}\Big)\Big| \leq \\ &\quad\leq Ch^2\|a\|_{2,\infty,K}\|p'\|_{2,K}|p|_{1,K}.\end{aligned} \tag{4.4.58}$$

By adding up inequalities (4.4.58), we find that the expression $E_K(a\partial_i p'\partial_j p)$ (cf. (4.4.47)) satisfies an inequality similar to (4.4.58), and the proof is complete. □

Theorem 4.4.5. *Let there be given a regular isoparametric family of n-simplices K of type* (2), *let the quadrature scheme over the reference* *finite element be such that*

$$\forall \hat{\varphi} \in P_2(\hat{K}), \quad \hat{E}(\hat{\varphi}) = 0, \tag{4.4.59}$$

and finally, let $q \in [1, \infty]$ *be any number which satisfies the inequality*

$$2 - \frac{n}{q} > 0. \tag{4.4.60}$$

Then there exists a constant C *independent of* K *such that*

$$\forall f \in W^{2,q}(K), \quad \forall p \in P_K,$$

$$|E_K(fp)| \leq Ch_{\tilde{K}}^2(\text{meas}(\tilde{K}))^{1/2-1/q}\|f\|_{2,q,K}\|p\|_{1,K}, \tag{4.4.61}$$

where, for each K, $\tilde{K}$ *denotes the* n*-simplex with the same vertices as those of* K.

Proof. We have, for all $f \in W^{2,q}(K)$ and all $p \in P_K$,

$$E_K(fp) = \hat{E}(\hat{f}\hat{p}J_F). \tag{4.4.62}$$

It follows from the proof of Theorem 4.1.5 (cf. (4.1.54) and (4.1.55)) that there exists a constant C such that

$$\forall \hat{g} \in W^{2,q}(\hat{K}), \quad \forall \hat{p} \in P_2(\hat{K}),$$

$$|\hat{E}(\hat{g}\hat{p})| \leq C((|\hat{g}|_{1,q,\hat{K}} + |\hat{g}|_{2,q,\hat{K}})|\hat{p}|_{1,\hat{K}} + |\hat{g}|_{2,q,\hat{K}}|\hat{p}|_{0,\hat{K}}). \tag{4.4.63}$$

By letting

$$\hat{g} = \hat{f}J_F \tag{4.4.64}$$

in the above inequality and by making use of inequalities (4.1.42), (4.4.43), (4.4.44) and

$$|\hat{f}|_{\mu,q,\hat{K}} \leq C(\text{meas}(\tilde{K}))^{-1/q}h^{\mu}\|f\|_{\mu,q,K}, \quad \mu = 0, 1, 2$$

(cf. Theorems 4.3.2 and 4.3.3), we obtain

$$|\hat{f}J_F|_{l,q,\hat{K}} \leq C\Big(\sum_{j=0}^{l} |J_F|_{j,\infty,\hat{K}}|\hat{f}|_{l-j,q,\hat{K}}\Big)$$

$$\leq C(\text{meas}(\tilde{K}))^{-1/q}h^{n+l}\|f\|_{l,q,K}, \; l = 1, 2.$$

These last inequalities, coupled with relations (4.4.62), (4.4.63), (4.4.64) and the first inequalities of (4.4.51) with $l = 0, 1$, yield inequality (4.4.61). ☐

Estimate of the error $\|\tilde{u} - u_h\|_{1,\Omega_h}$

Combining the previous theorems, we are in a position to prove the main result of this section, which the reader would profit from comparing with Theorem 4.1.6. We recall that u is the solution of the variational problem corresponding to the data (4.4.1). For references concerning the existence of extensions such as $\tilde{u}$ and $\tilde{a}_{ij}$ below, see LIONS (1962, chapter 2), NEČAS (1967, chapter 2).

Theorem 4.4.6. *Let $n \leq 5$, let (V_h) be a family of finite element spaces made up of isoparametric n-simplices of type (2), and let there be given a quadrature scheme on the reference finite element such that*

$$\forall \hat{\varphi} \in P_2(\hat{K}), \quad \hat{E}(\hat{\varphi}) = 0. \tag{4.4.65}$$

Let $\tilde{\Omega}$ be an open set such that the inclusions

$$\Omega \subset \tilde{\Omega}, \quad \text{and} \quad \Omega_h \subset \tilde{\Omega} \text{ for all } h, \tag{4.4.66}$$

hold, and such that the functions u and a_{ij}, $1 \leq i, j \leq n$, possess extensions $\tilde{u}$ and $\tilde{a}_{ij}$, $1 \leq i, j \leq n$, which satisfy

$$\tilde{u} \in H^3(\tilde{\Omega}), \quad \tilde{a}_{ij} \in W^{2,\infty}(\tilde{\Omega}), \quad 1 \leq i, j \leq n, \tag{4.4.67}$$

$$\tilde{f} = \sum_{i,j=1}^{n} \partial_j(\tilde{a}_{ij}\partial_i\tilde{u}) \in W^{2,q}(\tilde{\Omega}) \quad \text{for some} \quad q \geq 2 \text{ with } 2 > \frac{n}{q}. \tag{4.4.68}$$

Then, if hypothesis (H1) *holds, there exists a constant C independent of h such that*

$$\|\tilde{u} - u_h\|_{1,\Omega_h} \leq Ch^2\Big(\|\tilde{u}\|_{3,\tilde{\Omega}} + \sum_{i,j=1}^{n} \|\tilde{a}_{ij}\|_{2,\infty,\tilde{\Omega}}\|\tilde{u}\|_{3,\tilde{\Omega}} + \|\tilde{f}\|_{2,q,\tilde{\Omega}}\Big), \tag{4.4.69}$$

where $h = \max_{K \in \mathcal{T}_h} h_K$.

Proof. By Theorem 4.4.2, the approximate bilinear forms are uniformly V_h-elliptic and therefore, we can use the abstract error estimate (4.4.21) of Theorem 4.4.1.

(i) Since $n \leq 5$, the inclusion $H^3(\tilde{\Omega}) \hookrightarrow \mathscr{C}^0(\bar{\tilde{\Omega}})$ holds, and by Theorem 4.4.3, the function $\Pi_h\tilde{u}$ belongs to the space X_{0h} since on the boundary Γ, we have $u = \tilde{u} = 0$. Thus we may let $v_h = \Pi_h\tilde{u}$ in the term $\inf_{v_h \in V_h}\{\ldots\}$

which appears in the abstract error estimate. In this fashion we obtain

$$\|\tilde{u} - u_h\|_{1,\Omega_h} \leq C\Big(\|\tilde{u} - \Pi_h\tilde{u}\|_{1,\Omega_h} + \sup_{w_h \in V_h} \frac{|\tilde{a}_h(\Pi_h\tilde{u}, w_h) - a_h(\Pi_h\tilde{u}, w_h)|}{\|w_h\|_{1,\Omega_h}} + \sup_{w_h \in V_h} \frac{|\tilde{a}_h(\tilde{u}, w_h) - f_h(w_h)|}{\|w_h\|_{1,\Omega_h}}\Big). \tag{4.4.70}$$

By Theorem 4.4.3, we know that

$$\|\tilde{u} - \Pi_h\tilde{u}\|_{1,\Omega_h} \leq Ch^2\|\tilde{u}\|_{3,\Omega_h} \leq Ch^2\|\tilde{u}\|_{3,\tilde{\Omega}}. \tag{4.4.71}$$

(ii) To evaluate the two consistency errors, a specific choice must be made for the functions $\tilde{a}_{ij}$ which appear in the bilinear form $\tilde{a}_h(.,.)$: *We shall choose precisely the functions given in* (4.4.67). Notice that, since the inclusion $W^{2,\infty}(\tilde{\Omega}) \hookrightarrow \mathscr{C}^1(\bar{\tilde{\Omega}})$ holds, the functions $\tilde{a}_{ij}$ are in particular defined everywhere on the set $\tilde{\Omega}$. Then we have, for all $w_h \in V_h$,

$$\tilde{a}_h(\Pi_h\tilde{u}, w_h) - a_h(\Pi_h\tilde{u}, w_h) = \int_{\Omega_h} \sum_{i,j=1}^{n} \tilde{a}_{ij}\partial_i\Pi_h\tilde{u}\partial_j w_h \, dx - \sum_{K \in \mathcal{T}_h} \sum_{l=1}^{L} \omega_{l,K} \sum_{i,j=1}^{n} (a_{ij}\partial_i\Pi_h\tilde{u}\partial_j w_h)(b_{l,K}),$$

and, since all the quadrature nodes $b_{l,K}$ belong to the set $\bar{\Omega}$, we have $a_{ij}(b_{l,K}) = \tilde{a}_{ij}(b_{l,K})$. Consequently, we can rewrite the above expression as

$$\tilde{a}_h(\Pi_h\tilde{u}, w_h) - a_h(\Pi_h\tilde{u}, w_h) = \sum_{K \in \mathcal{T}_h} \sum_{i,j=1}^{n} E_K(\tilde{a}_{ij}\partial_i\Pi_h\tilde{u}\partial_j w_h).$$

Using the estimates of Theorem 4.4.4 and the Cauchy–Schwarz inequality, we obtain

$$|\tilde{a}_h(\Pi_h\tilde{u}, w_h) - a_h(\Pi_h\tilde{u}, w_h)| \leq C \sum_{K \in \mathcal{T}_h} h_K^2 \sum_{i,j=1}^{n} \|\tilde{a}_{ij}\|_{2,\infty,K}\|\Pi_K\tilde{u}\|_{2,K}|w_h|_{1,K}$$
$$\leq Ch^2\Big(\sum_{i,j=1}^{n} \|\tilde{a}_{ij}\|_{2,\infty,\tilde{\Omega}}\Big)\Big(\sum_{K \in \mathcal{T}_h} \|\Pi_K\tilde{u}\|_{2,K}^2\Big)^{1/2} |w_h|_{1,\Omega_h}.$$

By another application of Theorem 4.4.3,

$$\Big(\sum_{K\in\mathcal{T}_h}\|\Pi_K\tilde{u}\|_{2,K}^2\Big)^{1/2}\leqslant\|\tilde{u}\|_{2,\Omega_h}+\Big(\sum_{K\in\mathcal{T}_h}\|\tilde{u}-\Pi_K\tilde{u}\|_{2,K}^2\Big)^{1/2}$$
$$\leqslant\|\tilde{u}\|_{2,\Omega_h}+Ch\|\tilde{u}\|_{3,\Omega_h}\leqslant C\|\tilde{u}\|_{3,\tilde{\Omega}},$$

and thus, we have shown that

$$\sup_{w_h\in V_h}\frac{|\tilde{a}_h(\Pi_h\tilde{u},w_h)-a_h(\Pi_h\tilde{u},w_h)|}{\|w_h\|_{1,\Omega_h}}\leqslant Ch^2\sum_{i,j=1}^n\|\tilde{a}_{ij}\|_{2,\infty,\tilde{\Omega}}\|\tilde{u}\|_{3,\tilde{\Omega}}. \tag{4.4.72}$$

(iii) Let us next examine the expression which appears in the numerator of the second consistency error. First it is easily verified that assumptions (4.4.67) imply in particular that the functions $(a_{ij}\partial_i\tilde{u})$ belong to the space $H^1(\Omega)$.

Therefore Green's formula yields

$$\forall w_h\in V_h\subset H_0^1(\Omega_h),\quad \tilde{a}_h(\tilde{u},w_h)=\int_{\Omega_h}\sum_{i,j=1}^n\tilde{a}_{ij}\partial_i\tilde{u}\partial_jw_h\,\mathrm{d}x$$
$$=-\int_{\Omega_h}\sum_{i,j=1}^n\partial_j(\tilde{a}_{ij}\partial_i\tilde{u})w_h\,\mathrm{d}x$$
$$=\int_{\Omega_h}\tilde{f}w_h\,\mathrm{d}x.$$

Since we have

$$-\sum_{i,j=1}^n\partial_j(\tilde{a}_{ij}\partial_i\tilde{u})=-\sum_{i,j=1}^n\partial_j(a_{ij}\partial_iu)=f\quad\text{on}\quad\Omega,$$

the function $\tilde{f}$ given in (4.4.68) *is an extension of the function f*. Besides, using once again the fact that all integration nodes $b_{l,K}$ belong to the set $\bar{\Omega}$, we obtain $f(b_{l,K})=\tilde{f}(b_{l,K})$ and consequently, we can write

$$\tilde{a}_h(\tilde{u},w_h)-f_h(w_h)=\int_{\Omega_h}\tilde{f}w_h\,\mathrm{d}x-\sum_{K\in\mathcal{T}_h}\sum_{l=1}^L\omega_{l,K}(fw_h)(b_{l,K})$$
$$=\sum_{K\in\mathcal{T}_h}E_K(\tilde{f}w_h).$$

Using the estimates of Theorem 4.4.5, we get

$$|\tilde{a}_h(\tilde{u},w_h)-f_h(w_h)|\leqslant C\sum_{K\in\mathcal{T}_h}h_K^2(\operatorname{meas}(\tilde{K}))^{1/2-1/q}\|\tilde{f}\|_{2,q,K}\|w_h\|_{1,K}$$
$$\leqslant Ch^2\Big(\sum_{K\in\mathcal{T}_h}\operatorname{meas}(\tilde{K})\Big)^{1/2-1/q}\|\tilde{f}\|_{2,q,\Omega_h}\|w_h\|_{1,\Omega_h}.$$

By construction, the interiors of the n-simplices do not overlap and therefore the quantity $\Sigma_{K\in\mathcal{T}_h}\operatorname{meas}(\tilde{K}) = \operatorname{meas}(\bigcup_{K\in\mathcal{T}_h}\tilde{K})$ is clearly bounded independently of h. Thus, we have shown that

$$\sup_{w_h\in V_h}\frac{|\tilde{a}_h(\tilde{u}, w_h) - f_h(w_h)|}{\|w_h\|_{1,\Omega_h}} \leq Ch^2\|\tilde{f}\|_{2,q,\tilde{\Omega}}, \tag{4.4.73}$$

and inequality (4.4.69) follows from inequalities (4.4.70), (4.4.71), (4.4.72) and (4.4.73). □

We have therefore reached a *remarkable conclusion*: *In order to retain the same order of convergence as in the case of polygonal domains (when only straight finite elements are used), the same quadrature scheme should be used, whether it be for straight or for isoparametric finite elements.* Thus, if $n = 2$ for instance, we can use the quadrature scheme of (4.1.17), which is exact for polynomials of degree ≤ 2.

Remark 4.4.4. (i) As one would expect, it is of course true that, in the absence of numerical integration, the order of convergence is the same, i.e., one has $\|\tilde{u} - \tilde{u}_h\|_{1,\Omega_h} = O(h^2)$, where $\tilde{u}$ is now the solution of the discrete problem (4.4.6). To show this is the object of Exercise 4.4.3.

(ii) To make the analysis even more complete, it would remain to show that for a given domain with a curved boundary (irrespectively of whether or not numerical integration is used), isoparametric n-simplices of type (2) yield better estimates than their straight counterparts! Indeed, STRANG & BERGER (1971) and THOMÉE (1973b) have shown that one gets in the latter case an $O(h^{3/2})$ convergence. In this direction, see Exercise 4.4.4. □

Remark 4.4.5. By contrast with the case of straight finite elements (cf. Remark 4.1.8), *the integrals* $\int_K a_{ij}\partial_i u_h \partial_j v_h \,\mathrm{d}x$ *are no longer computed exactly when the coefficients* a_{ij} *are constant functions.* If K is an isoparametric n-simplex of type (2), we have

$$\forall p', p \in P_K, \quad \int_K \partial_i p' \partial_j p \,\mathrm{d}x = \int_{\hat{K}} J_F(\partial_i p')^\wedge(\partial_j p)^\wedge \,\mathrm{d}\hat{x},$$

and (cf. (4.4.46)),

$$\begin{aligned} J_F(\hat{x})(\partial_i p')^{\wedge}(\hat{x}) &= \sum_{k=1}^{n} \partial_k \hat{p}'(\hat{x}) \det(\partial_1 F(\hat{x}), \ldots, \\ &\quad \times \partial_{k-1} F(\hat{x}), e_i, \partial_{k+1} F(\hat{x}), \ldots, \partial_n F(\hat{x})) \\ &= \{\text{polynomial of degree} \leq n \text{ in } \hat{x}\}, \end{aligned}$$

$$(\partial_i p)^{\wedge}(x) = (J_F(\hat{x}))^{-1} \times \{\text{polynomial of degree} \leq n \text{ in } \hat{x}\}.$$

Since

$$\begin{aligned} J_F(\hat{x}) &= \det(\partial_1 F(\hat{x}), \ldots, \partial_n F(\hat{x})) \\ &= \{\text{polynomial of degree} \leq n \text{ in } \hat{x}\}, \end{aligned}$$

we eventually find that

$$\int_K \partial_i p' \partial_j p \, \mathrm{d}x = \int_{\hat{K}} \frac{\{\text{polynomial of degree} \leq 2n \text{ in } \hat{x}\}}{\{\text{polynomial of degree} \leq n \text{ in } \hat{x}\}} \, \mathrm{d}\hat{x}.$$

Therefore the exact computation of such integrals would require a quadrature scheme which is exact for rational functions of the form N/D with $N \in P_{2n}(\hat{K})$, $D \in P_n(\hat{K})$. □

Exercises

4.4.1. With the notations of Fig. 4.4.3, show that the image $b_K = F_K(\hat{b})$ of any point $\hat{b} \in \hat{K}$ belongs to the set $\bar{\Omega} \cap K$ provided h_K is small enough.

4.4.2. With the same assumptions as in Theorem 4.4.4, show that the estimates

$$|E_K(a \partial_i p' \partial_j p)| \leq C h_K \|a\|_{2,\infty,K} \|p'\|_{1,K} |p|_{1,K}$$

hold. Deduce from these another proof of the uniform V_h-ellipticity of the approximate bilinear forms (this type of argument is used by ZLÁMAL (1974)).

4.4.3. Analyze the case where isoparametric n-simplices of type (2) are used *without* numerical integration, i.e., the discrete problem is defined as in (4.4.6).

[Hint: After defining appropriate extensions of the functions a_{ij} so that the discrete bilinear forms are uniformly V_h-elliptic, use the abstract error estimates of Theorem 4.4.1. This type of analysis is carried out in SCOTT (1973a).]

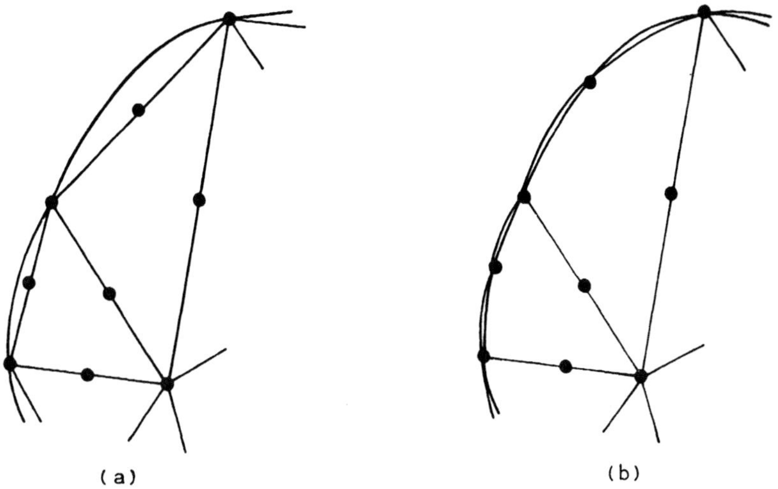

Fig. 4.4.4

4.4.4. Assume that the set Ω is a bounded convex domain in $\mathbf{R}^2$. Given a triangulation $\mathcal{T}_h$ made up of triangles *with straight sides only*, let X_h denote the finite element space whose generic finite element is the triangle of type (2), and let $V_h = \{v_h \in X_h;\ v_h = 0 \text{ on } \Gamma_h\}$, where Γ_h is the boundary of the set $\bigcup_{K\in\mathcal{T}_h} K$ (cf. Fig. 4.4.4(a)).

Show that (cf. STRANG & BERGER (1971), THOMÉE (1973b); see also STRANG & FIX (1973, Chapter 4))

$$\|u - u_h\|_{1,\Omega_h} = \mathrm{O}(h^{3/2}),$$

where $u_h \in V_h$ is the solution of the equations

$$\forall v_h \in V_h, \quad \int_\Omega \sum_{i,j=1}^n a_{ij}\partial_i u_h \partial_j v_h \,\mathrm{d}x = \int_\Omega f v_h \,\mathrm{d}x$$

(one should notice that in this case, the X_h-interpolant of the solution u does *not* belong to the space V_h). In other words, triangulations of type (b) are asymptotically better than triangulations of type (a) (cf. Fig. 4.4.4).

4.4.5. Analyze the case where n-simplices of type (1) are used, with or without numerical integration, over a curved domain Ω. It is assumed that all the vertices which are on the boundary of the set $\bar{\Omega}_h = \bigcup_{K\in\mathcal{T}_h} K$ are also on the boundary Γ.

Bibliography and comments

4.1. The content of this section is essentially based on CIARLET & RAVIART (1972c, 1975) and RAVIART (1972).

The abstract error estimate of Theorem 4.1.1 is based on STRANG (1972b). The proof of the uniform V_h-ellipticity given in Theorem 4.1.2 is based on, and generalizes, an idea of G. Strang (STRANG & FIX (1973, Section 4.3)).

Theorem 4.1.3 is due to BRAMBLE & HILBERT (1970). It is recognized as an important tool in getting error estimates in numerical integration and interpolation theory (although we did not use it in Section 3.1).

In CIARLET & RAVIART (1975), the content of this section is given a general treatment so as to comprise as special cases the inclusions $P_k(\hat{K}) \subset \hat{P} \subset P_{k'}(\hat{K})$ (cf. Exercise 4.1.6), the case of quadrilateral elements (cf. Exercise 4.1.7), etc... As regards in particular the error estimate in the norm $|.|_{0,\Omega}$ (cf. the abstract error estimate of Exercise 4.1.3), the following is proved: Assuming that the adjoint problem is regular and that $\hat{P} = P_k(\hat{K})$, one has $|u - u_h|_{0,\Omega} = O(h^{k+1})$ if the quadrature scheme is exact for the space $P_{2k-2}(\hat{K})$ if $k \geq 2$, or if the quadrature scheme is exact for the space $P_1(\hat{K})$ if $k = 1$.

For other references concerning the effect of numerical integration, see BABUŠKA & AZIZ (1972, Ch. 9), FIX (1972a, 1972b), HERBOLD (1968) where this problem was studied for the first time, HERBOLD, SCHULTZ & VARGA (1969), HERBOLD & VARGA (1972), ODEN & REDDY (1976a, Section 8.8), SCHULTZ (1972), STRANG & FIX (1973, Section 4.3).

Comparisons between finite element methods (with or without numerical integration) and finite-difference methods are found in BIRKHOFF & GULATI (1974), TOMLIN (1972), WALSH (1971).

Examples of numerical quadrature schemes used in actual computations are found in the book of ZIENKIEWICZ (1971, Section 8.10).

For general introductions to the subject of numerical integration (also known as: *numerical quadrature, approximate integration, approximate quadrature*), see the survey of HABER (1970), and the books of DAVIS & RABINOWITZ (1974), STROUD (1971).

For studies of numerical integration along the lines developed here, see also MANSFIELD (1971, 1972a). In ARCANGELI & GOUT (1976) and MEINGUET (1975), the constants appearing in the quadrature error estimates are evaluated.

4.2. The abstract error estimate of Theorem 4.2.2 is due to STRANG

(1972b). The description of Wilson's brick is given in WILSON & TAYLOR (1971).

In analyzing the consistency error, we have followed the method set up in CIARLET (1974a) for studying nonconforming methods, the main idea being to obtain two polynomial invariances in the functions $D_K(.,.)$ so as to apply the bilinear lemma. For the specific application of this method to Wilson's brick, we have extended to the three-dimensional case the analysis which LESAINT (1976) has made for Wilson's rectangle. P. Lesaint has considered the use of this element for approximating the system of plane elasticity, for which he was able to show the uniform ellipticity of the corresponding approximate bilinear forms. In this fashion, P. Lesaint obtains an $O(h)$ convergence in the norm $\|.\|_h$ and an $O(h^2)$ convergence in the norm $|.|_{0,\Omega}$ (the corresponding technique is indicated in Exercise 4.2.3). Also, the idea of introducing the degrees of freedom $\int_K \partial_{jj} p \, dx$ is due to P. Lesaint.

In his pioneering work on the mathematical analysis of nonconforming methods, G. Strang (cf. STRANG (1972b), and also STRANG & FIX (1973, Section 4.2) where the study of Wilson's brick is sketched) has shown in particular the importance of the patch test of B. Irons (cf. IRONS & RAZZAQUE (1972a)). For more recent developments on the connection with the patch test, see OLIVEIRA (1976).

There are other ways of generating nonconforming finite element methods. See for example RACHFORD & WHEELER (1974). In NITSCHE (1974), several types of such methods are analyzed in a systematic way. See also CÉA (1976).

References more specifically concerned with nonconforming methods for fourth-order problems are postponed till Section 6.2.

4.3. This section is based on CIARLET & RAVIART (1972b), where an attempt was made to establish an interpolation theory for general isoparametric finite elements (in this direction see Exercises 4.3.1, 4.3.8 and 4.3.9). A survey is given in CIARLET (1973).

To see that our description indeed coincides with the one used by the Engineers, let us consider for example the isoparametric triangle of type (2) as described by FELIPPA & CLOUGH (1970, p. 224): Given six points $a_i = (a_{1i}, a_{2i})$ $1 \leq i \leq 6$, in the plane (the points a_4, a_5 and a_6 play momentarily the role of the points which we usually call a_{12}, a_{23} and a_{13}, respectively), a "natural" coordinate system is defined, whereby the following relation (written in matrix form) should hold between the Cartesian coordinates x_1 and x_2 describing the finite element and the "new" coordinates λ_1, λ_2 and λ_3:

$$\begin{pmatrix} x_1 \\ x_2 \\ 1 \end{pmatrix} = \begin{pmatrix} a_{11} & a_{12} & a_{13} & a_{14} & a_{15} & a_{16} \\ a_{21} & a_{22} & a_{23} & a_{24} & a_{25} & a_{26} \\ 1 & 1 & 1 & 1 & 1 & 1 \end{pmatrix} \begin{pmatrix} \lambda_1(2\lambda_1 - 1) \\ \lambda_2(2\lambda_2 - 1) \\ \lambda_3(2\lambda_3 - 1) \\ 4\lambda_1\lambda_2 \\ 4\lambda_2\lambda_3 \\ 4\lambda_3\lambda_1 \end{pmatrix}$$

Then we observe that the first two lines of the above relation precisely represent relation (4.3.7), with $F(\hat{x}) = (F_1(\hat{x}), F_2(\hat{x}))$ now denoted (x_1, x_2). The last line of the above matrix equation implies either $\lambda_1 + \lambda_2 + \lambda_3 = 1$ or $\lambda_1 + \lambda_2 + \lambda_3 = -\frac{1}{2}$, so that the solution $\lambda_1 + \lambda_2 + \lambda_3 = 1$ is the only one which is acceptable if we impose the restriction that $\lambda_i \geqslant 0$, $1 \leqslant i \leqslant 3$.

Therefore, the "natural" coordinates λ_1, λ_2 and λ_3 are nothing but the barycentric coordinates with respect to a fixed triangle $\hat{K}$, and the isoparametric finite element associated with the points a_i, $1 \leqslant i \leqslant 6$, is in this formulation the set of those points (x_1, x_2) given by the first two lines of the above matrix equation when the "natural" coordinates λ_i (also known as "*curvilinear*" *coordinates*) satisfy the inequalities $0 \leqslant \lambda_i \leqslant 1$, $1 \leqslant i \leqslant 3$, and the equality $\sum_{i=1}^{3} \lambda_i = 1$.

A general description of isoparametric finite elements along these lines is also found in ZIENKIEWICZ (1971, chapter 8). The first references where such finite elements are found are ARGYRIS & FRIED (1968) and ERGATOUDIS, IRONS & ZIENKIEWICZ (1968).

In case of isoparametric quadrilateral elements, JAMET (1976b) has significantly contributed to the interpolation error analysis, by relaxing some assumptions of CIARLET & RAVIART (1972b).

Curved finite elements of other than isoparametric type have also been considered, notably by ZLÁMAL (1970, 1973a, 1973b, 1974) and SCOTT (1973a). Both authors begin by constructing a curved face K' by approximating a smooth surface through an $(n-1)$-dimensional interpolation process. This interpolation serves to define a mapping F_K which in turn allows to define a finite element with K' as a curved face. Then the corresponding interpolation theory follows basically the same pattern as here. In particular, R. Scott constructs in this fashion a curved finite element which resembles the isoparametric triangle of type (3) and for which an interpolation theory can be developed which requires weaker assumptions than those indicated in Exercise 4.3.8.

In ARCANGÉLI & GOUT (1976), a polynomial interpolation process over a curved domain is analyzed. For curved finite elements based on the so-called *blending function* interpolation process, see notably CAVENDISH, GORDON & HALL (1976), GORDON & HALL (1973), the paper of BARNHILL (1976a) and the references therein. WACHSPRESS (1971, 1973, 1975) uses rational functions for constructing general polygonal finite elements in the plane with straight or curved sides. For additional references, see LEAF, KAPER & LINDEMAN (1976), LUKÁŠ (1974), MCLEOD & MITCHELL (1972, 1976), MITCHELL (1976), MITCHELL & MARSHALL (1975).

4.4. The error analysis developed in this section follows the general approach set up in CIARLET & RAVIART (1972c) (however it was thought at that time that more accurate quadrature schemes were needed for isoparametric elements), where an estimate of the error in the norm $|\cdot|_{0,\Omega}$ was also obtained.

An analogous study is made in ZLÁMAL (1974), where it is shown that, for two-dimensional curved elements for which $\hat{P} = P_k(\hat{K})$, k even, it is sufficient to use quadrature schemes exact for polynomials of degree $\leq 2k-2$, in order to retain the $O(h^k)$ convergence in the norm $\|\cdot\|_{1,\Omega_h}$. ZLÁMAL (1973b) has also evaluated the error in the absence of numerical integration. For complementary results, see VEIDINGER (1975). Likewise, SCOTT (1973a) has shown that quadrature schemes of higher order of accuracy are not needed when curved finite elements are used. However, the finite elements considered by M. Zlámal and R. Scott are not of the isoparametric type as understood here. For such elements, a general theory is yet to be developed, in particular for quadrilateral finite elements.

In spite of the absence of a uniform V_h-ellipticity condition, GIRAULT (1976a) has successfully studied the use of quadrilaterals of type (1) in conjunction with a one-point quadrature scheme.

Alternate ways of handling Dirichlet problems posed over domains with curved boundaries have been proposed, which rely on various alterations of the bilinear form of the given problem. In this direction, we notably mention

(i) penalty methods, as advocated by AUBIN (1969) and BABUŠKA (1973b), and later improved by KING (1974),

(ii) methods where the boundary condition is considered as a constraint and as such is treated via techniques from duality theory, as in BABUŠKA (1973a),

(iii) least square methods as proposed and studied in BRAMBLE & SCHATZ (1970, 1971), BRAMBLE & NITSCHE (1973), BAKER (1973),

(iv) methods where the domain is approximated by a polygonal domain, as in BRAMBLE, DUPONT & THOMÉE (1972),

(v) various methods proposed by NITSCHE (1971, 1972b).

For additional references for the finite element approximation of boundary value problems over curved boundaries, see BABUŠKA (1971b), BERGER (1973), BERGER, SCOTT & STRANG (1972), BLAIR (1976), BRAMBLE (1975), NITSCHE (1972b), SCOTT (1975), SHAH (1970), STRANG & BERGER (1971), STRANG & FIX (1973, Chapter 4), THOMÉE (1973a, 1973b). See also Chapter 6 for fourth-order problems.

We finally mention that, following the terminology of STRANG (1972b), we have perpetrated in this chapter three *variational crimes*: numerical integration, nonconforming methods, approximation of curved boundaries.

Additional bibliography and comments

Problems on unbounded domains

Let us consider one physical example: Given an electric conductor which occupies a bounded volume $\bar{\Omega}$ in $\mathbf{R}^3$, and assuming that the electric potential u_0 is known along the boundary Γ of the set Ω, the *electric conductor problem* consists in finding the space distribution of the electric potential u. This potential u is the solution of

$$\begin{cases} \Delta u = 0 & \text{in} \quad \Omega, \\ \Delta u = 0 & \text{in} \quad \Omega' = \complement\bar{\Omega}, \\ u = u_0 & \text{on} \quad \Gamma. \end{cases}$$

Thus, in addition to a standard problem on the set $\bar{\Omega}$, we have to solve a boundary value problem on the *unbounded* set $\bar{\Omega}'$. Classically, this problem is solved in the following fashion: Denoting by $\partial_\nu u$ the normal derivative of $u|_{\bar{\Omega}}$ across Γ and by $(\partial_\nu u)'$ the normal derivative of $u|_{\bar{\Omega}'}$ across Γ (both normals being oriented in the same direction), let

$$q = \partial_\nu u - (\partial_\nu u)'.$$

Then if the function q is known on Γ, the solution u is obtained in $\mathbf{R}^3$ as a *single layer potential* through the formula

$$\forall x \in \mathbf{R}^3, \quad u(x) = \frac{1}{4\pi} \int_\Gamma \frac{q(y)}{\|x-y\|} \, \mathrm{d}\gamma(y).$$

By specializing the points x to belong to the boundary Γ in the above formula, we are therefore led to solve the *integral equation*:

$$\forall x \in \Gamma, \quad u_0(x) = \frac{1}{4\pi} \int_\Gamma \frac{q(y)}{\|x-y\|} \, \mathrm{d}\gamma(y)$$

in the unknown q. For details about this classical approach, see for instance PETROVSKY (1954).

Interestingly, this integral equation can be given a variational formulation which, among other things, make it amenable to finite element approximations, as shown by NÉDÉLEC & PLANCHARD (1973). First we need a new *Sobolev space*, the space

$$H^{1/2}(\Gamma) = \{r \in L^2(\Gamma);\ \exists v \in H^1(\Omega);\ \operatorname{tr} v = r \quad \text{on} \quad \Gamma\},$$

which is dense in the space $L^2(\Gamma)$. It is a Hilbert space when it is equipped with the quotient norm

$$r \in H^{1/2}(\Gamma) \to \|r\|_{H^{1/2}(\Gamma)} = \inf\{\|v\|_{1,\Omega};\ v \in H^1(\Omega),\ \operatorname{tr} v = r \quad \text{on} \quad \Gamma\}.$$

We shall denote by $H^{-1/2}(\Gamma)$ its dual space, and by $\|\cdot\|_{H^{-1/2}(\Gamma)}$ the dual norm. Denoting by $\langle \cdot, \cdot \rangle_\Gamma$ the duality pairing between the spaces $H^{-1/2}(\Gamma)$ and $H^{1/2}(\Gamma)$, we note that

$$\forall r \in L^2(\Gamma) \subset H^{-1/2}(\Gamma), \quad \forall s \in H^{1/2}(\Gamma), \quad \langle r, s \rangle_\Gamma = \int_\Gamma rs \, \mathrm{d}\gamma.$$

For details about these spaces (and more generally about the spaces $H^t(\Gamma)$, $t \in \mathbf{R}$), see LIONS & MAGENES (1968).

The bilinear form

$$(q, r) \to \frac{1}{4\pi} \int_\Gamma \int_\Gamma \frac{q(x) r(y)}{\|x-y\|} \, \mathrm{d}\gamma(x) \, \mathrm{d}\gamma(y)$$

is well-defined over the space $\mathscr{D}(\Gamma) \times \mathscr{D}(\Gamma)$, and it is continuous when the space $\mathscr{D}(\Gamma)$ is equipped with the norm $\|\cdot\|_{H^{-1/2}(\Gamma)}$. Consequently, it has a unique extension over the space $H^{-1/2}(\Gamma) \times H^{-1/2}(\Gamma)$, which shall be denoted by $a(\cdot, \cdot)$, and one can show that this bilinear form is $H^{-1/2}(\Gamma)$-

elliptic. Therefore, the natural variational formulation of the problem posed above as an integral equation consists in finding the unique function q which satisfies

$$q \in H^{-1/2}(\Gamma) \quad \text{and} \quad \forall r \in H^{-1/2}(\Gamma), \quad a(q, r) = \langle u_0, r \rangle_\Gamma,$$

assuming the data u_0 belongs to the space $H^{1/2}(\Gamma)$.

Once the function q is found in this fashion, the solution u of the original problem is obtained as follows. Define the space $W_0^1(\mathbf{R}^3)$ as being the completion of the space $\mathcal{D}(\mathbf{R}^3)$ with respect to the norm $|\cdot|_{1,\mathbf{R}^3}$. This space (which does *not* coincide with the space $H^1(\mathbf{R}^3)$, i.e., the completion of the space $\mathcal{D}(\mathbf{R}^3)$ with respect to the norm $\|\cdot\|_{1,\mathbf{R}^3}$) can be equally characterized by (cf. Barros, Neto (1965), Deny & Lions (1953–1954))

$$\begin{aligned} W_0^1(\mathbf{R}^3) &= \{v \in L^6(\mathbf{R}^3);\ \partial_i v \in L^2(\mathbf{R}^3);\quad 1 \leq i \leq 3\} \\ &= \left\{v \in \mathcal{D}'(\mathbf{R}^3);\ \frac{v}{(1+\|x\|^2)^{1/2}} \in L^2(\mathbf{R}^3);\ \partial_i v \in L^2(\mathbf{R}^3),\right. \\ &\qquad \left. 1 \leq i \leq 3\right\}. \end{aligned}$$

Then for each function $q \in \mathcal{D}(\Gamma)$, the function

$$u: x \in \mathbf{R}^3 \to u(x) = \frac{1}{4\pi} \int_\Gamma \frac{q(y)}{\|x-y\|} \,\mathrm{d}\gamma(y)$$

belongs to the space $W_0^1(\mathbf{R}^3)$ and, besides, the mapping $q \in \mathcal{D}(\Gamma) \to u \in W_0^1(\mathbf{R}^3)$ defined in this fashion is continuous when the space $\mathcal{D}(\Gamma)$ is equipped with the norm $\|\cdot\|_{H^{-1/2}(\Gamma)}$. Therefore, it has a unique extension over the space $H^{-1/2}(\Gamma)$. In other words, we have solved the original problem via the mappings $u_0 \in H^{1/2}(\Gamma) \to q \in H^{-1/2}(\Gamma) \to u \in W_0^1(\mathbf{R}^3)$ (as indicated in Nédélec & Planchard (1973), one can also solve directly the problem $u_0 \in H^{1/2}(\Gamma) \to u \in W_0^1(\mathbf{R}^3)$). We mention that the related *perfect dielectric problem* can be also handled in an analogous manner (the boundary condition $u = u_0$ on Γ is then replaced by $\partial_\nu u - c(\partial_\nu u)' = u_1$ on Γ, $c > 0$).

J.C. Nédélec and J. Planchard then construct a general finite element approximation of the above problem. Given a subspace of the space $H^{-1/2}(\Gamma)$, they first derive an abstract error estimate: Let $u_{0h} \in V_h$ be an approximation of the function u_0 and let the discrete solution q_h be such that

$$\forall r_h \in V_h, \quad a(q_h, r_h) = \langle u_{0h}, r_h \rangle_\Gamma.$$

Then there exists a constant C independent of the subspace V_h such that

$$\|q - q_h\|_{H^{-1/2}(\Gamma)} \leq C\Big(\inf_{r_h \in V_h} \|q - r_h\|_{H^{-1/2}(\Gamma)} + \|u_0 - u_{0h}\|_{H^{1/2}(\Gamma)}\Big).$$

To apply this error estimate the authors assume that the boundary Γ is polygonal so that it may be triangulated in an obvious fashion, i.e., the set Γ is written as a union $\bigcup_{K \in \mathcal{T}_h} K$ of triangles K. Then they look for a discrete solution in either space

$$V_{0h} = \{v_h : \Gamma \to \mathbf{R};\ \forall K \in \mathcal{T}_h,\ v_h|_K \in P_0(K)\} \subset L^2(\Gamma) \subset H^{-1/2}(\Gamma),$$

$$V_{1h} = \{v_h \in \mathscr{C}^0(\Gamma);\ \forall K \in \mathcal{T}_h,\ v_h|_K \in P_1(K)\} \subset H^1(\Gamma) \subset H^{-1/2}(\Gamma)$$

(note that the functions in the space V_{0h} are discontinuous across adjacent triangles) and they show that

$$\|q - q_h\|_{H^{1/2}(\Gamma)} \leq \begin{cases} C(q)h^{3/2} + C\|u_0 - u_{0h}\|_{H^{1/2}(\Gamma)} & \text{in} \quad V_{0h}, \\ C(q)h^{5/2} + C\|u_0 - u_{0h}\|_{H^{1/2}(\Gamma)} & \text{in} \quad V_{1h}, \end{cases}$$

assuming the function q is smooth enough. To conclude their analysis, they compute the function

$$u_h : x \in \mathbf{R}^3 \to \frac{1}{4\pi} \int_\Gamma \frac{q_h(y)}{\|x - y\|} \, d\gamma(y),$$

and they obtain in both cases

$$\|u - u_h\|_{W_0^1(\mathbf{R}^3)} \leq C(q)h^{3/2} + \|u - u_{0h}\|_{H^{1/2}(\Gamma)}.$$

Of course, the major computational difficulty in this approach is the evaluation of the coefficients of the resulting linear system. For a review of the numerical aspects of such integral equation techniques for solving problems on unbounded domains arising in the study of 2- or 3-dimensional incompressible potential flows around obstacles, see HESS (1975a, 1975b).

NÉDÉLEC (1976) next considers the case of a curved surface Γ which needs therefore to be approximated by another surface Γ_h made up of finite elements of isoparametric type (such a construction is related to – and is of interest for – the surface approximation found in the shell problem; cf. Section 8.2). Again error estimates for the differences $(q - q_h)$ and $(u - u_h)$ are obtained in appropriate Hilbert spaces.

LE ROUX (1974, 1977) considers the finite element approximation of the analogous problem in dimension two. In this case, the kernel in the integral transform is $\ln \|x - y\|$ instead of $1/\|x - y\|$. A similar analysis is found in HSIAO & WENDLAND (1976).

There are other ways of handling problems on unbounded domains. In particular, there are methods where the unbounded domain is triangulated and then the triangulation is "truncated" in some fashion. In this spirit, BABUŠKA (1972c) considers the model problem: Find $u \in H^1(\mathbf{R}^n)$ such that $-\Delta u + u = f$ in $\mathbf{R}^n$, $f \in L^2(\mathbf{R}^n)$. Using an "abstract" variational approximation (cf. BABUŠKA (1970, 1971a)), he obtains orders of convergence on compact subsets of $\mathbf{R}^n$ which are arbitrarily close to the orders of convergence obtained in the case of bounded domains. By contrast with the method of FIX & STRANG (1969), the discrete solution is obtained via the solution of a linear system with a finite number of unknowns. In SILVESTER & HSIEH (1971), a bounded subdomain is triangulated in the usual way while the remaining unbounded part is represented by a single "finite element" of a special type.

As we shall mention in the section "Bibliography and Comments" of Section 5.1, problems on unbounded domains which typically arise in the study of 2-dimensional compressible flows may be reduced to variational inequalities, as in CIAVALDINI & TOURNEMINE (1977) and ROUX (1976).

The Stokes problem

Classically, the *Stokes problem* for an incompressible viscous fluid in a domain $\bar{\Omega} \subset \mathbf{R}^n$, $n = 2$ or 3, consists in finding functions $\boldsymbol{u} = (u_i)_{i=1}^n$ and p defined over the set $\bar{\Omega}$, which satisfy ($\Delta \boldsymbol{u} = (\Delta u_i)_{i=1}^n$)

$$\begin{cases} -\nu \Delta \boldsymbol{u} + \nabla p = f & \text{in} \quad \Omega, \\ \operatorname{div} \boldsymbol{u} = 0 & \text{in} \quad \Omega, \\ \boldsymbol{u} = \boldsymbol{0} & \text{on} \quad \Gamma. \end{cases}$$

The vector function $\boldsymbol{u}$ represents the velocity distribution, the scalar function p is the pressure, and the given vector function $\boldsymbol{f} = (f_i)_{i=1}^n \in (L^2(\Omega))^n$ represents the volumic forces per unit mass. The constant $\nu > 0$ is the *dynamic viscosity*, a constant which is inversely proportional to the *Reynolds number.*

In order to derive the variational formulation of this problem, we introduce the space

$$V = \{\boldsymbol{v} \in (H_0^1(\Omega))^n;\ \operatorname{div} \boldsymbol{v} = 0\}$$

provided with the norm

$$|\cdot|_{1,\Omega}: \boldsymbol{v} = (v_i)_{i=1}^n \to |\boldsymbol{v}|_{1,\Omega} = \left(\sum_{i=1}^n |v_i|_{1,\Omega}^2\right)^{1/2},$$

and we introduce the bilinear form

$$\boldsymbol{u}, \boldsymbol{v} \in (H^1(\Omega))^n \to a(\boldsymbol{u}, \boldsymbol{v}) = \sum_{i,j=1}^n \int_\Omega \partial_j u_i \partial_j v_i \, \mathrm{d}x,$$

which is clearly $\boldsymbol{V}$-elliptic. We shall use the notation

$$(u, v) = \int_\Omega uv \, \mathrm{d}x, \quad (\boldsymbol{u}, \boldsymbol{v}) = \int_\Omega \boldsymbol{u} \cdot \boldsymbol{v} \, \mathrm{d}x.$$

Since one has $(\nabla p, \boldsymbol{v}) = -(p, \operatorname{div} \boldsymbol{v})$ for all $v \in \mathcal{D}(\Omega)$ for smooth functions p, *the natural variational formulation of this problem consists in finding a pair* $(\boldsymbol{u}, p)$ *such that*

$$(\boldsymbol{u}, p) \in \boldsymbol{V} \times (L^2(\Omega)/P_0(\Omega)), \quad \text{and}$$
$$\forall \boldsymbol{v} \in (H_0^1(\Omega))^n, \quad \nu a(\boldsymbol{u}, \boldsymbol{v}) - (p, \operatorname{div} \boldsymbol{v}) = (\boldsymbol{f}, \boldsymbol{v})$$

(notice that the definition of the space $L^2(\Omega)/P_0(\Omega)$ reflects the fact that the unknown p can be determined up to additive constants only). Then we observe that the relations

$$\forall \boldsymbol{v} \in \boldsymbol{V}, \quad \nu a(\boldsymbol{u}, \boldsymbol{v}) = (\boldsymbol{f}, \boldsymbol{v})$$

determine uniquely the function $\boldsymbol{u}$ (a word of caution: Since the space $(\mathcal{D}(\Omega))^n$ is *not* contained in the space $\boldsymbol{V}$, the above variational problem *cannot* be interpreted in the usual way as a boundary value problem). Once the function $\boldsymbol{u}$ is known, it remains to find a function $p \in L^2(\Omega)/P_0(\Omega)$ such that

$$\forall \boldsymbol{v} \in (H_0^1(\Omega))^n, \quad (p, \operatorname{div} \boldsymbol{v}) = g(\boldsymbol{v}),$$

where the linear form

$$g: \boldsymbol{v} \in (H_0^1(\Omega))^n \to g(\boldsymbol{v}) = \nu a(\boldsymbol{u}, \boldsymbol{v}) - (\boldsymbol{f}, \boldsymbol{v})$$

is continuous over the space $(H_0^1(\Omega))^n$ and vanishes over its subspace $\boldsymbol{V}$, by definition of the function $\boldsymbol{u}$.

It then follows that there exists a function $p \in L^2(\Omega)$, unique up to an additive constant factor, such that the linear form can also be written as

$$\forall \boldsymbol{v} \in (H_0^1(\Omega))^n, \quad g(\boldsymbol{v}) = \int_\Omega p \operatorname{div} \boldsymbol{v} \, \mathrm{d}x.$$

This is a nontrivial fact whose proof may be found in de RHAM (1955) (the converse is clear).

Notice that if $n = 2$, the Stokes problem can be reduced to a familiar problem: Since $\operatorname{div} \boldsymbol{u} = 0$, there exists a *stream function* ψ such that $u_1 = \partial_2\psi$ and $u_2 = -\partial_1\psi$. Then a simple computation shows that $\nu\Delta^2\psi = f$ with $f = \partial_1 f_2 - \partial_2 f_1$. When the set Ω is simply connected, we may impose the boundary condition $\psi = 0$ on Γ, so that we also have $\partial_\nu\psi = 0$ on Γ as a consequence of the boundary condition $\boldsymbol{u} = \boldsymbol{0}$ on Γ. Therefore *the solution of the Stokes problem is reduced in this case to the solution of a biharmonic problem* (cf. Section 1.2). Observe that the *vorticity* $-\Delta\psi$ is then nothing but the value of the rotational of the velocity $\boldsymbol{u}$. Finite element methods for this problem will be described in Section 6.1 and Chapter 7.

As regards the finite element approximation of the general Stokes problem, it is realized that a major difficulty consists in taking properly into account the *incompressibility condition* $\operatorname{div} \boldsymbol{u} = 0$. A first approach is to use standard finite element spaces $\boldsymbol{V}_h$ in which the condition $\operatorname{div} \boldsymbol{v}_h = 0$ is exactly imposed. However, this process often results in sophisticated elements. Methods of this type have been extensively studied by FORTIN (1972a, 1972b).

In a second approach, whose applicability seems wider, the *incompressibility condition is approximated.* This is the method advocated by CROUZEIX & RAVIART (1973), who seek the discrete solution in a space of the form

$$\boldsymbol{V}_h = \left\{\boldsymbol{v}_h \in \boldsymbol{X}_{0h};\ \forall\phi_h \in \Phi_h,\ \sum_{K\in\mathcal{T}_h} \int_K \phi_h \operatorname{div} \boldsymbol{v}_h \,\mathrm{d}x = 0\right\},$$

where $\boldsymbol{X}_{0h}$ is a product of standard finite element spaces and Φ_h appears as an appropriate space of "Lagrange multipliers", following the terminology of duality theory (cf. the section "Additional Bibliography and Comments" in Chapter 7). For instance if the generic finite element in the space X_{0h} is the triangle – or tetrahedron – of type (k), the space Φ_h is the product $\Pi_{K\in\mathcal{T}_h} P_{k-1}(K)$. In their remarkable paper, M. Crouzeix and P.-A. Raviart construct both conforming and nonconforming finite elements of special type (cf. Exercise 2.3.9) and they obtain estimates for the error $(\Sigma_{K\in\mathcal{T}_h} |\boldsymbol{u} - \boldsymbol{u}_h|^2_{1,K})^{1/2}$, and for the error $(\Sigma_{i=1}^n |u_i - u_{ih}|^2_{0,\Omega})^{1/2}$ through an extension of the Aubin–Nitsche lemma. They also compute an approximation p_h of the pressure p and they evaluate the norm $\|p - p_h\|_{L^2(\Omega)/P_0(\Omega)}$. Finally, they briefly consider the case of the inhomogeneous boundary condition $\boldsymbol{u} = \boldsymbol{u}_0$ on Γ. As usual the error estimates depend upon the smoothness of the solution, a question studied in KELLOGG & OSBORN (1976), OSBORN (1976b), and TÉMAM (1973). It seems however that the most promising finite element approximations of the Stokes problem are of the so-called *mixed* type. For such methods, the reader is referred to the section "Additional Bibliography and Comments" at the end of Chapter 7.

Further references concerning the finite element approximation of the Stokes problem are FALK (1976a, 1976c), FALK & KING (1976), and the thorough treatment given by TÉMAM (1977). We also mention that CROUZEIX & LE ROUX (1976) have proposed and analyzed a finite element method for two-dimensional irrotational fluid flows, in which the unknown $\boldsymbol{u} = (u_1, u_2)$ satisfies

$$\begin{cases} \mathit{rot}\, \boldsymbol{u} = \boldsymbol{0} & \text{in} \quad \Omega, \\ \operatorname{div} \boldsymbol{u} = f & \text{in} \quad \Omega, \\ \boldsymbol{u} \cdot \boldsymbol{\nu} = g & \text{on} \quad \Gamma. \end{cases}$$

Eigenvalue problems

Given an elliptic operator $\mathscr{L}$ defined on a bounded open subset Ω of $\mathbf{R}^n$ and given the boundary condition $u = 0$ on Γ, the associated *eigenvalue problem* classically consists in finding real numbers λ and functions $u \neq 0$ such that

$$\begin{cases} \mathscr{L}u = \lambda u & \text{in} \quad \Omega, \\ u = 0 \quad \text{in} \quad \Gamma. \end{cases}$$

Indeed, eigenvalue problems may be associated with any other homogeneous boundary conditions but, for simplicity, we shall consider only the Dirichlet condition.

Such problems typically arise when one looks for *periodic* (in time) solutions of evolution problems of the form $\partial_{00}u + \mathscr{L}u = 0$, $u = 0$ on Γ, where ∂_{00} denotes the second partial derivative with respect to the time variable t. Such particular solutions being of the form $u(x)e^{i\mu t}$, $\mu \in \mathbf{R}$, the pair (λ, u), $\lambda = \mu^2$, is therefore obtained through the solution of an eigenvalue problem. This is why such problems are of fundamental importance, in the analysis of vibrations of structures for instance.

We shall in fact consider the *variational formulation of this eigenvalue problem*, which consists in finding pairs (λ, u), $\lambda \in \mathbf{R}$, $u \in V - \{0\}$, such that

$$\forall v \in V \quad a(u, v) = \lambda(u, v),$$

where $V = H_0^1(\Omega)$ or $H_0^2(\Omega)$, depending upon whether $\mathscr{L}$ is a second-order or fourth-order operator, $a(\cdot,\cdot)$ is the associated bilinear form (i.e., which satisfies $a(u, v) = (\mathscr{L}u, v)$ for all $v \in \mathscr{D}(\Omega)$), and $(\cdot,\cdot)$ is the inner-product in the space $L^2(\Omega)$. If (λ, u) is a solution, then u is called an *eigenfunction* associated with the *eigenvalue* λ.

Let us make the usual assumptions that the bilinear form is continuous and V-elliptic, so that for each $f \in L^2(\Omega)$, there exists a unique function $u \in V$ which satisfies $a(u, v) = (f, v)$ for all $v \in V$ (if we identify the function f with an element of V', we have $u = A^{-1}f$ with the notations of Theorem 1.1.3). In this fashion, we define a mapping

$$G: f \in L^2(\Omega) \to u = Gf \in V,$$

which is continuous (cf. Remark 1.1.3), and consequently, *the mapping*

$$G: V \to V$$

is compact, by Rellich theorem. Since $(u, v) = a(Gu, v)$ for all $u, v \in V$ by definition of the mapping G, *the eigenvalue problem amounts to finding the inverses of the eigenvalues of the mapping* $G: V \to V$ (clearly, zero cannot be an eigenvalue of the mapping G nor of the original problem).

If we finally add the assumption of *symmetry* of the bilinear form, then the problem is reduced to that of finding the eigenvalues and eigenfunctions of a compact symmetric operator in the Hilbert space V, considered as equipped with the inner-product $a(\cdot,\cdot)$ (the symmetry is a consequence of the equalities $a(Gu, v) = (u, v) = (v, u) = a(Gv, u) = a(u, Gv)$). Consequently, an application of the spectral theory of such operators (cf. e.g. Riesz & Nagy (1952)) yields the following result concerning the existence and characterizations of the solutions of the eigenvalue problem: *There exists an increasing sequence of strictly positive eigenvalues*:

$$0 < \lambda_1 \leq \lambda_2 \leq \cdots \leq \lambda_k \leq \lambda_{k+1} \leq \cdots \textit{ with } \lim_{k\to\infty} \lambda_k = \infty,$$

associated with eigenfunctions u_k, $k \geq 1$, *which can be orthonormalized in the sense that*

$$a(u_k, u_l) = \lambda_k \delta_{kl}, \quad (u_k, u_l) = \delta_{kl}, \quad k,l \geq 1,$$

and which form a complete system in both the Hilbert spaces V and $L^2(\Omega)$.

Moreover, if we introduce the *Rayleigh quotient*

$$R: v \in V - \{0\} \to R(v) = \frac{a(v, v)}{(v, v)},$$

the eigenvalues are characterized by the relations

$$\begin{cases} \lambda_1 = \inf\{R(v);\ v \in V\} = R(u_1), \\ \lambda_k = \inf\{R(v);\ v \in V,\ (v, u_l) = 0,\ 0 \leq l \leq k-1\} = R(u_k),\ k \geq 2. \end{cases}$$

Other characterizations (quite useful for the analysis of such problems and of their approximations) as well as further developments may be found in COURANT & HILBERT (1953, Chapter V).

The simplest discretization of such problems is called the *Rayleigh–Ritz method* and it is defined as follows: Given a subspace V_h, of dimension M, of the space V, find pairs $(\lambda_h, u_h) \in \mathbf{R} \times V$ such that

$$\forall v_h \in V_h, \quad a(u_h, v_h) = \lambda_h (u_h, v_h).$$

Equivalently, if we let w_k, $1 \leq k \leq M$, denote a basis in the space V_h, the problem consists in finding the solutions of the generalized matrix eigenvalue problem in $\mathbf{R}^M$:

$$\mathcal{A}_h \boldsymbol{u} = \lambda_h \mathcal{B}_h \boldsymbol{u},$$

where the coefficients of the symmetric and positive definite matrices $\mathcal{A}_h$ and $\mathcal{B}_h$ have respectively for expressions $a(w_h, w_l)$ and (w_k, w_l). In this fashion we obtain M strictly positive *approximate eigenvalues*

$$0 < \lambda_{1h} \leq \lambda_{2h} \leq \cdots \leq \lambda_{Mh}$$

and M *approximate eigenfunctions* u_{kh}, $1 \leq k \leq M$, which can be orthonormalized in the sense that

$$a(u_{kh}, u_{lh}) = \lambda_{kh} \delta_{kl}, \quad (u_{kh}, u_{lh}) = \delta_{kl}, \quad 1 \leq k, l \leq M.$$

One can then show that for any fixed integer l, one has $\lim_{h \to 0} \lambda_{kh} = \lambda_k$ (from above), $1 \leq k \leq l$, provided $\lim_{h \to 0} \inf_{v_h \in V_h} \|u_k - v_h\| = 0$, $1 \leq k \leq l$ (compare with Theorem 2.4.1). More precisely, one obtains inequalities of the form

$$\lambda_k \leq \lambda_{kh} \leq \lambda_k + C(l) \sum_{r=1}^{k} \inf_{v_h \in V_h} \|u_r - v_h\|^2, \quad 1 \leq k \leq l,$$

which show that the order of the error $|\lambda_{kh} - \lambda_k| = (\lambda_{kh} - \lambda_k)$ is the *square* of the order of the interpolation error (provided as usual the eigenfunctions are smooth enough). For the eigenfunctions one can show (again with appropriate smoothness assumptions) that the orders of the errors $\|u_{kh} - u_k\|$ are the same as those of the interpolation error (with additional difficulties in case of multiple eigenvalues, as expected). Such error estimates are found in BIRKHOFF, de BOOR, SWARTZ & WENDROFF (1966), CIARLET, SCHULTZ & VARGA (1968b), FIX (1969), PIERCE & VARGA (1972a, 1972b), CHATELIN & LEMORDANT (1975).

For treatments more directly connected with the finite element method, see BABUŠKA & AZIZ (1972, Chapter 10), FIX (1972a) (where in particular the effect of numerical integration is studied), FIX (1973), GRÉGOIRE, NÉDÉLEC & PLANCHARD (1976), STRANG & FIX (1973, Chapter 6).

Extensions to the nonsymmetric case have been obtained by BRAMBLE & OSBORN (1972, 1973). See also BRAMBLE (1972), OSBORN (1974). For an extension to a noncompact operator, see RAPPAZ (1976, 1977).

For the practical implementation of such methods, see for example BATHE & WILSON (1973), LINDBERG & OLSON (1970).

CHAPTER 5

APPLICATION OF THE FINITE ELEMENT METHOD TO SOME NONLINEAR PROBLEMS

Introduction

Consider the minimization problem: Find $u \in U \subset V$ such that $J(u) = \inf_{v \in U} J(v)$, with a functional J of the form $J(v) = F(v) - f(v)$, $f \in V'$. There are two ways in which this problem can become nonlinear (the nonlinearity is as usual that of the mapping $f \to u$): Either the functional is quadratic, i.e., $F(v) = \frac{1}{2}a(v, v)$ but the set U is not a vector space, or the functional is not quadratic, in which case the problem is nonlinear even if the set V is a vector space.

In the first case, we have shown (Section 1.1) that when U is a closed convex subset of the space V the minimization problem is equivalent to a set of *variational inequalities*. Several important physical problems correspond to this modeling, in particular the *obstacle problem*, which corresponds to the following data:

$$V = H_0^1(\Omega), \quad \Omega \subset \mathbf{R}^2,$$

$$U = \{v \in H_0^1(\Omega);\ v \geqslant \chi \text{ a.e. in } \Omega\},$$

$$a(u, v) = \int_\Omega \nabla u \cdot \nabla v \, \mathrm{d}x, \qquad f(v) = \int_\Omega fv \, \mathrm{d}x.$$

In Section 5.1, we consider the finite element approximation of this problem. Following an analysis of R.S. Falk, we show (Theorem 5.1.2) that the discrete solution u_h obtained with triangles of type (1) satisfies

$$\|u - u_h\|_{1,\Omega} = \mathrm{O}(h).$$

This result is itself a consequence of an abstract error estimate (Theorem 5.1.1) valid for a general class of variational inequalities.

In the second case, i.e., of non quadratic functionals, there are almost as many problems as there are non quadratic functionals and, consequently, no general theory is available. We have nevertheless considered two significant examples.

The first one, considered in Section 5.2, is the *minimal surface problem*, which consists in minimizing the functional

$$J(v) = \int_\Omega \sqrt{1 + \|\nabla v\|^2}\,\mathrm{d}x, \quad \Omega \subset \mathbf{R}^2,$$

over all functions v in the convex set

$$U = \{v \in H^1(\Omega);\ v = u_0 \quad \text{on} \quad \Gamma\}, \quad u_0 \in H^1(\Omega).$$

Following a recent paper of C. Johnson and V. Thomée, we show (Theorem 5.2.2) that

$$\|u - u_h\|_{1,\Omega} = \mathrm{O}(h),$$

where u_h is the discrete solution again obtained through the use of triangles of type (1).

The second problem, studied in Section 5.3, consists in minimizing the functional

$$J(v) = \frac{1}{p}\int_\Omega \|\nabla v\|^p\,\mathrm{d}x - f(v), \quad f \in V',$$

over the space

$$V = W_0^{1,p}(\Omega), \quad p \geq 2.$$

We show in particular that this problem has a unique solution (Theorem 5.3.1), which is also solution of the equation

$$J'(u) = Au - f = 0,$$

where the operator $A\colon V \to V'$ is an instance of so-called *strongly monotone operators*, i.e., which satisfy inequalities of the form

$$(Au - Av)(u - v) \geq \chi(\|u - v\|)\|u - v\|,$$

for some function $\chi\colon [0, +\infty[\to [0, +\infty[$ such that $\lim_{t\to\infty} \chi(t) = \infty$.

Following a recent work by R. Glowinski and A. Marrocco, we next consider a finite element approximation of this problem (for $n = 2$) using again triangles of type (1). We then prove the following convergence results (Theorems 5.3.2 and 5.3.5):

$$\lim_{h\to 0} \|u - u_h\|_{1,p,\Omega} = 0 \quad \text{if } u \in W_0^{1,p}(\Omega),$$

$$\|u - u_h\|_{1,p,\Omega} = \mathrm{O}(h^{1/(p-1)}) \quad \text{if } u \in W_0^{1,p}(\Omega) \cap W^{2,p}(\Omega).$$

The last error estimate is itself a corollary of an abstract error estimate valid for strongly monotone operators in general (Theorem 5.3.4).

Perhaps the most striking feature of nonlinear problems, by contrast with linear problems, is that their solutions *cannot be very smooth over the whole set* $\bar{\Omega}$ even if the data are very smooth. For example, the solution of the membrane problem is in general "only" in the space $H^2(\Omega)$, whatever the smoothness of the data χ, f and Γ. Consequently, *finite element "of low degree" (of the local polynomial spaces* P_K*) are sufficient for all practical purposes, a fact amply confirmed by numerical evidence.*

Finally, we mention that the three sections in this chapter can be read independently of each other.

5.1. The obstacle problem

Variational formulation of the obstacle problem

The *obstacle problem* consists in finding the equilibrium position of an elastic membrane, with tension τ, which

(i) passes through a curve Γ, i.e., the boundary of an open set Ω of the "horizontal" plane of coordinates (x_1, x_2),

(ii) is subjected to the action of a "vertical" force of density $F = \tau f$,

(iii) must lie over an "obstacle" which is represented by a function $\chi: \bar{\Omega} \to \mathbf{R}$, as illustrated in Fig. 5.1.1.

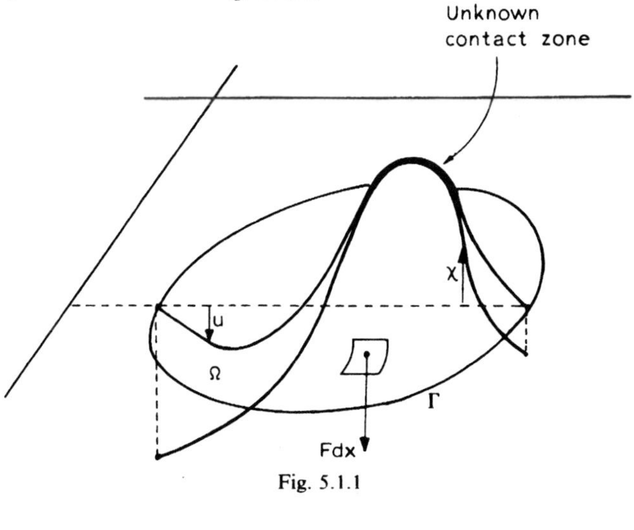

Fig. 5.1.1

Thus this is another *membrane problem* which, following the example given in Section 1.2, is associated with the following data:

$$\begin{cases} V = H_0^1(\Omega), \qquad n = 2, \\ U = \{v \in H_0^1(\Omega), \quad v \geqslant \chi \text{ a.e. in } \Omega\}, \\ a(u, v) = \int_\Omega \nabla u \cdot \nabla v \, dx, \\ f(v) = \int_\Omega fv \, dx. \end{cases} \tag{5.1.1}$$

Throughout this section, we shall make the following assumptions on the functions χ and f:

$$\chi \in H^2(\Omega), \quad \chi \leqslant 0 \quad \text{on} \quad \Gamma, \quad f \in L^2(\Omega). \tag{5.1.2}$$

The set U, which is not empty by virtue of the second assumption of (5.1.2), is easily seen to be convex. To show that it is closed, it suffices to notice that every convergent sequence in the space $L^2(\Omega)$ contains an a.e. pointwise convergent subsequence.

Thus we may apply Theorem 1.1.1: There exists a unique function $u \in U$ which minimizes the membrane energy

$$J: v \to J(v) = \frac{1}{2} \int_\Omega \|\nabla v\|^2 \, dx - \int_\Omega fv \, dx \tag{5.1.3}$$

over the set U, and it is also the unique solution of the variational inequalities

$$\forall v \in U, \quad \int_\Omega \nabla u \cdot \nabla(v - u) \, dx \geqslant \int_\Omega f(v - u) \, dx, \tag{5.1.4}$$

by Theorem 1.1.2.

This variational problem corresponds to the formal solution of a boundary value problem. See Exercise 5.1.1.

One should notice that *the region where the membrane touches the obstacle, i.e., the set* $\{x \in \Omega;\ u(x) = \chi(x)\}$, *is not known in advance.*

By contrast with the linear membrane problem of Section 1.2, *the solution of the obstacle problem is not smooth in general, even if the data are very smooth.* To be convinced of this phenomenon, consider the one-dimensional analog with $f = 0$, as shown in Fig. 5.1.2. In this case, the solution is affine in the region where it does not touch the obstacle and consequently, whatever the smoothness of the function χ, the

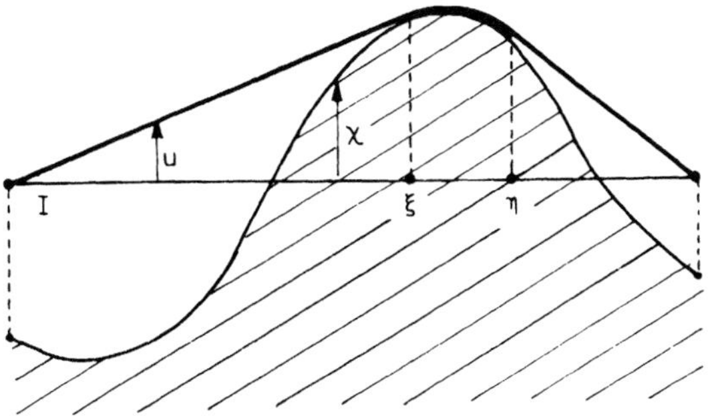

Fig. 5.1.2.

second derivative of u has discontinuities at points such as ξ and η. Therefore the solution u is "only" in the space $H^2(I)$.

These results carry over to the 2-dimensional case, but they are of course much less easy to prove. For example, it is known that if the function χ satisfies the assumptions of (5.1.2), $f = 0$, and $\bar{\Omega}$ is a convex polygon, the solution u belongs to the space $H_0^1(\Omega) \cap H^2(\Omega)$. If the set $\bar{\Omega}$ is convex with a boundary of class $\mathscr{C}^2$ and assumptions (5.1.2) hold then we have again $u \in H_0^1(\Omega) \cap H^2(\Omega)$. In both cases, the norm $\|u\|_{2,\Omega}$ can be estimated in terms of the norms $\|\chi\|_{2,\Omega}$ and $|f|_{0,\Omega}$ of the data. These results are proved in Brezis & Stampacchia (1968) and Lewy & Stampacchia (1969).

An abstract error estimate for variational inequalities

We next consider the approximation of such a problem. Following an analysis due to R.S. Falk, we shall first prove an abstract error estimate (Theorem 5.1.1) which is valid for a general class of approximation schemes for variational inequalities of the form (5.1.5) below, and then we shall apply this result to a particular finite element method, well adapted to the present problem (Theorem 5.1.2).

The *abstract setting* is the following: Let V be a Hilbert space, with norm $\|\cdot\|$, let $a(\cdot,\cdot)$: $V \times V \to \mathbf{R}$ be a continuous, symmetric and V-elliptic bilinear form (with the usual V-ellipticity and continuity constants α and M), let f: $V \to \mathbf{R}$ be a continuous linear form, and let U be a non empty closed convex subset of V. Then there is a unique element u which satisfies (cf. Theorem 1.1.2).

$$u \in U \quad \text{and} \quad \forall v \in U, \quad a(u, v-u) \geqslant f(v-u). \tag{5.1.5}$$

Let then V_h be a finite-dimensional subspace of the space V and let U_h be a non empty closed convex subset of V_h. Observe that, in general, *the set U_h is not a subset of U.*

Then, quite naturally, the *discrete problem* consists in finding an element u_h such that

$$u_h \in U_h \quad \text{and} \quad \forall v_h \in U_h, \quad a(u_h, v_h - u_h) \geq f(v_h - u_h), \tag{5.1.6}$$

and, again, this abstract variational problem has a unique solution u_h.

In the proof of the next theorem, we shall need the mapping $A \in \mathcal{L}(V; V')$ defined by the relations

$$\forall v, w \in V, \quad Av(w) = a(v, w), \tag{5.1.7}$$

and which we already used in the proof of Theorem 1.1.3. Notice that in the present situation we do *not* have $Au = f$ in general, as in the case of the linear problem ($U = V$). Also, we shall consider a Hilbert space H, with norm $|\cdot|$ and inner product $(\cdot\,.\,\cdot)$, such that

$$\bar{V} = H \quad \text{and} \quad V \hookrightarrow H. \tag{5.1.8}$$

The space H will be identified with its dual, so that it may be in turn identified with a subspace of the dual space of V, as we showed in Section 3.2.

We are now in a position to prove an *abstract error estimate* in the norm $\|\cdot\|$.

Theorem 5.1.1. *Assume that*

$$(Au - f) \in H. \tag{5.1.9}$$

Then there exists a constant C independent of the subspace V_h and of the set U_h such that

$$\|u - u_h\| \leq C\Big(\inf_{v_h \in U_h} \{\|u - v_h\|^2 + |Au - f|\,|u - v_h|\} + \\ + |Au - f| \inf_{v \in U} |u_h - v|\Big)^{1/2}. \tag{5.1.10}$$

Proof. We have

$$\begin{aligned} \alpha\|u - u_h\|^2 &\leq a(u - u_h, u - u_h) \\ &= a(u, u) + a(u_h, u_h) - a(u, u_h) - a(u_h, u), \end{aligned}$$

and, using (5.1.5) and (5.1.6),

$$\forall v \in U, \quad a(u, u) \leq a(u, v) + f(u - v),$$
$$\forall v_h \in U_h, \quad a(u_h, u_h) \leq a(u_h, v_h) + f(u_h - v_h).$$

Therefore we deduce that, for all $v \in U$ and all $v_h \in U_h$,

$$\begin{aligned}\alpha\|u - u_h\|^2 &\leq a(u, v - u_h) + a(u_h, v_h - u) + f(u - v) + f(u_h - v_h)\\ &= a(u, v - u_h) - f(v - u_h) + a(u, v_h - u) - f(v_h - u)\\ &\qquad + a(u_h - u, v_h - u)\\ &= (f - Au, u - v_h) + (f - Au, u_h - v)\\ &\qquad + a(u - u_h, u - v_h).\end{aligned}$$

We thus have, for all $v \in U$ and all $v_h \in U_h$,

$$\alpha\|u - u_h\|^2 \leq |f - Au|(|u - v_h| + |u_h - v|) + M\|u - u_h\|\,\|u - v_h\|.$$

Since

$$\|u - u_h\|\,\|u - v_h\| \leq \frac{1}{2}\left(\frac{\alpha}{M}\|u - u_h\|^2 + \frac{M}{\alpha}\|u - v_h\|^2\right),$$

we obtain, upon combining the two previous inequalities,

$$\frac{\alpha}{2}\|u - u_h\|^2 \leq |f - Au|(|u - v_h| + |u_h - v|) + \frac{M^2}{2\alpha}\|u - v_h\|^2, \quad (5.1.11)$$

from which inequality (5.1.10) follows. □

Remark 5.1.1. Several comments are in order about this theorem:

(i) The proof has been given in such a way that it includes the case where the bilinear form is not symmetric.

(ii) If $U = V$ then $Au - f = 0$, so that, with the obvious choice $U_h = V_h$, the error estimate of (5.1.10) reduces to the familiar error estimate of Céa's lemma.

(iii) If the inclusion $U_h \subset U$ holds, then of course the term $\inf_{v \in U} |u_h - v|$ (which can be expected to be the harder to evaluate) vanishes in the error estimate. For such an example, see Exercise 5.1.3. This is not the case, however, of the finite element approximation of the obstacle problem which we shall describe below.

(iv) Also, had we not introduced the space H in our argument, we would have found, instead of inequality (5.1.11), the inequality

$$\frac{\alpha}{2}\|u-u_h\|^2 \leq \|f-Au\|^\star(\|u-v_h\|+\|u_h-v\|)+\frac{M^2}{2\alpha}\|u-v_h\|^2,$$

where $\|\cdot\|^\star$ denotes as usual the norm of the dual space of V. However, this last inequality is likely to yield a poorer order of convergence, since the term $\inf_{v_h\in U_h}\{\|u-v_h\|^2+|Au-f|\,|u-v_h|\}$ can be anticipated to be of a higher order than the term $\inf_{v_h\in U_h}\{\|u-v_h\|^2+\|Au-f\|^\star\|u-v_h\|\}$. This will be confirmed in the proof of Theorem 5.1.2. □

Finite element approximation with triangles of type (1). Estimate of the error $\|u-u_h\|_{1,\Omega}$

Let us return to the obstacle problem. For simplicity, we shall assume that the set $\bar{\Omega}$ is a polygon, leaving the case of a curved boundary as a problem (Exercise 5.1.2). With a triangulation $\mathcal{T}_h$ of the set $\bar{\Omega}=\bigcup_{K\in\mathcal{T}_h}K$, we associate the finite element space X_h whose generic finite element is the triangle of type (1) and we let as usual

$$X_{0h}=\{v_h\in X_h;\ v_h=0 \quad \text{on} \quad \Gamma\}. \tag{5.1.12}$$

Letting $\mathcal{N}_h$ denote the set of the nodes of the space X_h, i.e., the set of all the vertices, we let

$$V_h=X_{0h}, \tag{5.1.13}$$

$$U_h=\{v_h\in V_h;\ \forall b\in\mathcal{N}_h,\ v_h(b)\geq\chi(b)\}\tag{5.1.14}$$

(as an element of the space $H^2(\Omega)$, the function χ is in the space $\mathscr{C}^0(\bar{\Omega})$ and, therefore, its point values are well defined).

Notice that *the set U_h is not in general contained in the set U*, as the one-dimensional case considered in Fig. 5.1.3 exemplifies.

Let us now apply the abstract error estimate of Theorem 5.1.1.

Theorem 5.1.2. *Assume that the solution u is in the space $H^2(\Omega)$. Then, given a regular family of triangulations, there exists a constant $C(u,f,\chi)$ independent of h such that*

$$\|u-u_h\|_{1,\Omega}\leq C(u,f,\chi)h. \tag{5.1.15}$$

Proof. We shall let

$$H=L^2(\Omega),$$

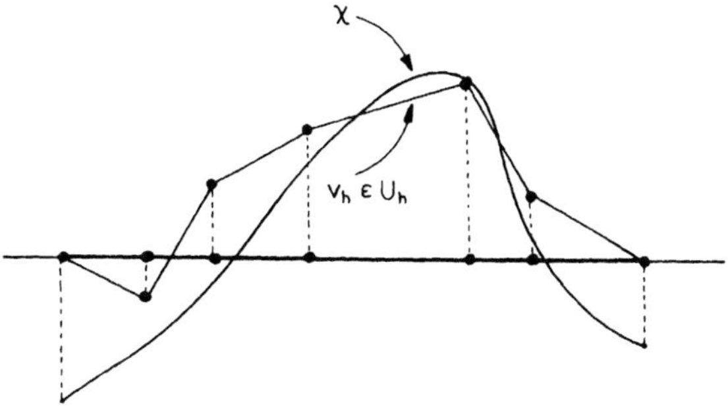

Fig. 5.1.3

so that we need to verify that $Au \in L^2(\Omega)$($f \in L^2(\Omega)$ by assumption). Since the solution u is assumed to be in the space $H^2(\Omega)$, we have

$$\forall v \in V, \quad Au(v) = \int_\Omega \nabla u \cdot \nabla v \, \mathrm{d}x = -\int_\Omega \Delta u \, v \, \mathrm{d}x$$

and thus

$$\forall v \in V, \; |Au(v)| \leq |\Delta u|_{0,\Omega} |v|_{0,\Omega},$$

so that Au is indeed an element of the space H.

Let $\Pi_h u$ denote as usual the X_h-interpolant of the function u, which is in the space X_{0h}. Since

$$\forall b \in \mathcal{N}_h, \; \Pi_h u(b) = u(b) \geq \chi(b),$$

it is also an element of the set U_h. Thus,

$$\begin{aligned}\inf_{v_h \in U_h} \{\|u - v_h\|_{1,\Omega}^2 + |Au - f|_{0,\Omega} |u - v_h|_{0,\Omega}\} &\leq \\ \leq \|u - \Pi_h u\|_{1,\Omega}^2 + (|\Delta u|_{0,\Omega} + |f|_{0,\Omega}) |u - \Pi_h u|_{0,\Omega} \\ \leq C(|u|_{2,\Omega}^2 + (|\Delta u|_{0,\Omega} + |f|_{0,\Omega}) |u|_{2,\Omega}) |u|_{2,\Omega}) h^2. \qquad (5.1.16)\end{aligned}$$

In order to evaluate the term $\inf_{v \in U} |u_h - v|_{0,\Omega}$, it is convenient to introduce the function (Fig. 5.1.4)

$$u_h^* = \max\{u_h, \chi\},$$

so that the inequality $u_h^* \geq \chi$ holds in Ω. Both functions u_h and χ being in the space $H^1(\Omega)$, it follows that their maximum u_h^* is also in $H^1(\Omega)$ (this is a non-trivial fact, whose proof may be found in p. 169 of LEWY & STAMPACCHIA (1969)). Finally, the condition $\chi \leq 0$ on Γ implies

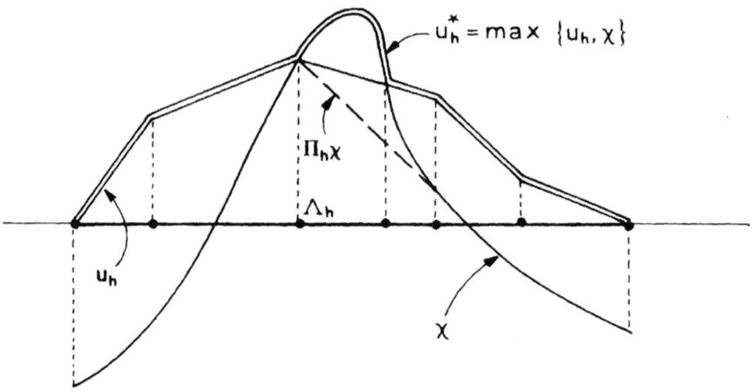

Fig. 5.1.4

that $u_h^* \in H_0^1(\Omega)$. Thus the function u_h^* is an element of the set U. Let

$$\Lambda_h = \{x \in \Omega;\ u_h < \chi\},$$

so that

$$|u_h - u_h^*|_{0,\Omega}^2 = \int_{\Lambda_h} |u_h - \chi|^2 \,\mathrm{d}x,$$

since $u_h - u_h^* = 0$ on $\Omega - \Lambda_h$. Let us introduce the X_h-interpolant $\Pi_h\chi$ of the function χ. Since

$$\forall b \in \mathcal{N}_h, \quad u_h(b) \geq \chi(b) = \Pi_h\chi(b),$$

it follows that

$$u_h - \Pi_h\chi \geq 0 \quad \text{in} \quad \Omega.$$

Consequently,

$$\forall x \in \Lambda_h, \quad 0 < |(\chi - u_h)(x)| = (\chi - u_h)(x) \leq (\chi - \Pi_h\chi)(x) = |(\chi - \Pi_h\chi)(x)|,$$

and thus,

$$|u_h - u_h^*|_{0,\Omega}^2 = \int_{\Lambda_h} |u_h - \chi|^2 \,\mathrm{d}x \leq \int_{\Lambda_h} |\chi - \Pi_h\chi|^2 \,\mathrm{d}x \leq |\chi - \Pi_h\chi|_{0,\Omega}^2.$$

Therefore, we obtain

$$\inf_{v\in U} |u_h - v|_{0,\Omega} \leq |u_h - u_h^*|_{0,\Omega} \leq |\chi - \Pi_h\chi|_{0,\Omega} \leq C|\chi|_{2,\Omega}h^2, \quad (5.1.17)$$

and the conclusion follows from inequalities (5.1.16) and (5.1.17). □

Exercises

5.1.1. Show that the solution of the variational problem associated with the data (5.1.1) corresponds to the formal solution of the following boundary value problem:

$$\begin{cases} -\Delta u \geqslant f \quad \text{in} \quad \Omega, \\ u \geqslant \chi \quad \text{in} \quad \Omega, \\ (-\Delta u - f)(u-\chi) = 0 \quad \text{in} \quad \Omega, \\ u = 0 \quad \text{on} \quad \Gamma, \\ u = \chi \quad \text{on} \quad \Gamma^*, \\ \partial_\nu u = \partial_\nu \chi \quad \text{on} \quad \Gamma^*, \end{cases}$$

where Γ^* is the "interface" between the sets $\{x \in \Omega;\ u(x) = \chi(x)\}$ and $\{x \in \Omega;\ u(x) > \chi(x)\}$ and ∂_ν is the normal derivative operator along Γ^*. Notice that the set Γ^* is an unknown of the problem: This is why such a problem is also called a *free surface problem.*

5.1.2. Show that the error estimate (5.1.15) of Theorem 5.1.2 holds unchanged in the following situation: The set Ω is convex with a sufficiently smooth boundary, so that $u \in H^2(\Omega)$. Then we let $\bar{\Omega}_h = \bigcup_{K \in \mathcal{T}_h} K$ denote a triangulation made up of triangles, in such a way that all the vertices of $\mathcal{T}_h$ which are on the boundary of the set Ω_h are also on Γ (Fig. 5.1.5).

With such a triangulation, we associate the finite element space X_h whose generic element is the triangle of type (1) and we let X_{0h} denote as usual the subspace of X_h whose functions vanish on the boundary of the set Ω_h. The space V_h then consists of the functions in the space X_{0h} prolongated by zero on the set $\bar{\Omega} - \bar{\Omega}_h$ (thus, the functions in the space V_h are defined over the set $\bar{\Omega}$).

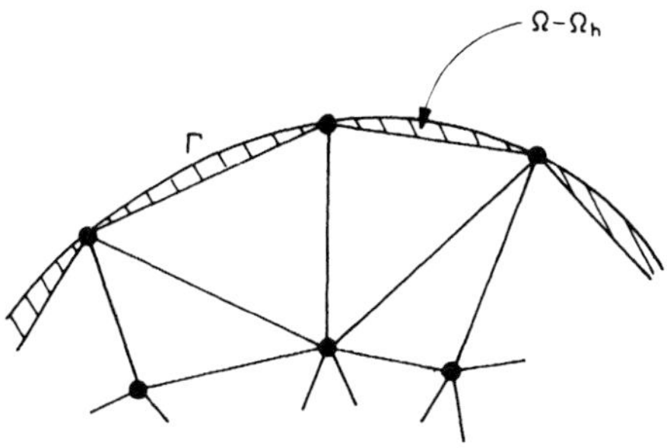

Fig. 5.1.5

[Hint: To prove the analog of inequality (5.1.16), show that, if $u \in H^2(\Omega) \cap H_0^1(\Omega)$, then

$$\|u\|_{m,\Omega-\Omega_h} \leq Ch^{2-m}\|u\|_{2,\Omega}, \quad m = 0, 1.$$

To prove the analog of inequality (5.1.17), assume for simplicity that $\chi = 0$ on Γ, and show that

$$|\max\{0, \chi\}|_{0,\Omega-\Omega_h} \leq |\chi|_{0,\Omega-\Omega_h} \leq Ch^2\|\chi\|_{2,\Omega}.]$$

5.1.3. Another problem which is modeled by variational inequalities is the *elastic-plastic torsion problem*, which arises in the following situation: Consider a cylindrical thin rod with a simply connected cross section $\bar{\Omega} \subset \mathbf{R}^2$, subjected to a torsion around the axis supporting the vector e_3. The torsion angle τ per unit length is assumed to be constant throughout the length of the rod (cf. Fig. 5.1.6, where the vertical scale should be considerably increased).

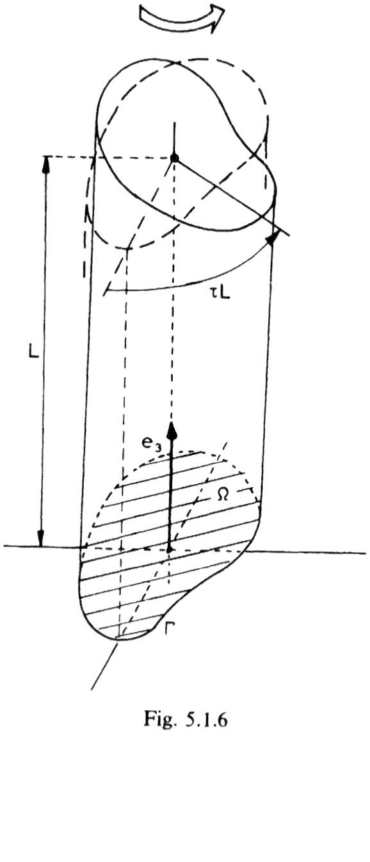

Fig. 5.1.6

Let us first assume that we are in the domain of validity of *linear* elasticity. Then certain simplifying assumptions (the weight of the rod is neglected among other things) imply that the components σ_{11}, σ_{12}, σ_{22} and σ_{33} of the stress tensor vanish everywhere in the rod, while the components σ_{13} and σ_{23} are functions of x_1, x_2 only, and are such that

$$\sigma_{13} = 2\mu\tau\partial_2 u, \quad \sigma_{23} = -2\mu\tau\partial_1 u,$$

where μ is the second Lamé coefficient of the constitutive material of the rod, and the *stress function* u satisfies

$$-\Delta u = 1 \quad \text{in} \quad \Omega \quad \text{and} \quad u = 0 \quad \text{on} \quad \Gamma.$$

Therefore the function u minimizes the functional

$$J: v \to J(v) = \frac{1}{2}\int_\Omega \|\nabla v\|^2 \, dx - \int_\Omega fv \, dx, \quad \text{with} \quad f = 1,$$

over the space $H_0^1(\Omega)$.

If we take into account the *plasticity* of the material, then the stresses cannot take arbitrary large values. A particular mathematical representation of this effect, known as the *von Mises criterion*, reduces in this case to the condition that the quantity $(|\sigma_{13}|^2 + |\sigma_{23}|^2)^{1/2}$, and consequently the norm $\|\nabla u\|$, cannot exceed a certain constant. Notice, however, that contrary to the linear case, it is not straightforward to recuperate the displacement field from the knowledge of the stress field, as shown by the discussion in DUVAUT & LIONS (1972, Chapter 5, Section 6).

Therefore this problem corresponds to the following data (where, for definiteness, the upper bound on $\|\nabla u\|$ has been set equal to one):

$$\begin{cases} V = H_0^1(\Omega), \qquad n = 2, \\ U = \{v \in H_0^1(\Omega);\ \|\nabla v\| \leq 1 \text{ a.e. in } \Omega\}, \\ a(u, v) = \int_\Omega \nabla u \cdot \nabla v \, dx, \\ f(v) = \int_\Omega fv \, dx, \quad f \in L^2(\Omega), \end{cases}$$

with $f = 1$ in this case.

(i) Show that U is a non empty closed convex subset of the space V and, consequently, that the variational problem associated with the above data has a unique solution u (which can be shown to be in the space $W^{2,p}(\Omega) \cap H_0^1(\Omega)$ for all $p < \infty$ if $f \in L^\infty(\Omega)$ and the boundary Γ is smooth enough; cf. BREZIS & STAMPACCHIA (1968)).

Show that this problem amounts to formally solving the following boundary value problem:

$$\begin{cases} -\Delta u \leq f \quad \text{in} \quad \Omega, \\ \|\nabla u\| \leq 1 \quad \text{in} \quad \Omega, \\ (-\Delta u - f)(1 - \|\nabla u\|) = 0 \quad \text{in} \quad \Omega, \\ u = 0 \quad \text{on} \quad \Gamma. \end{cases}$$

(ii) Consider the one-dimensional analogue of this problem and its finite element approximation, with

$$X_{0h} = \{v_h \in \mathscr{C}^0(\bar{I});\ v_{h|\bar{I}_i} \in P_1(\bar{I}_i),\ 1 \leq i \leq M,\ v_h(0) = v_h(1) = 0\},$$

where $\bar{I} = \bigcup_{i=1}^{M} \bar{I}_i$ is a partition of the interval $\bar{I} = [0, 1]$ with

$$\bar{I}_i = [x_{i-1}, x_i],\ 1 \leq i \leq M,\ h = \max_{1 \leq i \leq M} |x_i - x_{i-1}|,$$

and

$$U_h = \{v_h \in X_{0h};\ |v_h'| \leq 1 \text{ a.e.} \quad \text{in} \quad I\}.$$

Derive the error estimate

$$\|u - u_h\|_{1,I} = O(h).$$

(iii) Returning to a two-dimensional polygonal set $\bar{\Omega}$, let

$$U_h = \{v_h \in X_{0h};\ \|\nabla v_h\| \leq \text{ a.e.} \quad \text{in} \quad \Omega\},$$

the space X_{0h} being defined as in (5.1.12).

Show that the X_h-interpolant of a function $v \in U$ is not necessarily contained in the set U_h.

Assume that the solution u belongs to the space $W^{2,p}(\Omega)$ for some $p \in]2, \infty]$. Then show that there exist appropriate quantities $\epsilon(h) > 0$ with $\lim_{h \to 0} \epsilon(h) = 0$ such that the functions $(1 + \epsilon(h))^{-1} \Pi_h u$ belong to the set U_h. Using this result, show that

$$\|u - u_h\|_{1,\Omega} = O(h^{1/2 - 1/p}).$$

5.2. The minimal surface problem

A formulation of the minimal surface problem

Let Ω be a bounded open subset of the plane $\mathbf{R}^2$, and let u_0 be a function given on the boundary Γ of the set Ω.

The *minimal surface problem* consists in finding a function u which minimizes the functional

$$J: v \to J(v) = \int_\Omega \sqrt{1 + \|\nabla v\|^2}\, \mathrm{d}x \tag{5.2.1}$$

over an appropriate space of functions which equal u_0 on Γ. In other words, among all surfaces given by an equation $x_3 = v(x_1, x_2)$, $x = (x_1, x_2) \in \Omega$ (for which the area can be defined) and which pass through a given curve of the form $x_3 = u_0(x_1, x_2)$, $(x_1, x_2) \in \Gamma$, one looks for a surface whose area is minimal.

The mathematical analysis of this problem is not easy. In particular, it is not straightforward to decide which function space is more appropriate to insure existence and uniqueness of a solution. However, we shall not go here into such matters, refering instead the reader to the section "Bibliography and Comments" for additional information. See also Exercise 5.2.1.

In this section, we shall make the following hypotheses: The set Ω is *convex* and has a Lipschitz-continuous boundary, and the function u_0 is the trace over Γ of a function (still denoted u_0) of the space $H^2(\Omega)$. Then, for our subsequent analysis, it will be convenient to consider that the minimal surface problem consists in finding a function u such that

$$u \in U \quad \text{and} \quad J(u) = \inf_{v \in U} J(v), \tag{5.2.2}$$

where

$$U = \{v \in H^1(\Omega);\ (v - u_0) \in H_0^1(\Omega)\}. \tag{5.2.3}$$

Remark 5.2.1. The functional J of (5.2.1) is defined over any Sobolev space $W^{1,p}(\Omega)$, $1 \leq p \leq \infty$. One reason for the present choice $p = 2$ is that it is easily seen that the functional J is differentiable over the space $H^1(\Omega)$, as we next show. □

For notational simplicity in the subsequent computations, it will be convenient to introduce the function

$$f: x=(x_1, x_2) \in \mathbf{R}^2 \to f(x) = \sqrt{1+\|x\|^2} = \sqrt{1+x_1^2+x_2^2}. \tag{5.2.4}$$

Then, for all points $x \in \mathbf{R}^2$ and all vectors $\xi = (\xi_1, \xi_2) \in \mathbf{R}^2$,

$$\sum_{i,j=1}^{2} \partial_{ij} f(x) \xi_i \xi_j = \frac{\|\xi\|^2 + (x_2\xi_1 - x_1\xi_2)^2}{(1+\|x\|^2)^{3/2}} \leq \|\xi\|^2, \tag{5.2.5}$$

so that, for all $v, w \in H^1(\Omega)$,

$$J(v+w) - J(v) = \int_\Omega \frac{\nabla v \cdot \nabla w}{f(\nabla v)} \, dx + \mathcal{R}(v, w),$$

where

$$\mathcal{R}(v, w) \leq \frac{1}{2} \int_\Omega \|\nabla w\|^2 \, dx \leq \frac{1}{2} \|w\|_{1,\Omega}^2.$$

Therefore, the functional J is differentiable over the space $H^1(\Omega)$, and its derivative is given by

$$\forall v, w \in H^1(\Omega), \; J'(v)w = \int_\Omega \frac{\nabla v \cdot \nabla w}{f(\nabla v)} \, dx. \tag{5.2.6}$$

We also record the following result which has already been proved (cf. (1.1.9)):

Let V be a normed vector space, let U be a convex subset of V, let $J: V \to \mathbf{R}$ be a functional and, finally, let u be a point of the set U such that $J(u) = \inf_{v\in U} J(v)$ and such that the functional J is differentiable at the point u. Then the inequalities

$$\forall v \in U, \; J'(u)(v-u) \geq 0 \tag{5.2.7}$$

hold.

Finite element approximation with triangles of type (1). *Estimate of the error* $\|u - u_h\|_{1,\Omega_h}$

We next define the *discrete problem*: Let $\mathcal{T}_h$ be a triangulation made up of triangles K, $K \in \mathcal{T}_h$, in such a way that all the vertices situated on the boundary Γ_h of the set

$$\bar{\Omega}_h = \bigcup_{K \in \mathcal{T}_h} K \tag{5.2.8}$$

also belong to the boundary Γ (cf. Fig. 5.2.1). Notice that the inclusion $\bar{\Omega}_h \subset \bar{\Omega}$ holds, by virtue of the assumption of convexity for the set Ω.

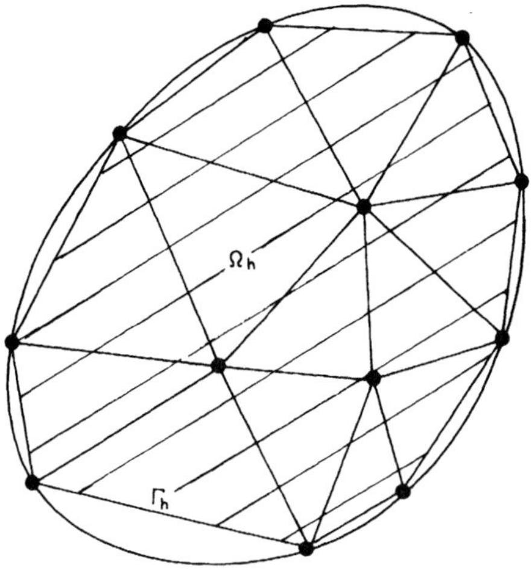

Fig. 5.2.1

With such a triangulation, we associate the finite element space X_h whose generic finite element is the triangle of type (1). The functions in the space X_h are therefore defined over the set $\bar{\Omega}_h$.

Letting $\mathcal{N}_h$ denote the set of nodes of the space X_h (which coincides with the set of vertices in this particular instance), we let

$$U_h = \{v_h \in X_h;\ \forall b \in \mathcal{N}_h \cap \Gamma, \quad v_h(b) = u_0(b)\} \tag{5.2.9}$$

(recall that, by assumption, $u_0 \in H^2(\Omega)$ and that $H^2(\Omega) \hookrightarrow \mathscr{C}^0(\bar{\Omega})$).

Then the discrete problem consists in finding a function u_h *such that*

$$u_h \in U_h \quad \text{and} \quad J_h(u_h) = \inf_{v_h \in U_h} J_h(v_h), \tag{5.2.10}$$

where

$$J_h(v_h) = \int_{\Omega_h} \sqrt{1 + \|\nabla v_h\|^2}\, \mathrm{d}x. \tag{5.2.11}$$

As usual, our first task is to examine the questions of existence and uniqueness for the discrete problem.

Theorem 5.2.1. *The discrete problem* (5.2.10) *has one and only one solution.*

Proof. If we define the norm

$$v_h \in X_h \to \|v_h\|_h = \max_{x \in \bar{\Omega}_h} |v_h(x)|,$$

it easily follows that

$$v_h \in U_h \quad \text{and} \quad \|v_h\|_h \to \infty \Rightarrow \lim J_h(v_h) = \infty. \tag{5.2.12}$$

To see this, it suffices to observe that

$$v_h \in U_h \quad \text{and} \quad \|v_h\|_h \to \infty \Rightarrow \max_{K \in \mathcal{T}_h} \|\nabla v_h|_K\| \to \infty$$

(argue by contradiction) and then to observe that

$$J_h(v_h) \geqslant \left(\sqrt{1 + \max_{K \in \mathcal{T}_h} \|\nabla v_h|_K\|^2}\right) \min_{K \in \mathcal{T}_h} \operatorname{meas}(K),$$

since the gradient of each function $v_h \in X_h$ is constant over each triangle K.

Let then $\tilde{v}_h$ denote a fixed function in the set U_h. Condition (5.2.12) implies that there exists a number r such that

$$v_h \in U_h \quad \text{and} \quad \|v_h\|_h > r \Rightarrow J_h(\tilde{v}_h) < J_h(v_h).$$

Therefore the solutions of the minimization problem (5.2.10) coincide with the solutions of an analogous minimization problem, with the set U_h replaced by the set

$$\tilde{U}_h = U_h \cap \{v_h \in X_h ; \|v_h\|_h \leqslant r\}.$$

Since the set $\tilde{U}_h$ is now compact, we have shown that problem (5.2.10) has at least one solution.

Let us next turn to the question of uniqueness. It follows from the equality of (5.2.5) that the function f defined in (5.2.4) is strictly convex. This will in turn imply that the function J_h: $U \to \mathbf{R}$ is also strictly convex: To prove this, let v_h and w_h be two distinct elements of the set U_h, let

$$\tilde{\Omega}_h = \left\{x \in \bigcup_{K \in \mathcal{T}_h} \mathring{K} ; \nabla v_h(x) \neq \nabla w_h(x)\right\},$$

and let θ be a given number in the interval $]0, 1[$. Then

$$J_h(\theta v_h + (1-\theta) w_h) = \int_{\Omega_h} f(\theta \nabla v_h(x) + (1-\theta) \nabla w_h(x))\, dx,$$

and the assertion follows by using the relations

$$\forall x \in \Omega_h - \bar{\tilde{\Omega}}_h, \quad f(\theta \nabla v_h(x) + (1-\theta)\nabla w_h(x)) \\ = \theta f(\nabla v_h(x)) + (1-\theta) f(\nabla w_h(x)),$$

$$\forall x \in \tilde{\Omega}_h, \quad f(\theta \nabla v_h(x) + (1-\theta)\nabla w_h(x)) \\ < \theta f(\nabla v_h(x)) + (1-\theta) f(\nabla w_h(x)),$$

$$\text{meas}(\tilde{\Omega}_h) > 0.$$

Since the set U_h is convex, the minimization problem (5.2.10) has a unique solution. □

Remark 5.2.2. The same argument shows that the minimization problem (5.2.2) has at most one solution. □

We next obtain an error estimate in the norm $\|\cdot\|_{1,\Omega_h}$.

Theorem 5.2.2. *Assume that the solution u of the minimization problem (5.2.2) exists and is in the space $H^2(\Omega) \cap W^{1,\infty}(\Omega)$. Then, given a regular family of triangulations, there exists a constant $C(u)$ independent of h such that*

$$\|u - u_h\|_{1,\Omega_h} \leqslant C(u)h. \tag{5.2.13}$$

Proof. In what follows, the notation $C(u)$ stands for various constants solely dependent upon the solution u. For clarity, the proof has been subdivided in five steps. The first four steps consist in establishing that $|u - u_h|_{1,\Omega_h} = O(h)$.

(i) Let us first record some *relations which are consequences of the minimizing properties of the functions u_h and u.*

Using (5.2.7), we know that

$$\forall v_h \in U_h, \quad J_h'(u_h)(v_h - u_h) \geqslant 0.$$

But in view of the particular form of the set U_h (cf. (5.2.9)), these inequalities are equivalent to the equations

$$\forall w_h \in X_{0h}, \quad J_h'(u_h) w_h = 0,$$

where, as usual,

$$X_{0h} = \{v_h \in X_h;\ v_h = 0 \quad \text{on} \quad \Gamma_h\}. \tag{5.2.14}$$

By a computation similar to that which led to (5.2.6), we deduce that

$$\forall w_h \in X_{0h}, \quad J_h'(u_h)w_h = \int_{\Omega_h} \frac{\nabla u_h \cdot \nabla w_h}{\sqrt{1+\|\nabla u_h\|^2}}\,\mathrm{d}x. \tag{5.2.15}$$

Using again (5.2.7) and the particular form of the set U (cf. (5.2.3)), we see that

$$\forall w \in H_0^1(\Omega), \quad J'(u)w = 0,$$

and therefore, by (5.2.6),

$$\forall w_h \in X_{0h}, \quad \int_{\Omega_h} \frac{\nabla u \cdot \nabla w_h}{\sqrt{1+\|\nabla u\|^2}}\,\mathrm{d}x = 0. \tag{5.2.16}$$

Clearly, this application of (5.2.6) supposes that each function $w_h \in X_{0h}$ be identified with its extension to the space $H_0^1(\Omega)$ obtained by prolongating it by zero on the set $\Omega - \Omega_h$.

(ii) Let us next show that, with the assumption that the solution u belongs to the space $H^2(\Omega) \cap W^{1,\infty}(\Omega)$, *there exists a constant* $C(u)$ *such that the quantity*

$$\Delta_h = \left(\int_{\Omega_h} \frac{\|\nabla(u-u_h)\|^2}{\sqrt{1+\|\nabla u_h\|^2}}\,\mathrm{d}x\right)^{1/2} \tag{5.2.17}$$

satisfies an inequality of the form

$$\Delta_h \leqslant C(u)h. \tag{5.2.18}$$

Let v_h be an arbitrary function in the set U_h, so that the function $w_h = v_h - u_h$ belongs to the space X_{0h}. Then, using relations (5.2.15) and (5.2.16) established in step (i), we can write ($f(x) = \sqrt{1+\|x\|^2}$; cf. (5.2.4)):

$$\Delta_h^2 = \int_{\Omega_h} \frac{\nabla(u-u_h)\cdot\nabla(u-v_h)}{f(\nabla u_h)}\,\mathrm{d}x + \\ + \int_{\Omega_h} \left(\frac{1}{f(\nabla u_h)} - \frac{1}{f(\nabla u)}\right)\nabla u \cdot \nabla w_h\,\mathrm{d}x. \tag{5.2.19}$$

The first integral can be bounded as follows:

$$\left|\int_{\Omega_h} \frac{\nabla(u-u_h)\cdot\nabla(u-v_h)}{f(\nabla u_h)}\,\mathrm{d}x\right| \leqslant \int_{\Omega_h} \left(\frac{\|\nabla(u-u_h)\|}{\sqrt{f(\nabla u_h)}}\right)\|\nabla(u-v_h)\|\,\mathrm{d}x \\ \leqslant \Delta_h |u - v_h|_{1,\Omega_h}. \tag{5.2.20}$$

In order to get an estimate for the second integral, we observe that

$$\frac{1}{f(\nabla u_h)} - \frac{1}{f(\nabla u)} = \frac{\nabla(u-u_h)\cdot\nabla(u+u_h)}{f(\nabla u_h)f(\nabla u)(f(\nabla u_h)+f(\nabla u))},$$

and thus,

$$\left|\frac{1}{f(\nabla u_h)} - \frac{1}{f(\nabla u)}\right| \leqslant \frac{\|\nabla(u-u_h)\|}{f(\nabla u_h)f(\nabla u)}.$$

Therefore,

$$\begin{aligned}
&\left|\int_{\Omega_h}\left(\frac{1}{f(\nabla u_h)} - \frac{1}{f(\nabla u)}\right)\nabla u\cdot\nabla w_h\,\mathrm{d}x\right| \leqslant \\
&\quad\leqslant \int_{\Omega_h}\frac{\|\nabla u\|}{f(\nabla u)}\frac{\|\nabla(u-u_h)\|}{\sqrt{f(\nabla u_h)}}\frac{\|\nabla w_h\|}{\sqrt{f(\nabla u_h)}}\,\mathrm{d}x \\
&\quad\leqslant \gamma(u)\Delta_h\left(\int_{\Omega_h}\frac{\|\nabla w_h\|^2}{f(\nabla u_h)}\,\mathrm{d}x\right)^{1/2} \\
&\quad\leqslant \gamma(u)\Delta_h\left(\Delta_h + \left(\int_{\Omega_h}\frac{\|\nabla(u-v_h)\|^2}{f(\nabla u_h)}\,\mathrm{d}x\right)^{1/2}\right) \\
&\quad\leqslant \gamma(u)\Delta_h(\Delta_h + |u-v_h|_{1,\Omega_h}),
\end{aligned} \tag{5.2.21}$$

where ($u\in W^{1,\infty}(\Omega)$ by assumption)

$$\gamma(u) = \left|\frac{\|\nabla u\|}{\sqrt{1+\|\nabla u\|^2}}\right|_{0,\infty,\Omega}. \tag{5.2.22}$$

Combining relations (5.2.19), (5.2.20) and (5.2.21), we obtain

$$\forall v_h\in U_h,\quad \Delta_h \leqslant \gamma(u)\Delta_h + (1+\gamma(u))|u-v_h|_{1,\Omega_h}.$$

Since the constant $\gamma(u)$ of (5.2.22) is strictly less than one, it follows that

$$\Delta_h = \left(\int_{\Omega_h}\frac{\|\nabla(u-u_h)\|^2}{\sqrt{1+\|\nabla u_h\|^2}}\,\mathrm{d}x\right)^{1/2} \leqslant C(u)\inf_{v_h\in U_h}|u-v_h|_{1,\Omega_h}, \tag{5.2.23}$$

with $C(u) = (1+\gamma(u))/(1-\gamma(u))$.

Since the function u belongs to the space $H^2(\Omega)$ by assumption, its X_h-interpolant is well defined, and it belongs to the set U_h. Thus

$$\inf_{v_h\in U_h}|u-v_h|_{1,\Omega_h} \leqslant |u|_{2,\Omega}h, \tag{5.2.24}$$

and inequality (5.2.18) is a consequence of inequalities (5.2.23) and (5.2.24).

(iii) Let us show that

$$|u_h|_{1,\infty,\Omega_h} \leqslant C(u). \tag{5.2.25}$$

Let K be an arbitrary triangle in the triangulation. Using step (ii) (cf. inequality (5.2.18)), one obtains

$$\begin{aligned} \left(\int_K \frac{\|\nabla u_h\|^2}{\sqrt{1+\|\nabla u_h\|^2}}\,\mathrm{d}x\right)^{1/2} &\leqslant \Delta_h + \left(\int_K \frac{\|\nabla u\|^2}{\sqrt{1+\|\nabla u_h\|^2}}\,\mathrm{d}x\right)^{1/2} \qquad (5.2.26)\\ &\leqslant C(u)h + |u|_{1,\infty,\Omega}(\operatorname{meas}(K))^{1/2} \leqslant C(u)h. \end{aligned}$$

Because the restriction $\nabla u_h|_K$ is constant over the triangle K, we may write

$$\begin{aligned} \int_K \frac{\|\nabla u_h\|^2}{\sqrt{1+\|\nabla u_h\|^2}}\,\mathrm{d}x &= \frac{\|\nabla u_h|_K\|^2}{\sqrt{1+\|\nabla u_h|_K\|^2}}\operatorname{meas}(K)\\ &\geqslant C\frac{\|\nabla u_{h|K}\|^2}{\sqrt{1+\|\nabla u_{h|K}\|^2}}h^2, \end{aligned} \tag{5.2.27}$$

for some constant C independent of h. Then the conjunction of inequalities (5.2.26) and (5.2.27) implies that

$$\forall K \in \mathcal{T}_h, \quad \forall h, \quad \frac{\|\nabla u_h|_K\|^2}{\sqrt{1+\|\nabla u_h|_K\|^2}} \leqslant C(u).$$

Therefore, the norms $\|\nabla u_{h|K}\|$ are bounded independently of $K \in \mathcal{T}_h$ and $h(\lim_{x\to\infty}(x^2/\sqrt{1+x^2}) = \infty)$ and thus property (5.2.25) is proved.

(iv) Combining steps (ii) and (iii), we obtain

$$\begin{aligned} |u-u_h|_{1,\Omega_h} &= \left(\int_{\Omega_h} \frac{\|\nabla(u-u_h)\|^2}{\sqrt{1+\|\nabla u_h\|^2}}\sqrt{1+\|\nabla u_h\|^2}\,\mathrm{d}x\right)^{1/2}\\ &\leqslant \left(\max_{K\in\mathcal{T}_h}\sqrt{1+\|\nabla u_h|_K\|^2}\right)^{1/2}\Delta_h \leqslant C(u)h. \end{aligned} \tag{5.2.28}$$

(v) Let us add triangles $K \in \mathcal{T}_h^{\partial}$ to each triangulation $\mathcal{T}_h$ as indicated in Fig. 5.2.2, i.e., in such a way that, for all h,

$$\bar{\Omega} \subset \bar{\Omega}_h^* = \bar{\Omega}_h \cup \left(\bigcup_{K\in\mathcal{T}_h^{\partial}} K\right),$$

so that the triangulations $\mathcal{T}_h \cup \mathcal{T}_h^{\partial}$ again constitute a regular family (such a construction is certainly possible).

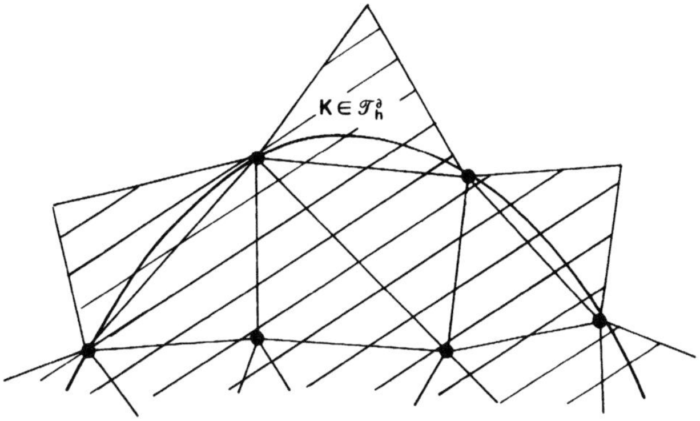

Fig. 5.2.2

Because the boundary Γ is Lipschitz-continuous, there exists (cf. LIONS (1962, Chapter 2) or NEČAS (1967, Chapter 2)) an *extension operator* $E: H^2(\Omega) \to H^2(\mathbf{R}^n)$, i.e., such that for all $v \in H^2(\Omega)$, the function $Ev \in H^2(\mathbf{R}^n)$ satisfies $Ev|_\Omega = v$ and, besides, this operator is continuous: There exists a constant $C(\Omega)$ such that

$$\forall v \in H^2(\Omega), \quad \|Ev\|_{2,\mathbf{R}^n} \leq C(\Omega)\|v\|_{2,\Omega}. \tag{5.2.29}$$

Let then $Eu = u^*$. We define an extension $u_h^*: \bar{\Omega}_h^* \to \mathbf{R}$ of the function u_h by letting

$$\begin{cases} u_h^* = u_h \quad \text{on} \quad \bar{\Omega}_h, \\ \forall K \in \mathcal{T}_h^\partial, \quad u_h^* = \Pi_K(u^*), \end{cases} \tag{5.2.30}$$

where Π_K denotes the $P_1(K)$-interpolant associated with triangles of type (1). Observe that, since the function u_h belongs to the set U_h as defined in (5.2.9), the function u_h^* is continuous over the set $\bar{\Omega}_h^*$ by virtue of the second condition (5.2.30) and thus, it is in the space $H^1(\Omega_h^*)$.

Finally, we shall use the following inequality, due to Friedrichs (cf. NEČAS (1967), Theorem 1.9): There exists a constant $C(\Omega)$ such that

$$\forall v \in H^1(\Omega), \quad \|v\|_{1,\Omega} \leq C(\Omega)(|v|_{1,\Omega} + \|v\|_{L^2(\Gamma)}).$$

Let then $v = u^* - u_h^*$ in this inequality. We obtain, upon combining with inequality (5.2.28),

$$
\begin{aligned}
\|u-u_h\|_{1,\Omega_h} &\leq \|u^*-u_h^*\|_{1,\Omega} \\
&\leq C(\Omega)(|u-u_h|_{1,\Omega_h}+|u-u_h^*|_{1,\Omega-\Omega_h}+\|u-u_h^*\|_{L^2(\Gamma)}) \\
&\leq C(\Omega,u)h+C(\Omega)(|u^*-u_h^*|_{1,\Omega_h^*-\Omega_h}+\|u^*-u_h^*\|_{L^2(\Gamma)}).
\end{aligned}
\tag{5.2.31}
$$

Using inequality (5.2.29), we get

$$
|u^*-u_h^*|_{1,\Omega_h^*-\Omega_h} \leq Ch|u^*|_{2,\Omega_h^*-\Omega_h} \leq Ch\|u\|_{2,\Omega}, \tag{5.2.32}
$$

$$
\begin{aligned}
|u^*-u_h^*|_{0,\infty,\Omega_h^*-\Omega_h} &= \max_{K\in\mathcal{T}_h^\partial} |u^*-u_h^*|_{0,\infty,K} \\
&\leq Ch \max_{K\in\mathcal{T}_h^\partial} |u^*|_{2,K} \leq Ch|u^*|_{2,\Omega_h^*-\Omega_h} \\
&\leq Ch\|u\|_{2,\Omega},
\end{aligned}
\tag{5.2.33}
$$

through two applications of Theorem 3.1.5. Inequality (5.2.33) implies that

$$
\|u^*-u_h^*\|_{L^2(\Gamma)} \leq |u^*-u_h^*|_{0,\infty,\Omega_h^*-\Omega_h}\left(\int_\Gamma \mathrm{d}\gamma\right)^{1/2} \leq Ch\|u\|_{2,\Omega}, \tag{5.2.34}
$$

and inequality (5.2.13) follows from inequalities (5.2.31), (5.2.32) and (5.2.34). □

Exercises

5.2.1. Let $\Omega=\{x\in\mathbf{R}^2;\ 1<\|x\|<2\}$, $u_0=\gamma$ for $\|x\|=1$ and $u_0=0$ for $\|x\|=2$, where γ is a constant. Show that the associated minimal surface problem has a solution if γ is smaller than a quantity γ^* while there is no solution if $\gamma>\gamma^*$ (cf. Fig. 5.2.3).
[Hint: Reduce this problem to a minimization problem for functions in one variable.]

This is a very simple example of a general phenomenon that R. Témam has analyzed through the introduction of "generalized solutions" (cf. the section "Bibliography and Comments").

5.2.2. (i) Show that the minimal surface problem amounts to formally solving a boundary value problem of the form

$$
(*) \qquad \begin{cases} -\sum_{i,j=1}^{2} \partial_j(a_{ij}(\nabla u)\partial_i u)=0 & \text{in} \quad \Omega, \\ u=u_0 \quad \text{sur} \quad \Gamma, \end{cases}
$$

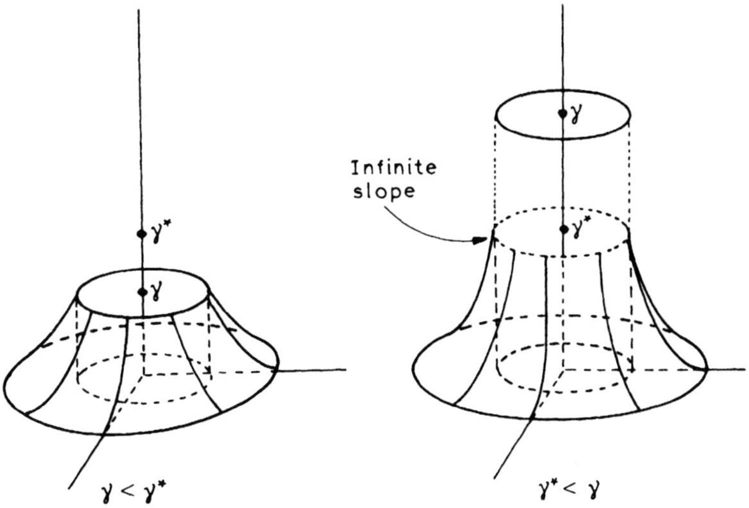

Fig. 5.2.3

i.e., a nonhomogeneous Dirichlet problem for the nonlinear operator

$$u \to -\sum_{i,j=1}^{2} \partial_j(a_{ij}(\nabla u)\partial_i u)$$

and that this operator satisfies an ellipticity condition in the sense that, for any smooth enough function u,

$$\exists \beta(u) > 0, \quad \forall \xi_i, \quad i = 1, 2, \sum_{i,j=1}^{2} a_{ij}(\nabla u)\xi_i\xi_j \geqslant \beta(u) \sum_{i=1}^{2} \xi_i^2.$$

However, the constant $\beta(u)$ cannot be bounded below away from zero independently of u.

(ii) Show that, for smooth functions, the boundary value problem (*) can also be written

$$\begin{cases} (1 + (\partial_2 u)^2)\partial_{11}u - 2\partial_1 u \partial_2 u \partial_{12} u + (1 + (\partial_1 u)^2)\partial_{22} u = 0 & \text{in} \quad \Omega, \\ u = u_0 & \text{on} \quad \Gamma. \end{cases}$$

5.3. Nonlinear problems of monotone type

A minimization problem over the space $W_0^{1,p}(\Omega)$, $2 \leq p$, *and its finite element approximation with n-simplices of type* (1)

Let there be given a convex open subset Ω of $\mathbf{R}^n$ and let p be a number such that

$$2 \leq p \tag{5.3.1}$$

(for the case where $1 < p < 2$, see Exercise 5.3.2). We consider the *minimization problem*: Find a function u such that

$$u \in W_0^{1,p}(\Omega) \quad \text{and} \quad J(u) = \inf_{v \in W_0^{1,p}(\Omega)} J(v) \tag{5.3.2}$$

where the functional J is given by

$$J(v) = \frac{1}{p} \int_\Omega \|\nabla v\|^p \, dx - f(v), \tag{5.3.3}$$

for some given element f of the dual space of the space $W_0^{1,p}(\Omega)$. We use the standard notation

$$\|\nabla v\| = \left(\sum_{i=1}^n (\partial_i v)^2 \right)^{1/2}.$$

For computational convenience, we shall consider throughout this section that the space $W_0^{1,p}(\Omega)$ is equipped with the norm

$$v \to \|v\| = \left(\int_\Omega \|\nabla v\|^p \, dx \right)^{1/p}, \tag{5.3.4}$$

which is clearly equivalent to the standard semi-norm $|\cdot|_{1,p,\Omega}$, itself a norm equivalent to the norm $\|\cdot\|_{1,p,\Omega}$ over the space $W_0^{1,p}(\Omega)$. Finally, we shall use the notation $\|\cdot\|^*$ for the norm in the dual space $(W_0^{1,p}(\Omega))'$ of the space $W_0^{1,p}(\Omega)$.

Remark 5.3.1. For $p = 2$, this minimization problem reduces to the familiar homogeneous Dirichlet problem $-\Delta u = f$ in Ω, $u = 0$ on Γ. □

Our proof of the existence of a solution of the minimization problem (5.3.2) (cf. Theorem 5.3.1) uses the simplest finite element approximation of this problem, which we now proceed to describe: We consider triangulations $\mathcal{T}_h$ made up of n-simplices $K \in \mathcal{T}_h$, in such a way that all the vertices situated on the boundary Γ_h of the set $\bar{\Omega}_h = \bigcup_{K \in \mathcal{T}_h} K$ also belong to the boundary Γ of the set Ω (a similar situation was considered in the previous section; see in particular Fig. 5.2.1 for $n = 2$). Then with each such triangulation, we associate the finite element space X_h whose generic finite element is the n-simplex of type (1) (notice that the functions in the space X_h are defined only on the set $\bar{\Omega}_h$), and we let as usual

$$X_{0h} = \{v_h \in X_h;\ v_{h|\Gamma_h} = 0\}.$$

Then we denote by V_h the space formed by the extensions of the functions of the space X_{0h} which vanish over the set $\bar{\Omega} - \bar{\Omega}_h$. In fact we shall not distinguish between the functions in X_{0h} and their corresponding extensions in the space V_h.

Notice that, because the set $\bar{\Omega}$ was assumed to be convex, the inclusion

$$V_h \subset W_0^{1,p}(\Omega) \tag{5.3.5}$$

and the relations

$$v \in \mathscr{C}^0(\bar{\Omega}) \quad \text{and} \quad v = 0 \quad \text{on} \quad \Gamma \Rightarrow \Pi_h v \in V_h \tag{5.3.6}$$

hold.

Then the *discrete problem* consists in finding a function u_h such that

$$u_h \in V_h \quad \text{and} \quad J(u_h) = \inf_{v_h \in V_h} J(v_h), \tag{5.3.7}$$

where the functional J is defined as in (5.3.3).

Theorem 5.3.1. *The minimization problems* (5.3.2) *and* (5.3.7) *both have one and only one solution. Their respective solutions* $u \in W_0^{1,p}(\Omega)$ *and* $u_h \in V_h$ *are also the unique solutions of the variational equations*

$$\forall v \in W_0^{1,p}(\Omega), \quad \int_\Omega \|\nabla u\|^{p-2} \nabla u \cdot \nabla v \, dx = f(v), \tag{5.3.8}$$

$$\forall v_h \in V_h, \quad \int_\Omega \|\nabla u_h\|^{p-2} \nabla u_h \cdot \nabla v_h \, dx = f(v_h), \tag{5.3.9}$$

respectively.

Proof. We begin by proving several properties of the functional J of (5.3.3).

(i) Since

$$J(v) = \frac{1}{p}\|v\|^p - f(v) \geq \frac{1}{p}\|v\|^p - \|f\|^*\|v\|,$$

we deduce that

$$\lim_{\|v\|\to\infty} J(v) = \infty. \tag{5.3.10}$$

(ii) Let us next establish the *strict convexity* of the functional J. The functional f being convex, it suffices to establish the strict convexity of the mapping

$$v \in W_0^{1,p}(\Omega) \to \int_\Omega F(\nabla v(x))\,dx, \quad \text{with} \quad F: \xi \in \mathbf{R}^n \to \frac{1}{p}\|\xi\|^p. \tag{5.3.11}$$

Let u and v be two different elements in the space $W_0^{1,p}(\Omega)$ such that

$$\operatorname{meas} \tilde{\Omega} > 0, \quad \text{where} \quad \tilde{\Omega} = \{x \in \Omega; \nabla u \neq \nabla v\},$$

and let $\theta \in]0, 1[$ be given. Then write

$$\int_\Omega F(\theta\nabla u + (1-\theta)\nabla v)\,dx = \int_{\tilde{\Omega}} F(\theta\nabla u + (1-\theta)\nabla v)\,dx$$
$$+ \int_{\Omega - \tilde{\Omega}} F(\theta\nabla u + (1-\theta)\nabla v)\,dx,$$

so that the conclusion follows by making use of the strict convexity of the mapping F (which is itself a straightforward consequence of the strict convexity of the mapping $t \in \mathbf{R} \to |t|^p$). For a similar argument, see the proof of Theorem 5.2.1.

Notice at this stage that the property of strict convexity implies the *uniqueness* of the solution of both minimization problems (5.3.2) and (5.3.7).

(iii) We then show that *the functional J is differentiable*, and in so doing, we compute its derivative. Clearly, it suffices to examine the differentiability properties of the mapping considered in (5.3.11). We first observe that the mapping F is twice differentiable, with

$$\partial_i F(\xi) = \|\xi\|^{p-2}\xi_i, \quad 1 \leq i \leq n,$$

$$\partial_{ij} F(\xi) = (p-2)\|\xi\|^{p-4}\xi_i\xi_j + \|\xi\|^{p-2}\delta_{ij}, \quad 1 \leq i, j \leq n.$$

Consequently, we can write

$$F(\xi+\eta)-F(\xi)=\|\xi\|^{p-2}\xi\cdot\eta+R(\xi,\eta),$$

with

$$|R(\xi,\eta)|\leqslant C(p)(\|\xi\|+\|\eta\|)^{p-2}\|\eta\|^2,$$

and thus

$$\int_\Omega F(\nabla(u+v)(x))\,dx-\int_\Omega F(\nabla u(x))\,dx=$$
$$=\int_\Omega \|\nabla u\|^{p-2}\nabla u\cdot\nabla v\,dx+\mathcal{R}(u,v),$$

with

$$|\mathcal{R}(u,v)|\leqslant C(p)\int_\Omega(\|\nabla u\|+\|\nabla v\|)^{p-2}\|\nabla v\|^2\,dx.$$

On the one hand, we have

$$\left|\int_\Omega \|\nabla u\|^{p-2}\nabla u\cdot\nabla v\,dx\right|\leqslant\|u\|^{p-1}\|v\|,$$

and thus the linear mapping

$$v\in W_0^{1,p}(\Omega)\to\int_\Omega\|\nabla u\|^{p-2}\nabla u\cdot\nabla v\,dx$$

is continuous for a fixed $u\in W_0^{1,p}(\Omega)$. On the other, we have

$$\int_\Omega(\|\nabla u\|+\|\nabla v\|)^{p-2}\|\nabla v\|^2\,dx\leqslant(\|u\|+\|v\|)^{p-2}\|v\|^2,$$

and thus the mapping of (5.3.11) is differentiable. Let us then record for future uses the expression of the derivative of the mapping J:

$$\forall v\in W_0^{1,p}(\Omega),\quad J'(u)v=\int_\Omega\|\nabla u\|^{p-2}\nabla u\cdot\nabla v\,dx-f(v). \quad (5.3.12)$$

This shows in particular that the solutions u and u_h of the minimization problems (5.3.2) and (5.3.7) (assuming at this stage their existence) must satisfy relations (5.3.8) and (5.3.9), respectively. In view of the strict convexity of the functional J (step (ii)), these relations are also sufficient for the existence of a unique minimum.

(iv) We next show that *the approximate minimization problem* (5.3.7) *always has a solution*: This is simply a consequence of the strict convexity of the functional J (step (ii)) and of the property

$$\lim_{\substack{\|v_h\|\to\infty \\ v_h\in V_h}} J(v_h)=\infty$$

(step (i)) (the argument has already been given in the proof of Theorem 5.2.1 and shall not be repeated here).

We also remark that *the discrete solutions u_h are bounded independently of the subspace V_h*: Letting $v_h = u_h$ in (5.3.9), we obtain $\|u_h\|^p = f(u_h)$ and thus,

$$\|u_h\| \leq (\|f\|^*)^{1/(p-1)}. \tag{5.3.13}$$

(v) We are now in a position to show the *existence of a solution of the minimization problem* (5.3.2): We consider from now on a family V_h of finite element spaces (of the type described at the beginning of this section) associated with a regular family of triangulations.

The space $W_0^{1,p}(\Omega)$ being reflexive, the uniform boundedness of the discrete solutions u_h, as shown in (5.3.13), implies that there exists a sequence $(u_{h_k})_{k=1}^{\infty}$ which weakly converges to some element $u \in W_0^{1,p}(\Omega)$.

Let then ϕ be an arbitrary function in the space $\mathcal{D}(\Omega)$. By definition of the discrete problems, we have, in particular

$$\forall k \geq 1, \quad J(u_{h_k}) \leq J(\Pi_{h_k}\phi).$$

Since the functional J is continuous and convex, it is weakly lower semicontinuous. Consequently,

$$J(u) \leq \liminf_{k\to\infty} J(u_{h_k}) \leq \liminf_{k\to\infty} J(\Pi_{h_k}\phi). \tag{5.3.14}$$

Because the support of the function ϕ is a compact subset of the set Ω, it is easily seen that there exists an integer k_0 such that

$$k \geq k_0 \Rightarrow \operatorname{supp}\phi \subset \bar{\Omega}_{h_k}.$$

Using Theorem 3.1.6, we obtain for any $k \geq k_0$,

$$\|\Pi_{h_k}\phi - \phi\|_{1,p,\Omega} \leq Ch_k(\text{meas}\,\Omega)^{1/p}|\phi|_{2,\infty,\Omega},$$

and therefore,

$$\lim_{k\to\infty} \|\Pi_{h_k}\phi - \phi\| = 0.$$

This last relation and the continuity of the functional J imply that

$$\lim_{k\to\infty} J(\Pi_{h_k}\phi) = J(\phi). \tag{5.3.15}$$

Combining (5.3.14) and (5.3.15), we have thus proved that

$$\forall \phi \in \mathcal{D}(\Omega), \quad J(u) \leq J(\phi).$$

The space $\mathcal{D}(\Omega)$ being dense in the space $W_0^{1,p}(\Omega)$, we deduce that

$$\forall v \in W_0^{1,p}(\Omega), \quad J(u) \leq J(v),$$

and therefore the function u is the (unique as observed in step (ii)) solution of the minimization problem (5.3.2). □

Remark 5.3.2. From relations (5.3.8), it is immediately seen that the minimization problem (5.3.2) is formally equivalent to the homogeneous Dirichlet problem

$$\begin{cases} -\sum_{i=1}^{n} \partial_i(\|\nabla u\|^{p-2}\partial_i u) = f & \text{in } \Omega, \\ u = 0 & \text{on } \Gamma, \end{cases}$$

where the operator $u \to -\sum_{i=1}^{n} \partial_i(\|\nabla u\|^{p-2}\partial_i u)$ is nonlinear for $p > 2$. □

Sufficient condition for $\lim_{h\to 0} \|u - u_h\|_{1,p,\Omega} = 0$

Using the last part of the proof of the above theorem, we are in addition able to prove the convergence of the discrete solutions towards the solution u, as we now show.

Theorem 5.3.2. *Let there be given a family of finite element spaces as previously described, i.e., made up of n-simplices of type* (1), *associated with a regular family of triangulations. Then with the sole assumption that the solution u is in the space* $W_0^{1,p}(\Omega)$ *we have*

$$\lim_{h\to 0} \|u - u_h\|_{1,p,\Omega} = 0. \tag{5.3.16}$$

Proof. We continue the argument used in part (v) of the proof of the previous theorem. Since the weak limit u is unique, we deduce that the whole family (u_h) weakly converges to the solution u. Thus,

$$f(u) = \lim_{h\to 0} f(u_h).$$

On the other hand, we have

$$\forall \phi \in \mathcal{D}(\Omega), \quad \limsup_{h\to 0} J(u_h) \leq \limsup_{h\to 0} J(\Pi_h\phi) = \lim_{h\to 0} J(\Pi_h\phi) = J(\phi).$$

Since the functions ϕ can be chosen arbitrarily close to the solution u (in the norm of the space $W_0^{1,p}(\Omega)$), we deduce from the above relations that

$$J(u) = \lim_{h\to 0} J(u_h),$$

i.e., in view of the expression for the functional J, that

$$\|u\| = \lim_{h\to 0} \|u_h\|, \tag{5.3.17}$$

since $\lim_{h\to 0} f(u_h) = f(u)$.

The space $W_0^{1,p}(\Omega)$ being uniformly convex, the weak convergence and the convergence (5.3.17) imply the convergence in the norm. □

The equivalent problem $Au = f$. Two properties of the operator A

In order to have an approach similar to that of the linear case, let us introduce, for any function $u \in W_0^{1,p}(\Omega)$, the element $Au \in (W_0^{1,p}(\Omega))'$ defined by (cf. the proof of Theorem 5.3.1)

$$\forall v \in W_0^{1,p}(\Omega), \quad Au(v) = \int_\Omega \|\nabla u\|^{p-2}\nabla u \cdot \nabla v \,\mathrm{d}x. \tag{5.3.18}$$

Notice that the element Au is nothing but the derivative of the mapping of (5.3.11), so that relation (5.3.12) may be equivalently written as

$$J'(u) = Au - f. \tag{5.3.19}$$

In other words, the original minimization problem (5.3.2) is equivalent to the solution of the (nonlinear if $p > 2$) equation $Au = f$. Our next task is to establish (cf. Theorem 5.3.3) two properties of the operator

$$A\colon W_0^{1,p}(\Omega) \to (W_0^{1,p}(\Omega))'$$

defined in (5.3.18) and whose bijectivity has been proved in Theorem 5.3.1. The first property (cf. (5.3.20)) is a generalization of the usual ellipticity condition in the linear case, while the second property (cf. (5.3.21)) is a generalization of the continuity of the operator A in the linear case (cf. the inequality $\|A\|_{\mathcal{L}(V;V')} \leq M$ established in (1.1.21)). In order to simplify the exposition, we shall henceforth assume that $n = 2$ (the extension to higher dimensions is indeed possible, but at the expense of additional technicalities).

Theorem 5.3.3. *For a given number p in the interval* $[2, \infty[$, *let* $A: W_0^{1,p}(\Omega) \to (W_0^{1,p}(\Omega))'$ *be the operator as defined in* (5.3.18). *Then,*

$$\exists \alpha > 0, \quad \forall u, v \in W_0^{1,p}(\Omega), \quad \alpha \|u - v\|^p \leqslant (Au - Av)(u - v), \tag{5.3.20}$$

$$\exists M > 0, \quad \forall u, v \in W_0^{1,p}(\Omega), \quad \|Au - Av\|^* \leqslant \leqslant M(\|u\| + \|v\|)^{p-2}\|u - v\|. \tag{5.3.21}$$

Proof. Let us introduce the auxiliary function

$$\phi: (\xi, \eta) \in \mathcal{O} = \{(\xi, \eta) \in \mathbf{R}^2 \times \mathbf{R}^2;\ \xi \neq \eta\} \to$$
$$\to \phi(\xi, \eta) = \frac{(\|\xi\|^{p-2}\xi - \|\eta\|^{p-2}\eta) \cdot (\xi - \eta)}{\|\xi - \eta\|^p}, \tag{5.3.22}$$

where $\cdot$ denotes as usual the Euclidean inner-product in the space $\mathbf{R}^2$. We shall show that

$$\exists \alpha > 0, \quad \forall (\xi, \eta) \in \mathcal{O}, \quad \alpha \leqslant \phi(\xi, \eta), \tag{5.3.23}$$

a property which is easily seen to imply inequality (5.3.20). First, we notice that since

$$\forall \eta \neq 0, \quad \phi(0, \eta) = 1, \tag{5.3.24}$$

it suffices to consider the case where $\xi \neq 0$. Next, we prove that

$$\forall (\xi, \eta) \in \mathcal{O}, \quad \phi(\xi, \eta) > 0. \tag{5.3.25}$$

This follows from the relations

$$\begin{aligned}
&(\|\xi\|^{p-2}\xi - \|\eta\|^{p-2}\eta) \cdot (\xi - \eta) = \\
&\quad = \|\xi\|^p - (\|\xi\|^{p-2} + \|\eta\|^{p-2})(\xi \cdot \eta) + \|\eta\|^p \\
&\quad \geqslant \|\xi\|^p - \|\xi\|^{p-1}\|\eta\| - \|\eta\|^{p-1}\|\xi\| + \|\eta\|^p \\
&\quad = (\|\xi\|^{p-1} - \|\eta\|^{p-1})(\|\xi\| - \|\eta\|) \\
&\quad > 0 \quad \text{unless} \quad \|\xi\| = \|\eta\|.
\end{aligned}$$

Since the penultimate inequality is an equality if and only if $\eta = \mu\xi$ for some $\mu \in \mathbf{R}$, the only remaining case is that where $\eta = -\xi$. But then

$$(\|\xi\|^{p-2}\xi - \|\eta\|^{p-2}\eta) \cdot (\xi - \eta) = 4\|\xi\|^p > 0.$$

Finally, we observe that we may restrict ourselves to the case where $\xi = \bar{\xi} = (1, 0)$ since $\phi(\lambda\xi, \lambda\eta) = \phi(\xi, \eta)$ for all $\lambda > 0$ on the one hand and since the Euclidean inner product is invariant through rotations around the origin on the other. Because

$$\lim_{\|\eta\| \to \infty} \phi(\bar{\xi}, \eta) = 1, \tag{5.3.26}$$

it remains to study the behavior of the function $\eta = (\eta_1, \eta_2) \in (\mathbf{R}^2 - \bar{\xi}) \to \phi(\bar{\xi}, \eta)$ in the neighborhood of the point $\bar{\xi}$. For this purpose, let

$$\eta_1 = 1 + \rho \cos\theta, \quad \eta_2 = \rho \sin\theta.$$

Then a simple computation shows that

$$\phi(\bar{\xi}, \eta) = \frac{1 + (p-2)\cos^2\theta + \epsilon(\rho, \theta)}{\rho^{p-2}},$$

with $\lim_{\rho \to 0} \epsilon(\rho, \theta) = 0$ uniformly with respect to $\theta \in [0, 2\pi[$. Therefore,

$$\lim_{\eta \to \bar{\xi}} \phi(\bar{\xi}, \eta) = \begin{cases} 1 & \text{if} \quad p = 2, \\ \infty & \text{if} \quad p > 2, \end{cases} \tag{5.3.27}$$

and relation (5.3.23) follows from the conjunction of relations (5.3.24) through (5.2.27).

To prove the second relation (5.3.21), we introduce the auxiliary function

$$\psi\colon (\xi, \eta) \in \mathcal{O} = \{(\xi, \eta) \in \mathbf{R}^2 \times \mathbf{R}^2, x \neq y\} \to$$
$$\to \psi(\xi, \eta) = \frac{\|\,\|\eta\|^{p-2}\eta - \|\xi\|^{p-2}\xi\|}{\|\eta - \xi\|(\|\eta\| + \|\xi\|)^{p-2}}, \tag{5.3.28}$$

and we shall show that

$$\exists M > 0, \quad \forall (\xi, \eta) \in \mathcal{O}, \quad \psi(\xi, \eta) \leq M. \tag{5.3.29}$$

Since

$$\forall \eta \neq 0, \quad \psi(0, \eta) = 1, \tag{5.3.30}$$

we may assume that $\xi \neq 0$. In fact, it suffices to consider the case where $\xi = \bar{\xi} = (1, 0)$ since $\psi(\lambda\xi, \lambda\eta) = \psi(\xi, \eta)$ for all $\lambda > 0$ on the one hand, and since the Euclidean norm is invariant through rotations around the origin on the other. We also have

$$\lim_{\|\eta\|\to\infty} \psi(\bar{\xi}, \eta) = 1. \tag{5.3.31}$$

To study the behavior of the function $\eta = (\eta_1, \eta_2) \in (\mathbf{R}^2 - \bar{\xi}) \to \psi(\bar{\xi}, \eta)$ in the neighborhood of the point $\bar{\xi}$, we let $\eta_1 = 1 + \rho \cos\theta$, $\eta_2 = \rho \sin\theta$ as before. In this fashion we obtain

$$\psi(\bar{\xi}, \eta) = 2^{2-p}(1 + p(p-2)\cos^2\theta)^{1/2} + \epsilon(\rho, \theta),$$

with $\lim_{\rho\to 0} \epsilon(\rho, \theta) = 0$ uniformly with respect to $\theta \in [0, 2\pi[$ and therefore,

$$\limsup_{\eta\to\bar{\xi}} \psi(\bar{\xi}, \eta) < \infty. \tag{5.3.32}$$

Then relation (5.3.29) follows from relations (5.3.30) to (5.3.32). As a consequence, we have

$$\forall \xi, \eta \in \mathbf{R}^2, \quad \| \|\eta\|^{p-2}\eta - \|\xi\|^{p-2}\xi\| \leqslant M\|\eta - \xi\|(\|\eta\| + \|\xi\|)^{p-2}. \tag{5.3.33}$$

To prove inequality (5.3.21), we shall use the characterization

$$\|Au - Av\|^* = \sup_{w\in V} \frac{|(Au - Av)w|}{\|w\|}. \tag{5.3.34}$$

By making use of inequality (5.3.33), we infer that

$$\begin{aligned}
|(Au - Av)(w)| &= \left|\int_\Omega (\|\nabla u\|^{p-2}\nabla u - \|\nabla v\|^{p-2}\nabla v) \cdot \nabla w \,\mathrm{d}x\right| \\
&\leqslant \int_\Omega \| \|\nabla u\|^{p-2}\nabla u - \|\nabla v\|^{p-2}\nabla v\| \,\|\nabla w\| \,\mathrm{d}x \\
&\leqslant M \int_\Omega \|\nabla(u - v)\|(\|\nabla u\| + \|\nabla v\|)^{p-2}\|\nabla w\| \,\mathrm{d}x \\
&\leqslant M\|u - v\|\left\{\int_\Omega (\|\nabla u\| + \|\nabla v\|)^p \,\mathrm{d}x\right\}^{(p-2)/p} \|w\| \\
&\leqslant M\|u - v\|(\|u\| + \|v\|)^{p-2}\|w\|,
\end{aligned}$$

and inequality (5.3.21) follows from the above inequality coupled with characterization (5.3.34). □

Strongly monotone operators. Abstract error estimate

We are now in a position to describe an *abstract setting* particularly appropriate for this type of problem and its approximation: We are given a (generally nonlinear) mapping

$$A: V \to V'$$

acting from a space V, with norm $\|\cdot\|$, into its dual space V', with norm $\|\cdot\|^*$, which possesses the two following properties:

(i) The mapping A is *strongly monotone*, i.e., there exists a strictly increasing function $\chi: [0, +\infty[\to \mathbf{R}$ such that

$$\chi(0) = 0 \quad \text{and} \quad \lim_{t\to\infty} \chi(t) = \infty, \tag{5.3.35}$$

$$\forall u, v \in V, \quad (Au - Av)(u - v) \geq \chi(\|u - v\|)\|u - v\|. \tag{5.3.36}$$

In particular, the operator A as defined in (5.3.18) is strongly monotone, with (cf. Theorem 5.3.3)

$$\chi(t) = \alpha t^{p-1}. \tag{5.3.37}$$

(ii) The mapping A is *Lipschitz-continuous for bounded arguments* in the sense that, for any ball $B(0; r) = \{v \in V; \|v\| \leq r\}$, there exists a constant $\Gamma(r)$ such that

$$\forall u, v \in B(0; r), \quad \|Au - Av\|^* \leq \Gamma(r)\|u - v\|. \tag{5.3.38}$$

Thus, the operator A as defined in (5.3.18) is Lipschitz-continuous for bounded arguments, with (cf. Theorem 5.3,3)

$$\Gamma(r) = M(2r)^{p-2}. \tag{5.3.39}$$

Let there be given an element $f \in V'$. For operators which satisfy assumptions (i) and (ii), we are able to obtain in the next theorem an abstract estimate for the error $\|u - u_h\|$, where u and u_h are respectively the solutions of the equations

$$\forall v \in V, \quad Au(v) = f(v) \quad \text{(equivalently,} \quad Au = f), \tag{5.3.40}$$

$$\forall v_h \in V_h, \quad Au_h(v_h) = f(v_h), \tag{5.3.41}$$

where V_h is a (finite-dimensional in practice) subspace of the space V (we showed in Theorem 5.3.1 that, with the operator A of (5.3.18), problems (5.3.40) and (5.3.41) have solutions; for general existence results, see "Bibliography and Comments").

Theorem 5.3.4. *Let there be given a mapping $A: V \to V'$ which is strongly monotone and Lipschitz-continuous for bounded arguments.*

Then there exists a constant C independent of the subspace V_h such that

$$\chi(\|u - u_h\|) \leq C \inf_{v_h \in V_h} \|u - v_h\|. \tag{5.3.42}$$

Proof. To begin with, we show that the assumption of strong monotonicity for the operator A implies that *the same a priori bound holds for both solutions u and u_h*: The conjunction of inequality (5.3.36) and relations (5.3.40) implies that

$$\begin{aligned}\chi(\|u\|)\|u\| &\leq (Au - AO)u \\ &= f(u) - (AO)u \leq (\|f\|^* + \|AO\|^*)\|u\|,\end{aligned}$$

and a similar inequality holds with u replaced by u_h. Therefore, the function χ being strictly increasing with $\chi(0) = 0$ and $\lim_{t\to\infty} \chi(t) = \infty$ by assumption, we have

$$\|u\|, \quad \|u_h\| \leq \chi^{-1}(\|f\|^* + \|AO\|^*). \tag{5.3.43}$$

Next, let v_h be an arbitrary element in the space V_h. Using the inclusion $V_h \subset V$ and relations (5.3.40) and (5.3.41), we obtain $(Au - Au_h)w_h = 0$ for all $w_h \in V_h$ so that, in particular,

$$(Au - Au_h)(u_h - v_h) = 0.$$

Combining the above equations with inequalities (5.3.36), (5.3.38) and the *a priori* bound (5.3.43), we obtain

$$\begin{aligned}\chi(\|u - u_h\|)\|u - u_h\| &\leq (Au - Au_h)(u - u_h) \\ &= (Au - Au_h)(u - v_h) \\ &\leq \|Au - Au_h\|^*\|u - v_h\| \\ &\leq \Gamma(\chi^{-1}(\|f\|^* + \|AO\|^*))\|u - u_h\|\,\|u - v_h\|,\end{aligned}$$

and thus inequality (5.3.42) is proved, with $C = \Gamma(\chi^{-1}(\|f\|^* + \|AO\|^*))$. □

Remark 5.3.3. The abstract error estimate of the previous theorem is another generalization of Céa's lemma, since in the linear case one has $\chi(t) = \alpha t$. □

Remark 5.3.4. In the particular case of the operator A of (5.3.18), we have $AO = O$, so that with the function χ of (5.3.37), we obtain

$$\|u\|, \quad \|u_h\| \leq \left(\frac{\|f\|^*}{\alpha}\right)^{1/(p-1)}$$

If we argue as in part (iv) of the proof of Theorem 5.3.1, however, we obtain the improved *a priori* bound

$$\|u\|,\quad \|u_h\| \leqslant (\|f\|^*)^{1/(p-1)}. \qquad \square$$

Estimate of the error $\|u - u_h\|_{1,p,\Omega}$

Let us now return to the minimization problem (5.3.2) and its finite element approximation as described at the beginning of this section. For simplicity, *we shall assume that the set* $\bar{\Omega}$ *is polygonal.* Then we get as an application of Theorem 5.3.4:

Theorem 5.3.5. *Let there be given a family of finite element spaces made up of triangles of type* (1), *associated with a regular family of triangulations. Then, if the solution* $u \in W_0^{1,p}(\Omega)$ *of the minimization problem* (5.3.2) *is in the space* $W^{2,p}(\Omega)$, *there exists a constant* $C(\|f\|^*, |u|_{2,p,\Omega})$ *such that*

$$\|u - u_h\|_{1,p,\Omega} \leqslant C(\|f\|^*, |u|_{2,p,\Omega})h^{1/(p-1)}. \tag{5.3.44}$$

Proof. Since $AO = O$, the constant which appears in inequality (5.3.42) is a function of $\|f\|^*$ only. Next, for some constants C independent of the subspace V_h, we have

$$\inf_{v_h \in V_h} \|u - v_h\| \leqslant C|u - \Pi_h u|_{1,p,\Omega} \leqslant Ch|u|_{2,p,\Omega}.$$

It then remains to apply inequality (5.3.42) with the function $\chi(t) = \alpha t^{p-1}$.

$\square$

One should be aware that the above error estimate may be somehow illusive in that the solution u need not be in the space $W^{2,p}(\Omega)$ even with very smooth data (cf. Exercise 5.3.1). This is why it was worth proving convergence with the minimal assumption that $u \in W_0^{1,p}(\Omega)$ (Theorem 5.3.2). This is also why we did not consider the (otherwise straightforward) case where the generic finite element in the spaces V_h would be for example the triangle of type (k).

Exercises

5.3.1. Following GLOWINSKI & MARROCCO (1975), consider the one-dimensional analog of the minimization problem (5.3.2), where

$$\begin{cases} \Omega =]-1, +1[, \\ J(v) = \dfrac{1}{p}\displaystyle\int_\Omega |v'|^p \, dx - \gamma \int_\Omega v \, dx, \quad \gamma \in \mathbf{R}. \end{cases}$$

Show that the unique solution $u \in W_0^{1,p}(\Omega)$ of this problem is given by

$$u(x) = \left(1 - \frac{1}{p}\right)\gamma^{1/(p-1)}(1 - |x|^{p/(p-1)}),$$

and that

$$u \in W^{2,p}(\Omega) \quad \text{if} \quad 1 < p < \frac{3+\sqrt{5}}{2},$$

$$u \notin W^{2,p}(\Omega) \quad \text{if} \quad \frac{3+\sqrt{5}}{2} \leqslant p.$$

5.3.2. The object of this problem is to study the minimization problem (5.3.2) (with the functional J as in (5.3.3)) when $1 < p < 2$.

(i) Let V be a reflexive Banach space, and let $J: V \to \mathbf{R}$ be a continuous and convex functional such that $\lim_{\|v\| \to \infty} J(v) = \infty$. Show that there exists at least one element $u \in V$ such that $J(u) = \inf_{v \in V} J(v)$ (cf. CÉA (1971) or LIONS (1968, 1969)).

(ii) Deduce from this result the existence of a unique solution of the minimization problem (5.3.2). Show that this problem is equivalent to solving the equation $Au = f$, where the mapping $A: V \to V'$ is defined as in (5.3.18) and $V = W_0^{1,p}(\Omega)$.

(iii) Following GLOWINSKI & MARROCCO (1975), show that

$$\exists \alpha > 0, \quad \forall u, v \in V, \quad \alpha\|u - v\|^2 \leqslant (\|u\| + \|v\|)^{2-p}(Au - Av)(u - v),$$

$$\exists M > 0, \quad \forall u, v \in V, \quad \|Au - Av\|^* \leqslant M\|u - v\|^{p-1}.$$

(iv) Deduce from (iii) that (GLOWINSKI & MARROCCO (1975))

$$\|u - u_h\|_{1,p,\Omega} \leqslant C(\|f\|^*, |u|_{2,p,\Omega}) h^{1/(3-p)},$$

if $u \in W_0^{1,p}(\Omega) \cap W^{2,p}(\Omega)$.

Bibliography and comments

5.1. The content of this section is based on the analysis of FALK (1974). The abstract error estimate of Theorem 5.1.1 has been independently rediscovered by ROUX (1976) in the study of the numerical approximation of a two-dimensional compressible flow problem which can be reduced to variational inequalities, using the method of BREZIS & STAMPACCHIA (1973). Incidentally, the functional setting for this problem is interesting in itself in that the corresponding space V is a weighted Sobolev space, and the domain of definition of its functions is unbounded. The same problem is similarly studied via variational inequalities by CIAVALDINI & TOURNEMINE (1977), who have extended the abstract error estimate of Theorem 5.1.1 so as to include the case where the bilinear and linear forms are approximated (through the process of numerical integration).

FALK (1975) has extended his results to the case of a non convex domain with a smooth boundary. For additional results concerning the approximation of the obstacle problem, see MOSCO & STRANG (1974), MOSCO & SCARPINI (1975). BREZZI, HAGER & RAVIART (1977) have given another proof of Theorem 5.1.2. They have also shown that $\|u - u_h\|_{1,\Omega} = O(h^{(3/2)-\epsilon})$, $\epsilon > 0$ arbitrarily small, when triangles of type (2) are used. NATTERER (1976) has studied the error in the norm $|\cdot|_{0,\Omega}$, using an argument based on the Aubin–Nitsche lemma. For another approach, see BERGER (1976). Finally, NITSCHE (1977) has been able to apply his method of weighted norms to this problem. In this fashion, he obtains an estimate of the form

$$|u - u_h^*|_{0,\infty,\Omega} \leq Ch^2|\ln h|(\|u\|_{2,\infty,\Omega} + \|\psi\|_{2,\infty,\Omega}).$$

However, the corresponding discrete solution u_h^* is found in the subset $U_h^* = U \cap X_{0h}$, instead of the present subset U_h.

FRÉMOND (1971a, 1972) has given a thorough treatment of the related problem of an elastic body lying on a support, the contact surface being unknown.

The elastic-plastic torsion problem (Exercise 5.1.3) is extensively studied in LANCHON (1972). Using techniques from duality theory, FALK & MERCIER (1977) have recently constructed a finite element method which yields directly an approximation of the stresses σ_{13} and σ_{23} with an $O(h)$ convergence in the norm $|\cdot|_{0,\Omega}$. In fact their formulation is more appropriate for this type of problem, where a direct knowledge of the stresses is more important than a knowledge of the stress function. For related results, see MERCIER (1975a, 1975b), GABAY & MERCIER (1976), and BREZZI, JOHNSON & MERCIER (1977), where elasto-plastic plates are considered.

A third type of problem which reduces to variational inequalities occurs with sets U of the form

$$U = \{v \in H^1(\Omega);\ v \geq \psi \text{ a.e. on } \Gamma\}.$$

Such problems with unilateral constraints occur in particular in elasticity, where they are known as *Signorini problems* (cf. Exercise 1.2.5). A finite element approximation of such problems is studied in SCARPINI & VIVALDI (1977).

An extension of the present setting consists in looking for the solution u of variational inequalities of the form (see DUVAUT & LIONS (1972)):

$$\forall v \in U, \quad a(u, v-u) + j(v) - j(u) \geq f(v-u),$$

where $j: V \to \mathbf{R}$ is a *non differentiable* functional. Such problems are found in particular in the study of *Bingham flows*, with $j(v) = \int_\Omega \|\nabla v\|\,dx$. Their finite element approximations have been analyzed in BRISTEAU (1975, Chapter 2), FORTIN (1972a), GLOWINSKI (1975).

An extensive treatment of variational inequalities and of their approximations is found in GLOWINSKI, LIONS & TRÉMOLIÈRES (1976a, 1976b). The reader who is also interested in the actual solution of the corresponding discrete problems should consult GLOWINSKI (1976b).

A crucial generalization consists in considering the *quasi-variational inequalities* introduced by BENSOUSSAN & LIONS (1973, 1974): Instead of a fixed set U, one considers a family $(U(v))_{v\in V}$ of nonempty closed convex subsets of V and one looks for an element u such that

$$u \in U(u) \quad \text{and} \quad \forall v \in U(u), \quad a(u, v-u) \geq f(v-u).$$

Introductions to such "quasi-variational" problems are given in LIONS (1975a, 1975b). A much more complete treatment is given in LIONS (1976).

A variety of *free surface problems* can be reduced to quasi-variational inequalities. In particular, problems of *flows through porous media* can be reduced to variational inequalities or quasi-variational inequalities, by a method due to BAIOCCHI (1971, 1972, 1974, 1975). See also BAIOCCHI, COMINCIOLI, MAGENES & POZZI (1973). Such problems may be also reduced to *optimal domain problems* as in BÉGIS & GLOWINSKI (1974, 1975), CÉA, GIOAN & MICHEL (1974).

5.2. There exist several approaches for analyzing the minimal surface problem. When $n = 2$, there is always a solution in the classical sense (i.e., of the associated boundary value problem; cf. Exercise 5.2.2) for continuous boundary data (RADÓ (1930)), when the set Ω is convex. In higher dimensions, JENKINS & SERRIN (1968) have shown that, rather than convexity, it is the positivity of the mean curvature of the boundary which insures existence of a unique solution in the classical sense (for sufficiently smooth boundary and boundary data). They also proved that, if the mean curvature of the boundary is not everywhere positive, there exist smooth boundary data for which the Dirichlet problem has no solution. MORREY (1966, Theorem 4.2.1) has shown that if the set Ω is strictly convex with a sufficiently smooth boundary and if the function u_0 belongs to the space $W^{2,q}(\Omega)$ for some $q > 2$ and satisfies a "bounded slope condition", then the minimization problem (5.2.2) has a unique solution in the space $W^{2,q}(\Omega)$. For an extensive treatment of the minimal surface problem, see the monumental work of J.C.C. NITSCHE (1975).

TÉMAM (1971) (see also EKELAND & TÉMAM (1974, Chapter V)) has extended the notion of solution so as to get existence for arbitrary bounded open sets Ω, when there is no solution in a more traditional sense (cf. Exercise 5.2.1). The main result is the following:

If $u_0 \in W^{1,1}(\Omega) \cap L^\infty(\Omega)$, there exists a *generalized solution* $\tilde{u} \in W^{1,1}(\Omega)$ (unique up to a constant additive factor) in the following sense:

(i) It is analytic in Ω and solution of the associated partial differential equation.

(ii) Any sequence (v_k) with $v_k \in W^{1,1}(\Omega)$ and $\lim_{k\to\infty} J(v_k) = \inf\{J(v);\ (v - u_0) \in W_0^{1,1}(\Omega)\}$ is such that $\lim_{k\to\infty} v_k = \tilde{u}$ in the space $L^1(\Omega)/\mathbf{R}$ and $\lim_{k\to\infty} |v_k - \tilde{u}|_{1,\tilde{\Omega}} = 0$ for any open set $\tilde{\Omega}$ with $\bar{\tilde{\Omega}} \subset \Omega$.

(iii) If there exists a point $x_0 \in \Gamma$ such that $\limsup_{\{x\to x_0,\ x\in\Omega\}} \|\nabla u(x)\| < \infty$, then the generalized solution is unique, and $\tilde{u} = u_0$ on the set $\{x \in \Gamma;\ \limsup_{\{y\to x,\ y\in\Omega\}} \|\nabla u(y)\| < \infty\}$. For recent developments of R. Témam's analysis, see LICHNEWSKY (1974a, 1974b).

There are relatively few references on the application of the finite element method to this problem. Let us first quote HINATA, SHIMASAKI & KIYONO (1974) where only numerical results are presented. The proof that $|u - u_h|_{1,\Omega_h} = O(h)$, i.e., the four first steps of the proof of Theorem 5.2.2, as well as the proof of Theorem 5.2.1, are given in JOHNSON & THOMÉE (1975). Using an adaptation of the Aubin–Nitsche lemma, C. Johnson and V. Thomée have in addition shown that, if $u \in W^{2,q}(\Omega)$ for some $q > 2$ and if u_0 is sufficiently smooth, then for any p with $1 \leqslant p < 2$, one has $|u - u_h|_{0,p,\Omega_h} = O(h^2)$.

More recently, RANNACHER (1977) has completed the results of C.

Johnson and V. Thomée by showing, under the same assumptions, that $|u - u_h|_{0,\Omega_h} = O(h^2)$. Especially, R. Rannacher has been able to adapt the method of weighted norms of J.A. Nitsche described in Section 3.3 so as to derive the error estimate

$$|u - u_h|_{0,\infty,\Omega_h} = O(h^{2-(2/q)}|\ln h|),$$

assuming $u \in W^{2,q}(\Omega)$ for some q with $2 < q \leq \infty$. See also FREHSE & RANNACHER (1976). For similar, but weaker, results, see MITTELMANN (1977). Finally, JOURON (1975) has made an interesting study of the approximation of the generalized solution in the sense of R. Témam.

5.3. We have followed in this section a paper of GLOWINSKI & MARROCCO (1975), where the case $1 < p < 2$ is also treated along the lines indicated in Exercise 5.3.2 (problems of this last type arise, with more complicated boundary conditions however, in the modeling of strains in ice; in this respect, see the thorough study of PÉLISSIER (1975)). Actual methods for solving the discrete problems are described and studied in the above paper by R. Glowinski and A. Marrocco.

In CÉA (1971) and LIONS (1968, 1969), several general theorems concerning the existence of solutions for a problem of the form $\inf_{v \in V} J(v)$ are proved for general functionals J. Another approach for obtaining existence results is to use the theory of monotone operators: A mapping $A: V \to V'$ is said to be *monotone* if $(Au - Av)(u - v) \geq 0$ for all $u, v \in V$. BROWDER (1965) and LERAY & LIONS (1965) have proved: Let V be a reflexive Banach space and let A be a monotone operator such that

(i) there exists a strictly increasing function χ with $\lim_{t\to\infty} \chi(t) = \infty$ such that $Av(v) \geq \chi(\|v\|)\|v\|$ for all $v \in V$ (a property implied by the definition of strongly monotone operators as given in the text),

(ii) given any finite-dimensional subspace W of the space V and given any sequence of elements $w_k \in W$ which converges to $w \in W$, one has $\lim_{k\to\infty} Aw_k(v) = Aw(v)$ for all $v \in V$ (a property implied by the Lipschitz-continuity for bounded arguments).

Then the mapping $A: V \to V'$ is a bijection. Therefore this result provides a more general method for proving existence.

One of the first systematic treatments of variational approximations of nonlinear problems of monotone type appears to be that of CIARLET (1966). This work was then extended in several directions, in CIARLET, SCHULTZ & VARGA (1967, 1968a, 1968c, 1969), CIARLET, NATTERER & VARGA (1970), MOCK (1975), NOOR & WHITEMAN (1976), SCHULTZ (1969a, 1971), LOUIS (1976). See also MELKES (1970) for an independent work. In particular, Theorem 5.3.4 is adapted from Theorem 2.1 of CIARLET, SCHULTZ & VARGA (1969). In this paper, one considers monotone operator equations of the general form

$$\forall v \in V, \quad \sum_{|\alpha| \leq m} \int_{\Omega} A_{\alpha}(x, u, Du, \ldots, D^m u) \partial^{\alpha} v \, dx = 0.$$

These contain as special case the equations $J'(u)v = 0$ associated with functionals of the type

$$v \in H_0^1(\Omega) \to J(v) = \frac{1}{2} \int_{\Omega} \|\nabla v\|^2 \, dx - \int_{\Omega} \left\{ \int_0^{v(x)} f(x, \eta) \, d\eta \right\} dx,$$

which correspond to *nonlinear Dirichlet problems* of the form $-\Delta u = f(x, u)$ in Ω, $u = 0$ on Γ. General approximate methods for problems with monotone operators are studied in BRÉZIS & SIBONY (1968). Techniques from duality theory can be applied to such problems, as in BERCOVIER (1976), SCHEURER (1977).

Additional bibliography and comments

Other nonlinear problems

We continue this review by mentioning nonlinear problems of various type, some of which are reminiscent of the problems considered in Chapter 5.

For example, the nonlinear boundary value problem

$$\begin{cases} -\sum_{i=1}^{2} \partial_i (a(\|\nabla u\|) \partial_i u) = 0 & \text{in} \quad \Omega \subset \mathbf{R}^2, \\ u = 0 & \text{on} \quad \Gamma, \end{cases}$$

where $a(r)$ is a strictly increasing function of its argument r, describes the magnetic state in the cross-section of an alternator. It is thoroughly studied, as well as its finite element approximation, in GLOWINSKI & MARROCCO (1974).

NITSCHE (1976c) and FREHSE & RANNACHER (1977) have applied the method of weighted norms of J.A. Nitsche (cf. Section 3.3) to nonlinear problems of the form

$$\begin{cases} -\sum_{i=1}^{n} \partial_i (a(x, u) \partial_i u) = f & \text{in} \quad \Omega, \\ u = 0 & \text{on} \quad \Gamma, \end{cases}$$

and

$$\begin{cases} -\sum_{i=1}^{n} \partial_i a_i(x, u, \nabla u) + a_0(x, u, \nabla u) = f & \text{in } \Omega, \\ u = 0 & \text{on } \Gamma, \end{cases}$$

respectively. See also DOUGLAS & DUPONT (1975) for another approach.

The method of weighted norms has also been applied to problems where the solution $u \in H_0^1(\Omega)$, $\Omega \subset \mathbf{R}^2$, minimizes an integral of the form $\int_\Omega F(x, v, \nabla v)\,dx$ over the space $H_0^1(\Omega)$. For uniformly convex functions $F(x, \xi, \cdot)$ (this is not the case of the minimal surface problem), FREHSE (1976) shows that $|u - u_h|_{0,\infty,\Omega_h} = O(h^2|\ln h|)$ for the piecewise linear approximations.

A wide class of nonlinear problems arises in nonlinear elasticity, particularly in the study of large strains. For a thorough treatment of the application of the finite element method to such problems, see the book of ODEN (1972a), and also ODEN (1973b, 1976b), CAREY (1974).

Problems in which the solution must satisfy various equality and inequality constraints arise in water pollution control. Their numerical solution, which combines finite element methods and linear programming methods, is considered by FUTAGAMI (1976). A challenging domain of study is the approximation of *bifurcation problems* (which arise in particular in elasticity). In this direction, we mention the pioneering work of KIKUCHI (1976b), who considers the problem

$$\begin{cases} -\Delta u = \lambda u - u^3 & \text{in } \Omega, \\ u = 0 & \text{on } \Gamma. \end{cases}$$

The Navier–Stokes problem

For large gradients of the velocity, a new term has to be added in the partial differential equation of the Stokes problem (cf. the section "Additional Bibliography and Comments" of Chapter 4), a process which gives rise to the *Navier–Stokes problem*:

$$\begin{cases} -\nu \Delta \boldsymbol{u} + \sum_{i=1}^{n} u_i \partial_i \boldsymbol{u} + \nabla p = \boldsymbol{f} & \text{in } \Omega, \\ \operatorname{div} \boldsymbol{u} = 0 & \text{in } \Omega, \\ \boldsymbol{u} = \boldsymbol{0} & \text{on } \Gamma. \end{cases}$$

One can show (cf. LADYŽENSKAJA (1963), LIONS (1969)) that if the ratio $|f|_{0,\Omega}/\nu^2$ is small enough, the associated variational problem (whose derivation follows the same line as for the Stokes problem) has one and only one solution in the space $V \times \{L^2(\Omega)/P_0(\Omega)\}$.

JAMET & RAVIART (1974) have extended to this problem the analysis which CROUZEIX & RAVIART (1973) developed for the Stokes problem, even adding the effect of numerical integration (which is especially needed for the computation of the integrals associated with the nonlinear term $\Sigma_{i=1}^{n} u_i\partial_i u$). Their results are related to those of FORTIN (1972a), who was the first to mathematically analyze the finite element approximation of the Navier-Stokes problem.

Further references are BERCOVIER (1976), GIRAULT (1976b), OSBORN (1976a) for the finite element approximation of the associated eigenvalue problem, TÉMAM & THOMASSET (1976), THOMASSET (1974) and especially, the extensive treatments given by TÉMAM (1973, 1977).

For a reference in the Engineering literature, see TAYLOR & HOOD (1973), GARTLING & BECKER (1976) where infinite domains are also considered.

CHAPTER 6

FINITE ELEMENT METHODS FOR THE PLATE PROBLEM

Introduction

In this chapter, we study two commonly used finite element approximations of the plate problem.

To begin with, we consider in Section 6.1 various *conforming methods*. Assuming for simplicity that the domain $\bar{\Omega}$ is polygonal, the elaboration of such methods requires the use of *straight finite elements of class* $\mathscr{C}^1$. Although such finite elements cannot be imbedded in affine families in general, we show that they form *almost-affine families*, in the sense that if the P_K-interpolation operator Π_K leaves invariant the space $P_k(K)$, there exists a constant C independent of K such that, for a regular family,

$$\forall v \in H^{k+1}(K), \quad |v - \Pi_K v|_{m,K} \leq C h_K^{k+1-m} |v|_{k+1,K},$$

for all integers $m \leq k+1$ for which $P_K \subset H^m(K)$. This is the case not only of the finite elements of class $\mathscr{C}^1$ introduced in Section 2.2, such as the *Argyris triangle*, but it is also the case of *composite finite elements* such as the *Hsieh–Clough–Tocher triangle*, or of *singular finite elements* such as the *singular Zienkiewicz triangle*.

For finite element spaces made up of such almost-affine families, we obtain (Theorem 6.1.6) error estimates of the form

$$\|u - u_h\|_{2,\Omega} \leq C \|u - \Pi_h u\|_{2,\Omega} = \mathrm{O}(h^{k-1}), \quad \text{with} \quad h = \max_{K \in \mathscr{T}_h} h_K,$$

by an application of Céa's lemma. We also show (Theorem 6.1.7) that the minimal assumptions "$u \in H^2(\Omega)$" and "$P_2(K) \subset P_K$, $K \in \mathscr{T}_h$" insure convergence, i.e., $\lim_{h \to 0} \|u - u_h\|_{2,\Omega} = 0$.

The actual implementation of conforming methods offers serious computational difficulties: Either the dimension of the "local" spaces P_K is fairly large (at least 18 for triangular polynomial elements) or the structure of the space P_K is complicated (cf. the Hsieh–Clough–Tocher triangle or the singular Zienkiewicz triangle for example). The basic source of these difficulties is of course the required continuity of the first order partial derivatives across adjacent finite elements.

It is therefore tempting to relax this continuity requirement, and this results in *nonconforming methods*: One looks for a discrete solution in a finite element space V_h which is no longer contained in the space $H^2(\Omega)$ (not even in the space $H^1(\Omega)$ in some cases). The discrete solution then satisfies $a_h(u_h, v_h) = f(v_h)$ for all $v_h \in V_h$, where

$$a_h(.\,,.) = \sum_{K \in \mathcal{T}_h} \int_K \{\cdots\}\,\mathrm{d}x,$$

the integrand $\{\cdots\}$ being the same as in the bilinear form of the original problem.

The analysis of such nonconforming methods follows exactly the same pattern as in the case of nonconforming methods for second-order problems (cf. Section 4.2). In Section 6.2, we concentrate on one example, where the generic finite element is the *Adini rectangle*. For this finite element, we show that (Theorem 6.2.3)

$$\Big(\sum_{K \in \mathcal{T}_h} |u - u_h|_{2,K}^2\Big)^{1/2} = \mathrm{O}(h),$$

if the solution u is in the space $H^3(\Omega)$.

6.1. Conforming methods

Conforming methods for fourth-order problems

In this section, we study several types of conforming finite element methods which are commonly used for approximating the solution of plate problems. For definiteness, we shall consider the *clamped plate problem*, which corresponds to the following data (cf. Section 1.2):

$$\begin{cases} V = H_0^2(\Omega), \quad \Omega \subset \mathbf{R}^2, \\ a(u, v) = \displaystyle\int_\Omega \{\Delta u \Delta v + \\ \qquad + (1-\sigma)(2\partial_{12}u\partial_{12}v - \partial_{11}u\partial_{22}v - \partial_{22}u\partial_{11}v)\}\,\mathrm{d}x, \\ f(v) = \displaystyle\int_\Omega fv\,\mathrm{d}x, \quad f \in L^2(\Omega), \end{cases} \tag{6.1.1}$$

where the constant σ (the Poisson coefficient of the material of which the plate is composed) lies in the interval $]0, \frac{1}{2}[$.

As a matter of fact, the methods which we shall describe apply equally well to *any* fourth-order boundary value problem posed over a space V such as $H_0^2(\Omega)$, $H^2(\Omega) \cap H_0^1(\Omega)$ or $H^2(\Omega)$, whose data $a(.,.)$ and $f(.)$ satisfy the assumptions of the Lax–Milgram lemma. For instance, we could likewise consider the simply supported plate (Exercise 1.2.7) or the biharmonic problem (Section 1.2).

Remark 6.1.1. By contrast, the nonconforming methods studied in the next section are specifically adapted to plate problems (cf. Remark 6.2.1). □

We shall assume that the set Ω is polygonal, so that it may be covered by triangulations composed of straight finite elements. Then in order to develop a conforming method, we face the problem of constructing subspaces of the space $H^2(\Omega)$. Since the functions found in standard finite element spaces are "locally regular" ($P_K \subset H^2(K)$ for all $K \in \mathcal{T}_h$), this construction amounts in practice to finding finite element spaces X_h which satisfy the inclusion $X_h \subset \mathscr{C}^1(\bar{\Omega})$ (Theorem 2.1.2), i.e., whose finite elements are *of class* $\mathscr{C}^1$.

We have already described three finite elements which meet this requirement, the *Argyris triangle*, the *Bell triangle* (cf. Theorem 2.2.13), and the *Bogner–Fox–Schmit rectangle* (cf. Theorem 2.2.15).

Almost-affine families of finite elements

As we pointed out in Section 2.3, Argyris triangles or Bell triangles *cannot* be imbedded in affine families in general, because normal derivatives at some nodes are used either as degrees of freedom (for the Argyris triangle) or in the definition of the space P_K (for the Bell triangle). This is in general the rule for finite elements of class $\mathscr{C}^1$, but there are exceptions. For instance, the Bogner–Fox–Schmit rectangle is a rectangular finite element of class $\mathscr{C}^1$ which *can* be imbedded in an affine family.

Nevertheless, if most finite elements of class $\mathscr{C}^1$ do not form affine families, we shall show that *their interpolation properties are quite similar to those of affine families*, and it is this similarity that motivates the following definition (compare with Theorem 3.1.6).

Consider a family of finite elements (K, P_K, Σ_K) of a given type, for which s denotes the greatest order of partial derivatives occuring in the definition of the set Σ_K. Then such a family is said to be *almost-affine* if, for any integers $k, m \geq 0$ and any numbers $p, q \in [1, \infty]$ compatible with the following inclusions:

$$W^{k+1,p}(K) \hookrightarrow \mathscr{C}^s(K), \tag{6.1.2}$$

$$W^{k+1,p}(K) \hookrightarrow W^{m,q}(K), \tag{6.1.3}$$

$$P_k(K) \subset P_K \subset W^{m,q}(K), \tag{6.1.4}$$

there exists a constant C independent of K such that

$$\forall v \in W^{k+1,p}(K), \quad \|v - \Pi_K v\|_{m,q,K} \leq \\ \leq C(\operatorname{meas}(K))^{1/q-1/p} h_K^{k+1-m} |v|_{k+1,p,K}, \tag{6.1.5}$$

where $h_K = \operatorname{diam}(K)$.

In order to simplify the exposition, we shall consider in the subsequent examples *only the highest possible value* of the integer k for which the inclusions $W^{k+1,p}(K) \hookrightarrow \mathscr{C}^s(K)$ and $P_k(K) \subset P_K$ are satisfied, but it is implicitly understood that any lower value of k compatible with these two inclusions is also admissible (a related observation was made in Remark 3.1.5).

As expected, *a regular affine family is almost-affine* (cf. Theorem 3.1.6). In particular, this is the case of *a regular family of Bogner–Fox–Schmit rectangles* (cf. Fig. 2.2.20), for which the set K is a rectangle with vertices a_i, $1 \leq i \leq 4$, $P_K = Q_3(K)$, and

$$\Sigma_K = \{p(a_i), \quad \partial_1 p(a_i), \quad \partial_2 p(a_i), \quad \partial_{12} p(a_i), \quad 1 \leq i \leq 4\}.$$

Hence, *for all* $p \in]1, \infty]$ (so as to guarantee the inclusion $W^{4,p}(K) \hookrightarrow \mathscr{C}^2(K) = \operatorname{dom} \Pi_K$) *and all pairs* (m, q) *with* $m \geq 0$ *and* $q \in [1, \infty]$ *compatible with the inclusion*

$$W^{4,p}(K) \hookrightarrow W^{m,q}(K), \tag{6.1.6}$$

there exists a constant C independent of K such that

$$\forall v \in W^{4,p}(K), \quad \|v - \Pi_K v\|_{m,q,K} \leq C(\operatorname{meas}(K))^{1/q-1/p} h_K^{4-m} |v|_{4,p,K}. \tag{6.1.7}$$

A "polynomial" finite element of class $\mathscr{C}^1$: The Argyris triangle

Let us next examine the *Argyris triangle* (the case of Bell's triangle is left as a problem; cf. Exercise 6.1.1). We recall that this finite element is a triple

(K, P_K, Σ_K) where the set K is a triangle with vertices a_i, $1 \leq i \leq 3$, and mid-points $a_{ij} = (a_i + a_j)/2$, $1 \leq i < j \leq 3$, of the sides, the space P_K is the space $P_5(K)$, and the set Σ_K (whose $P_5(K)$-unisolvence has been proved in Theorem 2.2.11) can be chosen in the form (cf. Fig. 2.2.17)

$$\Sigma_K = \{\partial^\alpha p(a_i),\quad 1 \leq i \leq 3,\quad |\alpha| \leq 2,\quad \partial_\nu p(a_{ij}),\quad 1 \leq i < j \leq 3\}.$$

Theorem 6.1.1. *A regular family of Argyris triangles is almost-affine: For all $p \in [1, \infty]$ and all pairs (m, q) with $m \geq 0$ and $q \in [1, \infty]$ compatible with the inclusion*

$$W^{6,p}(K) \hookrightarrow W^{m,q}(K), \tag{6.1.8}$$

there exists a constant C independent of K such that

$$\forall v \in W^{6,p}(K),\quad \|v - \Pi_K v\|_{m,q,K} \leq C(\text{meas}(K))^{1/q-1/p} h_K^{6-m} |v|_{6,p,K}, \tag{6.1.9}$$

where Π_K denotes the associated $P_5(K)$-interpolation operator.

Proof. The key idea is to introduce a finite element similar to the Argyris triangle, but which *can* be imbedded in an affine family, and which will play a crucial intermediary role in obtaining the interpolation error estimate. Inasmuch as it is the presence of the degrees of freedom $\partial_\nu p(a_{ij})$, $1 \leq i < j \leq 3$, which prevents the property of affine-equivalence, we are naturally led to introduce the *Hermite triangle of type* (5), whose associated data are indicated in Fig. 6.1.1. For notational convenience, we shall henceforth denote by b_i the mid-point of the side which does not contain the vertex a_i, $1 \leq i \leq 3$.

It is easily seen that the set Ξ_K is $P_5(K)$-unisolvent and that this is a finite element of class $\mathscr{C}^0$, but not of class $\mathscr{C}^1$ (cf. Exercise 2.3.5). In addition, it is clear that two arbitrary Hermite triangles of type (5) are affine-equivalent. Therefore, if we denote by Λ_K the associated $P_5(K)$-interpolation operator, for all $p \in [1, \infty]$ and all pairs (m, q) with $0 \leq m \leq 6$ and $q \in [1, \infty]$ such that $W^{6,p}(K) \hookrightarrow W^{m,q}(K)$, there exists a constant C independent of K such that, for all functions $v \in W^{6,p}(K)$,

$$|v - \Lambda_K v|_{m,q,K} \leq C(\text{meas}(K))^{1/q-1/p} h_K^{6-m} |v|_{6,p,K}. \tag{6.1.10}$$

It therefore remains to evaluate the semi-norms $|\Pi_K v - \Lambda_K v|_{m,q,K}$. For a given function $v \in W^{6,p}(K)$, the difference

$$\Delta = \Pi_K v - \Lambda_K v \tag{6.1.11}$$

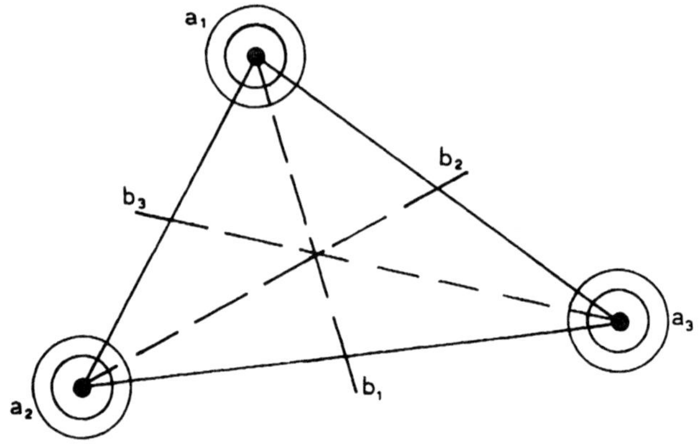

Hermite triangle of type (5)
$P_K = P_5(K)$; dim $P_K = 21$; $\Xi_K = \{\partial^\alpha p(a_i),\ \ 1 \leqslant i \leqslant 3,\ \ |\alpha| \leqslant 2;\ Dp(b_i)(a_i - b_i),\ \ 1 \leqslant i \leqslant 3\}$

Fig. 6.1.1

is a polynomial of degree $\leqslant 5$ which satisfies

$$\partial^\alpha \Delta(a_i) = 0, \quad |\alpha| \leqslant 2, \quad 1 \leqslant i \leqslant 3, \tag{6.1.12}$$

since $\partial^\alpha \Pi_K v(a_i) = \partial^\alpha \Lambda_K v(a_i) = \partial^\alpha v(a_i), \quad |\alpha| \leqslant 2, \quad 1 \leqslant i \leqslant 3,$ and

$$\partial_\nu \Delta(b_i) = \partial_\nu (v - \Lambda_K v)(b_i), \quad 1 \leqslant i \leqslant 3, \tag{6.1.13}$$

since $\partial_\nu \Pi_K v(b_i) = \partial_\nu v(b_i)$, $1 \leqslant i \leqslant 3$. For $1 \leqslant i \leqslant 3$, let ν_i and τ_i be the unit outer normal and tangential vectors along the side opposite to the vertex a_i, as indicated in Fig. 6.1.2.

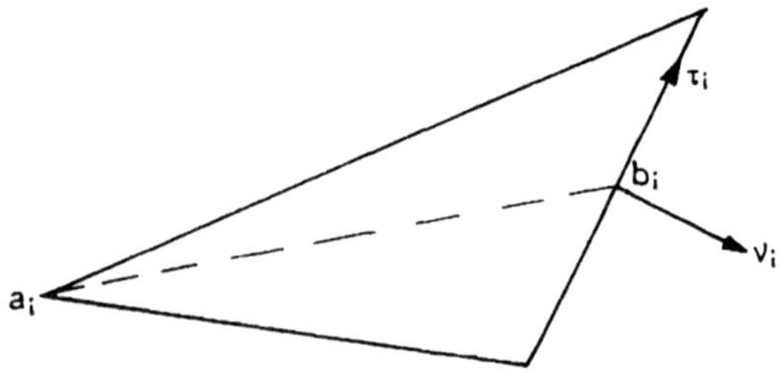

Fig. 6.1.2.

Denoting by $\cdot$ the Euclidean inner product in $\mathbf{R}^2$, we can write, for $1 \leq i \leq 3$,

$$D\Delta(b_i)(a_i - b_i) = \partial_\nu \Delta(b_i)\{(a_i - b_i) \cdot \nu_i\}, \tag{6.1.14}$$

since on the one hand $D\Delta(b_i)\nu_i = \partial_\nu \Delta(b_i)$, and since on the other $D\Delta(b_i)\tau_i = 0$ as a consequence of relations (6.1.12), which imply that the difference Δ vanishes along each side of the triangle. Combining relations (6.1.13) and (6.1.14), we obtain

$$D\Delta(b_i)(a_i - b_i) = \partial_\nu(v - \Lambda_K v)(b_i)\{(a_i - b_i) \cdot \nu_i\}, \quad 1 \leq i \leq 3. \tag{6.1.15}$$

Let q_i denote the basis functions of the Hermite triangle of type (5) which are associated with the degrees of freedom $Dp(b_i)(a_i - b_i)$, $1 \leq i \leq 3$. Then using relations (6.1.12) and (6.1.15), we can write

$$\begin{aligned} \Delta = \Pi_K v - \Lambda_K v &= \sum_{i=1}^{3} \{D\Delta(b_i)(a_i - b_i)\} q_i \\ &= \sum_{i=1}^{3} \partial_\nu(v - \Lambda_K v)(b_i)\{(a_i - b_i) \cdot \nu_i\} q_i. \end{aligned} \tag{6.1.16}$$

Applying Theorem 3.1.5 with $m = 1$, $q = \infty$ and $k = 5$, we obtain

$$\begin{aligned} |\partial_\nu(v - \Lambda_K v)(b_i)| &\leq \sqrt{2}|v - \Lambda_K v|_{1,\infty,K} \\ &\leq C(\text{meas}(K))^{-1/p} \frac{h_K^6}{\rho_K} |v|_{6,p,K}, \quad 1 \leq i \leq 3. \end{aligned} \tag{6.1.17}$$

Next, it is clear that

$$|(a_i - b_i) \cdot \nu_i| \leq h_K, \quad 1 \leq i \leq 3. \tag{6.1.18}$$

Finally, let $\hat{q}_i$ be the basis functions of a reference Hermite triangle of type (5) associated in the usual correspondence with the basis functions q_i. From Theorems 3.1.2 and 3.1.3, we infer that

$$|q_i|_{m,q,K} \leq C \frac{(\text{meas}(K))^{1/q}}{\rho_K^m} |\hat{q}_i|_{m,q,\hat{K}}. \tag{6.1.19}$$

Relations (6.1.16), (6.1.17), (6.1.18) and (6.1.19) then imply that

$$\begin{aligned} |\Pi_K v - \Lambda_K v|_{m,q,K} &\leq C(\text{meas}(K))^{1/q - 1/p} \frac{h_K^7}{\rho_K^{m+1}} |v|_{6,p,K} \\ &\leq C(\text{meas}(K))^{1/q-1/p} h_K^{6-m} |v|_{6,p,K}, \end{aligned} \tag{6.1.20}$$

since we are considering a regular family. Inequality (6.1.9) is therefore a consequence of inequalities (6.1.10) and (6.1.20). □

A composite finite element of class $\mathscr{C}^1$: The Hsieh–Clough–Tocher triangle

In our next examples, we shall for the first time leave the realm of "purely polynomial" finite elements.

As we already pointed out (cf. Bibliography and Comments of Sections 2.2 and 2.3), Bell's triangle is optimal among triangular polynomial finite elements of class $\mathscr{C}^1$ in the sense that for such finite elements, one has necessarily $\dim P_K \geq 18$, as a consequence of Ženíšek's result. Therefore, *a smaller dimension of the space P_K for triangular finite elements of class $\mathscr{C}^1$ requires that functions other than polynomials be used.*

For example, one can use piecewise polynomials *inside* the set K, a process which results in so-called *composite finite elements*, also named *macroelements*. Or one can add some judiciously selected rational functions to a space of polynomials, a process which results in so-called *singular finite elements* (singular in the sense that some functions in the space P_K or some of their derivatives become infinite and/or are not defined at some points of K). We shall describe and study one example of each type. Other examples of composite and singular finite elements are suggested as problems (cf. Exercises 6.1.3, 6.1.4, 6.1.5, 6.1.6 and 6.1.7).

The *Hsieh–Clough–Tocher triangle*, sometimes abbreviated as the *HCT triangle*, is defined as follows: The set K is a triangle subdivided into three triangles K_i with vertices a, a_{i+1}, a_{i+2}, $1 \leq i \leq 3$ (Fig. 6.1.3), the point a being in the interior of the set K (here and subsequently, the indices are counted modulo 3 when necessary). The space P_K and the set Σ_K are indicated in Fig. 6.1.3. For convenience, we again denote by b_i, $1 \leq i \leq 3$, the mid-point of the side which does not contain the vertex a_i. Our first task is as usual to prove the P_K-unisolvence of the set Σ_K. Since $\dim P_3(K_i) = 10$, it is necessary to find 30 equations to define the three polynomials $P_{|K_i}$, $1 \leq i \leq 3$. First, it is easily seen that the data of the degrees of freedom of the set Σ_K amounts to the data of 21 equations. To see that the condition "$p \in \mathscr{C}^1(K)$" yields 9 additional equations, it suffices to write the continuity of the functions p, $\partial_1 p$ and $\partial_2 p$ at the point a (6 equations) and the continuity of the normal derivatives across the mid-points of the sides $[a, a_i]$ (3 equations).

It therefore remains to show that the 30×30 matrix of the corresponding linear system is invertible, and this is the object of the next theorem (another proof is suggested in Exercise 6.1.2).

Theorem 6.1.2. *With the definitions of Fig.* 6.1.3, *the set* Σ_K *is* P_K-*unisolvent.*

The resulting Hsieh–Clough–Tocher triangle is a finite element of class $\mathscr{C}^1$.

Proof. It suffices to show that a function p in the space P_K vanishes if

$$p(a_i) = \partial_1 p(a_i) = \partial_2 p(a_i) = \partial_\nu p(b_i) = 0, \quad 1 \leq i \leq 3. \qquad (6.1.21)$$

For $1 \leq i \leq 3$, let μ_i denote the unique function which satisfies

$$\mu_i \in P_1(K_i), \quad \mu_i(a) = 1, \quad \mu_i(a_{i+1}) = \mu_i(a_{i+2}) = 0,$$

so that the function $\mu: K \to \mathbf{R}$ defined by

$$\mu_{|K_i} = \mu_i, \quad 1 \leq i \leq 3,$$

is continuous. Since over each triangle K_i, the function $p_{|K_i}$ is a polynomial of degree ≤ 3, assumptions (6.1.21) imply that there exist func-

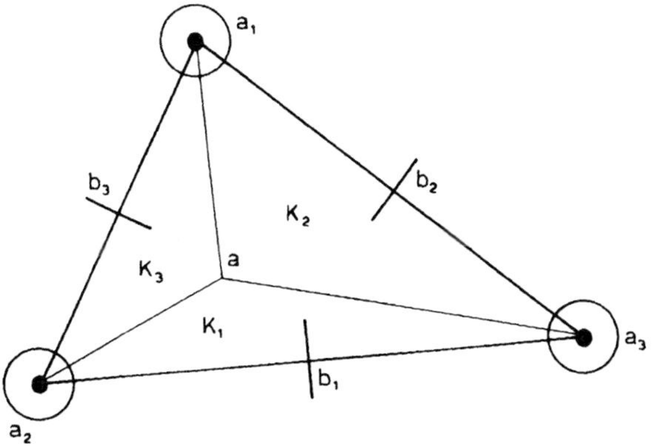

Hsieh–Clough–Tocher triangle
$P_K = \{p \in \mathscr{C}^1(K);\ p_{\mid K_i} \in P_3(K_i),\ 1 \leq i \leq 3\}$; $\dim P_K = 12$; $\Sigma_K = \{p(a_i),\ \partial_1 p(a_i),\ \partial_2 p(a_i),\ \partial_\nu p(b_i),\ 1 \leq i \leq 3\}$.

Fig. 6.1.3

tions ν_i such that

$$\nu_i \in P_1(K_i), \quad p_{|K_i} = \nu_i \mu_i^2, \quad 1 \leq i \leq 3.$$

Since the functions $p: K \to \mathbf{R}$ and $\mu: K \to \mathbf{R}$ are continuous, the function $\nu: K \to \mathbf{R}$ defined by

$$\nu_{|K_i} = \nu_i, \quad 1 \leq i \leq 3,$$

is also continuous (the function μ does not vanish in the interior of K).

On each segment $[a, a_{i+2}]$, the gradient ∇p is well-defined since the function p is continuously differentiable, and it is given by either expressions

$$\nabla p\,|_{[a,a_{i+2}]} = \begin{cases} (2\nu_i\mu_i \nabla \mu_i + \mu_i^2 \nabla \nu_i)|_{[a,a_{i+2}]}, \\ (2\nu_{i+1}\mu_{i+1} \nabla \mu_{i+1} + \mu_{i+1}^2 \nabla \nu_{i+1})|_{[a,a_{i+2}]}, \end{cases}$$

so that we deduce ($\mu \neq 0$ in $\mathring{K}$)

$$2\nu\nabla(\mu_{i+1} - \mu_i) + \mu\nabla(\nu_{i+1} - \nu_i) = 0 \quad \text{along} \quad [a, a_{i+2}].$$

Since $\mu(a_{i+2}) = 0$ and $\nabla(\mu_{i+1} - \mu_i) \neq 0$ (otherwise the lines $\mu_i = 0$ and $\mu_{i+1} = 0$ would be parallel), we conclude that $\nu(a_{i+2}) = \nu_i(a_{i+2}) = 0$. A similar argument would show that $\nu_i(a_{i+1}) = 0$. Consequently each function $\nu_i \in P_1(K_i)$ is of the form

$$\nu_i = C_i\mu_i \quad \text{with} \quad C_i = \text{constant}.$$

The function ν being continuous, we have

$$\nu(a) = \nu_i(a) = C_i, \quad 1 \leq i \leq 3.$$

Denoting by C the common value of the constants C_i, we conclude that

$$\nu_{|K_i} = \nu_i = C\mu_i, \quad 1 \leq i \leq 3,$$

and therefore that

$$p_{|K_i} = C\mu_i^3, \quad \text{whence} \quad \nabla p_{|K_i} = 3C\mu^2\nabla\mu_i, \quad 1 \leq i \leq 3.$$

Then the constant C is necessarily zero for otherwise the function p would not be continuously differentiable along the segment $[a, a_{i+2}]$ since $\nabla\mu_i \neq \nabla\mu_{i+1}$.

That the Hsieh–Clough–Tocher triangle is of class $\mathscr{C}^1$ follows by an argument analogous to the proof of Theorem 2.2.13. □

Remark 6.1.2. The normal derivatives at the mid-point of the sides can be eliminated by requiring that the normal derivative vary linearly along the sides. This elimination results in a finite element of class $\mathscr{C}^1$ for which dim $P_K = 9$ (cf. Exercise 6.1.3). □

There are two reasons that prevent the Hsieh–Clough–Tocher triangle from being imbedded in an affine family. As for the Argyris triangle, one reason is the presence of the normal derivatives $\partial_\nu p(b_i)$ as degrees of freedom. The additional reason is that the point a may be allowed to vary inside the set K. This is why we must adapt to this element the notion of a regular family:

We shall say that *a family of Hsieh–Clough–Tocher triangles K is regular* if the following three conditions are satisfied:

(i) There exists a constant σ such that

$$\forall K, \quad \frac{h_K}{\rho_K} \leq \sigma.$$

(ii) The quantities h_K approach zero.

(iii) Let $\hat{K}$ be any fixed triangle with vertices $\hat{a}_i$, $1 \leq i \leq 3$. For each Hsieh–Clough–Tocher triangle K with vertices $a_{i,K}$, $1 \leq i \leq 3$, let F_K denote the unique affine mapping which satisfies $F_K(\hat{a}_i) = a_{i,K}$, $1 \leq i \leq 3$.

Then (Fig. 6.1.4) *the points $\hat{a}_K = F_K^{-1}(a_K)$ all belong to some compact subset $\hat{B}$ of the interior of the triangle $\hat{K}$* (clearly, the compact subset $\hat{B}$

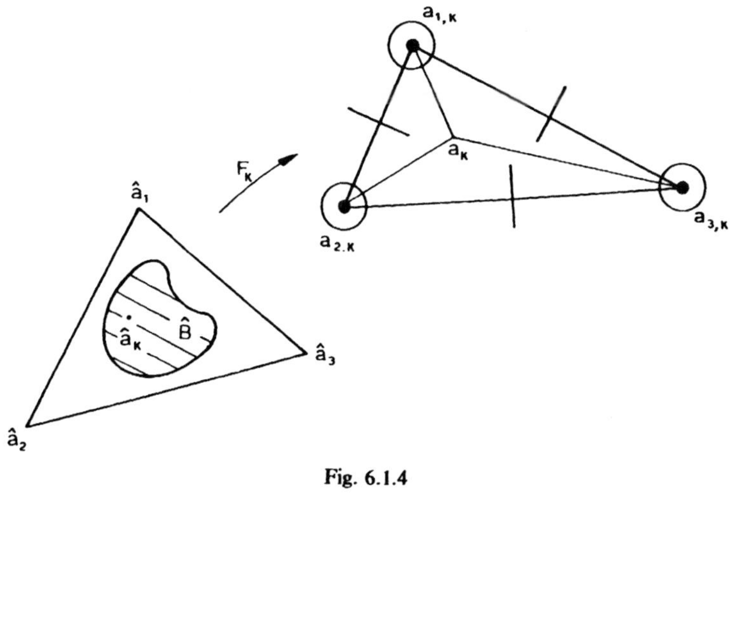

Fig. 6.1.4

may vary from one regular family to another). Notice that the $\hat{K}$ is here simply understood as being a triangle, not a finite element.

Remark 6.1.3. Conditions (i) and (ii) are the familiar ones for a regular family of finite elements. Condition (iii) expresses precisely in which sense the points a_K may vary inside the triangle K, so as to guarantee that the family under consideration is almost-affine, as we next show. □

Theorem 6.1.3. *A regular family of Hsieh–Clough–Tocher triangles is almost affine: For all $p \in [1, \infty]$ and all pairs (m, q) with $m \geq 0$ and $q \in [1, \infty]$ such that*

$$\begin{cases} W^{4,p}(K) \hookrightarrow W^{m,q}(K), \\ P_K \subset W^{m,q}(K), \end{cases} \tag{6.1.22}$$

there exists a constant C independent of K such that

$$\forall v \in W^{4,p}(K), \quad \|v - \Pi_K v\|_{m,q,K} \leq C(\text{meas}(K))^{1/q - 1/p} h_K^{4-m} |v|_{4,p,K}. \tag{6.1.23}$$

Proof. We first observe that the inclusion $W^{4,p}(K) \hookrightarrow \mathscr{C}^1(K) = \text{dom}\, \Pi_K$ holds for all $p \geq 1$. The proof of the theorem consists of three steps.

(i) As expected, we shall introduce a finite element which is similar to the Hsieh–Clough–Tocher triangle but which *can* be imbedded in an affine family. This will be achieved through the replacement of the normal derivatives by appropriate directional derivatives and through a restriction on the position of the points a_K. More precisely, with each Hsieh–Clough–Tocher triangle (K, P_K, Σ_K), we associate the finite element (K, P_K, Ξ_K), where

$$\begin{aligned} \Xi_K = \{&p(a_i), \quad Dp(a_i)(a_{i+1} - a_i), \quad Dp(a_i)(a_{i-1} - a_i), \\ &Dp(b_i)(a - b_i), \quad 1 \leq i \leq 3\} \end{aligned} \tag{6.1.24}$$

(the proof of the P_K-unisolvence of the set Ξ_K is similar to the proof of the P_K-unisolvence of the set Σ_K as given in Theorem 6.1.2), and we denote by Λ_K the P_K-interpolation operator associated with each finite element (K, P_K, Ξ_K).

For each point $\hat{a} \in \hat{B}$, let $\mathscr{K}(\hat{a})$ denote the (possibly empty) subfamily of Hsieh–Clough–Tocher triangles for which $a_K = F_K(\hat{a})$. Then, for each $\hat{a} \in \hat{B}$, the subfamily (K, P_K, Ξ_K), $K \in \mathscr{K}(\hat{a})$ is affine, and consequently, the inclusion

$$P_3(K) \subset P_K \tag{6.1.25}$$

implies that there exists a constant $C(\hat{a}, \hat{K})$ such that

$$\begin{aligned} &\forall v \in W^{4,p}(K), \quad \forall K \in \mathcal{K}(\hat{a}), \\ &|v - \Lambda_K v|_{m,q,K} \leq C(\hat{a}, \hat{K})(\text{meas}(K))^{1/q-1/p} h_K^{4-m} |v|_{4,p,K}, \end{aligned} \tag{6.1.26}$$

for all pairs (m, q) compatible with the inclusions (6.1.22).

(ii) We next show that, when the points $\hat{a}$ vary in the compact set $\hat{B}$, the constants $C(\hat{a}, \hat{K})$ which appear in the last inequality are bounded. To prove this, we recall that in the proof of Theorem 3.1.4, we found that these constants are of the form (cf. (3.1.33)):

$$C(\hat{a}, \hat{K}) = C(\hat{K}) \| I - \hat{\Lambda}(\hat{a}) \|_{\mathcal{L}(W^{4,p}(\hat{K});\, W^{m,q}(\hat{K}))}, \tag{6.1.27}$$

where, for each $\hat{a} \in \hat{B}$, $\hat{\Lambda}(\hat{a})$ denotes the $P_{\hat{K}}$-interpolation operator associated with the corresponding reference finite element $(\hat{K}, \hat{P}(\hat{a}), \hat{\Xi}(\hat{a}))$.

With self-explanatory notations, we have, for all functions $\hat{v} \in W^{4,p}(\hat{K})$,

$$\begin{aligned} \hat{\Lambda}(\hat{a})\hat{v} = &\sum_{i=1}^{3} \hat{v}(\hat{a}_i)\hat{p}_i(\hat{a}, .) \\ &+ \sum_{\substack{i=1 \\ |j-i|=1}}^{3} \{D\hat{v}(\hat{a}_i)(\hat{a}_j - \hat{a}_i)\}\hat{q}_{ij}(\hat{a}, .) \\ &+ \sum_{i=1}^{3} \{D\hat{v}(\hat{b}_i)(\hat{a} - \hat{b}_i)\}\hat{r}_i(\hat{a}, .), \end{aligned} \tag{6.1.28}$$

and

$$\begin{aligned} &|\hat{v}(\hat{a}_i)| \leq |\hat{v}|_{0,\infty,\hat{K}} \leq C(\hat{K})\|\hat{v}\|_{4,p,\hat{K}}, \\ &\left.\begin{aligned} &|\{D\hat{v}(\hat{a}_i)(\hat{a}_j - \hat{a}_i)\}| \\ &|\{D\hat{v}(\hat{b}_i)(\hat{a} - \hat{b}_i)\}| \end{aligned}\right\} \leq \sqrt{2}\, \text{diam}(\hat{K}) |\hat{v}|_{1,\infty,\hat{K}} \leq C(\hat{K})\|\hat{v}\|_{4,p,\hat{K}}, \end{aligned} \tag{6.1.29}$$

where the constants $C(\hat{K})$ are independent of $\hat{a}$.

Let us then consider the norm $\|.\|_{m,q,K}$ of any one of the basis functions $\hat{p}_i(\hat{a}, .)$, $\hat{q}_{ij}(\hat{a}, .)$ and $\hat{r}_i(\hat{a}, .)$. On each of the triangles $\hat{K}_i(\hat{a})$, $1 \leq i \leq 3$, which subdivide the triangle $\hat{K}$, the restriction of any one of these basis functions is a polynomial of degree ≤ 3, whose coefficients are obtained through the solution of a linear system with an invertible matrix (the set $\hat{\Xi}(\hat{a})$ is $\hat{P}(\hat{a})$-unisolvent as long as the point $\hat{a}$ belongs to the interior of the set $\hat{K}$). This matrix depends continuously on the point $\hat{a}$ since its coefficients are polynomial functions of the coordinates of the point $\hat{a}$. Consequently, each coefficient is in turn a continuous function of the point $\hat{a}$ and there exists a constant $\hat{C}$ such that

$$\sup_{\hat{a}\in\hat{B}} \{\|\hat{p}_i(\hat{a},.)\|_{m,q,\hat{K}},\ \|\hat{q}_{ij}(\hat{a},.)\|_{m,q,\hat{K}},\ \|\hat{r}_i(\hat{a},.)\|_{m,q,\hat{K}}\} \leq \hat{C}, \quad (6.1.30)$$

since the set $\hat{B}$ is compact. Then it follows from relations (6.1.27) to (6.1.30) that

$$\sup_{\hat{a}\in\hat{B}} C(\hat{a}, \hat{K}) = C(\hat{B}, \hat{K}) < \infty.$$

Combining this result with inequality (6.1.26), we obtain

$$\forall v \in W^{4,p}(K),\quad \forall K,$$
$$|v - \Lambda_K v|_{m,q,K} \leq C(\hat{B}, \hat{K})(\mathrm{meas}(K))^{1/q-1/p} h_K^{4-m} |v|_{4,p,K}. \quad (6.1.31)$$

(iii) By an argument similar to that used in the proof of Theorem 6.1.1 (cf. (6.1.16)), we find that

$$\Pi_K v - \Lambda_K v = \sum_{i=1}^{3} \partial_\nu(v - \Lambda_K v)(b_i)\{(a - b_i)\cdot \nu_i\} r_i, \quad (6.1.32)$$

where the functions r_i, $1 \leq i \leq 3$, are the basis functions associated with the degrees of freedom $\{Dp(b_i)(a - b_i)\}$ in the finite element (K, P_K, Ξ_K). Applying Theorem 3.1.5 with $m = 1$, $q = \infty$ and $k = 3$, we find that

$$|\partial_\nu(v - \Lambda_K v)(b_i)| \leq \sqrt{2}|v - \Lambda_K v|_{1,\infty,K} \leq C(\mathrm{meas}(K))^{-1/p} \frac{h_K^4}{\rho_K} |v|_{4,p,K}. \quad (6.1.33)$$

Next we have

$$|\{(a - b_i)\cdot \nu_i\}| \leq h_K, \quad (6.1.34)$$

$$|r_i|_{m,q,K} \leq C \frac{(\mathrm{meas}(K))^{1/q}}{\rho_K^m} |\hat{r}_i(\hat{a},.)|_{m,q,\hat{K}}, \quad (6.1.35)$$

and we deduce from relations (6.1.30) and (6.1.32) to (6.1.35) that

$$|\Pi_K v - \Lambda_K v|_{m,q,K} \leq C(\mathrm{meas}(K))^{1/q-1/p} \frac{h_K^5}{\rho_K^{m+1}} |v|_{4,p,K}$$
$$\leq C(\mathrm{meas}(K))^{1/q-1/p} h_K^{4-m} |v|_{4,p,K}. \quad (6.1.36)$$

Then the proof is completed by combining the above inequality with inequality (6.1.31). □

Remark 6.1.4. When $q = 2$, it is easily seen that the highest admissible value for the integer m compatible with the inclusion $P_K \subset H^m(K)$ is $m = 2$ (that the integer m is at least 2 follows from an application of Theorem 2.1.2 to the partitionned triangle $K = \bigcup_{i=1}^{3} K_i$; to prove that the integer m cannot exceed 2 requires an argument which shall be used later, cf. Theorem 6.2.1). Notice that this is the first instance of a restriction on the possible inclusions $P_K \subset W^{m,q}(K)$. The next finite element under study will be another instance. Fortunately, the inclusion $P_K \subset H^2(K)$ is precisely that which is needed to insure convergence, as we shall show at the end of this section. □

A singular finite element of class $\mathscr{C}^1$: The singular Zienkiewicz triangle

Let us next turn to an example of a triangular finite element, which is of class $\mathscr{C}^1$ as a result of the addition of appropriate rational functions to a familiar space of polynomials.

The *singular Zienkiewicz triangle* is defined as follows (Fig. 6.1.5): The set K is a triangle with vertices a_i, $1 \leqslant i \leqslant 3$, the space P_K is the space $P_3''(K)$ of the Zienkiewicz triangle (cf. (2.2.39)) to which are added three

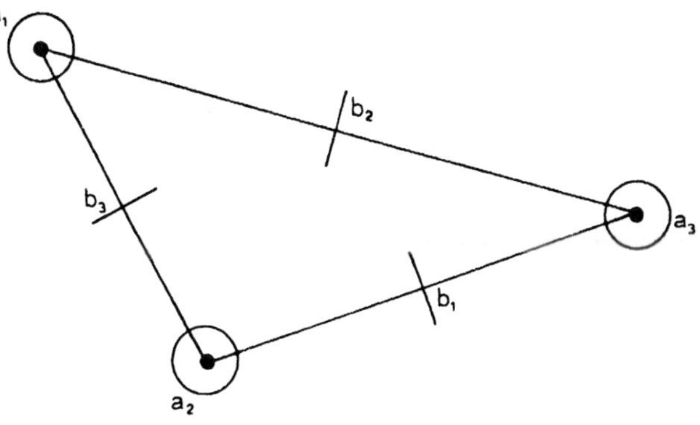

Singular Zienkiewicz triangle

$$P_K = P_3''(K) \oplus \bigvee_{i=1}^{3} \left\{ \frac{4\lambda_i \lambda_{i+1}^2 \lambda_{i+2}^2}{(\lambda_i + \lambda_{i+1})(\lambda_i + \lambda_{i+2})} \right\} \text{ (cf. (2.2.39)); } \dim P_K = 12;$$

$$\Sigma_K = \{p(a_i),\ \partial_1 p(a_i),\ \partial_2 p(a_i),\ \partial_\nu p(b_i),\quad 1 \leqslant i \leqslant 3\}.$$

Fig. 6.1.5

functions $q_i: K \to \mathbf{R}$, $1 \leq i \leq 3$, defined by

$$\begin{cases} q_i = \dfrac{4\lambda_i\lambda_{i+1}^2\lambda_{i+2}^2}{(\lambda_i + \lambda_{i+1})(\lambda_i + \lambda_{i+2})} & \text{for } 0 \leq \lambda_i \leq 1,\ 0 \leq \lambda_{i+1}, \lambda_{i+2} < 1, \\ q_i(a_{i+1}) = q_i(a_{i+2}) = 0, \end{cases} \tag{6.1.37}$$

where the functions λ_i, $1 \leq i \leq 3$, denote the barycentric coordinates in the triangle K (notice that the function given in the first line of definition (6.1.37) is not defined for $\lambda_i + \lambda_{i+1} = 0$ or $\lambda_i + \lambda_{i+2} = 0$, i.e., for $\lambda_{i+2} = 1$ or $\lambda_{i+1} = 1$; this is why we have to assign values to the function q_i at the vertices a_{i+1} and a_{i+2}). Finally the set Σ_K is the same as for the Hsieh–Clough–Tocher triangle.

As usual, we begin by examining the question of unisolvence. Observe that this finite element is an instance where the validity of the inclusions $P_K \subset \mathscr{C}^1(K)$ (which is part of the definition of elements of class $\mathscr{C}^1$) and $P_K \subset H^2(K)$ requires a proof.

Theorem 6.1.4. *With the definitions of Fig.* 6.1.5, *the set Σ_K is P_K-unisolvent.*

The resulting singular Zienkiewicz triangle is a finite element of class $\mathscr{C}^1$, and the inclusion $P_K \subset H^2(K)$ holds.

Proof. (i) To begin with, let us verify the inclusions $P_K \subset \mathscr{C}^1(K)$ and $P_K \subset H^2(K)$. Since such properties are invariant through affine transformations, we may consider the case where the set K is the unit triangle $\hat{K}$ with vertices $\hat{a}_1 = (1, 0)$, $\hat{a}_2 = (0, 1)$, and $\hat{a}_3 = (0, 0)$. Then it suffices to study the behavior of the function $\hat{q}_1: \hat{K} \to \mathbf{R}$ in a neighborhood of the origin in $\hat{K}$. We have

$$\begin{cases} \hat{q}_1(x) = \dfrac{x_2 x_2^2}{(x_1 + x_2)} f(x) & \text{for } x \neq 0, \\ \hat{q}_1(0) = 0, \end{cases} \tag{6.1.38}$$

where the function $f(x) = 4(1 - x_1 - x_2)^2/(1 - x_2)$ and its derivatives have no singularity at the origin. Since $\lim_{x_1, x_2 \to 0^+} x_1 x_2^2/(x_1 + x_2) = 0$, we deduce that $\lim_{x_1, x_2 \to 0^+} \hat{q}_1(x) = 0$. Therefore the function $\hat{q}_1$ is continuous at the origin. For $x_1, x_2 \geq 0$ and $x \neq 0$, we have:

$$\begin{cases} \partial_1 \hat{q}_1(x) = \dfrac{x_2^3}{(x_1 + x_2)^2} f(x) + \dfrac{x_1 x_2^2}{x_1 + x_2} \partial_1 f(x), \\ \partial_2 \hat{q}_1(x) = \dfrac{x_1 x_2 (2x_1 + x_2)}{(x_1 + x_2)^2} f(x) + \dfrac{x_1 x_2^2}{x_1 + x_2} \partial_2 f(x), \end{cases} \tag{6.1.39}$$

and thus we conclude that

$$\lim_{x_1,x_2\to 0^+} \partial_j\hat{q}_1(x) = 0 = \partial_j\hat{q}_1(0), \quad j = 1, 2, \tag{6.1.40}$$

which proves that the function $\hat{q}_1$ is continuously differentiable at the origin.

Arguing analogously with the vertex $\hat{a}_2$, and next with the functions $\hat{q}_2$ and $\hat{q}_3$, we conclude that the inclusion

$$P_K \subset \mathscr{C}^1(K)$$

holds. This inclusion implies the inclusion $P_K \subset H^1(K)$ and thus, to obtain the inclusion $P_K \subset H^2(K)$, it remains to show that the second partial derivatives of the function $\hat{q}_1$ are square integrable around the origin. For $x \neq 0$, we find

$$\begin{cases} \partial_{11}\hat{q}_1(x) = -\dfrac{2x_2^3}{(x_1+x_2)^3} f(x) + g_{11}(x), \\ \partial_{12}\hat{q}_1(x) = \dfrac{x_2^2(3x_1+x_2)}{(x_1+x_2)^3} f(x) + g_{12}(x), \\ \partial_{22}\hat{q}_1(x) = \dfrac{2x_1^3}{(x_1+x_2)^3} f(x) + g_{22}(x), \end{cases} \tag{6.1.41}$$

where the functions g_{11}, g_{12} and g_{22} are continuous around the origin. Since the three functions factoring the function $f(x)$ are bounded on the set $\hat{K}$, the inclusion

$$P_K \subset H^2(K)$$

follows.

(ii) The inclusion $P_K \subset \mathscr{C}^1(K)$ proved in (i) guarantees that the degrees of freedom of the set Σ_K are well-defined for the functions in the space P_K. The P_K-unisolvence of the set Σ_K will be an easy consequence of the P_K-unisolvence of the set

$$\Xi_K = \{p(a_i),\quad Dp(a_i)(a_j - a_i),\quad Dp(b_i)(a_i - b_i),\\ 1 \leq i, j \leq 3,\quad |j - i| = 1\}, \tag{6.1.42}$$

which we proceed to show.

Let us denote by p_i, $1 \leq i \leq 3$, and p_{ij}, $1 \leq i, j \leq 3$, $|j - i| = 1$, the basis functions of the space $P_3''(K)$ as given in (2.2.39). By definition, they satisfy

$$p_i(a_k) = \delta_{ik}, \quad Dp_i(a_k)(a_l - a_k) = 0, \tag{6.1.43}$$

for $1 \leq i, k, l \leq 3$, $|k - l| = 1$, and

$$p_{ij}(a_k) = 0, \quad Dp_{ij}(a_k)(a_l - a_k) = \delta_{ik}\delta_{jl}, \tag{6.1.44}$$

for $1 \leq i, j, k, l \leq 3$, $|j-i| = |k-l| = 1$. We next show that they satisfy

$$Dp_i(b_k)(a_k - b_k) = -\tfrac{1}{4} + \tfrac{3}{4}\delta_{ik}, \quad 1 \leq i, k \leq 3, \tag{6.1.45}$$

$$Dp_{ij}(b_k)(a_k - b_k) = -\tfrac{1}{4} + \tfrac{3}{8}\delta_{ik} + \tfrac{5}{8}\delta_{jk}, \quad 1 \leq i, j, k \leq 3, \quad |j-i| = 1. \tag{6.1.46}$$

For the purpose of proving these relations, it is convenient to compute the directional derivatives $Dp(b_i)(a_i - b_i)$ for a function $p: K \to \mathbf{R}$ expressed in terms of barycentric coordinates (the computation below is not restricted to $n = 2$). Let then $p(x_1, x_2) = q(\lambda_1, \lambda_2, \lambda_3)$ be such a function. Denoting as usual by $B = (b_{ij})$ the inverse matrix of the matrix A of (2.2.4), we find that

$$\partial_j p = \sum_{k=1}^{3} \partial_k q \partial_j \lambda_k = \sum_{k=1}^{3} b_{kj} \partial_k q, \quad j = 1, 2.$$

Let us compute for example the quantity

$$Dp(b_1)(a_1 - b_1) = \sum_{j=1}^{2} \sum_{k=1}^{3} b_{kj} \partial_k q\left(0, \frac{1}{2}, \frac{1}{2}\right)\left(a_{j1} - \frac{a_{j2} + a_{j3}}{2}\right),$$

where a_{ji}, $j = 1, 2$, denote the coordinates of the vertex a_i. By definition of the matrices B and A,

$$\sum_{j=1}^{2} b_{kj} a_{jl} = \delta_{kl} - b_{k3}, \quad 1 \leq k, l \leq 3,$$

so that

$$Dp(b_1)(a_1 - b_1) = \partial_1 q(0, \tfrac{1}{2}, \tfrac{1}{2}) - \tfrac{1}{2}(\partial_2 q(0, \tfrac{1}{2}, \tfrac{1}{2}) + \partial_3 q(0, \tfrac{1}{2}, \tfrac{1}{2})). \tag{6.1.47}$$

Then relations (6.1.45) and (6.1.46) follow from the above result (and analogous computations for $Dp(b_i)(a_i - b_i)$, $i = 2, 3$) and the following expressions of the basis functions p_i and p_{ij} (which are easily derived from relations (2.2.37) and (2.2.38)):

$$p_i = -2\lambda_i^3 + 3\lambda_i^2 + 2\lambda_1\lambda_2\lambda_3, \tag{6.1.48}$$

$$p_{ij} = \frac{\lambda_i \lambda_j}{2}(\lambda_i - \lambda_j + 1). \tag{6.1.49}$$

On the other hand, the functions q_i as defined in (6.1.37) satisfy

$$\begin{cases} q_i(a_k) = 0, \quad 1 \leq i, k \leq 3, \\ Dq_i(a_k)(a_l - a_k) = 0, \quad 1 \leq i, k, l \leq 3, \quad |k-l| = 1, \\ Dq_i(b_k)(a_k - b_k) = \delta_{ik}, \quad 1 \leq i, k \leq 3. \end{cases} \tag{6.1.50}$$

The second equalities have been obtained in (6.1.40). The last ones are obtained through another application of relations of the form (6.1.47).

Then it follows from relations (6.1.43) to (6.1.50) that the functions (which all belong to the space P_K):

$$\begin{cases} \left\{p_i + \frac{1}{4}\left(-2q_i + \sum_{|j-i|=1} q_j\right)\right\}, & 1 \leqslant i \leqslant 3, \\ \left\{p_{ij} - \frac{1}{8}(q_i + 3q_j - 2q_l)\right\}, & 1 \leqslant i, j, l \leqslant 3, \quad \{i, j, l\} = \{1, 2, 3\}, \\ q_i, \quad 1 \leqslant i \leqslant 3, \end{cases} \tag{6.1.51}$$

form a basis of the space P_K, corresponding to the degrees of freedom of the set Ξ_K of (6.1.42). Thus this set is a P_K-unisolvent set.

It remains to prove that the set Σ_K is also P_K-unisolvent. To prove this, we make the following observation: Along each side K' of the triangle K, the restrictions $p_{|K'}$, $p \in P_K$, are polynomials of degree $\leqslant 3$ in one variable, while the restrictions $Dp(.)\xi_{|K'}$, $p \in P_K$, of any directional derivative are polynomials of degree $\leqslant 2$ in one variable. This is clearly true for the functions in the space $P''_3(K)$, and it is a straightforward consequence of the definition for the functions q_i. Notice in particular that this property implies that the finite element is of class $\mathscr{C}^1$.

Let then $p \in P_K$ be a function which satisfies

$$p(a_i) = \partial_1 p(a_i) = \partial_2 p(a_i) = \partial_\nu p(b_i) = 0, \quad 1 \leqslant i \leqslant 3.$$

The conjunction of these relations and of the above property implies that the normal derivative and the tangential derivative vanish along any side of the triangle K. Consequently, the directional derivatives $Dp(b_i)(a_i - b_i)$, $1 \leqslant i \leqslant 3$, vanish, and therefore the function p is identically zero since the set Ξ_K is P_K-unisolvent. □

Remark 6.1.5. Just as for the Hsieh–Clough–Tocher triangle, the normal derivatives at the mid-point of the sides can be eliminated by requiring that the normal derivatives vary linearly along the sides. Then we obtain in this fashion another finite element of class $\mathscr{C}^1$ for which $\dim(P_K) = 9$ (cf. Exercise 6.1.6). □

Theorem 6.1.5. *A regular family of singular Zienkiewicz triangles is almost affine: For all $p \in]1, \infty]$ and all pairs (m, q) with $m \geqslant 0$ and*

$q \in [1, \infty]$ *such that*

$$\begin{cases} W^{3,p}(K) \hookrightarrow W^{m,q}(K), \\ P_K \subset W^{m,q}(K), \end{cases} \tag{6.1.52}$$

there exists a constant C independent of K such that

$$\forall v \in W^{3,p}(K), \quad \|v - \Pi_K v\|_{m,q,K} \leq C(\text{meas}(K))^{1/q-1/p} h_K^{3-m} |v|_{3,p,K}. \tag{6.1.53}$$

Proof. We shall simply give some indications. The proof of inequality (6.1.53) rests on the inclusion

$$P_2(K) \subset P_K$$

(notice that the inequality $p > 1$ is required so as to guarantce the inclusion $W^{3,p}(K) \hookrightarrow \mathscr{C}^1(K) = \text{dom}\, \Pi_K$). One first argues with the finite element (K, P_K, Ξ_K), with Ξ_K as in (6.1.42), which *can* be imbedded in an affine family. Then one uses the same device as in the proofs of Theorems 6.1.1 and 6.1.3. □

Remark 6.1.6. The second partial derivatives of the basis function $\hat{q}_1$ (as given in (6.1.41)) are not defined at the origin. In fact, for each slope $t > 0$, an easy computation shows that

$$\lim_{\substack{x_2 = tx_1 \\ x_1 \to 0^+}} \partial_{11}\hat{q}_1(x) = \frac{-8t^3}{(1+t)^3}.$$

This phenomenon is observed in Zienkiewicz (1971, p. 199), where it is stated that "second-order derivatives have non-unique values at nodes". Hopefully, this observation carries no consequence since it does not prevent the function q_1 from being in the space $\mathscr{C}^1(K) \cap H^2(K)$. □

Estimate of the error $\|u - u_h\|_{2,\Omega}$

Let us now return to the finite element approximation of the clamped plate problem (6.1.1). We shall consider families of finite element spaces X_h, with the same generic finite element (K, P_K, Σ_K), for which we shall need the following assumptions:

(H1*) *The family* (K, P_K, Σ_K), $K \in \mathscr{T}_h$, *for all h, is an almost-affine family.*

(H2*) *The generic finite element is of class* $\mathscr{C}^1$.

If we assume (as in the subsequent theorems) that the inclusion $P_K \subset H^2(K)$ holds, the inclusion $X_h \subset H^2(\Omega)$ is then a consequence of hypothesis (H2*). This being the case, we let

$$V_h = X_{00h} = \{v_h \in X_h;\ \ v_h = \partial_\nu v_h = 0 \quad \text{on} \quad \Gamma\}. \tag{6.1.54}$$

Notice that the X_h-interpolation operator associated with any one of the finite elements of class $\mathscr{C}^1$ considered in this section satisfy the implication

$$v \in \operatorname{dom} \Pi_h \quad \text{and} \quad v = \partial_\nu v = 0 \quad \text{on} \quad \Gamma \Rightarrow \Pi_h v \in X_{00h}, \tag{6.1.55}$$

which will accordingly be an implicit assumption in the remainder of this section.

To begin with, we derive an error estimate in the norm $\|.\|_{2,\Omega}$. As usual, the letter C represents any constant independent of h and of all the functions appearing in a given inequality.

Theorem 6.1.6. *In addition to* (H1*) *and* (H2*), *assume that there exists an integer* $k \geq 2$ *such that the following inclusions are satisfied*:

$$P_k(K) \subset P_K \subset H^2(K), \tag{6.1.56}$$

$$H^{k+1}(K) \hookrightarrow \mathscr{C}^s(K), \tag{6.1.57}$$

where s *is the maximal order of partial derivatives occuring in the definition of the set* Σ_K.

Then if the solution $u \in H_0^2(\Omega)$ *of the clamped plate problem is also in the space* $H^{k+1}(\Omega)$, *there exists a constant* C *independent of* h *such that*

$$\|u - u_h\|_{2,\Omega} \leq Ch^{k-1}|u|_{k+1,\Omega}, \tag{6.1.58}$$

where $u_h \in V_h$ *is the discrete solution.*

Proof. Using Céa's lemma, inequality (6.1.7) and relation (6.1.55), we obtain

$$\begin{aligned}\|u - u_h\|_{2,\Omega} &\leq C \inf_{v_h \in V_h} \|u - v_h\|_{2,\Omega} \\ &\leq C\|u - \Pi_h u\|_{2,\Omega} = C\Big(\sum_{K \in \mathscr{T}_h} \|u - \Pi_K u\|_{2,K}^2\Big)^{1/2} \\ &\leq Ch^{k-1}\Big(\sum_{K \in \mathscr{T}_h} |u|_{k+1,K}^2\Big)^{1/2} = Ch^{k-1}|u|_{k+1,\Omega}.\end{aligned}$$

□

Remark 6.1.7. By the previous theorem, the least assumptions which insure an $0(h)$ convergence in the norm $\|.\|_{2,\Omega}$ are the inclusions $P_2(K) \subset P_K$ on the one hand, and the fact that the solution u of the plate problem is in the space $H^3(\Omega)$ on the other. It is remarkable that this last regularity result is precisely obtained if the right-hand side f is in the space $L^2(\Omega)$, and if $\bar{\Omega}$ is a convex polygon, an assumption often satisfied for plates. Therefore, since one cannot expect better regularity *in general*, the choice $P_K = P_2(K)$ appears optimal from the point of view of convergence. However, by Ženíšek's result, this choice is not compatible with the inclusion $X_h \subset \mathscr{C}^1(\bar{\Omega})$. □

Sufficient conditions for $\lim_{h\to 0} \|u - u_h\|_{2,\Omega} = 0$

We next obtain convergence in the norm $\|.\|_{2,\Omega}$ under minimal assumptions (cf. (6.1.59) below).

Theorem 6.1.7. *In addition to* (H1*) *and* (H2*), *assume that the inclusions*

$$P_2(K) \subset P_K \subset H^2(K) \tag{6.1.59}$$

are satisfied, and that the maximal order s of partial derivatives found in the set Σ_K satisfies $s \leq 2$.

Then we have

$$\lim_{h\to 0} \|u - u_h\|_{2,\Omega} = 0. \tag{6.1.60}$$

Proof. The argument is the same as in the proof of Theorem 3.2.3 and, for this reason, will be only sketched. Using inequality (6.1.5) with $k = 2$, $p = \infty$, $m = 2$ and $q = 2$, one first shows that the space

$$\mathscr{V} = W^{3,\infty}(\Omega) \cap H_0^2(\Omega)$$

is dense in the space $H_0^2(\Omega)$. Then it suffices to use the inequality

$$\inf_{v_h \in V_h} \|u - v_h\|_{2,\Omega} \leq \|u - v\|_{2,\Omega} + \|v - \Pi_h v\|_{2,\Omega},$$

valid for any function $v \in \mathscr{V}$. □

Conclusions

In the following tableau (Fig. 6.1.6), we have summarized the application of Theorem 6.1.6 to various finite elements of class $\mathscr{C}^1$.

Finite element	dim P_K	$P_k(K) \subset P_K$	$\|u - u_h\|_{2,\Omega}$	Assumed regularity
Argyris triangle	21	$P_5(K) = P_K$	$O(h^4)$	$u \in H^6(\Omega)$
Bell's triangle	18	$P_4(K) \subset P_K$	$O(h^3)$	$u \in H^5(\Omega)$
Bogner–Fox–Schmit rectangle	16	$P_3(K) \subset P_K$	$O(h^2)$	$u \in H^4(\Omega)$
Hsieh–Clough–Tocher triangle	12	$P_3(K) \subset P_K$	$O(h^2)$	$u \in H^4(\Omega)$
Reduced Hsieh–Clough–Tocher triangle (cf. Exercise 6.1.3)	9	$P_2(K) \subset P_K$	$O(h)$	$u \in H^3(\Omega)$
Singular Zienkiewicz triangle	12	$P_2(K) \subset P_K$	$O(h)$	$u \in H^3(\Omega)$
Reduced singular Zienkiewicz triangle (cf. Exercise 6.1.6)	9	$P_2(K) \subset P_K$	$O(h)$	$u \in H^3(\Omega)$

Fig. 6.1.6

One should notice that, if the reduced Hsieh–Clough–Tocher triangle and the reduced singular Zienkiewicz triangle are optimal in that the dimension of the corresponding spaces P_K is the smallest, this reduction in the dimension of the spaces P_K is obtained at the expense of an increased complexity in the *structure* of the functions $p \in P_K$.

Remark 6.1.8. In order to get an $O(h^{k+1})$ convergence in the norm $|.|_{0,\Omega}$, it would be necessary to assume that, for *any* $g \in L^2(\Omega)$, the corresponding solution φ_g of the plate problem belongs to the space $H^4(\Omega) \cap H_0^2(\Omega)$ and that there exists a constant C such that $\|\varphi_g\|_{4,\Omega} \leq C|g|_{0,\Omega}$ for *all* $g \in L^2(\Omega)$. However, *this regularity property is no longer true for convex polygons in general.* It is true only if the boundary Γ is sufficiently smooth: For example, this is the case if the boundary Γ is of class $\mathscr{C}^4$. But then this regularity of the boundary becomes incompatible with our assumption that $\bar{\Omega}$ be a polygonal set. □

Exercises

6.1.1. Show that a regular family of Bell's triangles (cf. Fig. 2.2.18) is almost affine, with the value $k = 4$ in the corresponding inequalities of the form (6.1.5).

6.1.2. The purpose of this problem is to give another proof of unisolvence for the Hsieh–Clough–Tocher triangle (as originally proposed in CIARLET (1974c)). Without loss of generality, it can be assumed that $a = (0, 0)$. Denoting by (x_i, y_i) the coordinates of the vertex a_i, let

$$\alpha_i = \det\begin{pmatrix} x_{i+1} & y_{i+1} \\ x_{i+2} & y_{i+2} \end{pmatrix}, \qquad \alpha = \det\begin{pmatrix} x_1 & x_2 & x_3 \\ y_1 & y_2 & y_3 \\ 1 & 1 & 1 \end{pmatrix} = \sum_{i=1}^{3} \alpha_i.$$

For definiteness, it shall be assumed that $\alpha = 1$.

Given a function $p \in P_K$ whose degrees of freedom are all zero, let

$$\delta_0 = p(a), \quad \delta_i = \alpha_i\{Dp(a)(a_i - a)\}, \quad 1 \leq i \leq 3.$$

(i) Show that

$$p_{|K_i} = \mu_i^2\{(-2\mu_i + 3)\delta_0 + (\mu_{i+1} - \mu_i)\delta_{i+1} + (\mu_{i+2} - \mu_i)\delta_{i+2}\},$$
$$1 \leq i \leq 3,$$

where for each i we denote by μ_i the unique function which satisfies

$$\mu_i \in P_1(K_i), \quad \mu_i(a) = 1, \qquad \mu_i(a_{i+1}) = \mu_i(a_{i+2}) = 0.$$

(ii) Show that

$$\sum_{i=1}^{3} \delta_i = 0,$$

$$\delta_0 - \sum_{k=1}^{3} \delta_k + \frac{2}{\alpha_j}\delta_j = 0, \quad 1 \leq j \leq 3,$$

and conclude that $\delta_i = 0$, $0 \leq i \leq 3$ (the first equality expresses that the function p is differentiable at the point a, while the other relations express the equalities

$$D(p_{i+1} - p_i)\left(\frac{a + a_{i+2}}{2}\right) = 0, \quad 1 \leq i \leq 3).$$

6.1.3. The *reduced Hsieh–Clough–Tocher triangle* is a triangular finite element whose corresponding data P_K and Σ_K are indicated in Fig. 6.1.7.

Show that the set Σ_K is P_K-unisolvent and that a regular family of reduced Hsieh–Clough–Tocher triangles is almost affine, with the value $k = 2$ in the corresponding inequalities of the form (6.1.5).

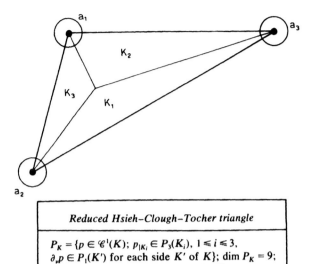

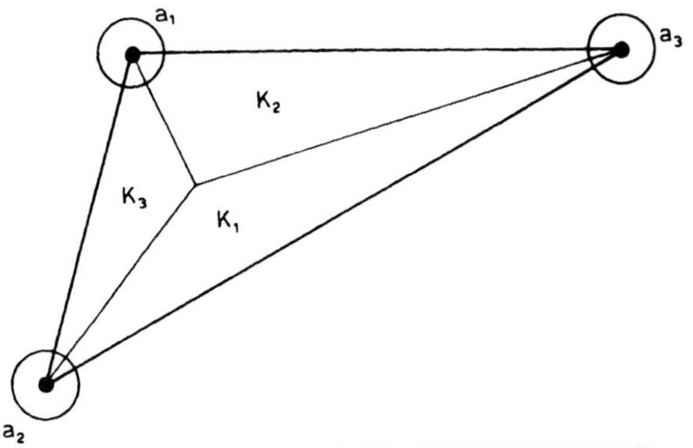

Reduced Hsieh–Clough–Tocher triangle
$P_K = \{p \in \mathscr{C}^1(K);\ p_{\mid K_i} \in P_3(K_i),\ 1 \leqslant i \leqslant 3,$ $\partial_\nu p \in P_1(K')$ for each side K' of $K\}$; $\dim P_K = 9$; $\Sigma_K = \{p(a_i),\ \partial_1 p(a_i),\ \partial_2 p(a_i),\ 1 \leqslant i \leqslant 3\}$.

Fig. 6.1.7

6.1.4. Following PERCELL (1976), one may define a triangular finite element of class $\mathscr{C}^1$ analogous to the Hsieh–Clough–Tocher triangle, as follows: With an identical subdivision $K = \bigcup_{i=1}^3 K_i$, let

$$P_K = \{P \in \mathscr{C}^1(K);\ p_{\mid K_i} \in P_4(K_i),\quad 1 \leqslant i \leqslant 3\},$$

$$\Sigma_K = \{p(a_i),\quad \partial_1 p(a_i),\quad \partial_2 p(a_i),\quad 1 \leqslant i \leqslant 3;$$

$$p(a_{ij}),\quad 1 \leqslant i < j \leqslant 3;\ \partial_\nu p(a_{iij}),\quad 1 \leqslant i, j \leqslant 3,\quad i \neq j;$$

$$p(a),\quad \partial_1 p(a),\quad \partial_2 p(a)\},$$

where

$$a_{ij} = \frac{a_i + a_j}{2},\quad a_{iij} = \frac{2a_i + a_j}{3}.$$

Then show that the set Σ_K is P_K-unisolvent.

6.1.5. *The Fraeijs de Veubeke–Sander quadrilateral* is a finite element (K, P_K, Σ_K) for which the set K is a convex nondegenerate quadrilateral with vertices a_i, $1 \leqslant i \leqslant 4$, and mid-points of the sides b_i, $1 \leqslant i \leqslant 4$. As indicated in Fig. 6.1.8, let K_1 denote the triangle with vertices a_1, a_2 and a_4, and let K_2 denote the triangle with vertices a_1, a_2 and a_3. The space P_K and the set Σ_K are indicated in Fig. 6.1.8.

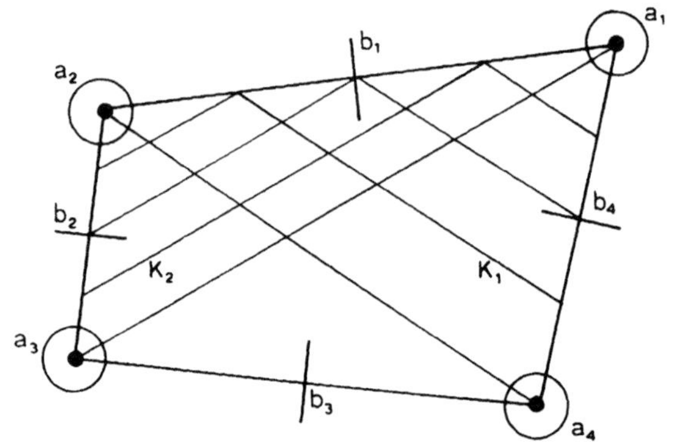

Fraeijs de Veubeke–Sander quadrilateral

$P_K = R_1(K) + R_2(K)$, where
$R_i(K) = \{p \in \mathscr{C}^1(K);\ p_{|K_i} \in P_3(K_i),\ p_{|K-K_i} \in P_3(K - K_i)\}$,
$i = 1, 2$; $\dim P_K = 16$;
$\Sigma_K = \{p(a_i),\ \partial_1 p(a_i),\ \partial_2 p(a_i),\ \partial_\nu p(b_i),\ 1 \leq i \leq 4\}$.

Fig. 6.1.8

(i) Show that the set Σ_K is P_K-unisolvent (CIAVALDINI & NÉDÉLEC (1974)).

(ii) We shall say that *a family of Fraeijs de Veubeke–Sander quadrilaterals is regular* if it is a regular family of finite elements in the usual sense and if, in addition, the following condition is satisfied: For each quadrilateral K in the family, let F_K denote the unique affine mapping which satisfies $F_K(0) = a_K$, $F_K(\hat{a}_1) = a_{1,K}$ and $F_K(\hat{a}_2) = a_{2,K}$, where a_K is the intersection of the two diagonals of the quadrilateral K, and where $\hat{a}_1 = (1, 0)$, $\hat{a}_2 = (0, 1)$ (cf. Fig. 6.1.9). *Then there exist compact intervals $\hat{I}_3$ and $\hat{I}_4$ contained in the half-axes*

$$\{(x_1, x_2) \in \mathbf{R}^2;\ x_1 < 0, x_2 = 0\}, \quad \{(x_1, x_2) \in \mathbf{R}^2;\ x_1 = 0, x_2 < 0\},$$

respectively, such that the points $\hat{a}_{j,K} = F_K^{-1}(a_{j,K})$ belong to the intervals $\hat{I}_j$, for $j = 3$ and 4. In other words, the quadrilateral $F_K^{-1}(K)$ is in between the two extremal quadrilaterals $\hat{K}_0$ and $\hat{K}_1$ indicated in Fig. 6.1.9. Then, following CIAVALDINI & NÉDÉLEC (1974), show that a regular family of Fraeijs de Veubeke–Sander quadrilaterals is almost affine, with the value $k = 3$ in the corresponding inequalities of the form (6.1.5).

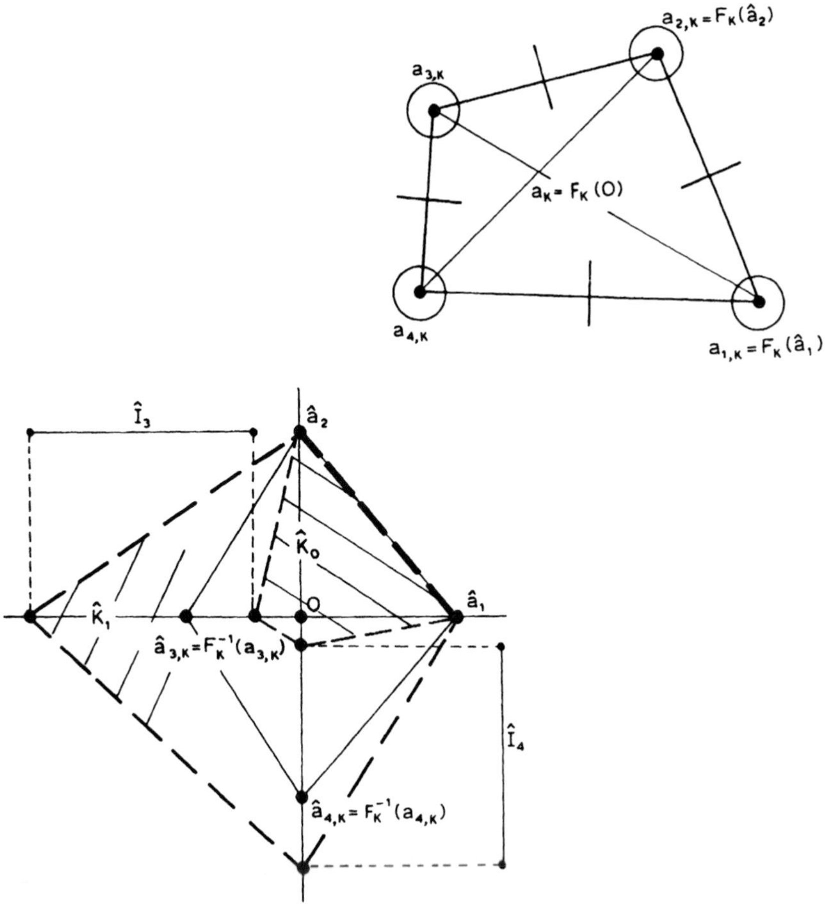

Fig. 6.1.9

(iii) Carry out a similar analysis (unisolvence, interpolation error) for the *reduced Fraeijs de Veubeke–Sander quadrilateral*, whose characteristics are indicated in Fig. 6.1.10 (for the definition of the spaces $R_1(K)$ and $R_2(K)$, see Fig. (6.1.8)).

6.1.6. The *reduced singular Zienkiewicz triangle* is a triangular finite element whose corresponding data P_K and Σ_K are indicated in Fig. 6.1.11.

Show that the set Σ_K is P_K-unisolvent and that a regular family of

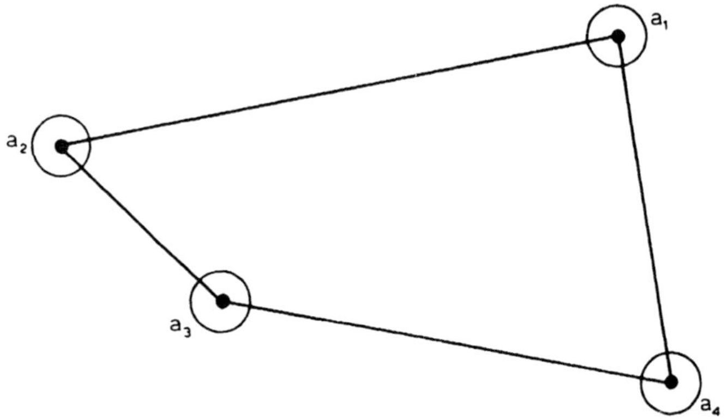

Reduced Fraeijs de Veubeke–Sander quadrilateral
$P_K = \{p \in R_1(K) + R_2(K);\ \partial_\nu p \in P_1(K')$ along each side K' of $K\}$; dim $P_K = 12$; $\Sigma_K = \{p(a_i), \partial_1 p(a_i), \partial_2 p(a_i), 1 \leq i \leq 4\}$.

Fig. 6.1.10

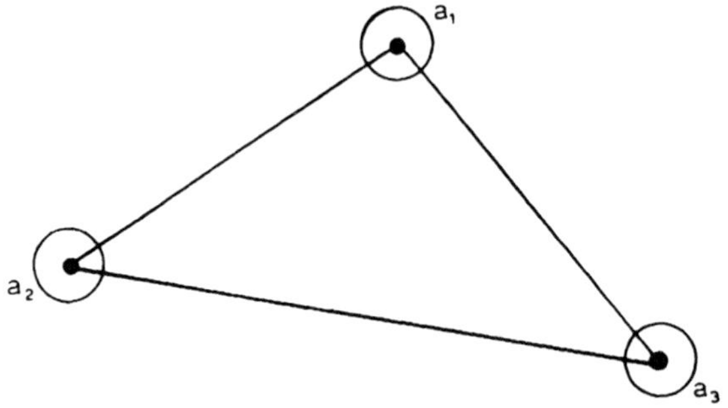

Reduced singular Zienkiewicz triangle
$P_K = \{p \in P_3''(K) \oplus \bigvee_{i=1}^{3} \{q_i\};\ \partial_\nu p \in P_1(K')$ for each side K' of $K\}$ (cf. (2.2.39) and (6.1.37)); dim $P_K = 9$; $\Sigma_K = \{p(a_i), \partial_1 p(a_i), \partial_2 p(a_i), 1 \leq i \leq 3\}$.

Fig. 6.1.11

reduced singular Zienkiewicz triangles is almost affine, with the value $k = 2$ in the corresponding inequalities of the form (6.1.5).

6.1.7. The purpose of this problem is to describe another instance where rational functions are added to a polynomial space so as to obtain a singular finite element of class $\mathscr{C}^1$. An analogous process yielded the singular Zienkiewicz triangle.

(i) Following BIRKHOFF (1971), let $T_3(K)$ denote, for any triangle K, the space of all polynomials whose restrictions along each parallel to any side of K are polynomials of degree ≤ 3 in one variable. Show that the space $T_3(K)$, of so-called *tricubic polynomials*, is the space $P_3(K)$ to which are added linear combinations of the three functions $\lambda_1^2\lambda_2\lambda_3$, $\lambda_1\lambda_2^2\lambda_3$ and $\lambda_1\lambda_2\lambda_3^2$ (which are not linearly independent). Show that $\dim P_3(K) = 12$.

(ii) Following BIRKHOFF & MANSFIELD (1974), we define the *Birkhoff–Mansfield triangle* as indicated in Fig. 6.1.12 (as usual, $\partial_{\nu\tau}p(b_i) = D^2p(b_i)(\nu, \tau)$ where τ is the unit tangential vector at the point b_i).

Show that, along each side of the triangle K, the functions in the space P_K are polynomials of degree ≤ 3 in one variable and that any directional derivative $Dp(.)\xi$, where ξ is any fixed vector in $\mathbf{R}^2$, is also a polynomial of degree ≤ 3 in one variable along each side of the triangle K.

Show that the set Σ_K is P_K-unisolvent.

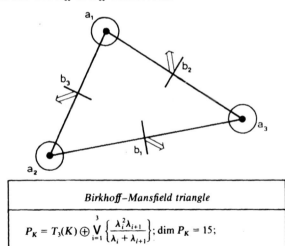

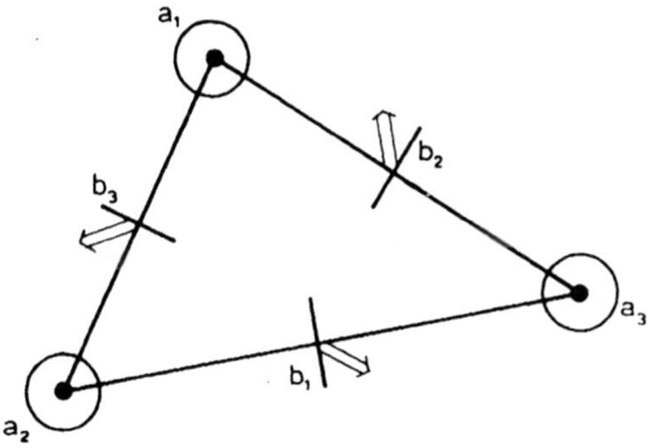

Birkhoff–Mansfield triangle

$$P_K = T_3(K) \oplus \bigvee_{i=1}^{3} \left\{ \frac{\lambda_i^2 \lambda_{i+1}}{\lambda_i + \lambda_{i+1}} \right\};\ \dim P_K = 15;$$

$$\Sigma_K = \{p(a_i),\ \partial_1 p(a_i),\ \partial_2 p(a_i),\ \partial_\nu p(b_i),\ \partial_{\nu\tau} p(b_i),\ 1 \leq i \leq 3\}.$$

Fig. 6.1.12

Show that the resulting finite element is of class $\mathscr{C}^1$ and that the inclusion $P_K \subset H^2(K)$ holds.

(iii) Show that a regular family of Birkhoff–Mansfield triangles is almost affine, with the value $k = 3$ in the corresponding inequalities of the form (6.1.5).

(iv) Carry out a similar analysis (unisolvence and interpolation error) for the *reduced Birkhoff–Mansfield triangle*, whose characteristics are indicated in Fig. 6.1.13.

6.2. Nonconforming methods

Nonconforming methods for the plate problem

To begin with, we shall give the general definition of a *nonconforming method* for solving the clamped plate problem (corresponding to the data (6.1.1)). Assuming the set $\bar{\Omega}$ polygonal, so that it may be exactly covered with triangulations, we construct *a finite element space X_h whose generic finite element is not of class $\mathscr{C}^1$*. Then the space X_h will not be a subspace of the space $H^2(\Omega)$, as a consequence of the next theorem (which is the converse of Theorem 2.1.2), whose proof is left to the reader (Exercise 6.2.1).

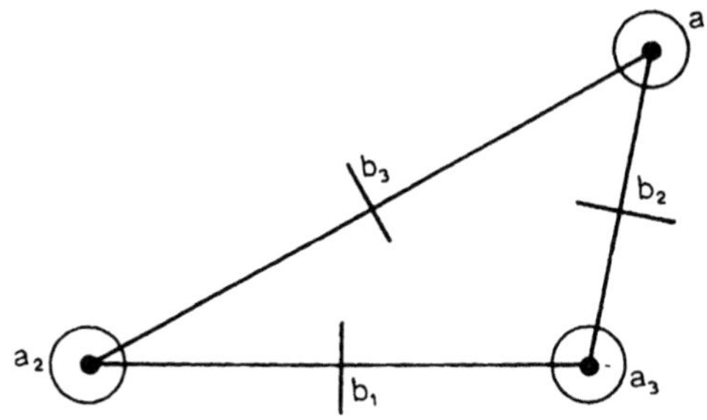

Reduced Birkhoff–Mansfield triangle
$P_K = \left\{p \in T_3(K) \oplus \bigvee_{i=1}^{3} \left\{\dfrac{\lambda_i^2\lambda_{i+1}}{\lambda_i + \lambda_{i+1}}\right\};\right.$ $\forall \xi \in \mathbf{R}^2,\ Dp(.)\xi \in P_2(K')$ for each side K' of $K\}$; $\dim P_K = 12$; $\Sigma_K = \{p(a_i),\ \partial_1 p(a_i),\ \partial_2 p(a_i),\ \partial_\nu p(b_i),\ 1 \leqslant i \leqslant 3\}$.

Fig. 6.1.13

Theorem 6.2.1. *Assume that the inclusions $P_K \subset \mathscr{C}^1(K)$ for all $K \in \mathscr{T}_h$ and $X_h \subset H^2(\Omega)$ hold. Then the inclusion*

$$X_h \subset \mathscr{C}^1(\bar{\Omega})$$

holds. □

Let us henceforth assume that we have

$$\forall K \in \mathscr{T}_h, \quad P_K \subset H^2(K), \tag{6.2.1}$$

so that, in particular, we have

$$X_h \subset L^2(\Omega). \tag{6.2.2}$$

After defining an appropriate subspace X_{00h} of X_h, so as to take into account the boundary conditions $v = \partial_\nu v = 0$ along Γ as well as possible (this will be illustrated on one example), we define the approximate bilinear form:

$$\begin{aligned} a_h(u_h, v_h) &= \sum_{K \in \mathscr{T}_h} \int_K \{\Delta u_h \Delta v_h + (1-\sigma)(2\partial_{12}u_h\partial_{12}v_h - \partial_{11}u_h\partial_{22}v_h \\ &\qquad - \partial_{22}u_h\partial_{11}v_h)\}\,\mathrm{d}x \\ &= \sum_{K \in \mathscr{T}_h} \int_K \{\sigma\Delta u_h \Delta v_h + (1-\sigma)(\partial_{11}u_h\partial_{11}v_h + \partial_{22}u_h\partial_{22}v_h \\ &\qquad + 2\partial_{12}u_h\partial_{12}v_h)\}\,\mathrm{d}x. \end{aligned} \tag{6.2.3}$$

Observe that this definition is justified by the inclusions (6.2.1). Then the *discrete problem* consists in finding a function $u_h \in V_h = X_{00h}$ such that

$$\forall v_h \in V_h, \quad a_h(u_h, v_h) = f(v_h) \tag{6.2.4}$$

(the linear form need not be approximated in view of the inclusion (6.2.2)). In analogy with the norm $|\cdot|_{2,\Omega}$ of the space $V = H_0^2(\Omega)$, we introduce the *semi-norm*

$$v_h \to \|v_h\|_h = \Big(\sum_{K \in \mathscr{T}_h} |v_h|_{2,K}^2\Big)^{1/2} \tag{6.2.5}$$

over the space V_h. Next we extend the domains of definition of the mappings $a_h(\cdot,\cdot)$ and $\|\cdot\|_h$ to the space $V_h + V$. Thus there exists a constant $\tilde{M}$ independent of the space V_h such that

$$\forall u, v \in (V_h + V), \quad |a_h(u, v)| \leq \tilde{M}\|u\|_h\|v\|_h. \tag{6.2.6}$$

An example of a nonconforming finite element: Adini's rectangle

In the remainder of this section, we shall essentially concentrate on one example of a *nonconforming finite element*, in the sense that it yields a nonconforming method when it is used in the approximation of the plate problem. This element, known as *Adini's rectangle*, corresponds to the following data K, P_K and Σ_K: The set K is a rectangle whose vertices a_i, $1 \leq i \leq 4$, are counted as in Fig. 6.2.1.

The space P_K is composed of all polynomials of the form

$$P: x = (x_1, x_2) \to p(x) = \sum_{\alpha_1 + \alpha_2 \leq 3} \gamma_{\alpha_1 \alpha_2} x_1^{\alpha_1} x_2^{\alpha_2} + \gamma_{13} x_1 x_2^3 + \gamma_{31} x_1^3 x_2,$$

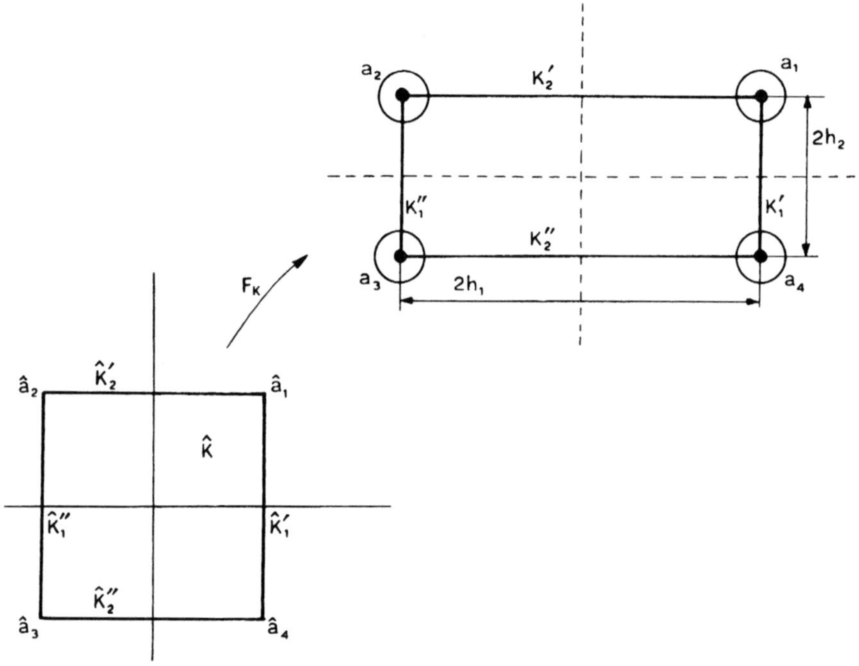

Adini's rectangle
$P_K = P_3(K) \oplus \bigvee \{x_1 x_2^3, x_1^3 x_2\}$; $\dim P_K = 12$; $\Sigma_K = \{p(a_i), \partial_1 p(a_i), \partial_2 p(a_i), 1 \leq i \leq 4\}$.

Fig. 6.2.1

i.e., we have

$$P_K = P_3(K) \oplus \bigvee \{x_1x_2^3, x_1^3x_2\}. \tag{6.2.7}$$

Notice that the inclusion

$$P_3(K) \subset P_K \tag{6.2.8}$$

holds, and that

$$\dim(P_K) = 12. \tag{6.2.9}$$

To see that *the set*

$$\Sigma_K = \{p(a_i), \partial_1 p(a_i), \partial_2 p(a_i), \quad 1 \leq i \leq 4\} \tag{6.2.10}$$

is a P_K*-unisolvent set*, let us argue on the square $\hat{K} = [-1, +1]^2$. Then we can write

$$\forall \hat{p} \in P_{\hat{K}}, \quad \hat{p} = \sum_{i=1}^{4} \hat{p}(\hat{a}_i)\hat{p}_i + \sum_{\begin{cases} |j-i| = 1 \\ (\mathrm{mod.}\ 4) \end{cases}} D\hat{p}(\hat{a}_i)(\hat{a}_j - \hat{a}_i)\hat{p}_{ij}, \tag{6.2.11}$$

with

$$\begin{cases} \hat{p}_1(x) = \dfrac{(1+x_1)(1+x_2)}{4}\left(1 + \dfrac{x_1+x_2}{2} - \dfrac{x_1^2+x_2^2}{2}\right), \\ \hat{p}_{12}(x) = \dfrac{(1+x_1)(1+x_2)^2(1-x_2)}{8}, \\ \hat{p}_{14}(x) = \dfrac{(1+x_2)(1+x_1)^2(1-x_1)}{8}, \quad \text{etc.}\ldots \end{cases} \tag{6.2.12}$$

Let us assume that the set $\bar{\Omega}$ is rectangular, so that it may be covered by triangulations made up of rectangles. With such a triangulation $\mathcal{T}_h$, we associate a finite element space X_h whose functions v_h are defined as follows:

(i) For each rectangle $K \in \mathcal{T}_h$, the restrictions $v_{h|K}$ span the space P_K of (6.2.7).

(ii) Each function $v_h \in X_h$ is defined by its values and the values of its first derivatives at all the vertices of the triangulation.

Along each side K' of an Adini's rectangle K, the restrictions $p_{|K'}$, $p \in P_K$, are polynomials of degree ≤ 3 in one variable. Since such polynomials are uniquely determined by their values and the values of their first derivative at the end points of K', we conclude that *Adini's rectangle is a finite element of class* $\mathscr{C}^0$. It is *not* of class $\mathscr{C}^1$, however:

Along the side $K_1' = [a_4, a_1]$ for instance (cf. Fig. 6.2.1), the normal derivative is a polynomial of degree ≤ 3 in the variable x_2 on the one hand, and on the other the only degrees of freedom that are available for the normal derivative along the side K_1' are its two values at the end points.

We let $V_h = X_{00h}$, where X_{00h} denotes the space of all functions $v_h \in X_h$ such that $v_h(b) = \partial_1 v_h(b) = \partial_2 v_h(b) = 0$ at all the boundary nodes b. Then the functions $v_h \in V_h$ vanish along the boundary Γ, but their derivatives $\partial_\nu v_h$ do not vanish along the boundary Γ in general, although they vanish at the boundary nodes. To sum up, we have constructed a finite element space V_h whose functions v_h satisfy

$$\begin{cases} v_h \in H_0^1(\Omega) \cap \mathscr{C}^0(\bar{\Omega}), & v_{h|K} \in H^2(K) \text{ for all } K \in \mathscr{T}_h, \\ \partial_\nu v_h(b) = 0 & \text{at the boundary nodes.} \end{cases} \tag{6.2.13}$$

Observe that the associated X_h-interpolation operator Π_h is such that

$$v \in H_0^2(\Omega) \cap \operatorname{dom} \Pi_h \Rightarrow \Pi_h v \in X_{00h} = V_h. \tag{6.2.14}$$

We shall use this implication in particular for functions in the space $H^3(\Omega) \cap H_0^2(\Omega) \subset \mathscr{C}^1(\bar{\Omega}) = \operatorname{dom} \Pi_h$.

Prior to the error analysis, we must examine whether the mapping $\|.\|_h$ of (6.2.5) is indeed a norm.

Theorem 6.2.2. *The mapping*

$$v_h \to \|v_h\|_h = \Big(\sum_{K \in \mathscr{T}_h} |v_h|_{2,K}^2\Big)^{1/2}$$

is a norm over the space V_h.

Proof. Let v_h be a function in the space V_h such that $\|v_h\|_h = 0$. Then the functions $\partial_j(v_{h|K})$, $j = 1, 2$, are constant over each rectangle $K \in \mathscr{T}_h$. Since they are continuous at the vertices, the functions $\partial_j v_h$, $j = 1, 2$, are therefore constant over the set $\bar{\Omega}$, and since they vanish at the boundary nodes, they are identically zero. Thus the function $v_h \in V_h$ is identically zero, as a consequence of the inclusion $V_h \subset H_0^1(\Omega) \cap \mathscr{C}^0(\Omega)$. □

Notice that *the approximate bilinear forms $a_h(.,.)$ are uniformly V_h-elliptic*, since one has (cf. (6.2.3))

$$\forall v_h \in V_h, \quad (1-\sigma)\|v_h\|_h^2 \leq a_h(v_h, v_h), \tag{6.2.15}$$

and the Poisson coefficient σ lies in the interval $]0, \frac{1}{2}[$ (for physical reasons).

Remark 6.2.1. Had we tried to use nonconforming finite element methods for the biharmonic problem (in which case the approximate bilinear form reduces to $\sum_{K\in\mathcal{T}_h}\int_K \Delta u_h \Delta v_h \,dx$), the uniform V_h-ellipticity is no longer automatic, and this is essentially why we restrict ourselves to plate problems. In contrast, conforming methods as described in the previous section apply equally well to any fourth-order elliptic boundary value problem. □

Consistency error estimate. Estimate of the error $(\sum_{K\in\mathcal{T}_h}|u-u_h|^2_{2,K})^{1/2}$

We are now in a position to apply the abstract error estimate of Theorem 4.2.2, which we recall here for convenience:

$$\|u-u_h\|_h \leqslant C\Big(\inf_{v_h\in V_h}\|u-v_h\|_h + \sup_{w_h\in V_h}\frac{|a_h(u,w_h)-f(w_h)|}{\|w_h\|_h}\Big). \tag{6.2.16}$$

In what follows, *the solution u will be assumed to be in the space* $H^3(\Omega)\cap H_0^2(\Omega)$ (this is true for any $f\in L^2(\Omega)$ if $\bar{\Omega}$ is a convex polygon, i.e., a rectangle in the present case). Observing that any family of Adini's rectangles is affine, we obtain for a regular family of triangulations,

$$\inf_{v_h\in V_h}\|u-v_h\|_h \leqslant \Big(\sum_{K\in\mathcal{T}_h}|u-\Pi_K u|^2_{2,K}\Big)^{1/2} \leqslant Ch|u|_{3,\Omega}, \tag{6.2.17}$$

and this estimate takes care of the first term in the right-hand side of inequality (6.2.16). The estimate of the second term, i.e., the *consistency error estimate*, rests on a careful decomposition of the difference

$$D_h(u,w_h) = a_h(u,w_h) - f(w_h), \quad w_h\in V_h. \tag{6.2.18}$$

Let us first show that the term $f(w_h)=\int_\Omega f w_h \,dx$ can be rewritten in the form

$$f(w_h) = -\int_\Omega \nabla(\Delta u)\cdot\nabla w_h \,dx \quad \text{for all} \quad w_h\in V_h \tag{6.2.19}$$

(this equality is obvious if $u\in H^4(\Omega)\cap H_0^2(\Omega)$, in which case $f(w_h)=\int_\Omega \Delta^2 u\, w_h\,dx$, but we only assume here that $u\in H^3(\Omega)\cap H_0^2(\Omega)$). To see this, let $w_h\in V_h$ be given, and let (w_h^k) be a sequence of functions $w_h^k\in\mathcal{D}(\Omega)$ such that $\lim_{k\to\infty}\|w_h^k-w_h\|_{1,\Omega}=0$ (recall that $w_h\in V_h\subset$

$H_0^1(\Omega)$). By making use of Green's formulas (1.2.5) and (1.2.9), we obtain for all integers k,

$$\int_\Omega \Delta u \Delta w_h^k \,\mathrm{d}x = -\int_\Omega \nabla(\Delta u)\cdot\nabla w_h^k \,\mathrm{d}x,$$

$$\int_\Omega \{2\partial_{12}u\partial_{12}w_h^k - \partial_{11}u\partial_{22}w_h^k - \partial_{22}u\partial_{11}w_h^k\}\,\mathrm{d}x = 0,$$

since $\partial_\nu w_h^k = \partial_\tau w_h^k = 0$ along Γ, and thus, by definition of the abstract problem (cf. (6.1.1)),

$$\int_\Omega f w_h^k \,\mathrm{d}x = -\int_\Omega \nabla(\Delta u)\cdot\nabla w_h^k \,\mathrm{d}x.$$

Therefore,

$$\begin{aligned}\int_\Omega f w_h \,\mathrm{d}x &= \lim_{k\to\infty}\int_\Omega f w_h^k \,\mathrm{d}x = \lim_{k\to\infty}\int_\Omega \Delta u \Delta w_h^k \,\mathrm{d}x\\ &= \lim_{k\to\infty}\left\{-\int_\Omega \nabla(\Delta u)\cdot\nabla w_h^k\,\mathrm{d}x\right\} = -\int_\Omega \nabla(\Delta u)\cdot\nabla w_h\,\mathrm{d}x,\end{aligned}$$

and equality (6.2.19) is proved.

Using the same Green's formulas as above, we obtain (cf. (1.2.9) for the notation).

$$\begin{aligned}&\forall K\in\mathcal{T}_h,\quad \forall w_h\in V_h,\\ &\int_K \{\Delta u\Delta w_h + (1-\sigma)(2\partial_{12}u\partial_{12}w_h - \partial_{11}u\partial_{22}w_h - \partial_{22}u\partial_{11}w_h)\}\,\mathrm{d}x\\ &\quad = -\int_K \nabla(\Delta u)\cdot\nabla w_h\,\mathrm{d}x + \int_{\partial K}\Delta u\partial_{\nu_K}w_h\,\mathrm{d}\gamma\\ &\qquad +(1-\sigma)\int_{\partial K}\{-\partial_{\tau\tau_K}u\partial_{\nu_K}w_h + \partial_{\nu\tau_K}u\partial_{\tau_K}w_h\}\,\mathrm{d}\gamma.\end{aligned}$$

When the above expressions are added up so as to form the approximate bilinear form of (6.2.3), we first find that

$$\sum_{K\in\mathcal{T}_h}\left\{-\int_K \nabla(\Delta u)\cdot\nabla w_h\,\mathrm{d}x\right\} = -\int_\Omega \nabla(\Delta u)\cdot\nabla w_h\,\mathrm{d}x = f(w_h),$$

using the inclusion $V_h\subset H^1(\Omega)$ and equality (6.2.19), and next we shall find that

$$\sum_{K\in\mathcal{T}_h}\int_{\partial K}\partial_{\nu\tau_K}u\partial_{\tau_K}(w_{h|K})\,\mathrm{d}\gamma = 0.$$

To prove this last relation, consider separately the case where $K' \subset \partial K$ is a side common to two adjacent rectangles K_1 and K_2, and the case where $K' \subset \partial K$ is a portion of the boundary Γ. In the first case the two corresponding integrals cancel because $u \in H^3(\Omega)$ and $w_h \in \mathscr{C}^0(\bar{\Omega})$, and in the second case the integral vanishes because $w_h = 0$ along Γ.

To sum up, we have found that

$$\forall w_h \in V_h, \quad D_h(u, w_h) = a_h(u, w_h) - f(w_h)$$
$$= \sum_{K \in \mathcal{T}_h} \int_{\partial K} (\Delta u - (1-\sigma)\partial_{\tau\tau_K} u)\partial_{\nu_K}(w_{h|K})\, \mathrm{d}\gamma, \tag{6.2.20}$$

i.e., we have obtained *one* decomposition of the expression $D_h(u, w_h)$ as a sum

$$D_h(u, w_h) = \sum_{K \in \mathcal{T}_h} D_K(u_{|K}, w_{h|K}),$$

where each mapping $D_K(.\,,.)$ appears as a bilinear form over the space $H^3(K) \times P_K$. Just as in the proof of Theorem 4.2.6, the key argument will consist in obtaining another decomposition of the form (6.2.20) (cf. (6.2.23)), which in this case takes into account the "conforming" part of the first order partial derivatives of the functions in the space V_h (for related ideas, cf. Remark 4.2.5). This will in turn allow us to obtain appropriate estimates of the difference $D_h(u, w_h)$, as we shall show in the proof of the next theorem.

Theorem 6.2.3. *Assume that the solution u of the plate problem is in the space $H_0^2(\Omega) \cap H^3(\Omega)$. Then, for any regular family of triangulations, there exists a constant C independent of h such that*

$$\|u - u_h\|_h = \Big(\sum_{K \in \mathcal{T}_h} |u - u_h|_{2,K}^2\Big)^{1/2} \leq Ch|u|_{3,\Omega}. \tag{6.2.21}$$

Proof. In view of the decomposition (6.2.20), we are naturally led to study the bilinear form

$$D_h(.\,,.): (v, w_h) \in H^3(\Omega) \times V_h \to$$
$$D_h(v, w_h) = \sum_{K \in \mathcal{T}_h} \int_{\partial K} (\Delta v - (1-\sigma)\partial_{\tau\tau_K} v)\partial_{\nu_K}(w_{h|K})\, \mathrm{d}\gamma$$
$$= D_h^1(v, \partial_1 w_h) + D_h^2(v, \partial_2 w_h), \tag{6.2.22}$$

with

$$D_h^j(v, \partial_j w_h) = \sum_{K \in \mathcal{T}_h} \left\{ \int_{K'_j} (\Delta v - (1-\sigma)\partial_{\tau\tau_K} v)\partial_j(w_{h|K})\,\mathrm{d}\gamma \right.$$
$$\left. - \int_{K''_j} (\Delta v - (1-\sigma)\partial_{\tau\tau_K} v)\partial_j(w_{h|K})\,\mathrm{d}\gamma \right\}, \quad j = 1, 2,$$

where, for each $K \in \mathcal{T}_h$, the sides K'_j and K''_j, $j = 1, 2$, are defined as in Fig. 6.2.1.

For each triangulation $\mathcal{T}_h$, we let Y_h denote the finite element space whose generic finite element is the rectangle of type (1) and we let $Z_h = Y_{0h}$ denote the space of all functions $w_h \in Y_h$ which vanish at the boundary nodes. Clearly, the inclusion

$$Z_h \subset \mathscr{C}^0(\bar{\Omega}) \cap H_0^1(\Omega)$$

implies that

$$\forall v \in H^3(\Omega), \quad \forall z_h \in Z_h, \quad D_h^j(v, z_h) = 0, \quad j = 1, 2,$$

with

$$D_h^j(v, z_h) = \sum_{K \in \mathcal{T}_h} \left\{ \int_{K'_j} (\Delta v - (1-\sigma)\partial_{\tau\tau_K} v) z_h\,\mathrm{d}\gamma \right.$$
$$\left. - \int_{K''_j} (\Delta v - (1-\sigma)\partial_{\tau\tau_K} v) z_h\,\mathrm{d}\gamma \right\}, \quad j = 1, 2.$$

Consequently, if, for each $K \in \mathcal{T}_h$, Λ_K denotes the $Q_1(K)$-interpolation operator, we can also write

$$\forall v \in H^3(\Omega), \quad \forall w_h \in V_h, \quad D_h(v, w_h) = \sum_{K \in \mathcal{T}_h} D_K(v, w_h), \tag{6.2.23}$$

where, for each $K \in \mathcal{T}_h$, the bilinear form $D_K(.,.)$ is given by

$$\forall v \in H^3(K), \quad \forall p \in P_K, \quad D_K(v, p) = \Delta_{1,K}(v, \partial_1 p) + \Delta_{2,K}(v, \partial_2 p), \tag{6.2.24}$$

with

$$\Delta_{j,K}(v, \partial_j p) = \int_{K'_j} (\Delta v - (1-\sigma)\partial_{\tau\tau_K} v)(\partial_j p - \Lambda_K \partial_j p)\,\mathrm{d}\gamma$$
$$- \int_{K''_j} (\Delta v - (1-\sigma)\partial_{\tau\tau_K} v)(\partial_j p - \Lambda_K \partial_j p)\,\mathrm{d}\gamma,$$
$$j = 1, 2. \tag{6.2.25}$$

Using the definition of the operator Λ_K, we find a *first polynomial invariance*:

$$\forall v \in H^3(K), \quad \forall q \in Q_1(K), \quad \Delta_{j,K}(v, q) = 0, \quad j = 1, 2, \quad (6.2.26)$$

with

$$\Delta_{j,K}(v, q) = \int_{K'_j} (\Delta v - (1-\sigma)\partial_{\tau\tau_K} v)(q - \Lambda_K q)\,\mathrm{d}\gamma$$

$$- \int_{K''_j} (\Delta v - (1-\sigma)\partial_{\tau\tau_K} v)(q - \Lambda_K q)\,\mathrm{d}\gamma, \quad j = 1, 2.$$

We next proceed to obtain the *second polynomial invariance*:

$$\forall v \in P_2(K), \quad \forall q \in \partial_j P_K, \quad \Delta_{j,K}(v, q) = 0, \quad j = 1, 2, \quad (6.2.27)$$

where the spaces

$$\partial_j P_K = \{\partial_j p\,;\, p \in P_K\}, \quad j = 1, 2, \quad (6.2.28)$$

both contain the space $Q_1(K)$. To see this, it suffices to show that

$$\forall q \in \partial_j P_K, \quad \int_{K'_j} (q - \Lambda_K q)\,\mathrm{d}\gamma = \int_{K''_j} (q - \Lambda_K q)\,\mathrm{d}\gamma, \quad j = 1, 2. \quad (6.2.29)$$

Let us prove this equality for $j = 1$, for instance. Each function $q \in \partial_1 P_K$ is of the form

$$q = \gamma_0(x_1) + \gamma_1(x_1)x_2 + \gamma_2 x_2^2 + \gamma_3 x_2^3,$$

where γ_0 and γ_1 are polynomials of degree ≤ 2 in the variable x_1. Given any function r defined on a side K', let $\lambda_{K'} r$ denote the linear function along K' which assumes the same values as the function r at the end points of K'. Then we have

$$(q - \Lambda_K q)_{|K'_1}(x_2) = \gamma_2 x_2^2 + \gamma_3 x_2^3 - \lambda_{K'_1}(\gamma_2 x_2^2 + \gamma_3 x_2^3)$$

and therefore

$$(q - \Lambda_K q)_{|K'_1}(x_2) = (q - \Lambda_K q)|_{K''_1}(x_2),$$

which proves (6.2.29). Consequently, the polynomial invariance of (6.2.27) holds.

To estimate the quantities $\Delta_{j,K}(v, \partial_j p)$ of (6.2.25), it suffices to estimate the similar expressions

$$\delta_{j,K}(\varphi, q) = \int_{K'_j} \varphi(q - \Lambda_K q)\,\mathrm{d}\gamma - \int_{K''_j} \varphi(q - \Lambda_K q)\,\mathrm{d}\gamma \quad (6.2.30)$$

for $\varphi \in H^1(K)$, $q \in \partial_j P_K$, $j = 1, 2$. Using the standard correspondences between the functions $\hat{v}: \hat{K} \to \mathbf{R}$ and $v: K \to \mathbf{R}$, we obtain

$$\delta_{1,K}(\varphi, q) = h_2 \delta_{1,\hat{K}}(\hat{\varphi}, \hat{q}), \quad \delta_{2,K}(\varphi, q) = h_1 \delta_{2,\hat{K}}(\hat{\varphi}, \hat{q}), \tag{6.2.31}$$

and we shall also take into account the fact that a function $\hat{q}$ belong to the space $\partial_j P_{\hat{K}}$ when the function q belongs to the space $\partial_j P_K$.

Paralleling the polynomial invariances (6.2.26) and (6.2.27), we now have:

$$\begin{cases} \forall \hat{\varphi} \in H^1(\hat{K}), \quad \forall \hat{q} \in P_0(\hat{K}), \quad \delta_{j,\hat{K}}(\hat{\varphi}, \hat{q}) = 0, \\ \forall \hat{\varphi} \in P_0(\hat{K}), \quad \forall \hat{q} \in \partial_j P_{\hat{K}}, \quad \delta_{j,\hat{K}}(\hat{\varphi}, \hat{q}) = 0, \quad j = 1, 2. \end{cases} \tag{6.2.32}$$

Then if we equip the spaces $\partial_j P_K$ with the norm $\|.\|_{1,K}$, we obtain $\forall \hat{\varphi} \in H^1(\hat{K})$, $\forall \hat{q} \in \partial_j P_{\hat{K}}$,

$$|\delta_{j,\hat{K}}(\hat{\varphi}, \hat{q})| \leq \hat{C} \|\hat{\varphi}\|_{L^2(\partial \hat{K})} \|\hat{q}\|_{L^2(\partial \hat{K})} \leq \hat{C} \|\hat{\varphi}\|_{1,\hat{K}} \|\hat{q}\|_{1,\hat{K}},$$

and thus each bilinear from $\delta_{j,\hat{K}}(.\,,.)$ is continuous over the space $H^1(\hat{K}) \times \partial_j P_{\hat{K}}$. Using the bilinear lemma (Theorem 4.2.5), there exists another constant $\hat{C}$ such that

$$\forall \hat{\varphi} \in H^1(\hat{K}), \quad \forall \hat{q} \in \partial_j P_{\hat{K}},$$
$$|\delta_{j,\hat{K}}(\hat{\varphi}, \hat{q})| \leq \hat{C} |\hat{\varphi}|_{1,\hat{K}} |\hat{q}|_{1,\hat{K}}. \tag{6.2.33}$$

By Theorem 3.1.2 and the regularity assumption, there exists a constant C such that

$$|\hat{\varphi}|_{1,\hat{K}} \leq C |\varphi|_{1,K}, \quad |\hat{q}|_{1,\hat{K}} \leq C |q|_{1,K}. \tag{6.2.34}$$

Combining relations (6.2.31), (6.2.33) and (6.2.34), we conclude that

$$\forall \varphi \in H^1(K), \quad \forall q \in \partial_j P_K,$$
$$|\delta_{j,K}(\varphi, q)| \leq C h_K |\varphi|_{1,K} |q|_{1,K}, \quad j = 1, 2. \tag{6.2.35}$$

Let then $v \in H^3(K)$ and $p \in P_K$ be two given functions, so that the functions $\varphi = \Delta v - (1 - \sigma)\partial_{22} v$ and $q = \partial_1 p$ belong to the spaces $H^1(K)$ and $\partial_1 P_K$, respectively. Then we have

$$|\Delta_{1,K}(v, p)| = |\delta_{1,K}(\Delta v - (1 - \sigma)\partial_{22} v, \partial_1 p)| \leq C h_K |v|_{3,K} |p|_{2,K}.$$

Arguing analogously with the term $|\Delta_{2,K}(v, p)|$, we obtain

$$\forall v \in H^3(K), \quad \forall p \in P_K,$$
$$|D_K(v, p)| \leq \sum_{j=1}^{2} |\Delta_{j,K}(v, \partial_j p)| \leq C h_K |v|_{3,K} |p|_{2,K}.$$

Then we are able to estimate the second term in the abstract error estimate (6.2.16): We find that, for all $w_h \in V_h$,

$$|a_h(u, w_h) - f(w_h)| \leq \sum_{K \in \mathcal{T}_h} |D_K(u, w_h)| \leq Ch|u|_{3,\Omega}\|w_h\|_h$$

and the proof is complete. □

Further results

The error estimate (6.2.21) can be improved *when all the rectangles* $K \in \mathcal{T}_h$ *are equal.* In this case, LASCAUX & LESAINT (1975) have shown that $\|u - u_h\|_h \leq Ch^2|u|_{4,\Omega}$ if the solution u is in the space $H^4(\Omega)$.

For an error estimate in the norm $\|.\|_{1,\Omega}$, see Exercise 6.2.2.

Another nonconforming finite element for solving the plate problem is the *Zienkiewicz triangle* (cf. BAZELEY, CHEUNG, IRONS & ZIENKIEWICZ (1965)) which was described in Section 2.2 (cf. Fig. 2.2.16). Through a refinement of the argument used in the proof of Theorem 6.2.3, LASCAUX & LESAINT (1975) have shown that the necessary polynomial invariances in the difference $D_h(u, w_h)$ (which in turn imply convergence) are obtained if and only if *all sides of all the triangles found in the triangulation are parallel to three directions only.* In this case, one gets $\|u - u_h\|_h \leq Ch|u|_{3,\Omega}$ and $\|u - u_h\|_{1,\Omega} \leq Ch^2|u|_{3,\Omega}$ assuming the solution u is the space $H^3(\Omega)$. This is therefore an answer to the *Union Jack problem*: As pointed out in ZIENKIEWICZ (1971, p. 188–189), the engineers had empirically discovered that configuration (a) systematically yields poorer results than configuration (b) (Fig. 6.2.2).

The reason why the degree of freedom $p(a_{123})$ (which is normally found in the Hermite triangle of type (3)) should be eliminated is that the presence of the associated basis function $\lambda_1\lambda_2\lambda_3$ (cf. (2.2.37)) would destroy the required polynomial invariances.

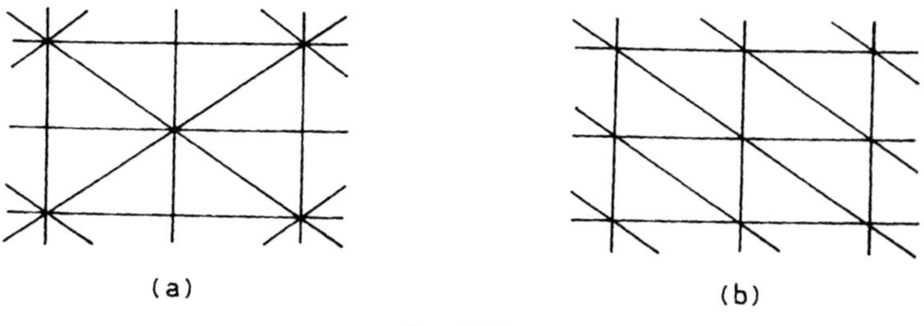

Fig. 6.2.2

Whereas Zienkiewicz triangles yield finite element spaces which satisfy the inclusion $V_h \subset \mathscr{C}^0(\bar{\Omega}) \cap H_0^1(\Omega)$ (just as Adini's rectangles), there exist nonconforming finite elements for the plate problem which are not even of class $\mathscr{C}^0$. Two such finite elements, the *Morley triangle* and the *Fraeijs de Veubeke triangle*, are analyzed in Exercise 6.2.3.

Exercises

6.2.1. Prove Theorem 6.2.1 (cf. Theorem 4.2.1 for a similar argument).
6.2.2. Using the abstract error estimate of Exercise 4.2.3, show that (Lascaux & Lesaint (1975))

$$\|u - u_h\|_{1,\Omega} \leq Ch^2 |u|_{3,\Omega},$$

for finite element spaces whose generic element is the Adini rectangle.
6.2.3. Following Lascaux & Lesaint (1975), the object of this problem is the study of two nonconforming finite elements which are not of class $\mathscr{C}^0$. The first element, known as *Morley's triangle* (cf. Morley (1968)) corresponds to the data indicated in Fig. 6.2.3.

The second element, known as *Fraeijs de Veubeke triangle* (cf. Fraeijs de Veubeke (1974)) is an example of a finite element where some degrees of freedom are *averages* (another related instance is Wilson's

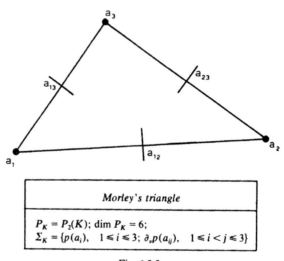

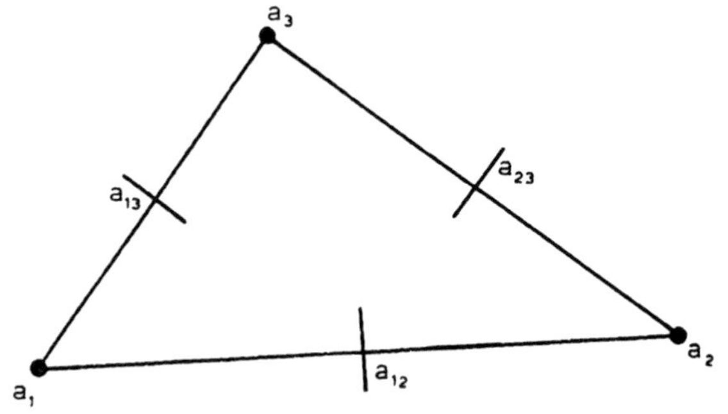

Morley's triangle
$P_K = P_2(K)$; $\dim P_K = 6$; $\Sigma_K = \{p(a_i),\ 1 \leq i \leq 3;\ \partial_\nu p(a_{ij}),\ 1 \leq i < j \leq 3\}$

Fig. 6.2.3.

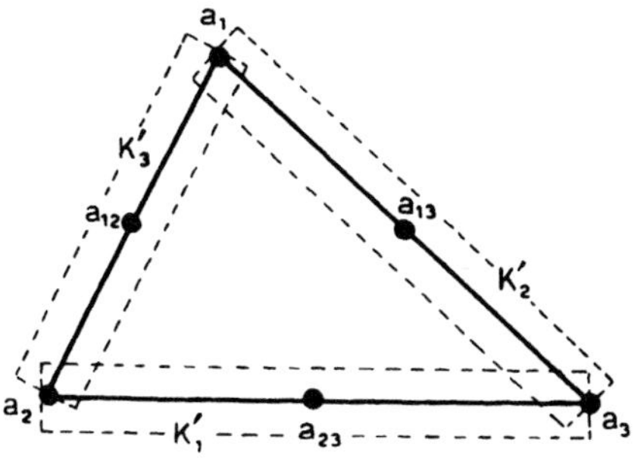

Fraeijs de Veubeke triangle

$$P_K = \{p \in P_3(K);\ \phi(p) = 0\};\ \dim P_K = 9;$$

$$\phi(p) = 27p(a_{123}) - \sum_{i=1}^{3} p(a_i) - 8 \sum_{1 \leq i < j \leq 3} p(a_{ij})$$

$$+ 3 \sum_{i=1}^{3} \frac{1}{|K_i'|} \int_{K_i'} Dp(.)(a_{123} - a_i)\, d\gamma;$$

$$\Sigma_K = \Big\{ p(a_i),\ 1 \leq i \leq 3;\ p(a_{ij}),\quad 1 \leq i \leq j \leq 3;$$

$$\frac{1}{|K_i'|} \int_{K_i'} \partial_\nu p\, d\gamma,\ 1 \leq i \leq 3 \Big\}.$$

Fig. 6.2.4

brick; cf. Section 4.2). All the relevant data are indicated in Fig. 6.2.4 where, for each $i = 1, 2, 3$, $|K_i'|$ denotes the length of the side K_i'.

(i) In each case, prove the P_K-unisolvence of the given sets Σ_K and that, for regular families, one has

$$\forall v \in H^3(K) \subset \operatorname{dom} \Pi_K, \quad |v - \Pi_K v|_{m,K} \leq C h^{3-m} |v|_{3,K},$$
$$0 \leq m \leq 3,$$

i.e., *regular families of Morley's triangles or Fraeijs de Veubeke triangles are almost-affine.* Prove in particular that the space P_K corresponding to the Fraeijs de Veubeke triangle contains the space $P_2(K)$.

(ii) For each finite element, describe the associated finite element space X_h, and then let $V_h = X_{00h}$, where X_{00h} is composed of the functions in X_h whose degrees of freedom vanish along the boundary Γ.

Show that neither element is of class $\mathscr{C}^0$. However, show that in each case the averages of the first order partial derivatives are the same across any side common to two adjacent finite elements, while the same averages vanish along a side included in Γ.

(iii) Show that for both elements the semi-norm $\|.\|_h$ of (6.2.5) is a norm over the space V_h.

(iv) Show that if the solution u belongs to the space $H^4(\Omega)$, the error estimates

$$\|u - u_h\|_h \leqslant C(h|u|_{3,\Omega} + h^2|u|_{4,\Omega})$$

holds. Therefore, contrary to the Zienkiewicz triangle, no restriction need to be imposed on the geometry of the triangulations so as to obtain convergence.

[Hint: The decomposition (6.2.20) is here to be replaced by

$$\forall w_h \in V_h,\quad D_h(u, w_h) = a_h(u, w_h) - f(w_h)$$
$$= \sum_{K \in \mathcal{T}_h} \int_{\partial K} (\Delta u - (1-\sigma)\partial_{\tau\tau_K} u)\partial_{\nu_K}(w_{h|K})\,\mathrm{d}\gamma$$
$$- \sum_{K \in \mathcal{T}_h} \int_{\partial K} \partial_{\nu_K}\Delta u\ w_{h|K} + (1-\sigma)\partial_{\nu\tau_K} u\,\partial_{\tau_K}(w_{h|K})\,\mathrm{d}\gamma,$$

and the key idea is again to subtract off appropriate "conforming" parts in the above expression. Then it is possible to apply the bilinear lemma (one side at a time rather than one element at a time, as in the case of Wilson's brick or Adini's rectangle).]

Bibliography and comments

6.1 and 6.2. The first interpolation error estimates for the Argyris triangle are due to Zlámal (1968), who obtained estimates in the spaces $\mathscr{C}^m(K)$. The results and methods of M. Zlámal were extended by Ženíšek (1970) to finite elements which yield inclusions of the form $X_h \subset \mathscr{C}^m(\bar{\Omega})$. Bramble & Zlámal (1970) have obtained estimates in Sobolev norms, which are contained in the estimates of Theorem 6.1.1.

The Hsieh–Clough–Tocher triangle appeared in Clough & Tocher (1965). It is also named after Hsieh who was the first to conceive in 1962 the idea of matching three polynomials so as to get a finite element of class $\mathscr{C}^1$. The interpolation theory given in Theorem 6.1.3 is based on Ciarlet (1974c) where a proof of unisolvence was also given along the lines indicated in Exercise 6.1.2. The proof of unisolvence given in

Theorem 6.1.2 is due to PERCELL (1976). See also DOUGLAS, DUPONT, PERCELL & SCOTT (1976). The Fraeijs de Veubeke–Sander quadrilateral (cf. Exercise 6.1.5) is due to SANDER (1964) and FRAEIJS DE VEUBEKE (1965a, 1968), and it has been theoretically studied by CIAVALDINI & NÉDÉLEC (1974).

The singular Zienkiewicz triangle is found in Section 10.10 of ZIENKIEWICZ (1971), where alternate singular finite elements are also described. Since the second derivatives of the functions in the space P_K have singularities at the vertices (Remark 6.1.6), very accurate quadrature schemes are used in practical computations. IRONS & RAZZAQUE (1972b) (see also RAZZAQUE (1973)) obviate this computational difficulty by "smoothing" the second derivatives. Other ways of adding rational functions are mentioned in BIRKHOFF & MANSFIELD (1974) (cf. Exercise 6.1.7), MANSFIELD (1974, 1976b), DUPUIS & GOEL (1970a). Boolean sum interpolation theory can also be used to derive *blending polynomial interpolants*, which interpolate a function $v \in \mathscr{C}^m(K)$ and all its derivatives of order $\leq m$ on the (possibly curved) boundary of a triangle K. In this direction, see BARNHILL (1976a, 1976b), BARNHILL, BIRKHOFF & GORDON (1973), BARNHILL & GREGORY (1975a, 1975b).

For a discussion about the use of finite elements of class $\mathscr{C}^1$ from the engineering viewpoint, see ZIENKIEWICZ (1971, chapter 10). There, finite elements of class $\mathscr{C}^1$ are called "compatible" while finite elements which are not of class $\mathscr{C}^1$ are called "incompatible", and rational functions such as those which are used in the singular Zienkiewicz triangle are called "singular shape functions". The Bogner–Fox–Schmit rectangle is not the only rectangular finite element of class $\mathscr{C}^1$ that may be used in practice. See for example GOPALACHARYULU (1973, 1976).

The general approach followed in Section 6.2 is that of CIARLET (1974a, 1974b). In LASCAUX & LESAINT (1975), a thorough study is made not only of Adini's rectangle, but of other nonconforming finite elements for the plate problem, such as the *Zienkiewicz triangle*, *Morley's triangle* (cf. Exercise 6.2.3) and various instances of *Fraeijs de Veubeke triangles* (an example of which is given in Exercise 6.2.3).

A survey of the use of such nonconforming elements, from an Engineering viewpoint, is found in ZIENKIEWICZ (1971, chapter 10). Adini's rectangle is due to ADINI & CLOUGH (1961) and MELOSH (1963) and, for this reason, it is sometimes called the *ACM rectangle*. The convergence of Adini's rectangle has also been studied by KIKUCHI (1975d, 1976a) and MIYOSHI (1972). KIKUCHI (1975d) considers in addition the use of this element for the approximation of the eigenvalue problem. This last problem is also considered from a numerical standpoint by LINDBERG & OLSON (1970) for conforming and nonconforming finite elements. An extension to the case of *curved nonconforming elements* is considered in BARNHILL & BROWN (1975).

Although some of the references given in Section 4.2 were more specifically concerned with second-order problems, some of them are also relevant in the present situation, notably CÉA (1976), NITSCHE (1974), OLIVEIRA (1976).

There are alternate definitions of nonconforming methods. For example, let us assume that we are given a finite element space V_h which satisfies the inclusion $V_h \subset \mathscr{C}^0(\bar{\Omega}) \cap H_0^1(\Omega)$. Assuming as usual that the functions in the spaces P_K are smooth, the conformity would require the additional conditions that $\partial_\nu(v_{h|K_1}) + \partial_\nu(v_{h|K_2}) = 0$ along any side K' common to two adjacent finite elements K_1 and K_2, and that $\partial_\nu v_h = 0$ along Γ. If these conditions cannot be exactly fulfilled, they may be considered as *constraints*, and accordingly, they may be dealt with either by a *penalty method* or by *duality techniques*.

In the first approach, one minimizes a functional of the form

$$J_h^*(v_h) = \frac{1}{2} a_h(v_h, v_h) - f(v_h) + \frac{1}{\epsilon(h)} \Phi(v_h)$$

where

$$\Phi(v_h) = \sum_{\substack{K_1, K_2 \in \mathcal{T}_h \\ K_1 \neq K_2}} \int_{K_1 \cap K_2} (\partial_\nu(v_{h|K_1}) + \partial_\nu(v_{h|K_2}))^2 \, \mathrm{d}\gamma + \int_\Gamma (\partial_\nu v_h)^2 \, \mathrm{d}\gamma,$$

and $\epsilon(\cdot)$ is a function of h which approaches zero with h. The function $\epsilon(\cdot)$ is usually of the form $\epsilon(h) = Ch^\sigma$, $C > 0$, where the exponent $\sigma > 0$ is to be chosen so as to maximize the order of convergence. A method of this type has been studied by BABUŠKA & ZLÁMAL (1973) who have shown that the use of the Hermite triangle of type (3) results in the error estimates

$$\|u - u_h\|_h \leq C\sqrt{h}\|u\|_{3,\Omega}, \quad \|u - u_h\|_{1,\Omega} \leq Ch\|u\|_{3,\Omega}$$

if $u \in H^3(\Omega)$, with the optimal choice $\epsilon(h) = Ch^2$, and

$$\|u - u_h\|_h \leq Ch\|u\|_{4,\Omega}, \quad \|u - u_h\|_{1,\Omega} \leq Ch\|u\|_{4,\Omega},$$

if $u \in H^4(\Omega)$, with the optimal choice $\epsilon(h) = Ch^3$ (let us add however that this penalty method is analyzed in the case of the biharmonic problem instead of the plate problem). Such techniques are used in practice: See ZIENKIEWICZ (1974).

The second approach consists in introducing an appropriate Lagrangian. This is done for example by HARVEY & KELSEY (1971) who use the Hermite triangle of type (3) for solving the plate problem.

Let us next review further aspects of the finite element approximation of the plate problem and of more general fourth-order problems. RANNACHER (1976a) has obtained error estimates in the norm $|\cdot|_{0,\infty,\Omega}$. The effect of numerical integration is analyzed in BERNADOU & DUCATEL (1976).

As regards the approximation of fourth-order problems on domains with curved boundaries, we mention MANSFIELD (1976b), who considers in addition the effect of numerical integration. Her approach parallels that given in CIARLET & RAVIART (1972c) for second-order problems. Curved isoparametric finite elements of a new type are suggested by ROBINSON (1973). In the case of the simply supported plate problem (cf. Exercise 1.2.7), we mention the *Babuška paradox* (cf. BABUŠKA (1963); see also BIRKHOFF (1969)): Contrary to second-order problems, no convergent approximation may be found if the curved boundary is replaced by a polygonal domain: This is because the boundary condition $\Delta u - (1-\sigma)\partial_{\tau\tau} u = 0$ on Γ (which is included in the variational formulation) is then replaced by the boundary condition $\partial_{\nu\nu} u = 0$.

Additional references concerning the handling of curved boundaries and/or boundary conditions for the plate problems are NITSCHE (1971, 1972b), CHERNUKA, COWPER, LINDBERG & OLSON (1972), and the survey of SCOTT (1976b).

Finite element approximation of variational inequalities of order four are considered by GLOWINSKI (1975, 1976b). See also GLOWINSKI, LIONS & TRÉMOLIÈRES (1976b, Chapter 4).

When large vertical displacements are considered, the plate problem amounts (cf. LANDAU & LIFSCHITZ (1967, Chapter 2)) to finding a pair $(u_1, u_2) \in (H_0^2(\Omega))^2$, solution of two coupled nonlinear equations, known as *von Karmann's equations*:

$$a_1 \Delta^2 u_1 - [u_1, u_2] = f, \quad f \in L^2(\Omega),$$
$$a_2 \Delta^2 u_2 + [u_1, u_1] = 0,$$

where

$$[v, w] = \partial_{11} v \partial_{22} w + \partial_{22} v \partial_{11} w - 2\partial_{12} v \partial_{12} w,$$

and a_1, a_2 are two strictly positive constants. The existence of a (possibly non unique) solution is proved in LIONS (1969, p. 53). For an analysis of a finite element approximation by a mixed method, see MIYOSHI (1976a, 1976b, 1976c, 1977). Another finite element method is proposed in BERGAN & CLOUGH (1973) to handle large displacements.

For yet other types of finite element approximation of the plate problem, see ALLMAN (1976), FRIED (1973c), FRIED & YANG (1973), IRONS (1974b), KIKUCHI (1975e), STRICKLIN, HAISLER, TISDALE & GUNDERSON (1969). Plates with cracks have been considered by YAMAMOTO & TOKUDA (1973), and YAMAMOTO & SUMI (1976). Further references are found in the next chapter, specially for the so-called mixed and hybrid methods.

CHAPTER 7

A MIXED FINITE ELEMENT METHOD

Introduction

In this chapter, we consider the problem of approximating the solution of the *biharmonic problem*: Find $u \in H_0^2(\Omega)$, $\Omega \subset \mathbf{R}^2$, such that

$$J(u) = \inf_{v \in H_0^2(\Omega)} J(v), \quad \text{with} \quad J(v) = \frac{1}{2}\int_\Omega |\Delta v|^2 \,\mathrm{d}x - \int_\Omega fv \,\mathrm{d}x.$$

Our objective is to study a method *based on a different variational formulation* of the biharmonic problem (it being implicitly understood that the above variational formulation is the standard one). Such methods fall themselves into several categories (cf. the discussion in the section "Additional Bibliography and Comments" at the end of this chapter), and it is the purpose of this chapter to study one of these, of the so-called *mixed* type. Basically, it corresponds to a variational formulation where the function u is the first argument of the minimum (u, φ) of a new functional. In this fashion, we shall directly get approximations not only of the solution u, but also of the second argument φ. Since this function φ turns out to be $-\Delta u$ in the present case, this approach is particularly appropriate for the study of two-dimensional steady-state flows, where $-\Delta u$ represents the *vorticity*.

Thus our first task in Section 7.1 is to construct a functional $\mathcal{J}$ and a space $\mathcal{V}$ such that (Theorem 7.1.2)

$$\mathcal{J}(u, -\Delta u) = \inf_{(v,\psi)\in\mathcal{V}} \mathcal{J}(v, \psi).$$

The space $\mathcal{V}$ consists of pairs $(v, \psi) \in H_0^1(\Omega) \times L^2(\Omega)$ which satisfy specific linear relations of the form $\beta((v, \psi), \mu) = 0$ for all functions $\mu \in H^1(\Omega)$.

Next, this problem is discretized in a natural way: Given a finite element space X_h contained in the space $H^1(\Omega)$, one looks for a pair

$(u_h, \varphi_h) \in \mathcal{V}_h$ such that

$$\mathcal{J}(u_h, \varphi_h) = \inf_{(v_h, \psi_h) \in \mathcal{V}_h} \mathcal{J}(v_h, \psi_h),$$

where the space $\mathcal{V}_h$ consists of those pairs $(v_h, \psi_h) \in X_{0h} \times X_h$ which satisfy linear relations of the form $\beta((v_h, \psi_h), \mu_h) = 0$ for all functions $\mu_h \in X_h$.

The major portion of Section 7.1 is then devoted to the study of convergence (Theorems 7.1.5 and 7.1.6): Our main conclusion is that, if the inclusions $P_k(K) \subset P_K$, $K \in \mathcal{T}_h$, hold, the error estimate

$$\|u - u_h\|_{1,\Omega} + |\Delta u + \varphi_h|_{0,\Omega} = O(h^{k-1})$$

holds. The main difficulty in this error analysis is that, in general, the space $\mathcal{V}_h$ is *not* a subspace of the space $\mathcal{V}$ (if this were the case, it would suffice to use the convergence analysis valid for conforming methods).

The advantages and drawbacks inherent in this method are easily understood: The main advantage is that *it suffices to use finite elements of class* $\mathscr{C}^0$, whereas finite elements of class $\mathscr{C}^1$ would be required for conforming methods. Another advantage (from the point of view of fluid mechanics) is that the present method not only yields a continuous approximation of the function u, but also of the vorticity $-\Delta u$, whereas a standard approximation using finite elements of class $\mathscr{C}^1$ would result in a discontinuous approximation $-\Delta u_h$ of the vorticity (which, in addition, needs to be computed).

The major drawback is that the computation of the discrete solution (u_h, φ_h) requires the solution of a *constrained* minimization problem, since the functions $v_h \in X_{0h}$ and $\psi_h \in X_h$ do not vary independently from one another. It is the object of Section 7.2 to show how such a problem may be solved, using *duality techniques*.

The basic idea consists in introducing an appropriate space $\mathcal{M}_h \subset X_h$ of "multipliers" and then in applying *Uzawa's method* for solving the saddle-point equations (cf. Theorem 7.2.2) of the Lagrangian associated with the present variational formulation. The convergence of Uzawa's method is established in Theorem 7.2.5.

In the process, we find an answer to a problem which has been often considered for the biharmonic problem and its various possible discretizations: We show (Theorem 7.2.4) that, *in this particular case, Uzawa's method amounts to solving a sequence of discrete Dirichlet problems for the operator* $-\Delta$.

Therefore we have at our disposal a method for approximating the solution of a fourth-order problem *which uses the same finite element programs as those needed for second-order problems.*

7.1. A mixed finite element method for the biharmonic problem

Another variational formulation of the biharmonic problem

Consider the variational problem which corresponds to the following data:

$$\begin{cases} V = H_0^2(\Omega), \\ a(u, v) = \int_\Omega \Delta u \Delta v \, dx, \\ f(v) = \int_\Omega f v \, dx, \end{cases} \tag{7.1.1}$$

where the set $\bar{\Omega}$ is a convex polygonal subset of $\mathbf{R}^2$ and the function f belongs to the space $L^2(\Omega)$. We recognize here the *biharmonic problem,* whose solution $u \in H_0^2(\Omega)$ also satisfies

$$J(u) = \inf_{v \in H_0^2(\Omega)} J(v), \tag{7.1.2}$$

with

$$J(v) = \frac{1}{2} \int_\Omega |\Delta v|^2 \, dx - \int_\Omega f v \, dx. \tag{7.1.3}$$

Thus we may equivalently consider that we are minimizing the functional

$$\mathcal{J}(v, \psi) = \frac{1}{2} \int_\Omega |\psi|^2 \, dx - \int_\Omega f v \, dx, \tag{7.1.4}$$

over those pairs $(v, \psi) \in H_0^2(\Omega) \times L^2(\Omega)$ whose elements v and ψ are related through the equality $-\Delta v = \psi$. This observation is the basis for another variational formulation of the biharmonic problem (Theorem 7.1.2), which depends on the fact that *the space*

$$\{(v, \psi) \in H_0^2(\Omega) \times L^2(\Omega); \; -\Delta v = \psi\}$$

can be described in an alternate way, as we now show.

Theorem 7.1.1. *Define the space*

$$\mathcal{V} = \{(v, \psi) \in H_0^1(\Omega) \times L^2(\Omega);\ \forall \mu \in H^1(\Omega),\quad \beta((v, \psi), \mu) = 0\}, \tag{7.1.5}$$

where

$$\beta((v, \psi), \mu) = \int_\Omega \nabla v \cdot \nabla \mu \,\mathrm{d}x - \int_\Omega \psi\mu \,\mathrm{d}x. \tag{7.1.6}$$

Then the mapping

$$(v, \psi) \in \mathcal{V} \to |\psi|_{0,\Omega}$$

is a norm over the space $\mathcal{V}$, *which is equivalent to the product norm* $(v, \psi) \in \mathcal{V} \to (|v|_{1,\Omega}^2 + |\psi|_{0,\Omega}^2)^{1/2}$, *and which makes* $\mathcal{V}$ *a Hilbert space. In addition, we have*

$$\mathcal{V} = \{(v, \psi) \in H_0^2(\Omega) \times L^2(\Omega);\ -\Delta v = \psi\}.$$

Proof. Equipped with the product norm, the space $\mathcal{V}$ is a Hilbert space since it is a closed subspace of the space $H_0^1(\Omega) \times L^2(\Omega)$.

Let (v, ψ) be any element of the space $\mathcal{V}$. The particular choice $\mu = v$ in the definition (7.1.5) of this space gives

$$|v|_{1,\Omega}^2 = \int_\Omega \psi v \,\mathrm{d}x \leq C(\Omega)|\psi|_{0,\Omega}|v|_{1,\Omega},$$

where $C(\Omega)$ is the constant appearing in the Poincaré–Friedrichs inequality (cf. (1.2.2)). Therefore,

$$(|v|_{1,\Omega}^2 + |\psi|_{0,\Omega}^2)^{1/2} \leq ((C(\Omega))^2 + 1)^{1/2}|\psi|_{0,\Omega},$$

and the first assertion is proved.

Since the set Ω has a Lipschitz-continuous boundary Γ, the Green formula

$$\forall v \in H^2(\Omega),\quad \forall \mu \in H^1(\Omega),$$

$$\int_\Omega \nabla v \cdot \nabla \mu \,\mathrm{d}x = -\int_\Omega \Delta v\, \mu \,\mathrm{d}x + \int_\Gamma \partial_\nu v\, \mu \,\mathrm{d}\gamma \tag{7.1.7}$$

holds. Let then the functions $v \in H_0^2(\Omega)$ and $\psi \in L^2(\Omega)$ be related through $-\Delta v = \psi$. For any function $\mu \in H^1(\Omega)$, an application of Green's formula (7.1.7) shows that $\beta((v, \psi), \mu) = 0$, since $\partial_\nu v = 0$ on Γ.

Conversely, let the functions $v \in H_0^1(\Omega)$ and $\psi \in L^2(\Omega)$ satisfy

$\beta((v, \psi), \mu) = 0$ for all $\mu \in H^1(\Omega)$. In particular then, we have

$$\forall \mu \in H_0^1(\Omega), \quad \int_\Omega \nabla v \cdot \nabla \mu \, dx = \int_\Omega \psi\mu \, dx,$$

so that v appears as the solution of a homogeneous Dirichlet problem for the operator $-\Delta$ on the set Ω. Since the set $\bar{\Omega}$ is convex, such a second-order boundary value problem is regular, i.e., the function v is in the space $H^2(\Omega)$. Using Green's formula (7.1.7) with functions μ in the space $H_0^1(\Omega)$, we first deduce that $-\Delta v = \psi$, and using the same Green formula with functions μ in the space $H^1(\Omega)$, we next deduce that $\partial_\nu v = 0$ along Γ. □

Theorem 7.1.2. *Let $u \in H_0^2(\Omega)$ denote the solution of the minimization problem* (7.1.2). *Then we also have*

$$\mathcal{J}(u, -\Delta u) = \inf_{(v,\psi)\in\mathcal{V}} \mathcal{J}(v, \psi), \tag{7.1.8}$$

where the functional $\mathcal{J}$ and the space $\mathcal{V}$ are defined as in (7.1.4) *and* (7.1.5), *respectively. In addition, the pair $(u, -\Delta u) \in \mathcal{V}$ is the unique solution of the minimization problem* (7.1.8).

Proof. The symmetric bilinear form

$$((u, \varphi), (v, \psi)) \in \mathcal{V} \times \mathcal{V} \to \int_\Omega \varphi\psi \, dx$$

is continuous and $\mathcal{V}$-elliptic (by Theorem 7.1.1), and the linear form

$$(v, \psi) \in \mathcal{V} \to \int_\Omega fv \, dx$$

is continuous. Therefore the minimization problem: Find an element $(u^*, \varphi) \in \mathcal{V}$ such that

$$\mathcal{J}(u^*, \varphi) = \inf_{(v,\psi)\in\mathcal{V}} \mathcal{J}(v, \psi), \tag{7.1.9}$$

has one and only one solution, also solution of the variational equations

$$\forall (v, \psi) \in \mathcal{V}, \quad \int_\Omega \varphi\psi \, dx = \int_\Omega fv \, dx. \tag{7.1.10}$$

Let us establish the relationship between this solution (u^*, φ) and the solution of problem (7.1.2). Since the pair (u^*, φ) is an element of the space $\mathcal{V}$, we deduce from Theorem 7.1.1 that the function u^* belongs to the space $H_0^2(\Omega)$ and that $-\Delta u^* = \varphi$. Applying again the same theorem in conjunction with relations (7.1.10), we find that

$$\forall v \in H_0^2(\Omega), \quad \int_\Omega \Delta u^* \Delta v \, \mathrm{d}x = \int_\Omega f v \, \mathrm{d}x,$$

and thus the function u^* coincides with the solution u of problem (7.1.2). □

The corresponding discrete problem. Abstract error estimate

We are now in a position to describe a discrete problem associated with this new variational formulation of the biharmonic problem.

Let there be given a finite element space X_h which satisfies the inclusion

$$X_h \subset H^1(\Omega). \tag{7.1.11}$$

We define as usual the finite element space

$$X_{0h} = \{v_h \in X_h;\ v_h = 0 \quad \text{on} \quad \Gamma\}, \tag{7.1.12}$$

and we let (compare with (7.1.5))

$$\mathcal{V}_h = \{(v_h, \psi_h) \in X_{0h} \times X_h;\ \forall \mu_h \in X_h, \quad \beta((v_h, \psi_h), \mu_h) = 0\}, \tag{7.1.13}$$

where the mapping $\beta((.,.),.)$ is defined as in (7.1.6).

Then, in analogy with (7.1.8), we define the *discrete problem* as follows: *Find an element* $(u_h, \varphi_h) \in \mathcal{V}_h$ *such that*

$$\mathcal{J}(u_h, \varphi_h) = \inf_{(v_h, \psi_h) \in \mathcal{V}_h} \mathcal{J}(v_h, \psi_h), \tag{7.1.14}$$

where $\mathcal{J}$ *is the functional defined in* (7.1.4).

Remark 7.1.1. It is thus realized that the *same* space X_h is used for the approximation of both spaces $H^1(\Omega)$ and $L^2(\Omega)$. It is indeed possible to develop a seemingly more general theory where another space, say Y_h, is used for approximating the space $L^2(\Omega)$, but eventually the advantage is nil: As shown in CIARLET & RAVIART (1974), one is naturally led, in the process of getting error estimates, to assume that the inclusion $Y_h \subset X_h$ holds, and this is precisely contrary to what one would have naturally expected. Besides, the assumption $X_h = Y_h$ yields significant simplifications in the developments to come. □

Theorem 7.1.3. *The discrete problem* (7.1.14) *has one and only one solution.*

Proof. Arguing as in the proof of Theorem 7.1.1, we deduce that the mapping

$$(v_h, \psi_h) \in \mathcal{V}_h \to |\psi_h|_{0,\Omega}$$

is a norm over the space $\mathcal{V}_h$. Thus, the existence and uniqueness of the solution of the discrete problem follows by an argument similar to that of Theorem 7.1.2. □

As a consequence of this result, the element $(u_h, \varphi_h) \in \mathcal{V}_h$ is also solution of the variational equations

$$\forall (v_h, \psi_h) \in \mathcal{V}_h, \quad \int_\Omega \varphi_h \psi_h \, dx = \int_\Omega f v_h \, dx. \tag{7.1.15}$$

We next begin our study of the convergence of this approximation process. As usual, we shall first establish an *abstract error estimate* (in two steps; cf. Theorems 7.1.4 and 7.1.5) and we shall then apply this to some typical finite element spaces (Theorem 7.1.6).

The abstract error estimate consists in getting an upper bound for the expression

$$|u - u_h|_{1,\Omega} + |\Delta u + \varphi_h|_{0,\Omega}$$

(recall that φ_h is an approximation of $-\Delta u$, whence the unusual sign in the second term). Notice that the above expression is a natural analogue in the present situation of the error in the norm $\|.\|_{2,\Omega}$ that arises in conforming methods.

As a first step towards getting the error estimate, we prove:

Theorem 7.1.4. *There exists a constant C independent of the space X_h such that*

$$|u - u_h|_{1,\Omega} + |\Delta u + \varphi_h|_{0,\Omega} \leqslant$$
$$\leqslant C\Big(\inf_{(v_h, \psi_h) \in \mathcal{V}_h} (|u - v_h|_{1,\Omega} + |\Delta u + \psi_h|_{0,\Omega}) + \inf_{\mu_h \in X_h} \|\Delta u + \mu_h\|_{1,\Omega}\Big). \tag{7.1.16}$$

Proof. Since the set Ω is convex, the solution u of problem (7.1.2) belongs to the space $H^3(\Omega)$. Thus we can write

$$\forall v \in \mathcal{D}(\Omega), \quad -\int_\Omega \nabla v \cdot \nabla(\Delta u)\,\mathrm{d}x = \int_\Omega \Delta v \Delta u\,\mathrm{d}x = \int_\Omega fv\,\mathrm{d}x,$$

and consequently

$$\forall v \in H_0^1(\Omega), \quad -\int_\Omega \nabla v \cdot \nabla(\Delta u)\,\mathrm{d}x = \int_\Omega fv\,\mathrm{d}x.$$

Using the definition (7.1.6) of the mapping $\beta((\cdot,\cdot),\cdot)$, we have therefore shown that, given any function $v \in H_0^1(\Omega)$ and any function $\psi \in L^2(\Omega)$, we have

$$\beta((v,\psi), -\Delta u) = \int_\Omega fv\,\mathrm{d}x + \int_\Omega \psi \Delta u\,\mathrm{d}x. \tag{7.1.17}$$

Let then (v_h, ψ_h) be an arbitrary element of the space $\mathcal{V}_h$ and let μ_h be an arbitrary element of the space X_h. Using the definition (7.1.5) of the space $\mathcal{V}$, the variational equations (7.1.15) and relation (7.1.17), we obtain

$$\int_\Omega (\Delta u + \varphi_h)(\varphi_h - \psi_h)\,\mathrm{d}x = -\beta((u_h - v_h, \varphi_h - \psi_h), \Delta u + \mu_h).$$

From this equality, we deduce that

$$\left|\int_\Omega (\Delta u + \varphi_h)(\varphi_h - \psi_h)\,\mathrm{d}x\right| \leq D(\Omega)|\varphi_h - \psi_h|_{0,\Omega}\|\Delta u + \mu_h\|_{1,\Omega},$$

where the constant $D(\Omega)$ depends solely on the constant $C(\Omega)$ of the Poincaré–Friedrichs inequality (argue as in the beginning of the proof of Theorem 7.1.1). Using this inequality, we get

$$\begin{aligned} |\varphi_h - \psi_h|_{0,\Omega}^2 &= \int_\Omega (\varphi_h - \psi_h)(\Delta u + \varphi_h)\,\mathrm{d}x \\ &\quad - \int_\Omega (\varphi_h - \psi_h)(\Delta u + \psi_h)\,\mathrm{d}x \\ &\leq D(\Omega)|\varphi_h - \psi_h|_{0,\Omega}\|\Delta u + \mu_h\|_{1,\Omega} \\ &\quad + |\varphi_h - \psi_h|_{0,\Omega}|\Delta u + \psi_h|_{0,\Omega}, \end{aligned}$$

and hence,

$$|\varphi_h - \psi_h|_{0,\Omega} \leq D(\Omega)\|\Delta u + \mu_h\|_{1,\Omega} + |\Delta u + \psi_h|_{0,\Omega}. \tag{7.1.18}$$

On the other hand, we have

$$\begin{aligned}|u-u_h|_{1,\Omega}+|\Delta u+\varphi_h|_{0,\Omega} &\leq |u-v_h|_{1,\Omega}+|v_h-u_h|_{1,\Omega}\\ &\quad+|\Delta u+\psi_h|_{0,\Omega}+|\psi_h-\varphi_h|_{0,\Omega}\\ &\leq |u-v_h|_{1,\Omega}+|\Delta u+\psi_h|_{0,\Omega}\\ &\quad+(1+C(\Omega))|\psi_h-\varphi_h|_{0,\Omega}. \qquad (7.1.19)\end{aligned}$$

Upon combining inequalities (7.1.18) and (7.1.19), we obtain

$$\begin{aligned}|u-u_h|_{1,\Omega}+|\Delta u+\varphi_h|_{0,\Omega} &\leq (|u-v_h|_{1,\Omega}+(2+C(\Omega))|\Delta u+\psi_h|_{0,\Omega})\\ &\quad+(1+C(\Omega))D(\Omega)\|\Delta u+\mu_h\|_{1,\Omega},\end{aligned}$$

and the inequality of (7.1.16) follows. □

To apply Theorem 7.1.4, we have to estimate on the one hand the expression $\inf_{\mu_h\in X_h}\|\Delta u+\mu_h\|_{1,\Omega}$, which is a standard problem. On the other hand, we also have to estimate the expression

$$\inf_{(v_h,\psi_h)\in\mathcal{V}_h}(|u-v_h|_{1,\Omega}+|\Delta u+\psi_h|_{0,\Omega}),$$

and this is no longer a standard problem, because the functions v_h and ψ_h do not vary independently in their respective spaces X_{0h} and X_h. Nevertheless, it is possible to estimate the above expression by means of the "unconstrained" terms $\inf_{v_h\in X_{0h}}|u-v_h|_{1,\Omega}$ and $\inf_{\mu_h\in X_h}|\Delta u-\mu_h|_{0,\Omega}$ (cf. (7.1.22)), provided we make use of an appropriate *inverse inequality*, as our next result shows.

Theorem 7.1.5. *Let $\alpha(h)$ be a strictly positive constant such that*

$$\forall\mu_h\in X_h,\quad |\mu_h|_{1,\Omega}\leq\alpha(h)|\mu_h|_{0,\Omega}. \qquad (7.1.20)$$

Then there exists a constant C independent of the space X_h such that

$$|u-u_h|_{1,\Omega}+|\Delta u+\varphi_h|_{0,\Omega}\leq$$

$$\leq C((1+\alpha(h))\inf_{v_h\in X_{0h}}|u-v_h|_{1,\Omega}+\inf_{\mu_h\in X_h}\|\Delta u-\mu_h\|_{1,\Omega}). \qquad (7.1.21)$$

Proof. Let (v_h,ψ_h) be an arbitrary element in the space $\mathcal{V}_h$, and let μ_h be an arbitrary element in the space X_h. The function $\nu_h=\mu_h+\psi_h$ belongs to the space X_h and thus,

$$\beta((v_h,\psi_h),\nu_h)=0.$$

Next, using the fact that $\partial_\nu u = 0$ on Γ, we get

$$\int_\Omega \Delta u\, \nu_h \,\mathrm{d}x = -\int_\Omega \nabla u \cdot \nabla \nu_h \,\mathrm{d}x,$$

so that, combining the two above equalities, we obtain

$$\int_\Omega (\Delta u + \psi_h)\nu_h \,\mathrm{d}x = \int_\Omega \nabla(v_h - u)\cdot \nabla \nu_h \,\mathrm{d}x.$$

Consequently, we get the inequality

$$\left|\int_\Omega (\Delta u + \psi_h)\nu_h \,\mathrm{d}x\right| \leq |u - v_h|_{1,\Omega}|\nu_h|_{1,\Omega}$$
$$\leq \alpha(h)|u - v_h|_{1,\Omega}|\nu_h|_{0,\Omega},$$

which in turn implies that

$$|\nu_h|_{0,\Omega}^2 = \int_\Omega (\mu_h - \Delta u)\nu_h \,\mathrm{d}x + \int_\Omega (\Delta u + \psi_h)\nu_h \,\mathrm{d}x$$
$$\leq |\Delta u - \mu_h|_{0,\Omega}|\nu_h|_{0,\Omega} + \alpha(h)|u - v_h|_{1,\Omega}|\nu_h|_{0,\Omega}.$$

From this inequality, we deduce that

$$|\Delta u + \psi_h|_{0,\Omega} \leq |\Delta u - \mu_h|_{0,\Omega} + |\nu_h|_{0,\Omega}$$
$$\leq 2|\Delta u - \mu_h|_{0,\Omega} + \alpha(h)|u - v_h|_{1,\Omega},$$

and thus,

$$\inf_{(v_h,\psi_h)\in \mathcal{V}_h} (|u - v_h|_{1,\Omega} + |\Delta u + \psi_h|_{0,\Omega}) \leq$$
$$\leq (1 + \alpha(h)) \inf_{v_h \in X_{0h}} |u - v_h|_{1,\Omega} + 2 \inf_{\mu_h \in X_h} |\Delta u - \mu_h|_{0,\Omega}. \tag{7.1.22}$$

To finish the proof, it suffices to combine inequalities (7.1.16) and (7.1.22). □

Estimate of the error $(|u - u_h|_{1,\Omega} + |\Delta u + \phi_h|_{0,\Omega})$

To apply the abstract error estimate proved in the previous theorem, we shall need the following standard assumptions on the family of finite element spaces X_h:

(H1) The associated family of triangulations $\mathcal{T}_h$ is regular.

(H2) All the finite elements (K, P_K, Σ_K), $K \in \bigcup_h \mathcal{T}_h$, are affine-equivalent to a single reference finite element $(\hat{K}, \hat{P}, \hat{\Sigma})$.

(H3) All the finite elements (K, P_K, Σ_K), $K \in \bigcup_h \mathcal{T}_h$, are of class $\mathscr{C}^0$.

(H4) The family of triangulations satisfies an inverse assumption (cf. 3.2.28)).

Theorem 7.1.6. *In addition to* (H1), (H2), (H3) *and* (H4), *assume that there exists an integer* $k \geq 2$ *such that the following inclusions are satisfied*:

$$P_k(\hat{K}) \subset \hat{P} \subset H^1(\hat{K}). \tag{7.1.23}$$

Then if the solution $u \in H_0^2(\Omega)$ *of the minimization problem* (7.1.2) *belongs to the space* $H^{k+2}(\Omega)$, *there exists a constant* C *independent of* h *such that*

$$|u - u_h|_{1,\Omega} + |\Delta u + \varphi_h|_{0,\Omega} \leq Ch^{k-1}(|u|_{k+1,\Omega} + |\Delta u|_{k,\Omega}). \tag{7.1.24}$$

Proof. In view of the inclusions (7.1.23), there exist constants C independent of h such that

$$\inf_{v_h \in X_{0h}} |u - v_h|_{1,\Omega} \leq Ch^k |u|_{k+1,\Omega}, \tag{7.1.25}$$

$$\inf_{\mu_h \in X_h} \|\Delta u - \mu_h\|_{1,\Omega} \leq Ch^{k-1} |\Delta u|_{k,\Omega}. \tag{7.1.26}$$

Next, the inverse assumption allows us to conclude (cf. (3.2.35)) that the constants $\alpha(h)$ in inequalities (7.1.20) may be taken of the form

$$\alpha(h) = \frac{C}{h}, \tag{7.1.27}$$

for another constant C independent of h. Then the conclusion follows by using relations (7.1.25), (7.1.26) and (7.1.27) in the error estimate (7.1.21). □

Remark 7.1.2. In principle, an inclusion such as $H^{k+1}(\hat{K}) \hookrightarrow \mathscr{C}^s(\hat{K})$ (where s is the maximal order of partial derivatives occurring in the definition of the set $\hat{\Sigma}$) should have been added, but it is always satisfied *in practice*: Since $s = 0$ or 1 for finite elements of class $\mathscr{C}^0$ and since $n = 2$, the inclusion $H^3(\hat{K}) \subset \mathscr{C}^s(\hat{K})$ holds. □

Concluding remarks

Let us briefly discuss the application of this theorem: The major conclusion is that *one can solve the biharmonic problem with the same* *finite element spaces that are normally used for solving second-order problems, provided the inclusions $P_2(K) \subset P_K$, $K \in \mathcal{T}_h$, hold.* If we are using in particular triangles of type (k), $k \geq 2$, we get

$$\|u - u_h\|_{1,\Omega} = O(h^{k-1}) \quad \text{and} \quad |\Delta u + \varphi_h|_{0,\Omega} = O(h^{k-1}).$$

We shall therefore retain two basic advantages of this method: First, we get a convergent approximation to the solution u (albeit in the norm $\|\cdot\|_{1,\Omega}$ instead of the norm $\|\cdot\|_{2,\Omega}$) with much less sophisticated finite element spaces than would be required in conforming methods. The second advantage is that we obtain a convergent approximation φ_h of the vorticity $-\Delta u$, a physical quantity of interest in steady-state flows.

Nevertheless, one should keep in mind that, in spite of the simplicity of the spaces X_h, there remains the practical problem of actually computing the pair (u_h, φ_h). This is the object of the next section.

Exercise

7.1.1. Following CIARLET & GLOWINSKI (1975), the object of this problem is to show that *the solution of the biharmonic problem can be reduced to the solution of a sequence of Dirichlet problems for the operator* $-\Delta$ (indeed, the analysis which shall be developed in the next section is nothing but the discrete analogue of what follows).

We recall that (cf. Theorem 7.1.2)

$$\mathcal{J}(u, -\Delta u) = \inf_{(v,\psi)\in\mathcal{V}} \mathcal{J}(v, \psi),$$

where the functional $\mathcal{J}$ and the space $\mathcal{V}$ are defined as in (7.1.4) and (7.1.5), respectively. Let there be given a subspace $\mathcal{M}$ of the space $H^1(\Omega)$ such that we may write the direct sum

$$H^1(\Omega) = H_0^1(\Omega) \oplus \mathcal{M}.$$

We next introduce the space

$$\mathcal{W} = \{(v, \psi) \in H_0^1(\Omega) \times L^2(\Omega);\ \forall \mu \in H_0^1(\Omega),\ \beta((v, \psi), \mu) = 0\},$$

where the mapping β is defined as in (7.1.6), and we define the *Lagrangian*

$$\mathcal{L}((v, \psi), \mu) = \mathcal{J}(v, \psi) + \beta((v, \psi), \mu).$$

(i) Show that, given a function $\lambda \in \mathcal{M}$, the problem: Find an element

$(u_\lambda, \varphi_\lambda) \in \mathcal{W}$ such that

$$\mathcal{L}((u_\lambda, \varphi_\lambda), \lambda) = \inf_{(v, \psi) \in \mathcal{W}} \mathcal{L}((v, \psi), \lambda),$$

has one and only one solution, which may also be obtained by solving the following Dirichlet problems for the operator $-\Delta$:

(*) Find a function $\varphi_\lambda \in H^1(\Omega)$ such that

$$(\varphi_\lambda - \lambda) \in H_0^1(\Omega),$$

$$\forall v \in H_0^1(\Omega), \quad \int_\Omega \nabla \varphi_\lambda \cdot \nabla v \, dx = \int_\Omega fv \, dx.$$

(**) Find a function $u_\lambda \in H_0^1(\Omega)$ such that

$$\forall v \in H_0^1(\Omega), \quad \int_\Omega \nabla u_\lambda \cdot \nabla v \, dx = \int_\Omega \varphi_\lambda v \, dx$$

(notice that since $\bar{\Omega}$ is a convex polygon, the function u_λ is in fact in the space $H^2(\Omega) \cap H_0^1(\Omega)$).

(ii) Let u denote the solution of problem (7.1.2), and let λ^* be that function in the space $\mathcal{M}$ which is such that the function $(\Delta u + \lambda^*)$ belongs to the space $H_0^1(\Omega)$. Show that $((u, -\Delta u), \lambda^*)$ is the unique *saddle-point* of the Lagrangian $\mathcal{L}$ over the space $\mathcal{W} \times \mathcal{M}$, in the sense that

$$\forall (v, \psi) \in \mathcal{W}, \quad \forall \mu \in \mathcal{M},$$
$$\mathcal{L}((u, -\Delta u), \mu) \leq \mathcal{L}((u, -\Delta u), \lambda^*) \leq \mathcal{L}((v, \psi), \lambda^*).$$

(iii) As a consequence of (ii), show that

$$\mathcal{L}((u, -\Delta u), \lambda^*) = \max_{\lambda \in \mathcal{M}} g(\lambda),$$

where the function $g: \mathcal{M} \to \mathbf{R}$ is defined by

$$g: \lambda \in \mathcal{M} \to g(\lambda) = \min_{(v, \psi) \in \mathcal{W}} \mathcal{L}((v, \psi), \lambda)$$
$$= \mathcal{L}((u_\lambda, \varphi_\lambda), \lambda) = \frac{1}{2} \int_\Omega |\varphi_\lambda|^2 \, dx$$

(this is a standard device in duality theory; see for example EKELAND & TEMAM (1974, chapter VI).

(iv) We next apply the *gradient method* to the maximization problem of question (iii) (a technique known as *Uzawa's method* for the original minimization problem, then called the *primal problem*): Given any function $\lambda_0 \in \mathcal{M}$ and a parameter $\rho > 0$ (to be specified in (v)), we define a sequence of functions $\lambda^n \in \mathcal{M}$ by the recurrence relation:

$$\forall \mu \in \mathcal{M}, \quad (\lambda^{n+1} - \lambda^n, \mu)_{\mathcal{M}} = -\rho\langle Dg(\lambda^n), \mu\rangle,$$

where $(.,.)_{\mathcal{M}}$ is an inner product in the space $\mathcal{M}$, whose associated norm is assumed to be equivalent to the norm $\|.\|_{1,\Omega}$, and $\langle .,. \rangle$ denotes the pairing between the spaces $\mathcal{M}'$ (= dual space of $\mathcal{M}$) and $\mathcal{M}$. Show that the function g is indeed everywhere differentiable over the space $\mathcal{M}$ and that one iteration of Uzawa's method consists of the following steps:

(*) Given a function $\lambda^n \in \mathcal{M}$, find the function $\varphi^n \in H^1(\Omega)$ which satisfies:

$$(\varphi^n - \lambda^n) \in H_0^1(\Omega),$$

$$\forall v \in H_0^1(\Omega), \quad \int_\Omega \nabla\varphi^n \cdot \nabla v \, dx = \int_\Omega fv \, dx.$$

(**) Find the function $u^n \in H_0^1(\Omega)$ which satisfies

$$\forall v \in H_0^1(\Omega), \quad \int_\Omega \nabla u^n \cdot \nabla v \, dx = \int_\Omega \varphi^n v \, dx.$$

(***) Find the function $\lambda^{n+1} \in \mathcal{M}$ which satisfies

$$\forall \mu \in \mathcal{M}, \quad (\lambda^{n+1} - \lambda^n, \mu)_{\mathcal{M}} = \rho\beta((u^n, \varphi^n), \mu).$$

(v) Show that the method described in question (iv) is convergent, in the sense that

$$\lim_{n\to\infty} \|u^n - u\|_{1,\Omega} = 0,$$

$$\lim_{n\to\infty} |\varphi^n + \Delta u|_{0,\Omega} = 0,$$

provided that

$$0 < \rho < 2c^2\sigma^2,$$

where the quantity σ is defined by

$$\sigma = \inf_{v \in H^2(\Omega)\cap H_0^1(\Omega)} \frac{|\Delta v|_{0,\Omega}}{\|\partial_\nu v\|_{L^2(\Gamma)}},$$

and c is any constant such that

$$\forall \mu \in H^1(\Omega), \quad c\|\mu\|_{L^2(\Gamma)} \leq \sqrt{(\mu, \mu)_{\mathcal{M}}}.$$

7.2. Solution of the discrete problem by duality techniques

Replacement of the constrained minimization problem by a saddle-point problem

Let us briefly review the definition of the discrete problem: We must find the unique element $(u_h, \varphi_h) \in \mathcal{V}_h$ which satisfies

$$\mathcal{J}(u_h, \varphi_h) = \inf_{(v_h, \psi_h) \in \mathcal{V}_h} \mathcal{J}(v_h, \psi_h), \tag{7.2.1}$$

with

$$\mathcal{V}_h = \{(v_h, \psi_h) \in X_{0h} \times X_h;\ \forall \mu_h \in X_h,\ \beta((v_h, \psi_h), \mu_h) = 0\}, \tag{7.2.2}$$

$$\beta((v_h, \psi_h), \mu_h) = \int_\Omega \nabla v_h \cdot \nabla \mu_h \,\mathrm{d}x - \int_\Omega \psi_h \mu_h \,\mathrm{d}x, \tag{7.2.3}$$

$$\mathcal{J}(v_h, \psi_h) = \frac{1}{2} \int_\Omega |\psi_h|^2 \,\mathrm{d}x - \int_\Omega f v_h \,\mathrm{d}x. \tag{7.2.4}$$

In the sequel, we assume that $\mathcal{M}_h$ is any supplementary subspace of the space X_{0h} in the space X_h, i.e., one has

$$X_h = X_{0h} \oplus \mathcal{M}_h \tag{7.2.5}$$

(practical choices of such subspaces $\mathcal{M}_h$ will be given later on). We also define the space

$$\mathcal{W}_h = \{(v_h, \psi_h) \in X_{0h} \times X_h;\ \forall \mu_h \in X_{0h},\ \beta((v_h, \psi_h), \mu_h) = 0\}, \tag{7.2.6}$$

and the *Lagrangian*

$$\mathcal{L}: (X_{0h} \times X_h) \times X_h \to \mathbf{R} \tag{7.2.7}$$

defined for all functions $v_h \in X_{0h}$, $\psi_h \in X_h$, $\mu_h \in X_h$ by

$$\mathcal{L}((v_h, \psi_h), \mu_h) = \mathcal{J}(v_h, \psi_h) + \beta((v_h, \psi_h), \mu_h), \tag{7.2.8}$$

the mappings β and $\mathcal{J}$ being given as in (7.2.3) and (7.2.4), respectively.

The next result is basic to the subsequent analysis.

Theorem 7.2.1. *Given a function $\lambda_h \in \mathcal{M}_h$, the minimization problem: Find an element $(u_{\lambda_h}, \varphi_{\lambda_h}) \in \mathcal{W}_h$ such that*

$$\mathcal{L}((u_{\lambda_h}, \varphi_{\lambda_h}), \lambda_h) = \inf_{(v_h, \psi_h) \in \mathcal{W}_h} \mathcal{L}((v_h, \psi_h), \lambda_h), \tag{7.2.9}$$

has one and only one solution, which may also be obtained through the consecutive solutions of the following problems:

(i) *Find a function* $\varphi_{\lambda_h} \in X_h$ *such that*

$$(\varphi_{\lambda_h} - \lambda_h) \in X_{0h}, \tag{7.2.10}$$

$$\forall v_h \in X_{0h}, \quad \int_\Omega \nabla \varphi_{\lambda_h} \cdot \nabla v_h \, dx = \int_\Omega f v_h \, dx. \tag{7.2.11}$$

(ii) *Find a function* $u_{\lambda_h} \in X_{0h}$ *such that*

$$\forall v_h \in X_{0h}, \quad \int_\Omega \nabla u_{\lambda_h} \cdot \nabla v_h \, dx = \int_\Omega \varphi_{\lambda_h} v_h \, dx. \tag{7.2.12}$$

Proof. Since the mapping

$$(v_h, \psi_h) \in \mathcal{W}_h \to |\psi_h|_{0,\Omega}$$

is a norm over the space $\mathcal{W}_h$ (argue as in the proof of Theorem 7.1.1), the minimization problem (7.2.9) has a unique solution.

Let us define a mapping

$$A_h : X_h \to X_{0h} \tag{7.2.13}$$

as follows: Given a function $\psi_h \in X_h$, the function $A_h\psi_h \in X_{0h}$ is the unique solution of the equations

$$\forall \mu_h \in X_{0h}, \quad \int_\Omega \nabla(A_h\psi_h) \cdot \nabla \mu_h \, dx = \int_\Omega \psi_h \mu_h \, dx.$$

Then the space $\mathcal{W}_h$ can also be written as

$$\mathcal{W}_h = \{(A_h\psi_h, \psi_h) \in X_{0h} \times X_h;\ \psi_h \in X_h\}.$$

Given a function $\lambda_h \in \mathcal{M}_h$, problem (7.2.9) consists in *minimizing the function* $\mathcal{L}((.,.), \lambda_h)$ *of the two variables* $v_h \in X_{0h}$ *and* $\psi_h \in X_h$ *when these two variables satisfy a relation of the form*

$$\Phi(v_h, \psi_h) = 0.$$

In the present case, the mapping $\phi: X_{0h} \times X_h \to X_{0h}$ is given by

$$\Phi : (v_h, \psi_h) \in X_{0h} \times X_h \to \Phi(v_h, \psi_h) = (A_h\psi_h - v_h) \in X_{0h}.$$

The functions $\mathcal{L}((.,.), \lambda_h)$ and Φ are both differentiable. Thus there necessarily exists a unique *Lagrange multiplier* $\Xi_{\lambda_h} \in X'_{0h}$ (X'_{0h} = dual space of X_{0h}) such that

$$D\mathscr{L}((u_{\lambda_h}, \varphi_{\lambda_h}), \lambda_h) = \Xi_{\lambda_h} \cdot D\Phi(u_{\lambda_h}, \varphi_{\lambda_h}).$$

By taking the partial derivatives with respect to each argument, the above equality is seen to be equivalent to the two relations

$$\forall v_h \in X_{0h}, \quad \langle \Xi_{\lambda_h}, v_h\rangle + \int_\Omega \nabla\lambda_h \cdot \nabla v_h \, dx = \int_\Omega f v_h \, dx, \tag{7.2.14}$$

$$\forall \psi_h \in X_h, \quad \langle \Xi_{\lambda_h}, A_h\psi_h\rangle + \int_\Omega \psi_h\lambda_h \, dx = \int_\Omega \varphi_{\lambda_h}\psi_h \, dx, \tag{7.2.15}$$

where $\langle .\, , .\rangle$ stands for the duality pairing between the spaces X'_{0h} and X_{0h}. Consequently, the two relations (7.2.14) and (7.2.15) and the equation

$$u_{\lambda_h} = A_h\varphi_{\lambda_h} \tag{7.2.16}$$

(which expresses that $(u_{\lambda_h}, \varphi_{\lambda_h})$ is an element of the space $\mathscr{W}_h$) allow for the determination of the functions u_{λ_h} and φ_{λ_h}.

In order to put relations (7.2.14) and (7.2.15) in a more convenient form, let us introduce the (unique) function $\xi_{\lambda_h} \in X_{0h}$ which satisfies

$$\forall v_h \in X_{0h}, \quad \langle \Xi_{\lambda_h}, v_h\rangle = \int_\Omega \nabla\xi_{\lambda_h} \cdot \nabla v_h \, dx.$$

Then relations (7.2.14) become

$$\forall v_h \in X_{0h}, \quad \int_\Omega \nabla(\xi_{\lambda_h} + \lambda_h) \cdot \nabla v_h \, dx = \int_\Omega f v_h \, dx, \tag{7.2.17}$$

while relations (7.2.15) become

$$\forall \psi_h \in X_h, \quad \int_\Omega (\xi_{\lambda_h} + \lambda_h - \varphi_{\lambda_h})\psi_h \, dx = 0,$$

and thus we deduce that

$$\varphi_{\lambda_h} - \lambda_h = \xi_{\lambda_h}. \tag{7.2.18}$$

Consequently, the proof is complete: The assertions (7.2.10), (7.2.11) and (7.2.12) have been proved in (7.2.18), (7.2.17) and (7.2.16), respectively. □

In the next theorem, we show that the Lagrangian $\mathscr{L}$ defined in (7.2.7) possesses a (unique) *saddle-point* (cf. (7.2.19) below) over the product space $\mathscr{W}_h \times \mathscr{M}_h$, *whose first argument is precisely the solution* (u_h, φ_h) *of the original minimization problem* (7.2.1).

Theorem 7.2.2. *Let* φ_{0h} *be the (unique) function in the space* X_{0h} *such that the function* $(\varphi_h - \varphi_{0h})$ *belongs to the space* $\mathscr{M}_h$. *Then the element* $((u_h, \varphi_h), \varphi_h - \varphi_{0h}) \in \mathscr{W}_h \times \mathscr{M}_h$ *is the unique saddle-point of the Lagrangian* $\mathscr{L}$ *over the space* $\mathscr{W}_h \times \mathscr{M}_h$, i.e., *one has*

$$\forall (v_h, \psi_h) \in \mathscr{W}_h, \quad \forall \mu_h \in \mathscr{M}_h,$$

$$\mathscr{L}((u_h, \varphi_h), \mu_h) \leq \mathscr{L}((u_h, \varphi_h), \varphi_h - \varphi_{0h}) \leq \mathscr{L}((v_h, \psi_h), \varphi_h - \varphi_{0h}). \tag{7.2.19}$$

Proof. Since the pair (u_h, φ_h) belongs to the space $\mathscr{V}_h$, we deduce that

$$\forall \mu_h \in \mathscr{M}_h, \quad \mathscr{L}((u_h, \varphi_h), \mu_h) = \mathscr{J}(u_h, \varphi_h),$$

and thus the first inequality of (7.2.19) is proved. The second inequality amounts to showing that

$$\mathscr{L}((u_h, \varphi_h), \varphi_h - \varphi_{0h}) = \inf_{(v_h, \psi_h) \in \mathscr{W}_h} \mathscr{L}((v_h, \psi_h), \varphi_h - \varphi_{0h}).$$

Thus, by Theorem 7.2.1, it suffices to verify that (cf. (7.2.11))

$$\forall v_h \in X_{0h}, \quad \int_\Omega \nabla \varphi_h \cdot \nabla v_h \, dx = \int_\Omega f v_h \, dx, \tag{7.2.20}$$

since the relations corresponding to (7.2.10) and (7.2.12) are clearly satisfied. Given any function $v_h \in X_{0h}$, let ψ_h denote the (unique) function in the space X_h such that $(v_h, \psi_h) \in \mathscr{V}_h$, i.e., which satisfies

$$\forall \mu_h \in X_h, \quad \int_\Omega \nabla v_h \cdot \nabla \mu_h \, dx = \int_\Omega \psi_h \mu_h \, dx. \tag{7.2.21}$$

Since, by (7.1.15),

$$\int_\Omega f v_h \, dx = \int_\Omega \varphi_h \psi_h \, dx,$$

an application of (7.2.21) with $\mu_h = \varphi_h$ yields (7.2.20).

Next, let $((u_h^*, \varphi_h^*), \lambda_h^*) \in \mathscr{W}_h \times \mathscr{M}_h$ be a saddle-point of the Lagrangian

$\mathcal{L}$ over the space $\mathcal{W}_h \times \mathcal{M}_h$. From Theorem 7.2.1, we deduce that

$$(\varphi_h^* - \lambda_h^*) \in X_{0h}, \tag{7.2.22}$$

$$\forall v_h \in X_{0h}, \quad \int_\Omega \nabla \varphi_h^* \cdot \nabla v_h \, dx = \int_\Omega f v_h \, dx, \tag{7.2.23}$$

$$\forall v_h \in X_{0h}, \quad \int_\Omega \nabla u_h^* \cdot \nabla v_h \, dx = \int_\Omega \varphi_h^* v_h \, dx, \tag{7.2.24}$$

since, by definition,

$$\mathcal{L}((u_h^*, \varphi_h^*), \lambda_h^*) = \inf_{(v_h, \psi_h) \in \mathcal{W}_h} \mathcal{L}((v_h, \psi_h), \lambda_h^*).$$

On the other hand, we have

$$\mathcal{L}((u_h^*, \varphi_h^*), \lambda_h^*) = \sup_{\mu_h \in \mathcal{M}_h} \mathcal{L}((u_h^*, \varphi_h^*), \mu_h),$$

so that

$$\forall \mu_h \in \mathcal{M}_h, \quad \int_\Omega \nabla u_h^* \cdot \nabla \mu_h \, dx = \int_\Omega \varphi_h^* \mu_h \, dx. \tag{7.2.25}$$

Since the space X_h is the direct sum of the subspaces X_{0h} and $\mathcal{M}_h$, we deduce from (7.2.24) and (7.2.25) that the pair (u_h^*, φ_h^*) is an element of the space $\mathcal{V}_h$.

Let (v_h, ψ_h) be an arbitrary element of the space $\mathcal{V}_h$. Using the definition of this space and relations (7.2.23), we have

$$\int_\Omega \varphi_h^* \psi_h \, dx = \int_\Omega \nabla \varphi_h^* \cdot \nabla v_h \, dx = \int_\Omega f v_h \, dx.$$

Therefore, we have shown that (cf. (7.1.15))

$$u_h^* = u_h \quad \text{and} \quad \varphi_h^* = \varphi_h.$$

Finally, it results from (7.2.22) that the function $\lambda_h^* \in \mathcal{M}_h$ is equal to the function $\varphi_h - \varphi_{0h}$. □

Use of Uzawa's method. Reduction to a sequence of Dirichlet problems for the operator $-\Delta$

Using a well-known result in duality theory (cf. Exercise 7.2.1), the property for $((u_h, \varphi_h), \varphi_h - \varphi_{0h})$ to be a saddle-point of the Lagrangian $\mathcal{L}$ implies that we also have

$$\mathcal{L}((u_h, \varphi_h), \varphi_h - \varphi_{0h}) = \max_{\lambda_h \in \mathcal{M}_h} g(\lambda_h), \tag{7.2.26}$$

where the function $g\colon \mathcal{M}_h \to \mathbf{R}$ is defined by

$$g\colon \lambda_h \in \mathcal{M}_h \to g(\lambda_h) = \min_{(v_h,\psi_h)\in \mathcal{W}_h} \mathcal{L}((v_h, \psi_h), \lambda_h) = \mathcal{L}((u_{\lambda_h}, \varphi_{\lambda_h}), \lambda_h)$$
$$= -\frac{1}{2}\int_\Omega |\varphi_{\lambda_h}|^2 \,\mathrm{d}x, \tag{7.2.27}$$

with the notations of Theorem 7.2.1.

The basic idea is then to apply the *gradient method* to the maximization problem (7.2.26), this technique for solving the so-called "*primal*" *problem* (7.2.1) being known in optimization theory as *Uzawa's method*. Thus we need to show that the function g is differentiable, and we need to compute its derivative: This is the object of the next theorem (as usual, $\mathcal{M}_h'$ denotes the dual space of the space $\mathcal{M}_h$ and $\langle .\, , .\rangle$ denotes the duality pairing between the spaces $\mathcal{M}_h'$ and $\mathcal{M}_h$).

Theorem 7.2.3. *At any point $\lambda_h \in \mathcal{M}_h$, the function g defined in* (7.2.27) *is differentiable, and its derivative $Dg(\lambda_h) \in \mathcal{M}_h'$ is defined by the relations*

$$\forall \mu_h \in \mathcal{M}_h, \quad \langle Dg(\lambda_h), \mu_h\rangle = \beta((u_{\lambda_h}, \varphi_{\lambda_h}), \mu_h). \tag{7.2.28}$$

Proof. The function $g\colon \mathcal{M}_h \to \mathbf{R}$ can be written as

$$g = g_1 \cdot g_0,$$

where the functions $g_1\colon X_h \to \mathbf{R}$ and $g_0\colon \mathcal{M}_h \to X_h$ are respectively given by

$$g_1\colon v_h \in X_h \to -\frac{1}{2}\int_\Omega |v_h|^2 \,\mathrm{d}x,$$
$$g_0\colon \lambda_h \in \mathcal{M}_h \to \varphi_{\lambda_h} \in X_h.$$

The mapping g_0 is affine (cf. (7.2.10) and (7.2.11)) and thus we may assume that $f = 0$ for computing its derivative, in which case we find that

$$\forall \mu_h \in \mathcal{M}_h, \quad Dg_0(\lambda_h)\mu_h = \varphi^\circ_{\mu_h}, \tag{7.2.29}$$

where the function $\varphi^\circ_{\mu_h}$ is such that

$$\begin{cases} (\varphi^\circ_{\mu_h} - \mu_h) \in X_{0h}, \\ \forall v_h \in X_{0h}, \quad \displaystyle\int_\Omega \nabla\varphi^\circ_{\mu_h} \cdot \nabla v_h \,\mathrm{d}x = 0. \end{cases} \tag{7.2.30}$$

On the other hand, we have

$$\forall v_h, w_h \in \mathcal{M}_h, \quad Dg_1(v_h)w_h = -\int_\Omega v_h w_h \,\mathrm{d}x, \tag{7.2.31}$$

so that, by (7.2.29) and (7.2.31),

$$\langle Dg(\lambda_h), \mu_h\rangle = -\int_\Omega \varphi_{\lambda_h}\varphi^\circ_{\mu_h}\,\mathrm{d}x.$$

Using (7.2.12) and (7.2.30), this last expression can be transformed into

$$-\int_\Omega \varphi_{\lambda_h}\varphi^\circ_{\mu_h}\,\mathrm{d}x = \int_\Omega \varphi_{\lambda_h}(\mu_h - \varphi^\circ_{\mu_h})\,\mathrm{d}x - \int_\Omega \varphi_{\lambda_h}\mu_h\,\mathrm{d}x \qquad (7.2.32)$$
$$= \beta((u_{\lambda_h}, \varphi_{\lambda_h}), \mu_h). \qquad \square$$

We recall that the gradient method as applied to the maximization problem (7.2.26) consists in defining a sequence $(\lambda_h^n)_{n=0}^\infty$ of functions $\lambda_h^n \in \mathcal{M}_h$ by the iterative scheme:

$$\forall \mu_h \in \mathcal{M}_h, \quad (\lambda_h^{n+1} - \lambda_h^n, \mu_h)_{\mathcal{M}_h} = -\rho\langle Dg(\lambda_h^n), \mu_h\rangle, \quad n \geq 0, \qquad (7.2.33)$$

where:

$(.\,,.)_{\mathcal{M}_h}$ = an arbitrary inner product in the space $\mathcal{M}_h$,
ρ = a strictly positive parameter, the admissible range of which will be determined later (Theorem 7.2.5),
λ_h° = an arbitrary function of the space $\mathcal{M}_h$.

Using Theorems 7.2.1 and 7.2.3, we can immediately convert one iteration (7.2.33) in a more explicit form:

Theorem 7.2.4. *One iteration of Uzawa's method amounts to consecutively solving the following three problems:*

(i) *Given the function $\lambda_h^n \in \mathcal{M}_h$, find the (unique) function $\varphi_h^n \in X_h$ which satisfies*

$$(\varphi_h^n - \lambda_h^n) \in X_{0h}, \qquad (7.2.34)$$

$$\forall v_h \in X_{0h}, \quad \int_\Omega \nabla\varphi_h^n \cdot \nabla v_h\,\mathrm{d}x = \int_\Omega f v_h\,\mathrm{d}x. \qquad (7.2.35)$$

(ii) *Find the function $u_h^n \in X_{0h}$ which satisfies*

$$\forall v_h \in X_{0h}, \quad \int_\Omega \nabla u_h^n \cdot \nabla v_h\,\mathrm{d}x = \int_\Omega \varphi_h^n v_h\,\mathrm{d}x. \qquad (7.2.36)$$

(iii) *Find the function $\lambda_h^{n+1} \in \mathcal{M}_h$ which satisfies*

$$\forall \mu_h \in \mathcal{M}_h, \quad (\lambda_h^{n+1} - \lambda_h^n, \mu_h)_{\mathcal{M}_h} = \rho\beta((u_h^n, \varphi_h^n), \mu_h). \qquad \square \qquad (7.2.37)$$

In other words, *the problem of approximating the solution of a fourth-order problem* (the biharmonic problem) *is reduced here to a sequence of "discrete second-order problems"*, namely problems (7.2.34)–(7.2.35) and (7.2.36), which correspond to the discretization of a nonhomogeneous and a homogeneous Dirichlet problems for the operator $-\Delta$, respectively. As will be explained later, the solution of problem (7.2.37) requires in principle a comparatively much smaller amount of work.

Convergence of Uzawa's method

Of course, these considerations implicitly assume that for some choices of the parameter ρ, Uzawa's method is convergent: This is what we shall prove in the next theorem.

First, we need to define a mapping

$$B_h: X_h \to \mathcal{M}_h, \tag{7.2.38}$$

as follows: For each function $\psi_h \in X_h$, the function $B_h\psi_h$ is the unique function in the space $\mathcal{M}_h$ which satisfies

$$\forall \mu_h \in \mathcal{M}_h, \quad (B_h\psi_h, \mu_h)_{\mathcal{M}_h} = \beta((A_h\psi_h, \psi_h), \mu_h), \tag{7.2.39}$$

where $A_h: X_h \to X_{0h}$ is the mapping of (7.2.13). Then we let

$$\|B_h\| = \sup_{v_h \in X_h} \frac{|B_h v_h|_{\mathcal{M}_h}}{|v_h|_{0,\Omega}}, \tag{7.2.40}$$

where $|.|_{\mathcal{M}_h}$ is the norm associated with the inner product $(.\,,.)_{\mathcal{M}_h}$.

Theorem 7.2.5. *If the parameter* ρ *satisfies*

$$0 < \rho < 2\sigma_h^2, \tag{7.2.41}$$

with (cf. (7.2.40))

$$\sigma_h = \frac{1}{\|B_h\|}, \tag{7.2.42}$$

Uzawa's method is convergent, in the sense that

$$\lim_{n\to\infty} u_h^n = u_h \quad in \quad X_{0h}, \tag{7.2.43}$$

$$\lim_{n\to\infty} \varphi_h^n = \varphi_h \quad in \quad X_h. \tag{7.2.44}$$

Proof. It suffices to show that $\lim_{n\to\infty} u_h^n = 0$ in X_{0h} and $\lim_{n\to\infty} \varphi_h^n = 0$ in X_h in the special case where $f = 0$. Using the definition (7.2.39) of the mapping B_h, the recurrence relation (7.2.37) takes the form

$$\lambda_h^{n+1} = \lambda_h^n + \rho B_h \varphi_h^n,$$

which, in conjunction with (7.2.32), yields

$$|\lambda_h^{n+1}|_{\mathcal{M}_h}^2 = |\lambda_h^n|_{\mathcal{M}_h}^2 - 2\rho|\varphi_h^n|_{0,\Omega}^2 + \rho^2 |B_h\varphi_h^n|_{\mathcal{M}_h}^2,$$

since $f = 0$. Therefore, we get the inequality

$$|\lambda_h^n|_{\mathcal{M}_h}^2 - |\lambda_h^{n+1}|_{\mathcal{M}_h}^2 \geq (2\rho - \rho^2\|B_h\|^2)|\varphi_h^n|_{0,\Omega}^2,$$

which in turn shows that

$$\lim_{n\to\infty} \varphi_h^n = 0,$$

provided the parameter ρ satisfies inequality (7.2.41). In addition, we deduce that

$$\lim_{n\to\infty} u_h^n = \lim_{n\to\infty} A_h \varphi_h^n = 0,$$

and the proof is complete. □

Concluding remarks

It is worth pointing out that the convergence of the present method is thus guaranteed for *any* choice of subspace $\mathcal{M}_h$ satisfying relation (7.2.5) and *any* choice of inner product $(\cdot,\cdot)_{\mathcal{M}_h}$ over the space $\mathcal{M}_h$. What is *not* independent of these data, however, is the quantity σ_h of (7.2.41) and it is of course desirable to get a concrete estimate of this quantity: This is the object of Exercise 7.2.2.

Although the space $\mathcal{M}_h$ is not uniquely determined by the sole equation $X_h = X_{0h} \oplus \mathcal{M}_h$, there is a "canonical" choice: Let us assume for definiteness that we are using Lagrange finite elements. Then the space $\mathcal{M}_h$ consists of those functions in the space X_h which are zero at the *interior nodes*, i.e., those nodes which are situated in the set Ω.

With the above choice for the space $\mathcal{M}_h$, assume that the inner product $(\cdot,\cdot)_{\mathcal{M}_h}$ is the inner product of the space $L^2(\Gamma)$. Then if we denote by M the dimension of the space X_{0h}, the solution of either problem (i) or (ii) (cf. (7.2.34)–(7.2.35) and (7.2.36)) requires the solution of a system of M linear equations, while the solution of problem (iii) (cf. (7.2.37)) amounts to solving a system of $0(\sqrt{M})$ linear equations. As a consequence, the amount of work required for solving problem (iii) is negligible compared with the total amount of work required in one iteration of Uzawa's method, at least asymptotically.

There remains in addition the possibility of reducing the computations involved in step (iii), simply by using *numerical integration* for computing the integrals over Γ, and this is precisely why Theorem 7.2.5 was proved with an arbitrary inner product over the space $\mathcal{M}_h$. In this direction, see Exercise 7.2.3.

Exercises

7.2.1. Let V and M be two arbitrary sets and let $L: V \times M \to \mathbf{R}$ be a given mapping. A pair $(v^*, \mu^*) \in V \times M$ is a *saddle-point* of the function L if

$$\sup_{\mu \in M} L(v^*, \mu) = L(v^*, \mu^*) = \inf_{v \in V} L(v, \mu^*).$$

Show that

$$\sup_{\mu \in M} \inf_{v \in V} L(v, \mu) = L(v^*, \mu^*) = \inf_{v \in V} \sup_{\mu \in M} L(v, \mu).$$

7.2.2. Let us assume that the finite element space X_h is made up of Lagrange finite elements and that $\mathcal{M}_h$ consists of those functions in the space X_h whose values are zero at all the nodes which belong to the set Ω. Assume in addition that the inner product $(\cdot\,,\cdot)_{\mathcal{M}_h}$ is the inner product of the space $L^2(\Gamma)$. The purpose of this problem is to show that (cf. CIARLET & GLOWINSKI (1975)) the quantity σ_h defined in (7.2.42) satisfies

$$\lim_{h \to 0} \sigma_h = \sigma,$$

where

$$\sigma = \inf_{v \in H^2(\Omega) \cap H_0^1(\Omega)} \frac{|\Delta v|_{0,\Omega}}{\|\partial_\nu v\|_{L^2(\Gamma)}}.$$

Such a quantity can be estimated for simple domains: See the section "Bibliography and Comments".

(i) Let ψ and μ be arbitrary functions in the space $H^1(\Omega)$. Show that

$$\limsup_{h \to 0} \sigma_h \leqslant \frac{|\psi|_{0,\Omega} \|\mu\|_{L^2(\Gamma)}}{\left| \int_\Gamma \partial_\nu v \, \mu \, \mathrm{d}\gamma \right|},$$

and deduce from this inequality that

$$\limsup_{h\to 0} \sigma_h \leqslant \sigma.$$

[Hint: Let, for each h, ψ_h and μ_h be two functions in the space X_h such that

$$\lim_{h\to 0} \|\psi_h - \psi\|_{1,\Omega} = 0, \quad \lim_{h\to 0} \|\mu_h - \mu\|_{1,\Omega} = 0.$$

Then prove and use the inequality

$$\left.\sigma_h \leqslant \frac{|\psi_h|_{0,\Omega}\|\mu_h\|_{L^2(\Gamma)}}{\int_\Gamma B_h\psi_h\, \mu_h \,\mathrm{d}\gamma}\right].$$

(ii) For each h, let ψ_h be an arbitrary function in the space X_h and let μ_h be an arbitrary function in the space $\mathcal{M}_h$. Show that there exists a constant C independent of h such that

$$\frac{1}{\sigma} \geqslant \frac{\left|\int_\Gamma B_h\psi_h\, \mu_h \,\mathrm{d}\gamma\right|}{|\psi_h|_{0,\Omega}\|\mu_h\|_{L^2(\Gamma)}} - Ch^{1/2},$$

and deduce from this inequality that

$$\liminf_{h\to 0} \sigma_h \geqslant \sigma.$$

[Hint: For each h, let $u_h = A_h\psi_h$ and let $\tilde{u}_h \in H^2(\Omega) \cap H_0^1(\Omega)$ satisfy $-\Delta\tilde{u}_h = \psi_h$. Prove and use the inequality

$$\left.\frac{1}{\sigma} \geqslant \frac{\left|\int_\Gamma B_h\psi_h\, \mu_h \,\mathrm{d}\gamma + \int_\Omega \nabla(\tilde{u}_h - u)\cdot\nabla\mu_h \,\mathrm{d}x\right|}{|\psi_h|_{0,\Omega}\|\mu_h\|_{L^2(\Gamma)}}\right].$$

7.2.3. Assume that the space X_h is made up of triangles of type (2), and let a quadrature scheme be given by

$$\int_\Gamma \varphi(x)\,\mathrm{d}x \sim \sum_{K'\subset\Gamma} \frac{\text{length}\,(K')}{6}\{\varphi(a_{K'}) + 4\varphi(b_{K'}) + \varphi(c_{K'})\},$$

where the symbol $\Sigma_{K'\subset\Gamma}$ means that the summation is extended over all sides K' (of the triangles) contained in the boundary Γ, and where, for each such side K', $a_{K'}$ and $c_{K'}$ denote the end-points of the segment K' while $b_{K'}$ stands for its mid-point.

Show that this quadrature scheme induces an inner product over the space $\mathcal{M}_h$. What is the corresponding structure of the matrix of the linear system found in the solution of problem (7.2.37)?

Bibliography and comments

7.1. As mentioned in the introduction of this chapter, a general discussion of analogous and related methods (equilibrium, mixed, hybrid methods) for second-order and fourth-order problems is postponed until the next section "Additional Bibliography and Comments". We shall discuss here only the particular mixed finite element approximation of the biharmonic problem considered in this chapter.

The content of this section is based on CIARLET & RAVIART (1974). The starting point was the work of GLOWINSKI (1973), who studied a related method. R. Glowinski obtained convergence, without orders of convergence, however, for piecewise polynomials of degree $\leqslant 1$ or $\leqslant 2$. In the first case (which is not covered by the present analysis), convergence holds provided a certain "patch test" is satisfied, which amounts to saying that there are only three admissible directions for the sides of the triangles (this condition is reminiscent of the analogous condition found when Zienkiewicz triangles are used for solving the plate problem). In addition, R. Glowinski made the interesting observation that for specific choices of subspaces, the method is identical to the usual 13-point difference approximation of the operator Δ^2. Likewise, MERCIER (1974) has also studied a similar method, again proving convergence without orders of convergence. For recent developments, see FALK (1976d).

7.2. The results contained in this section are proved in CIARLET & GLOWINSKI (1975). A further, and significant, step has been recently taken by GLOWINSKI & PIRONNEAU (1976a, 1976b, 1976c, 1976d) who reduced the approximation of the biharmonic problem to (i) a *finite* number of approximate Dirichlet problems for the operator $-\Delta$ and (ii) the solution of a linear system with a symmetric and positive definite matrix. The key idea consists in transforming the biharmonic problem into a variational problem posed over the boundary Γ, in which the unknown is $-\Delta u_{|\Gamma}$.

KESAVAN & VANNINATHAN (1977) have analyzed mathematically the effect of numerical integration, combined with the use of isoparametric finite elements, in the discrete second-order problems found in the method described in Section 7.2. We also mention that BOURGAT (1976) has implemented this method (with numerical integration and isoparametric finite elements). It turns out that the results compare favorably with those obtained with more familiar finite element methods. From a practical standpoint, it is clear that this is much less complex than a direct application of numerical integration and isoparametric finite elements to the more standard discretization of the biharmonic problem.

Several authors have considered either the problem of reducing the biharmonic problem to a sequence of Dirichlet problems for the Laplacian Δ, or its discrete counterpart. In this direction, we quote BOSSAVIT (1971), EHRLICH (1971), MCLAURIN (1974), SMITH (1968, 1973). In particular, it is shown in SMITH (1968) that the quantity σ (cf. Exercises 7.1.1 and 7.2.2) is equal to $2/R$ if $\bar{\Omega}$ is a disk of radius R and satisfies the inequalities $(2\sqrt{\pi}/\sqrt{ab}) \leqslant \sigma \leqslant (\pi^2(a^2+b^2)^2/4ab(a^3+b^3))$ if $\bar{\Omega}$ is a rectangle with sides a and b. For other estimates, see PAYNE (1970).

Although the exposition is self-contained, the reader who wishes to get a better acquaintance with optimization theory, in particular with the methods and techniques of duality theory referred to here (Lagrangian, saddle-point, gradient's method, Uzawa's method, etc. . .) may consult the books of AUSLENDER (1976), CÉA (1971), EKELAND & TÉMAM (1974), LAURENT (1972).

Additional bibliography and comments

Primal, dual and primal-dual formulations

As a preliminary step towards a better understanding of the various finite element methods which shall be described later, we must shed a new light on our familiar minimization problem: Find $u \in V$ such that $J(u) = \inf_{v\in V} J(v)$. We begin our discussion with the most illuminating example, the *elasticity problem*: With the notations of Section 1.2 (cf. Fig. 1.2.3 in particular), this problem consists in minimizing *the energy* (cf. (1.2.36))

$$\begin{aligned} J(\boldsymbol{v}) &= \frac{1}{2}\int_\Omega \left\{\lambda(\operatorname{div}\boldsymbol{v})^2 + 2\mu \sum_{i,j=1}^{3} (\epsilon_{ij}(\boldsymbol{v}))^2\right\} \mathrm{d}x \\ &\quad - \left(\int_\Omega \boldsymbol{f}\cdot\boldsymbol{v}\,\mathrm{d}x + \int_{\Gamma_1} \boldsymbol{g}\cdot\boldsymbol{v}\,\mathrm{d}\gamma\right) \\ &= \frac{1}{2}a(\boldsymbol{v},\boldsymbol{v}) - f(\boldsymbol{v}), \end{aligned}$$

over the space (cf. (1.2.30))

$$V = \{\boldsymbol{v} \in (H^1(\Omega))^3;\ \boldsymbol{v} = \mathbf{0} \quad \text{on} \quad \Gamma_0\}$$

of *admissible displacements*, and the associated boundary value problem is

$$\begin{cases} -\mu\Delta\boldsymbol{u} - (\lambda+\mu)\,\mathrm{grad}\,\mathrm{div}\,\boldsymbol{u} = \boldsymbol{f} \quad \text{in} \quad \Omega, \\ \boldsymbol{u} = \mathbf{0} \quad \text{on} \quad \Gamma_0, \\ \displaystyle\sum_{j=1}^{3} \sigma_{ij}(\boldsymbol{u})\nu_j = g_i \quad \text{on} \quad \Gamma_1, \quad 1 \leq i \leq 3. \end{cases}$$

We shall call this formulation of the elasticity problem the *displacement model*, or *formulation*, and we shall call *primal problem* the corresponding minimization, or variational, problem.

It turns out that in the analysis of actual structures, the knowledge of the *stress tensor* $(\sigma_{ij})_{i,j=1}^3$ is often of greater interest than the knowledge of the displacement $\boldsymbol{u}$. To make this tensor appear as the unknown of a new variational problem, we note from (1.2.33) that (with the convention that σ_{ij} denotes an unknown component of the stress tensor, while τ_{ij} is the corresponding "generic" component, $1 \leq i, j \leq 3$)

$$\begin{aligned} a(\boldsymbol{v}, \boldsymbol{v}) &= \int_\Omega \sum_{i,j=1}^{3} \tau_{ij}(\boldsymbol{v})\epsilon_{ij}(\boldsymbol{v})\,\mathrm{d}x \\ &= \int_\Omega \left\{\left(\frac{1+\sigma}{E}\right)\sum_{i,j=1}^{3} \tau_{ij}^2 - \frac{\sigma}{E}\left(\sum_{l=1}^{3}\tau_{ll}\right)^2\right\}\mathrm{d}x, \end{aligned}$$

since relations (1.2.32) can be inverted into

$$\epsilon_{ij} = \left(\frac{1+\sigma}{E}\right)\tau_{ij} - \frac{\sigma}{E}\left(\sum_{k=1}^{3}\tau_{kk}\right)\delta_{ij}, \quad 1 \leq i, j \leq 3,$$

where $E = \mu(3\lambda + 2\mu)/(\lambda + \mu)$ and $\sigma = (\lambda/2(\lambda + \mu))$ are respectively the Young modulus and the Poisson coefficient.

Using relations (cf. (1.2.39))

$$-\mu\Delta u_i - (\lambda + \mu)\partial_i\,\mathrm{div}\,\boldsymbol{u} = -\sum_{j=1}^{3}\partial_j\sigma_{ij}, \quad 1 \leq i \leq 3,$$

and the boundary conditions $\Sigma_{j=1}^3\,\sigma_{ij}(\boldsymbol{u})\nu_j = g_i$ on Γ_1, $1 \leq i \leq 3$, one finds (cf. DUVAUT & LIONS (1972, Chapter 3, Section 3.5)) that the tensor $\boldsymbol{\sigma} = (\sigma_{ij})_{i,j=1}^3$ minimizes the *complementary energy*

$$I(\boldsymbol{\tau}) = \frac{1}{2}\int_\Omega \left\{\left(\frac{1+\sigma}{E}\right)\sum_{i,j=1}^{3}\tau_{ij}^2 - \frac{\sigma}{E}\left(\sum_{l=1}^{3}\tau_{ll}\right)^2\right\}\mathrm{d}x$$

over the set

$$\mathcal{V}(f, g) = \Big\{ \tau \in \mathcal{W};\ -\sum_{j=1}^{3} \partial_j \tau_{ij} = f_i \quad \text{in} \quad \Omega,$$
$$\sum_{j=1}^{3} \tau_{ij}\nu_j = g_i \quad \text{on} \quad \Gamma_1, \quad 1 \leq i \leq 3 \Big\}$$

of *admissible stresses*, where

$$\mathcal{W} = \{\tau = (\tau_{ij})_{i,j=1}^{3} \in (L^2(\Omega))^9;\ \tau_{ij} = \tau_{ji}, \quad 1 \leq i, j \leq 3\}.$$

In the definition of the set $\mathcal{V}(f, g)$, the relations $-\Sigma_{j=1}^{3} \partial_j \tau_{ij} = f_i$ in Ω are to be understood in the sense of distributions, while the interpretation of the relations $\Sigma_{j=1}^{3} \tau_{ij}\nu_j = g_i$ on Γ_1 will be hinted at on a simpler problem. Then it is easily shown that the above minimization problem has one and only one solution σ, which is precisely related to the displacement u by relations (1.2.31) and (1.2.32). The reason the functional I is called the "complementary" energy is that $J(u) + I(\sigma) = 0$.

We shall call this formulation of the elasticity problem the *equilibrium model*, or *formulation* (it is called "equilibrium" model because the relations which define the set $\mathcal{V}(f, g)$ express the equilibrium of internal and boundary forces, respectively) and we shall call *dual problem* the corresponding minimization, or variational, problem.

It is therefore natural to conceive finite element approximations of this dual problem, but then the *major difficulty lies in the constraints used in the definition of the set* $\mathcal{V}(f, g)$. To obviate this difficulty, one key idea is to use techniques from *duality theory*. This is precisely the basis of the *mixed* and *hybrid* finite element methods, which we shall describe later.

For the sake of simplicity in the exposition, we shall often consider the model problem

$$\begin{cases} -\Delta u = f & \text{in} \quad \Omega \subset \mathbf{R}^n, \quad f \in L^2(\Omega), \\ u = 0 & \text{on} \quad \Gamma, \end{cases}$$

for which the *primal problem* consists in finding a function $u \in V = H_0^1(\Omega)$ such that

$$J(u) = \inf_{v \in V} J(v),$$

with

$$J(v) = \frac{1}{2}\int_\Omega \|\nabla v\|^2 \,\mathrm{d}x - \int_\Omega fv \,\mathrm{d}x,$$

or equivalently, such that

$$\forall v \in V, \quad \int_\Omega \nabla u \cdot \nabla v \, dx = \int_\Omega fv \, dx.$$

In this case, we shall consider that the unknowns which play the role of the "stresses" are the components of the gradient ∇u of the solution u (the "displacement") of the primal problem, and therefore, the objective is that these components be obtained by solving a minimization problem. In this direction, we introduce the space

$$H(\mathrm{div}; \Omega) = \{\boldsymbol{q} \in (L^2(\Omega))^n; \ \mathrm{div}\, \boldsymbol{q} \in L^2(\Omega)\},$$

a Hilbert space when equipped with the norm

$$\|\boldsymbol{q}\|_{H(\mathrm{div};\Omega)} = (|\boldsymbol{q}|_{0,\Omega}^2 + |\mathrm{div}\, \boldsymbol{q}|_{0,\Omega}^2)^{1/2}.$$

Then for any function $f \in L^2(\Omega)$, we define the affine hyperplane

$$\mathcal{V}(f) = \{\boldsymbol{q} \in H(\mathrm{div}; \Omega); \ \mathrm{div}\, \boldsymbol{q} + f = 0 \quad \text{in} \quad \Omega\}.$$

Using the $\mathcal{V}(0)$-ellipticity of the bilinear form $(\boldsymbol{p}, \boldsymbol{q}) \to \int_\Omega \boldsymbol{p} \cdot \boldsymbol{q} \, dx$, it is easily proved that there exists a unique function $\boldsymbol{p} \in \mathcal{V}(f)$ such that

$$I(\boldsymbol{p}) = \inf_{\boldsymbol{q} \in \mathcal{V}(f)} I(\boldsymbol{q}),$$

with

$$I(\boldsymbol{q}) = \frac{1}{2} \int_\Omega \|\boldsymbol{q}\|^2 \, dx,$$

or equivalently, such that

$$\forall \boldsymbol{q} \in \mathcal{V}(0), \quad \int_\Omega \boldsymbol{p} \cdot \boldsymbol{q} \, dx = 0.$$

Since in addition one has precisely

$$\boldsymbol{p} = \nabla u,$$

we have therefore constructed an adequate dual problem.

In order to get rid of the constraint $\mathrm{div}\, \boldsymbol{q} + f = 0$ which appears in the definition of the set $\mathcal{V}(f)$, we use a device standard in duality theory (cf. for example CÉA (1971), EKELAND & TÉMAM (1974)), which makes it possible to construct a problem in which the unknown $\boldsymbol{p}$ is no longer subjected to a constraint (i.e., it shall be simply required that $\boldsymbol{p} \in H(\mathrm{div}; \Omega)$). To achieve this goal, we have to find an appropriate space $\mathcal{M}$ of *Lagrange multipliers* and a *Lagrangian* $\mathcal{L}: H(\text{div}; \Omega) \times \mathcal{M} \to \mathbf{R}$, in such a way that *the unknown* $\boldsymbol{p}$ *is obtained as the first argument of the saddle-point* $(\boldsymbol{p}, \lambda)$ *of the Lagrangian* $\mathcal{L}$ over the space $H(\text{div}; \Omega) \times \mathcal{M}$ (this is the process that was followed in Section 7.2 for the solution of the discrete problem; see also Exercise 7.1.1 for the biharmonic problem itself).

In the present case, it turns out that we may choose

$$\mathcal{M} = L^2(\Omega) \quad \text{and} \quad \mathcal{L}(\boldsymbol{q}, \mu) = I(\boldsymbol{q}) + \int_\Omega \mu(\text{div}\, \boldsymbol{q} + f)\, \mathrm{d}x$$

(the particular form of the above Lagrangian is no coincidence; it is based on the fact that the functions $\boldsymbol{q}$ in the set $\mathcal{V}(f)$ may be equally characterized as those functions $\boldsymbol{q} \in H(\text{div}; \Omega)$ for which $\int_\Omega \mu(\text{div}\, \boldsymbol{q} + f)\, \mathrm{d}x = 0$ for all $\mu \in L^2(\Omega)$). Then one can show that the above Lagrangian has a unique saddle-point $(\boldsymbol{p}, \lambda)$ over the space $H(\text{div}; \Omega) \times L^2(\Omega)$, and that one has precisely

$$\boldsymbol{p} = \nabla u \quad \text{and} \quad \lambda = u$$

(that the second argument of the saddle-point is related to the solution of either the primal or dual problem is no coincidence either; cf. Section 7.2 for a similar circumstance). In other words, one has

$$\forall \boldsymbol{q} \in H(\text{div}; \Omega), \quad \forall \mu \in L^2(\Omega), \quad \mathcal{L}(\boldsymbol{p}, \mu) \leq \mathcal{L}(\boldsymbol{p}, \lambda) \leq \mathcal{L}(\boldsymbol{q}, \lambda),$$

or equivalently,

$$\forall \boldsymbol{q} \in H(\text{div}; \Omega), \quad \int_\Omega \boldsymbol{p} \cdot \boldsymbol{q}\, \mathrm{d}x + \int_\Omega \lambda\, \text{div}\, \boldsymbol{q}\, \mathrm{d}x = 0,$$

$$\forall \mu \in L^2(\Omega), \quad \int_\Omega \mu(\text{div}\, \boldsymbol{p} + f)\, \mathrm{d}x = 0$$

(these variational equations simply express the necessary, and sufficient in this case, conditions that the two partial derivatives of the Lagrangian vanish at the saddle-point). Notice in passing that the above variational equations mean that, at least formally, the given second-order problem has been replaced by a *first-order system*, namely,

$$\begin{cases} \boldsymbol{p} - \nabla \lambda = 0 & \text{in} \quad \Omega, \\ -\text{div}\, \boldsymbol{p} = f & \text{in} \quad \Omega, \\ \lambda = 0 & \text{on} \quad \Gamma. \end{cases}$$

For the finite element approximation of general first-order systems, see LESAINT (1973, 1975).

The verification of the above statements offers no difficulties. It relies in particular on the following result (LIONS & MAGENES (1968)). Given a function $\boldsymbol{q} \in H(\mathrm{div}; \Omega)$, one may define its "outer normal component" (denoted by definition) $\boldsymbol{q} \cdot \boldsymbol{\nu}$ along Γ as an element of the space $H^{-1/2}(\Gamma)$, in such a way that the Green formula

$$\forall v \in H^1(\Omega), \quad \int_\Omega \{\nabla v \cdot \boldsymbol{q} + v \operatorname{div} \boldsymbol{q}\} \, \mathrm{d}x = \langle \boldsymbol{q} \cdot \boldsymbol{\nu}, v \rangle_\Gamma$$

holds, where $\langle \cdot, \cdot \rangle_\Gamma$ denotes the duality pairing between the spaces $H^{-1/2}(\Gamma)$ and $H^{1/2}(\Gamma)$ (recall that $\langle \boldsymbol{q} \cdot \boldsymbol{\nu}, v \rangle_\Gamma = \int_\Gamma (\boldsymbol{q} \cdot \boldsymbol{\nu}) v \, \mathrm{d}\gamma$ if it so happens that $\boldsymbol{q} \cdot \boldsymbol{\nu} \in L^2(\Gamma) \subset H^{-1/2}(\Gamma)$; the spaces $H^{1/2}(\Gamma)$ and $H^{-1/2}(\Gamma)$ have been defined in connection with problems on unbounded domains; cf. the section "Additional Bibliography and Comments" in Chapter 4).

We shall call this formulation (of the model problem) the *primal-dual formulation*, and *primal-dual problem* the corresponding saddle-point, or variational, problem.

In the case of the elasticity problem, the corresponding Lagrangian is called the *Hellinger–Reissner energy*. It takes the form

$$\mathscr{L}(\boldsymbol{v}, \boldsymbol{\tau}) = -I(\boldsymbol{\tau}) + \int_\Omega \sum_{i,j=1}^{3} \epsilon_{ij}(\boldsymbol{v}) \tau_{ij} \, \mathrm{d}x - \left(\int_\Omega \boldsymbol{f} \cdot \boldsymbol{v} \, \mathrm{d}x + \int_{\Gamma_1} \boldsymbol{g} \cdot \boldsymbol{v} \, \mathrm{d}\gamma \right),$$

and it can be shown that the pair $(\boldsymbol{u}, \boldsymbol{\sigma})$ is a saddle-point of the Lagrangian $\mathscr{L}$ over the space $\boldsymbol{V} \times \mathscr{W}$.

The proper framework for justifying the previous considerations is of course that of *duality theory* (as was implicitly indicated by the use of the adjectives "primal" and "dual", for instance). For a thorough reference concerning duality theory in general, see EKELAND & TÉMAM (1974). For the application of duality theory to problems in elasticity, see FRÉMOND (1971a, 1971b, 1972, 1973), WASHIZU (1968), ODEN & REDDY (1974, 1976b), TONTI (1970). For applications to variational inequalities, see GLOWINSKI (1976a), GLOWINSKI, LIONS & TRÉMOLIÈRES (1976a, 1976b).

Displacement and equilibrium methods

Except in Chapter 7, the finite element methods described in this book are based on the *primal formulation* of a given problem. This explains why, by reference to the elasticity problem, such methods are sometimes known as *displacement methods.*

As we pointed out, it is also desirable to develop methods in which the "stresses" are directly computed. In this direction, the engineers have devised various ways of computing the stresses directly from the knowledge of the displacements, as in BARLOW (1976), HINTON & CAMPBELL (1974), STEIN & AHMAD (1974). Our primary interest, however, concerns methods directly based on a *dual formulation.*

To be more specific, consider the model problem $-\Delta u = f$ in Ω, $u = 0$ on Γ, whose dual problem consists in finding an element $\boldsymbol{p} \in \mathcal{V}(f)$ such that

$$\forall \boldsymbol{q} \in \mathcal{V}(0), \quad \int_\Omega \boldsymbol{p} \cdot \boldsymbol{q} \, \mathrm{d}x = 0,$$

where

$$\mathcal{V}(f) = \{\boldsymbol{q} \in H(\mathrm{div}; \Omega); \mathrm{div}\, \boldsymbol{q} + f = 0 \quad \text{in} \quad \Omega\}.$$

Let us assume that an element $\boldsymbol{p}_0 \in \mathcal{V}(f)$ is known, so that this problem is reduced to finding the (unique) element $\boldsymbol{p}^* = (\boldsymbol{p} - \boldsymbol{p}_0) \in \mathcal{V}(0)$ which satisfies

$$\forall \boldsymbol{q} \in \mathcal{V}(0), \quad \int_\Omega \boldsymbol{p}^* \cdot \boldsymbol{q} \, \mathrm{d}x = -\int_\Omega \boldsymbol{p}_0 \cdot \boldsymbol{q} \, \mathrm{d}x.$$

Since it is possible to construct subspaces $\mathcal{V}_h(0)$ of $\mathcal{V}(0)$ (as in the approximation of the Stokes problem; cf. the section "Additional Bibliography and Comments" of Chapter 4), the discrete problem consists in finding the (unique) element $\boldsymbol{p}_h^* \in \mathcal{V}_h(0)$ which satisfies

$$\forall \boldsymbol{q}_h \in \mathcal{V}_h(0), \quad \int_\Omega \boldsymbol{p}_h^* \cdot \boldsymbol{q}_h \, \mathrm{d}x = -\int_\Omega \boldsymbol{p}_0 \cdot \boldsymbol{q}_h \, \mathrm{d}x,$$

and one gets in this fashion an approximation $\boldsymbol{p}_h = \boldsymbol{p}_0 + \boldsymbol{p}_h^*$ of the solution $\boldsymbol{p}$ of the dual problem.

It is however exceptional that an element be known in the set $\mathcal{V}(f)$, so that the major difficulty is to take appropriately into account the constraint $\mathrm{div}\, \boldsymbol{p} + f = 0$ in Ω. There are essentially three ways to circumvent this difficulty.

First, *the constraint is approximated,* in such a way that the discrete solution $\boldsymbol{p}_h$ satisfies a relation of the form $\mathrm{div}\, \boldsymbol{p}_h + f_h = 0$ in Ω, where f_h is a typical finite element approximation of the function f (e.g. one has

$f_{h|K} \in P_k(K)$ for all $K \in \mathcal{T}_h$). Again by reference to the elasticity problem, such methods, which are directly based on the dual formulation, are known as *equilibrium methods* (of course, they include those where the constraint may be exactly satisfied). They have been first advocated by FRAEIJS DE VEUBEKE (1965b, 1973). Their numerical analysis is thoroughly made in THOMAS (1975, 1976, 1977). See also FALK (1976b) for a related method, and RAVIART (1975).

While the equilibrium methods are, by definition, based on a formulation where there is *only one unknown* (the gradient ∇u for the model problem, the stress tensor $\boldsymbol{\sigma}$ for the elasticity problem, etc...), one may use the *techniques of duality theory* to get rid of the constraint, a process which results in the addition of a *second unknown*, the Lagrange multiplier. This is in particular the basis of the *mixed* methods and the *dual hybrid* methods, which are other alternatives for handling the constraint.

Mixed methods

It is customary to call *mixed method* any finite element method based on the primal-dual formulation (notice that we shall later extend this definition; then we shall return to the mixed method described in Chapter 7, in the light of the present definitions).

Especially for second-order problems, the study of such methods may be based on a general approach of BREZZI (1974b) (notice that it does not directly apply to the method described in Section 7.2, however, even though several features are common to both analyses).

F. Brezzi considers the following variational problem (irrespective of whether it is obtained from a saddle-point problem): Find a pair $(p, \lambda) \in \mathcal{W} \times \mathcal{M}$ such that

$$\forall q \in \mathcal{W}, \quad a(p, q) + b(q, \lambda) = f(q),$$

$$\forall \mu \in \mathcal{M}, \quad b(p, \mu) = g(v),$$

where $\mathcal{W}$ and $\mathcal{M}$ are Hilbert spaces, $a(\cdot, \cdot)$ and $b(\cdot, \cdot)$ are continuous bilinear forms on $\mathcal{W} \times \mathcal{W}$ and $\mathcal{W} \times \mathcal{M}$ respectively, f and g are given elements in the dual spaces $\mathcal{W}'$ and $\mathcal{M}'$ respectively. First, F. Brezzi gives necessary and sufficient conditions on the bilinear forms $a(\cdot, \cdot)$ and $b(\cdot, \cdot)$, which insure that the above problem has one and only one solution for all $f \in \mathcal{W}'$, $g \in \mathcal{M}'$. Secondly, he considers the associated discrete problem: Find a pair $(p_h, \lambda_h) \in \mathcal{W}_h \times \mathcal{M}_h$ such that

$$\forall q_h \in \mathcal{W}_h, \quad a(p_h, q_h) + b(q_h, \lambda_h) = f(q_h),$$
$$\forall \mu_h \in \mathcal{M}_h, \quad b(p_h, \mu_h) = g(v_h),$$

where $\mathcal{W}_h$ and $\mathcal{M}_h$ are closed subspaces (finite-dimensional in practice) of the spaces $\mathcal{W}$ and $\mathcal{M}$, respectively. Then under suitable assumptions, F. Brezzi obtains an abstract error estimate for the quantity $\|p - p_h\|_{\mathcal{W}} + \|\lambda - \lambda_h\|_{\mathcal{M}}$.

To indicate the flavor of his results, let us return to our model problem. As we have seen, the primal-dual formulation consists in finding a pair $(\boldsymbol{p}, \lambda) \in H(\text{div}; \Omega) \times L^2(\Omega)$ such that

$$\forall \boldsymbol{q} \in H(\text{div}; \Omega), \int_\Omega \boldsymbol{p} \cdot \boldsymbol{q} \, \mathrm{d}x + \int_\Omega \lambda \, \text{div} \, \boldsymbol{q} \, \mathrm{d}x = 0,$$
$$\forall \mu \in L^2(\Omega), \quad \int_\Omega \mu(\text{div} \, \boldsymbol{p} + f) \, \mathrm{d}x = 0$$

(recall that $\boldsymbol{p} = \nabla u$ and $\lambda = u$). Accordingly, a *mixed finite element method* for solving this problem is defined as follows: Given two finite-dimensional spaces $\mathcal{W}_h$ and $\mathcal{M}_h$ which satisfy the inclusions

$$\mathcal{W}_h \subset H(\text{div}; \Omega), \quad \mathcal{M}_h \subset L^2(\Omega),$$

find a pair $(\boldsymbol{p}_h, \lambda_h) \in \mathcal{W}_h \times \mathcal{M}_h$ such that

$$\forall \boldsymbol{q}_h \in \mathcal{W}_h, \quad \int_\Omega \boldsymbol{p}_h \cdot \boldsymbol{q}_h \, \mathrm{d}x + \int_\Omega \lambda_h \, \text{div} \, \boldsymbol{q}_h \, \mathrm{d}x = 0,$$
$$\forall \mu_h \in \mathcal{M}_h, \quad \int_\Omega \mu_h(\text{div} \, \boldsymbol{p}_h + f) \, \mathrm{d}x = 0.$$

In this particular case, the spaces $\mathcal{W}_h$ and $\mathcal{M}_h$ should be related as follows: First, the implication

$$\boldsymbol{q}_h \in \mathcal{W}_h \quad \text{and} \quad \forall \mu_h \in \mathcal{M}_h, \quad \int_\Omega \mu_h \, \text{div} \, \boldsymbol{q}_h \, \mathrm{d}x = 0 \Rightarrow \text{div} \, \boldsymbol{q}_h = 0,$$

should hold. Secondly, *Brezzi's condition*:

$$0 < \beta = \inf_{\mu_h \in \mathcal{M}_h} \sup_{\boldsymbol{q}_h \in \mathcal{W}_h} \frac{\int_\Omega \mu_h \, \text{div} \, \boldsymbol{q}_h \, \mathrm{d}x}{|\mu_h|_{0,\Omega} \|\boldsymbol{q}_h\|_{H(\text{div};\Omega)}},$$

should hold. Under these assumptions, the discrete problem has a unique solution and there exists a constant C independent on the subspaces $\mathscr{W}_h$ and $\mathscr{M}_h$ such that

$$\|p - p_h\|_{H(\mathrm{div};\Omega)} + |\lambda - \lambda_h|_{0,\Omega} \leqslant$$
$$\leqslant C\left(\inf_{q_h \in \mathscr{W}_h} \|p - q_h\|_{H(\mathrm{div};\Omega)} + \inf_{\mu_h \in \mathscr{M}_h} |\lambda - \mu_h|_{0,\Omega}\right).$$

RAVIART & THOMAS (1977a) have constructed various finite element spaces $\mathscr{W}_h$ and $\mathscr{M}_h$ which satisfy Brezzi's condition and they have obtained the corresponding orders of convergence. See also MANSFIELD (1976a) for further results. SCHOLZ (1976, 1977), has obtained estimates of the error $(|\lambda - \lambda_h|_{0,\infty,\Omega} + h|p - p_h|_{0,\infty,\Omega})$, by adapting the method of weighted norms of J.A. Nitsche.

Let us also briefly review mixed finite element methods for the plate problem, as first proposed by HERMANN (1967) (for another mixed method for plates, see POCESKI (1975)). In this case, the dual formulation is defined as follows: Let

$$\mathscr{V}(f) = \{q = (q_i)_{i=1}^3 \in (L^2(\Omega))^3;$$
$$\partial_{11}q_1 + 2\partial_{12}q_2 + \partial_{22}q_3 = f \quad \text{in} \quad \Omega\},$$
$$I(q) = \frac{1}{2}\int_\Omega (q_1^2 + 2q_2^2 + q_3^2)\,\mathrm{d}x.$$

Then there exists a unique element $p \in \mathscr{V}(f)$ such that $I(p) = \inf_{q \in \mathscr{V}(f)} I(q)$, and one has precisely

$$p = (\partial_{11}u, \partial_{12}u, \partial_{22}u).$$

This is of particular interest for plates, where the second partial derivatives $\partial_{ij}u$ yield in turn the moments. In the primal-dual formulation, the triple $(\partial_{11}u, \partial_{12}u, \partial_{22}u)$ is the first argument of the saddle-point, while the second argument, i.e., the Lagrange multiplier, turns out to be the displacement u itself. For an analysis of such methods, see JOHNSON (1972, 1973), KIKUCHI & ANDO (1972a, 1973a), MIYOSHI (1973a), SAMUELSSON (1973).

Recently, BREZZI & RAVIART (1976) have developed a general theory of mixed methods for fourth-order problems, which contains in particular the analysis of Section 7.1, as well as the analyses of C. Johnson and T. Miyoshi quoted above. F. Brezzi and P.-A. Raviart also obtain optimal error estimates in the norm $|\cdot|_{0,\Omega}$.

As advocated in particular by TAYLOR & HOOD (1973), finite element methods of mixed type seem more and more popular for approximating the solutions of Stokes and Navier–Stokes problems (cf. the sections

"Additional Bibliography and Comments" at the end of Chapters 4 and 5). Such methods have been studied by BERCOVIER (1976), BERCOVIER & LIVNE (1976), FORTIN (1976), GIRAULT (1976c), RAVIART (1976).

Mixed methods are also increasingly used for solving nonlinear problems such as the von Karmann equations (cf. MIYOSHI (1976a, 1976c, 1977)), elastoplastic plates (cf. BREZZI, JOHNSON & MERCIER (1977)), nonlinear problems of monotone type as considered in Section 5.3 (cf. BERCOVIER (1976), SCHEURER (1977)).

To sum up, mixed methods yield simultaneous approximations of the solutions of both the primal and dual problems. Since the solution of the dual problem consists in practice of derivatives of the solution of the primal problem (e.g. ∇u for the model second-order problem), the terminology *mixed method* can also be used more generally for any approximation procedure in which an unknown *and* some of its derivatives are simultaneously approximated, irrespective of whether this is achieved through duality techniques. This is in particular the definition of J.T. Oden, who has done a thorough study of such methods. See ODEN (1972b, 1973c), ODEN & LEE (1975), ODEN & REDDY (1975, 1976a, Section 8.10, 1976c), REDDY (1973), REDDY & ODEN (1973), BABUŠKA, ODEN & LEE (1977).

In the light of the two above possible definitions of mixed methods, let us return to the method studied in the present chapter. The method described in Section 7.1 is mixed in the general sense only: The pair $(u, -\Delta u)$ is obtained through a minimization problem which, although not the standard one, is regarded as a primal problem. By contrast, the method described in Section 7.2 is mixed in the restricted sense, in that it is a natural discretization of a primal-dual problem (as described in Exercise 7.1.1).

For further references concerning the mathematical analysis of mixed methods, see KIKUCHI (1976a), HASLINGER & HLAVÁČEK (1975, 1976a, 1976b). The first abstract analyses of such methods are due to AUBIN & BURCHARD (1971) (see also AUBIN (1972)), and BABUŠKA (1971b). In particular, I. Babuška developed an abstract theory which resembles that of F. Brezzi and which is the basis of the paper of BABUŠKA, ODEN & LEE (1977).

Hybrid methods

A problem to be approximated by a mixed method is in practice formulated in such a way that the unknown u is a function together with some derivatives of this function *over the set* $\bar{\Omega}$, e.g. $(u, \nabla u)$ for the model problem, $(u, (\partial_{11}u, \partial_{12}u, \partial_{22}u))$ for the plate problem, etc. . . .

In another class of finite element methods, the unknown is a function together with some derivatives of this function *along the boundaries of appropriate subdomains of the set* $\bar{\Omega}$. Accordingly, an appropriate variational formulation of the given problem needs to be developed. To show how this is achieved, consider again the model problem

$$\begin{cases} -\Delta u = f & \text{in} \quad \Omega, \quad f \in L^2(\Omega), \\ u = 0 & \text{on} \quad \Gamma. \end{cases}$$

Assume that a triangulation $\mathcal{T}_h$ is established over the set $\bar{\Omega}$, in the sense that relations $(\mathcal{T}_h i)$, $1 \leq i \leq 4$ (cf. Section 2.1), are satisfied (at this stage, the sets K occurring in the decomposition $\bar{\Omega} = \bigcup_{K \in \mathcal{T}_h} K$ need not be related to actual finite elements), and assume that the objective is to compute not only the "displacement" u, but also the "stresses" along the boundaries ∂K of the sets K, understood here as the normal derivatives $\partial_{\nu_K} u$, $K \in \mathcal{T}_h$. We introduce the spaces

$$V(\mathcal{T}_h) = \prod_{K \in \mathcal{T}_h} H^1(K) \subset L^2(\Omega),$$

$$M(\mathcal{T}_h) = \Big\{ \mu \in \prod_{K \in \mathcal{T}_h} H^{-1/2}(\partial K);\ \exists \boldsymbol{q} \in H(\text{div}; \Omega), \\ \forall K \in \mathcal{T}_h,\ \boldsymbol{q} \cdot \boldsymbol{\nu}_K = \mu|_{\partial K} \Big\},$$

provided with the norms

$$\|v\|_{V(\mathcal{T}_h)} = \Big(\sum_{K \in \mathcal{T}_h} \|v\|_{1,K}^2 \Big)^{1/2},$$

$$\|\mu\|_{M(\mathcal{T}_h)} = \inf \Big\{ \|\boldsymbol{q}\|_{H(\text{div};\Omega)};\ \exists \boldsymbol{q} \in H(\text{div}; \Omega), \\ \forall K \in \mathcal{T}_h,\ \boldsymbol{q} \cdot \boldsymbol{\nu}_K = \mu|_{\partial K} \Big\},$$

respectively (recall that if a function $\boldsymbol{q}$ belongs to the space $H(\text{div}; K)$, its outer normal component $\boldsymbol{q} \cdot \boldsymbol{\nu}_K$ is well-defined as an element of the space $H^{-1/2}(\partial K)$). It can be verified that there exists a unique pair $(\bar{u}, \lambda) \in V(\mathcal{T}_h) \times M(\mathcal{T}_h)$ such that

$$\forall v \in V(\mathcal{T}_h), \quad \sum_{K \in \mathcal{T}_h} \int_K \nabla \bar{u} \cdot \nabla v \, dx - \sum_{K \in \mathcal{T}_h} \langle \lambda, v \rangle_{\partial K} = \int_\Omega f v \, dx,$$

$$\forall \mu \in M(\mathcal{T}_h), \qquad - \sum_{K \in \mathcal{T}_h} \langle \mu, v \rangle_{\partial K} = 0,$$

where $\langle\cdot,\cdot\rangle_{\partial K}$ denotes the duality between the spaces $H^{-1/2}(\partial K)$ and $H^{1/2}(\partial K)$. Moreover one has

$$\bar{u} = u, \quad \text{and} \quad \forall K \in \mathcal{T}_h, \quad \lambda|_{\partial K} = \partial_{\nu_K} u.$$

Notice that even though the unknown u lies in the space $\Pi_{K\in\mathcal{T}_h} H^1(K)$, where no continuity is *a priori* required, the function u is nevertheless automatically in the space $H_0^1(\Omega)$ (this is so because of the particular form of the constraint which appear in the definition of the space $M(\mathcal{T}_h)$).

Because the first unknown in such a formulation is the solution of the *primal* problem, we shall refer to this formulation as a *primal hybrid model*, or *formulation*, and the corresponding saddle-point, or variational, problem will be called *primal hybrid problem*. Accordingly, finite element methods based on this formulation shall be called *primal hybrid methods*.

Let us briefly describe such a method for the model problem. Assume that $n = 2$ and that the sets K are triangles. Then for some integer $k \geq 1$, we let $\bar{V}_h = \Pi_{K\in\mathcal{T}_h} P_k(K)$, so that the space $\bar{V}_h$ is contained in the space $V(\mathcal{T}_h)$. Next, for some integer $m \geq 0$ and for each triangle $K \in \mathcal{T}_h$, we let $S_m(\partial K) = \Pi_{i=1}^3 P_m(K_i')$ where K_i', $1 \leq i \leq 3$, denote the three sides of K. Then the other space M_h consists of those functions μ in the space $\Pi_{K\in\mathcal{T}_h} S_m(\partial K)$ which satisfy $\mu|_{K_1} + \mu|_{K_2} = 0$ along the side $K_1 \cap K_2$ whenever it happens that the triangles K_1 and K_2 are adjacent. In this fashion, we have constructed a subspace M_h of the space $M(\mathcal{T}_h)$. As in the case of mixed methods for second-order problems, the error analysis can be again based on the abstract approach of BREZZI (1974b). Once Brezzi's condition is verified (in the case of the above spaces $\bar{V}_h$ and M_h, it is satisfied if $k \geq m + 1$, m even, or $k \geq m + 2$ if m is odd), estimates are obtained for the error

$$\left(\sum_{K\in\mathcal{T}_h} \|\bar{u} - \bar{u}_h\|_{1,K}^2\right)^{1/2} + \left(\sum_{K\in\mathcal{T}_h} h_K \|\lambda - \lambda_h\|_{L^2(\partial K)}^2\right)^{1/2}.$$

These results are found in RAVIART & THOMAS (1977b), THOMAS (1977). See also the survey of RAVIART (1975).

Notice that since the first unknown is in a finite element space not contained in the space $H^1(\Omega)$, this method may be regarded as "nonconforming for the unknown u". In fact, the connection with nonconforming methods is deeper, as shown in the above references in particular. For related ideas, see BABUŠKA, ODEN & LEE (1977), CROUZEIX & RAVIART (1973), IRONS & RAZZAQUE (1972a).

Additional references for second-order problems are BREZZI (1974a) for the nonhomogeneous Neumann problem for the operator $-\Delta$, KIKUCHI (1973) for plane stress problems.

A complementary approach to the theory of mixed methods has been contributed by BABUŠKA, ODEN & LEE (1977), who have developed a general theory of *mixed-hybrid methods*, which includes results on both mixed and hybrid methods as special cases. In this direction, see also ODEN (1976a), ODEN & REDDY (1976a, Section 8.10), ODEN & LEE (1975, 1977).

One may analogously construct another formulation in which the first unknown is the solution of the dual problem while the second unknown is the trace along the boundaries ∂K of the solution of the primal problem. In the case of the model problem, we define the spaces

$$\mathcal{V}(f, \mathcal{T}_h) = \{\boldsymbol{q} \in (L^2(\Omega))^n;\ \forall K \in \mathcal{T}_h,\ \operatorname{div} \boldsymbol{q} + f = 0 \quad \text{in} \quad K\},$$

$$M'(\mathcal{T}_h) = \Big\{\mu \in \prod_{K \in \mathcal{T}_h} H^{1/2}(\partial K),\ \exists v \in H_0^1(\Omega),$$

$$\forall K \in \mathcal{T}_h,\ v|_{\partial K} = \mu|_{\partial K}\Big\},$$

provided with the norms $|\cdot|_{0,\Omega}$ and

$$\|\mu\|_{M'(\mathcal{T}_h)} = \inf\{\|v\|_{1,\Omega};\ \exists v \in H_0^1(\Omega),\ \forall K \in \mathcal{T}_h,\ v|_{\partial K} = \mu|_{\partial K}\},$$

respectively. Then there exists a unique pair $(\boldsymbol{p}, \lambda) \in \mathcal{V}(f, \mathcal{T}_h) \times M'(\mathcal{T}_h)$ such that

$$\forall \boldsymbol{q} \in \mathcal{V}(0, \mathcal{T}_h), \quad \int_\Omega \boldsymbol{p} \cdot \boldsymbol{q}\, \mathrm{d}x - \sum_{K \in \mathcal{T}_h} \langle \boldsymbol{q} \cdot \boldsymbol{\nu}_K, \lambda \rangle_{\partial K} = 0,$$

$$\forall \mu \in M'(\mathcal{T}_h), \quad -\sum_{K \in \mathcal{T}_h} \langle \boldsymbol{p} \cdot \boldsymbol{\nu}_K, \mu \rangle_{\partial K} = 0,$$

and one has in addition

$$\boldsymbol{p} = \nabla u \quad \text{and} \quad \forall K \in \mathcal{T}_h, \quad \lambda|_{\partial K} = u|_{\partial K}.$$

Such a formulation is called *dual hybrid model*, or *formulation*, and the corresponding saddle-point, or variational, problem, is called *dual hybrid problem*. The finite element approximation of such problems yields to *dual hybrid methods*, for an extensive study of which we refer to THOMAS (1976, 1977), in the case of second-order problems. Dual hybrid methods for the plate problem have been thoroughly studied by BREZZI (1975) and BREZZI & MARINI (1975). In this case, the first unknown is the triple $(\partial_{11}u, \partial_{12}u, \partial_{22}u)$ (i.e., the solution of the corresponding dual problem), while the second unknown are the triples $(u|_{\partial K}, \partial_1 u|_{\partial K}, \partial_2 u|_{\partial K})$, $K \in \mathcal{T}_h$. Hybrid methods for plates have been also studied by KIKUCHI (1973), KIKUCHI & ANDO (1972b, 1972c, 1972d, 1973b). REDDY (1976) has extended to fourth-order problems the mixed-hybrid approach of BABUŠKA, ODEN & LEE (1977).

Hybrid methods have been proposed and advocated by ALLMAN (1976), FRAEIJS DE VEUBEKE (1965b, 1973), HENSHELL (1973), JONES (1964), PIAN (1971, 1972), PIAN & TONG (1969a, 1969b), TORBE & CHURCH (1975), WOLF (1975).

In the same fashion as we extended the definition of mixed methods, we may define more generally as a *hybrid method* any finite element method based on a formulation where one unknown is a function, or some of its derivatives, on the set Ω, and the other unknown is the trace of some of the derivatives of the same function, or the trace of the function itself, along the boundaries of the set K. In other words, we ignore in this new acceptation that in practice, such methods are based on appropriate primal hybrid, or dual hybrid, formulations.

Even more general acceptations exist. For example FIX (1976) states that a finite element method is *hybrid* as soon as (any kind of) duality techniques are used for treating troublesome constraints. The use of Lagrange multipliers for handling boundary conditions, as proposed by BABUŠKA (1973a), is an example of such methods.

An attempt of general classification of finite element methods

Table 1 summarizes the previous considerations. For definiteness, we have formulated the problems as minimization or saddle-point problems, but they could have been equally expressed in the more general form of variational equations.

The reader will notice that notable omissions among the definitions of this table are those of *conforming* and *nonconforming methods*, amply illustrated in this book for displacement methods. The reason behind these omissions is that these make up *another* classification on their own. We shall simply illustrate by two examples the possible connections that may be established between the two classifications: First, mixed methods may be subdivided into conforming and nonconforming methods. For instance, the mixed method studied by JOHNSON (1972, 1973) for plates is "nonconforming with respect to the argument u",

Table 1.

Variational problem	Particular nomenclature for the elasticity problem	Special case of the model problem $-\Delta u = f$ in Ω, $u = 0$ on Γ	Name of the finite element methods based on the same formulations
Primal problem: Find $u \in V$ such that $J(u) = \inf_{v \in V} J(v)$	Displacement model u: displacement J: potential energy V: space of admissible displacements	$J(v) = \frac{1}{2}\int_\Omega \|\nabla v\|^2\,dx - \int_\Omega fv\,dx$ $V = H_0^1(\Omega)$	Displacement methods
Dual problem: Find $p \in \mathcal{V}(f)$ such that $I(p) = \inf_{q \in \mathcal{V}(f)} I(q)$	Equilibrium model p: stress tensor I: complementary energy $\mathcal{V}(f)$: set of admissible stresses	$I(q) = \frac{1}{2}\int_\Omega \|q\|^2\,dx$ $\mathcal{V}(f) = \{q \in H(\mathrm{div};\Omega);\ \mathrm{div}\,q + f = 0 \text{ in } \Omega\}$ $\boxed{p = \nabla u}$	Equilibrium methods
Primal-dual problem: Find: $(p, \lambda) \in \mathcal{W} \times \mathcal{M}$ such that $\forall q \in \mathcal{W}$, $\forall \mu \in \mathcal{M}$, $\mathcal{L}(p, \mu) \leq \mathcal{L}(p, \lambda) \leq \mathcal{L}(q, \lambda)$	$\mathcal{L}$: Hellinger–Reissner's energy	$\mathcal{L}(q, \mu) = I(q) + \int_\Omega \mu(\mathrm{div}\,q + f)\,dx$ $\mathcal{W} = H(\mathrm{div};\Omega)$, $\mathcal{M} = L^2(\Omega)$ $\boxed{p = \nabla u, \lambda = u}$	Mixed methods

Primal hybrid problem: Find $(\bar{u}, \lambda) \in V(\mathcal{T}_h) \times M(\mathcal{T}_h)$ such that $\forall v \in V(\mathcal{T}_h), \forall \mu \in M(\mathcal{T}_h)$, $L_h(\bar{u}, \mu) \leq L_h(\bar{u}, \lambda) \leq L_h(v, \lambda)$	$L_h(v, \mu) = \frac{1}{2}\sum_K \int_K \lVert\nabla v\rVert^2 \, dx - \int_\Omega fv \, dx - \sum_K \langle \mu, v\rangle_{\partial K}$; $V(\mathcal{T}_h) = \prod_K H^1(K)$; $M(\mathcal{T}_h) = \{\mu \in \prod_K H^{-1/2}(\partial K);\ \exists \boldsymbol{q} \in H(\mathrm{div}; \Omega),\ \forall K \in \mathcal{T}_h,\ \boldsymbol{q} \cdot \boldsymbol{\nu}_K = \mu\vert_{\partial K}\}$; $\boxed{\bar{u} = u,\ \lambda\vert_{\partial K} = \partial_{\nu_K} \text{ for all } K \in \mathcal{T}_h}$	**Primal hybrid methods**
Dual hybrid problem: Find $(\boldsymbol{p}, \lambda) \in \mathcal{V}(f, \mathcal{T}_h) \times M'(\mathcal{T}_h)$ such that $\forall \boldsymbol{q} \in \mathcal{V}(f, \mathcal{T}_h), \forall \mu \in M'(\mathcal{T}_h)$, $L_h'(\boldsymbol{p}, \mu) \leq L_h'(\boldsymbol{p}, \lambda) \leq L_h'(\boldsymbol{q}, \lambda)$	$L_h'(\boldsymbol{q}, \mu) = \frac{1}{2}\int_\Omega \lVert\boldsymbol{q}\rVert^2 \, dx - \sum_K \langle \boldsymbol{q} \cdot \boldsymbol{\nu}_K, \mu\rangle_{\partial K}$; $\mathcal{V}(f, \mathcal{T}_h) = \{\boldsymbol{q} \in \prod_K H(\mathrm{div}; K);\ \forall K \in \mathcal{T}_h,\ \mathrm{div}\, \boldsymbol{q} + f = 0 \text{ in } K\}$; $M'(\mathcal{T}_h) = \{\mu \in \prod_K H^{1/2}(\partial K);\ \exists v \in H_0^1(\Omega),\ \forall K \in \mathcal{T}_h,\ v\vert_{\partial K} = \mu\vert_{\partial K}\}$; $\boxed{\boldsymbol{p} = \nabla u,\ \lambda\vert_{\partial K} = u\vert_{\partial K} \text{ for all } K \in \mathcal{T}_h}$	**Dual hybrid methods**

which is not required to belong to a subspace of the space $H^2(\Omega)$. Secondly, primal hybrid methods as described in the case of the model problem are automatically "nonconforming for the argument u", which lies "only" in a subspace of the space $\Pi_{K\in\mathcal{T}_h} H^1(K)$.

We could likewise take into consideration the effect of *numerical integration* and/or the effect of the *approximation of the boundary* in the case of curved domains. We suggest that a classification according to such *variational crimes* make up a *secondary classification of finite element methods*, while the classification of the above table, i.e., based on the *formulation* of the problem, make up the *primary classification of finite element methods*.

CHAPTER 8

FINITE ELEMENT METHODS FOR SHELLS

Introduction

In Section 8.1, we give a description of a model of the shell problem, known as *Koiter's model*. The fact that a shell is a body with "small"' thickness makes it possible to use as the only unknown the displacement of the middle surface of the shell and, consequently, there are only two independent variables, namely the curvilinear coordinates of the middle surface of the shell.

Restricting ourselves to the linear theory, it can be shown that *the corresponding strain energy is elliptic*. However, since the proof of this fact is lengthy, we content ourselves to show the ellipticity of the strain energy of an arch (Theorem 8.1.2). This simplificationn is justified, inasmuch as the arch problem is a "model problem" for the shell problem.

In the following two sections, we examine the approximation of such problems by finite element methods. There are essentially five types of such approximations:

(i) The shell is considered as a *three-dimensional body* and, accordingly, three-dimensional isoparametric finite elements are used (therefore, the numerical analysis of such methods is known). Let us add that, in the engineers' experience, this method seems in some cases to be competitive with methods which are specifically based on a two-dimensional model.

(ii) The reduction of a three-dimensional model to a two-dimensional model is performed *not on the continuous problem*, but on the finite element model itself. The principle of this method is very attractive, but little seems to be known as regards its analysis.

(iii) The first example of a finite element method which uses only the two-dimensional model is the "ideal" one (and again its convergence analysis is known): In some instances, the strain energy (which involves partial derivatives of the mapping which defines the middle surface of the shell) and the potential energy of the exterior forces can be *exactly reproduced* in the finite element spaces. This happens in special cases where all coefficients in the energy of the shell are constant functions, such as when the shell is a portion of a right circular cylinder. Incidentally, in this case one may consider that the functions in the finite element spaces are piecewise polynomials expressed in terms of the curvilinear coordinates along the middle surface of the shell.

(iv) In general, the geometry of the shell has to be approximated. This approximation results in an approximate shell or, equivalently, in an approximate energy. In Section 8.2, this type of method is analyzed and, in so doing, we are led to the definition of *conforming finite element methods for shells.* A general convergence result is proved (Theorem 8.2.4), which depends upon a careful comparison between the exact and the approximate strain energies (Theorem 8.2.1).

(v) The last category of finite element methods for shells consists in approximating the geometry in too crude a manner, so that the method is no longer conforming. Following a recent work of C. Johnson, we present in Section 8.3 the corresponding analysis in the case of a circular arch. Here the arch is approximated by straight segments and, consequently, the strain energy is written as a sum of strain energies of "elementary" straight beams. It is proved that such a method is convergent, provided the functions in the finite element spaces satisfy appropriate *compatibility relations*, which essentially compensate for the inadequate approximation of the geometry.

8.1. The shell problem

Geometrical preliminaries. Koiter's model

Let Ω be a bounded subset in a plane $\mathscr{E}^2$, with boundary Γ. Then a *shell* $\mathscr{S}$ is the image of the set $\bar{\Omega}$ by a mapping $\boldsymbol{\varphi}$: $\bar{\Omega} \subset \mathscr{E}^2 \to \mathscr{E}^3$, where $\mathscr{E}^3$ is the usual Euclidean space. In fact, the surface $\mathscr{S}$ is the *middle surface* of the shell, but since we are only considering "thin" shells, *we shall constantly identify the shell with its middle surface.* The data Γ and $\boldsymbol{\varphi}$ are assumed to be sufficiently smooth for all subsequent purposes.

We will denote by $(\boldsymbol{a}, \boldsymbol{b}) \to \boldsymbol{a} \cdot \boldsymbol{b}$, $\|.\|$, and $\boldsymbol{e}^i$, $1 \leqslant i \leqslant 3$, the Euclidean scalar product, Euclidean norm, and an orthonormal basis of the space $\mathscr{E}^3$, respectively.

We shall assume that all points of the shell $\mathscr{S} = \boldsymbol{\varphi}(\bar{\Omega})$ are *regular*, in the sense that the two vectors

$$\boldsymbol{a}_\alpha = \partial_\alpha \boldsymbol{\varphi}, \quad \alpha = 1, 2, \tag{8.1.1}$$

are linearly independent, for all points $\xi = (\xi^1, \xi^2) \in \bar{\Omega}$.

As a rule, we shall use Greek letters: $\alpha, \beta, \tau, \ldots$, for indices which take their values in the set $\{1, 2\}$, while Latin letters: $i, j, k, \ldots$, will be used for indices which take their values in the set $\{1, 2, 3\}$. For these indices, we shall use Einstein's convention for summation. Finally, the usual symbols, such as ∂_α, $\partial_{\alpha\beta}$, etc..., shall be used also for partial derivatives of *vector-valued* functions of the form $\boldsymbol{\theta} = \theta_i \boldsymbol{e}^i : \bar{\Omega} \subset \mathscr{E}^2 \rightarrow \mathscr{E}^3$. Thus, for instance, one has $\partial_\alpha \boldsymbol{\theta} = \partial_\alpha \theta_i \boldsymbol{e}^i$, $\partial_{\alpha\beta} \boldsymbol{\theta} = \partial_{\alpha\beta} \theta_i \boldsymbol{e}^i$, etc...

The vectors $\boldsymbol{a}_\alpha$ are tangent to the *curvilinear coordinate lines* $\boldsymbol{\varphi}(\xi^\beta =$ constant), $\beta \neq \alpha$, and they define the tangent plane at the point $\boldsymbol{\varphi}(\xi)$. We introduce the vector (Fig. 8.1.1)

$$\boldsymbol{a}_3 = \boldsymbol{a}^3 = \frac{\boldsymbol{a}_1 \times \boldsymbol{a}_2}{\|\boldsymbol{a}_1 \times \boldsymbol{a}_2\|}. \tag{8.1.2}$$

The *first fundamental form* $(a_{\alpha\beta})$ of the surface is defined by

$$a_{\alpha\beta} = a_{\beta\alpha} = \boldsymbol{a}_\alpha \cdot \boldsymbol{a}_\beta = \partial_\alpha \boldsymbol{\varphi} \cdot \partial_\beta \boldsymbol{\varphi}. \tag{8.1.3}$$

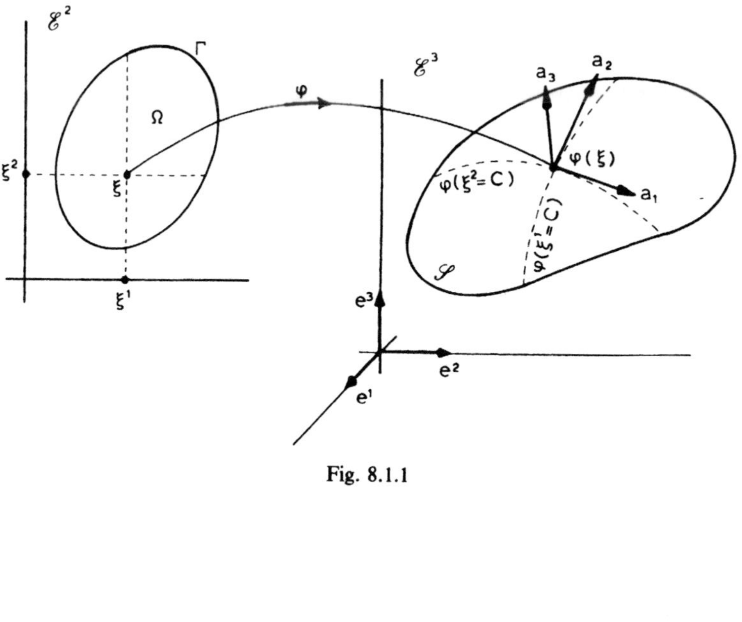

Fig. 8.1.1

With the *covariant basis* $(\boldsymbol{a}_\alpha)$ is associated (Fig. 8.1.2) the *contravariant basis* $(\boldsymbol{a}^\alpha)$ of the tangent plane, which is defined through the relations

$$\boldsymbol{a}^\alpha \cdot \boldsymbol{a}_\beta = \delta^\alpha_\beta \quad \text{(no summation if } \alpha = \beta), \tag{8.1.4}$$

where δ^α_β is the Kronecker symbol. We then have

$$\boldsymbol{a}_\alpha = a_{\alpha\beta}\boldsymbol{a}^\beta, \quad \boldsymbol{a}^\alpha = a^{\alpha\beta}\boldsymbol{a}_\beta, \quad a^{\alpha\beta} = \boldsymbol{a}^\alpha \cdot \boldsymbol{a}^\beta = a^{\beta\alpha}, \tag{8.1.5}$$

where the matrix $(a^{\alpha\beta})$ is the inverse of the matrix $(a_{\alpha\beta})$, which is always invertible since all points are regular, by assumption.

We recall that the *area measure* $\mathrm{d}S$ along the surface $\mathcal{S}$ is given by

$$\mathrm{d}S = \sqrt{a}\,\mathrm{d}\xi, \tag{8.1.6}$$

where

$$a = \det(a_{\alpha\beta}) = a_{11}a_{22} - (a_{12})^2. \tag{8.1.7}$$

We now come to the shell model, which is another example of a familiar problem: The solution $\boldsymbol{u}$, which will be defined below, minimizes the *shell energy*

$$J(\boldsymbol{v}) = \tfrac{1}{2}a(\boldsymbol{v}, \boldsymbol{v}) - f(\boldsymbol{v}), \tag{8.1.8}$$

when the functions vary over an appropriate space $\boldsymbol{V}$. We shall therefore successively define the bilinear form $a(.,.)$, the linear form f, and the space $\boldsymbol{V}$.

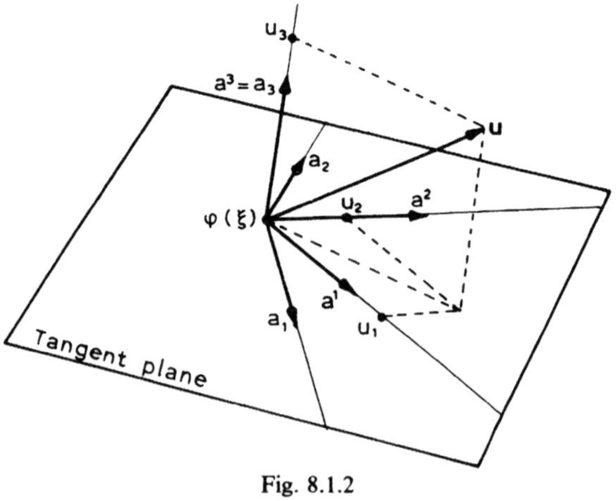

Fig. 8.1.2

The unknowns are the three functions

$$u_i: \xi \in \bar{\Omega} \to u_i(\xi) \in \mathbf{R},$$

which represent the *covariant components* of the displacement $\boldsymbol{u} = \boldsymbol{u}(\xi)$ of the point $\boldsymbol{\varphi}(\xi)$. In other words (Fig. 8.1.2), we have

$$\boldsymbol{u} = u_i \boldsymbol{a}^i. \tag{8.1.9}$$

Of course, it should be remembered that *the vectors $\boldsymbol{a}^i$ are also functions of $\xi \in \Omega$.*

The *strain energy* $\frac{1}{2}a(\boldsymbol{v}, \boldsymbol{v})$ *of the shell* is a surface integral:

$$\frac{1}{2}a(\boldsymbol{v}, \boldsymbol{v}) = \int_{\mathcal{S}} A(\cdots)\,\mathrm{d}S = \int_{\Omega} A(\cdots)\sqrt{a}\,\mathrm{d}\xi. \tag{8.1.10}$$

The function $A(\cdots)$ is given by (cf. KOITER (1970), eq. (3.16)):

$$A(\cdots) = \frac{Ee}{2(1-\sigma^2)}\{(1-\sigma)\gamma_\beta^\alpha\gamma_\alpha^\beta + \sigma\gamma_\alpha^\alpha\gamma_\beta^\beta\} + \frac{Ee^3}{24(1-\sigma^2)}\{(1-\sigma)\bar{\rho}_\beta^\alpha\bar{\rho}_\alpha^\beta + \sigma\bar{\rho}_\alpha^\alpha\bar{\rho}_\beta^\beta\}, \tag{8.1.11}$$

where e is the thickness of the shell, E is its Young modulus, σ is its Poisson coefficient, and the mixed tensors (γ_β^α) and $(\bar{\rho}_\beta^\alpha)$ are obtained from the doubly covariant *strain tensor* $(\gamma_{\alpha\beta})$ and *change of curvature tensor* $(\bar{\rho}_{\alpha\beta})$ through the tensorial operations

$$\gamma_\beta^\alpha = a^{\alpha\nu}\gamma_{\nu\beta}, \quad \bar{\rho}_\beta^\alpha = a^{\alpha\nu}\bar{\rho}_{\nu\beta}. \tag{8.1.12}$$

The tensors $(\gamma_{\alpha\beta})$ and $(\bar{\rho}_{\alpha\beta})$ are given by

$$\gamma_{\alpha\beta} = \gamma_{\beta\alpha} = \tfrac{1}{2}(v_{\alpha|\beta} + v_{\beta|\alpha}) - b_{\alpha\beta}v_3, \tag{8.1.13}$$

$$\bar{\rho}_{\alpha\beta} = \bar{\rho}_{\beta\alpha} = v_{3|\alpha\beta} - c_{\alpha\beta}v_3 + b_\alpha^\lambda v_{\lambda|\beta} + b_\beta^\lambda v_{\lambda|\alpha} + b_{\alpha|\beta}^\lambda v_\lambda, \tag{8.1.14}$$

where the various symbols occuring in these expressions will now be defined.

The *second fundamental form* $(b_{\alpha\beta})$ of the surface is given by

$$b_{\alpha\beta} = b_{\beta\alpha} = -\boldsymbol{a}_\alpha \cdot \partial_\beta \boldsymbol{a}_3 = \boldsymbol{a}_3 \cdot \partial_\beta \boldsymbol{a}_\alpha = \tfrac{1}{2}\boldsymbol{a}_3 \cdot (\partial_\beta \boldsymbol{a}_\alpha + \partial_\alpha \boldsymbol{a}_\beta), \tag{8.1.15}$$

from which the *third fundamental form* $(c_{\alpha\beta})$ of the surface is derived by letting

$$c_{\alpha\beta} = c_{\beta\alpha} = b_\alpha^\lambda b_{\lambda\beta}, \quad \text{where} \quad b_\alpha^\lambda = a^{\lambda\beta}b_{\alpha\beta}. \tag{8.1.16}$$

Then the *Christoffel symbols* $(\Gamma_{\beta\gamma}^\alpha)$ of the surface are defined by the formulas

$$\Gamma^{\alpha}_{\beta\gamma} = \boldsymbol{a}^{\alpha} \cdot \partial_{\beta}\boldsymbol{a}_{\gamma} = a^{\alpha\nu}\boldsymbol{a}_{\nu} \cdot \partial_{\beta}\boldsymbol{a}_{\gamma} = a^{\alpha\nu}\partial_{\nu}\boldsymbol{\varphi} \cdot \partial_{\gamma\beta}\boldsymbol{\varphi}. \tag{8.1.17}$$

These functions are symmetric with respect to the lower indices in the sense that

$$\Gamma^{\alpha}_{\beta\gamma} = \Gamma^{\alpha}_{\gamma\beta}, \tag{8.1.18}$$

and they satisfy

$$\Gamma^{\alpha}_{\beta\gamma} = a^{\alpha\lambda}\Gamma_{\lambda\beta\gamma}, \quad \text{with} \quad \Gamma_{\alpha\beta\gamma} = \tfrac{1}{2}(\partial_{\gamma}a_{\alpha\beta} + \partial_{\beta}a_{\alpha\gamma} - \partial_{\alpha}a_{\beta\gamma}). \tag{8.1.19}$$

Then the *covariant derivatives* $v_{\alpha|\beta}$, $v_{3|\alpha\beta}$, $b^{\delta}_{\alpha|\beta}$ are given by

$$v_{\alpha|\beta} = \partial_{\beta}v_{\alpha} - \Gamma^{\gamma}_{\alpha\beta}v_{\gamma}, \tag{8.1.20}$$

$$v_{3|\alpha\beta} = \partial_{\alpha\beta}v_3 - \Gamma^{\gamma}_{\alpha\beta}\partial_{\gamma}v_3, \tag{8.1.21}$$

$$b^{\delta}_{\alpha|\beta} = \partial_{\beta}b^{\delta}_{\alpha} + \Gamma^{\delta}_{\beta\lambda}b^{\lambda}_{\alpha} - \Gamma^{\lambda}_{\alpha\beta}b^{\delta}_{\lambda}. \tag{8.1.22}$$

Using the Mainardi–Codazzi and Ricci identities, one can show the equalities

$$b^{\delta}_{\alpha|\beta} = b^{\delta}_{\beta|\alpha}, \tag{8.1.23}$$

which in turn imply the symmetry of the tensor $\bar{\rho}_{\alpha\beta}$ of (8.1.14).

From all the previous formulas, it follows that in the integrand (8.1.11) appearing in the strain energy (8.1.10) of the shell, one finds the three functions v_i and some of their partial derivatives, which we shall sometimes record as the following twelve functions V_I, $1 \leq I \leq 12$:

$$(V_I)_{I=1}^{12} = \begin{cases} v_1,\ \partial_1 v_1,\ \partial_2 v_1, \\ v_2,\ \partial_1 v_2,\ \partial_2 v_2, \\ v_3,\ \partial_1 v_3,\ \partial_2 v_3,\ \partial_{11}v_3,\ \partial_{12}v_3,\ \partial_{22}v_3, \end{cases} \tag{8.1.24}$$

while the notation $\boldsymbol{v}$ is reserved for the triple (v_1, v_2, v_3).

Associating as in (8.1.24) twelve functions U_I, $1 \leq I \leq 12$, with another generic function $\boldsymbol{u} = (u_1, u_2, u_3)$, we are able to state the main properties of the bilinear form (8.1.10). The proof, which is a matter of lengthy verifications, is left as a problem (Exercise 8.1.1).

Theorem 8.1.1. *The bilinear form which occurs in the definition of the strain energy of the shell is of the following form:*

$$a(\boldsymbol{u}, \boldsymbol{v}) = \int_{\Omega} \sum_{I,J=1}^{12} A_{IJ}(\xi)U_I V_J \,\mathrm{d}\xi. \tag{8.1.25}$$

Denoting by φ_i the components of the mapping $\boldsymbol{\varphi} = \varphi_i \boldsymbol{e}^i$, we have, for each (I, J):

$$A_{IJ}(\xi) = f_{IJ}(\partial_\alpha \varphi_i(\xi), \partial_{\alpha\beta}\varphi_i(\xi), \partial_{\alpha\beta\gamma}\varphi_i(\xi)), \tag{8.1.26}$$

where the function f_{IJ} is a quotient between a polynomial in its arguments and a denominator which is an integer power of the expression

$$\sqrt{a} = \sqrt{\det(a_{\alpha\beta})} = \left(\left(\sum_{i=1}^{3} (\partial_1 \varphi_i)^2\right)\left(\sum_{j=1}^{3} (\partial_2 \varphi_j)^2\right) - \left(\sum_{i=1}^{3} \partial_1 \varphi_i \partial_2 \varphi_i\right)^2\right)^{1/2}. \tag{8.1.27}$$

The bilinear form is symmetric in its arguments $\boldsymbol{u}$ and $\boldsymbol{v}$ and, finally, it is defined and continuous over the space $H^1(\Omega) \times H^1(\Omega) \times H^2(\Omega)$. □

The *potential energy of the exterior forces* is another surface integral, of the form

$$f(\boldsymbol{v}) = \int_{\mathcal{S}} \boldsymbol{f} \cdot \boldsymbol{v} \, \mathrm{d}S = \int_\Omega f^i v_i \sqrt{a} \, \mathrm{d}\xi, \tag{8.1.28}$$

where the functions f^i represent the contravariant components, i.e., over the basis $(\boldsymbol{a}_i)$, of the reduced density per unit surface of the exterior forces. Clearly, such a linear form is also continuous over the space $H^1(\Omega) \times H^1(\Omega) \times H^2(\Omega)$.

Existence of a solution. Proof for the arch problem

Let then $\boldsymbol{V}$ be a space such as

$$\boldsymbol{V} = H_0^1(\Omega) \times H_0^1(\Omega) \times H_0^2(\Omega), \quad \text{or}$$

$$\boldsymbol{V} = H_0^1(\Omega) \times H_0^1(\Omega) \times (H^2(\Omega) \cap H_0^1(\Omega)), \tag{8.1.29}$$

which corresponds to the case of a *clamped shell*, and a *simply supported shell*, respectively. Then the problem of showing the existence of a displacement $\boldsymbol{u} \in \boldsymbol{V}$ such that $J(\boldsymbol{u}) = \inf_{\boldsymbol{v} \in \boldsymbol{V}} J(\boldsymbol{v})$, or equivalently such that $a(\boldsymbol{u}, \boldsymbol{v}) = f(\boldsymbol{v})$ for all $\boldsymbol{v} \in \boldsymbol{V}$, with J, $a(.,.)$ and $f(.)$ as in (8.1.8), (8.1.25) and (8.1.28) respectively, reduces to the problem of showing the $\boldsymbol{V}$-ellipticity of the bilinear form.

This is done in BERNADOU & CIARLET (1976), under the assumption that the mapping φ is of class $\mathscr{C}^3(\bar{\Omega})$. Rather than giving the lengthy and fairly intricate proof here, we shall instead focus our attention on a simpler problem, which nevertheless displays all the essential features of the general shell problem: the *arch problem*, where a single variable (instead of two) is needed.

The arch $\mathscr{A}$ is assumed to be in a plane just as the forces which act on it. Then following Fig. 8.1.3, which should be self explanatory as regards the various notations introduced, the *energy of the arch* $\mathscr{A}$ has the following form:

$$J(v) = \frac{1}{2}\int_I \left\{ EA\left(v_1' - \frac{v_2}{R}\right)^2 + EI\left(\left(v_2' + \frac{v_1}{R}\right)'\right)^2 \right\} \mathrm{d}s - \int_I \boldsymbol{f} \cdot \boldsymbol{v} \,\mathrm{d}s. \tag{8.1.30}$$

In this expression, *the parameter s is the curvilinear abcissa along the arch* and thus the vector $\boldsymbol{a}^1 = \boldsymbol{\varphi}'$ is a unit vector, the functions v_1 and v_2: $I \to \mathbf{R}$ are the tangential and normal components of the admissible displacements $\boldsymbol{v} = v_1\boldsymbol{a}^1 + v_2\boldsymbol{a}^2$, the function R: $\bar{I} \to \mathbf{R}$ is the (algebraically counted) radius of curvature, so that the function $1/R$: $\bar{I} \to \mathbf{R}$ is the *curvature* of the arch. Finally, the constant E is the Young modulus of the material of which the arch is composed, the constant A is the area of a cross-section of the arch and the constant I is the moment of inertia of a cross-section of the arch. Since these three constants are strictly

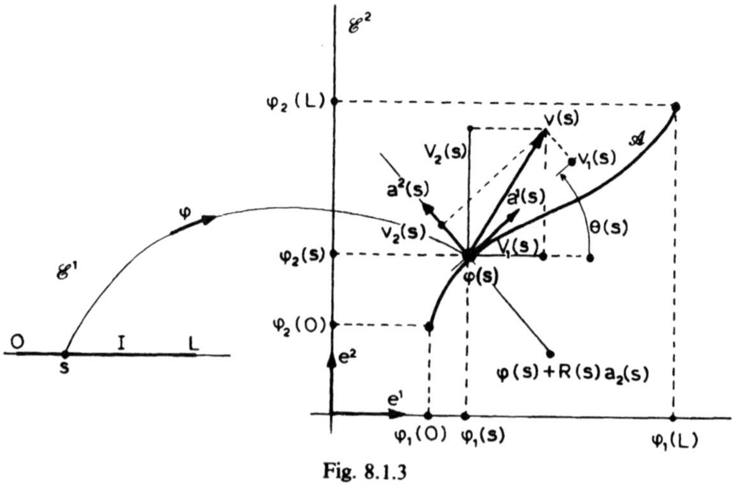

Fig. 8.1.3

positive, there is no loss of generality in assuming that $EA = EI = 1$, as will be henceforth assumed.

In the next theorem, the proof of the ellipticity of the bilinear form is given in a form which is similar to that given in BERNADOU & CIARLET (1976) for a shell.

Theorem 8.1.2. *If the function $1/R$ is continuously differentiable over the interval $\bar{I}$, the bilinear form defined by*

$$a(\boldsymbol{u}, \boldsymbol{v}) = \int_I \left\{\left(u_1' - \frac{u_2}{R}\right)\left(v_1' - \frac{v_2}{R}\right) + \left(u_2' + \frac{u_1}{R}\right)'\left(v_2' + \frac{v_1}{R}\right)'\right\} \mathrm{d}s \tag{8.1.31}$$

is $(H_0^1(I) \times (H^2(I) \cap H_0^1(I)))$-elliptic, and thus, it is a fortiori $(H_0^1(I) \times H_0^2(I))$-elliptic.

Proof. We shall equip the space

$$\boldsymbol{V} = H_0^1(I) \times (H^2(I) \cap H_0^1(I)) \tag{8.1.32}$$

with the norm

$$|\boldsymbol{v}| = (|v_1|_{1,I}^2 + |v_2|_{2,I}^2)^{1/2} \tag{8.1.33}$$

(it is easily verified that over the space $H^2(I) \cap H_0^1(I)$, the semi-norm $|\cdot|_{2,I}$ is a norm, equivalent to the norm $\|\cdot\|_{2,I}$).

The proof consists of three steps.

(i) *There exist a constant $\lambda > 0$ and a constant μ such that*

$$a(\boldsymbol{v}, \boldsymbol{v}) \geq \lambda |\boldsymbol{v}|^2 + \mu(|v_1|_{0,I}^2 + \|v_2\|_{1,I}^2),$$
$$\text{for all } \boldsymbol{v} = (v_1, v_2) \in H^1(I) \times H^2(I). \tag{8.1.34}$$

Let $\beta = |1/R|_{0,\infty,I}$. For all $\epsilon > 0$, we have

$$\int_I \left(v_1' - \frac{v_2}{R}\right)^2 \mathrm{d}s \geq |v_1|_{1,I}^2 - 2\beta|v_1|_{1,I}|v_2|_{0,I}$$
$$\geq (1 - \epsilon\beta)|v_1|_{1,I}^2 - \frac{\beta}{\epsilon}|v_2|_{0,I}^2,$$

and thus if we choose $\epsilon \in]0, 1/\beta[$, we have found a constant $\lambda_1 > 0$ and a constant μ_1 such that

$$\int_I \left(v_1' - \frac{v_2}{R}\right)^2 \mathrm{d}s \geq \lambda_1|v_1|_{1,I}^2 + \mu_1|v_2|_{0,I}^2. \tag{8.1.35}$$

Likewise, there exist a constant $\lambda_2 > 0$ and a constant μ_2' such that

$$\int_I \left(\left(v_2' + \frac{v_1}{R}\right)'\right)^2 \mathrm{d}s \geq \lambda_2 |v_2|_{2,I}^2 + \mu_2' \left|\left(\frac{v_1}{R}\right)'\right|_{0,I}^2.$$

Since

$$\left|\left(\frac{v_1}{R}\right)'\right|_{0,I} \leq \beta |v_1|_{1,I} + \beta^2 |R'|_{0,\infty,I} |v_1|_{0,I},$$

we have found a constant $\lambda_2 > 0$ and two constants μ_2 and ν_2 such that

$$\forall v_1 \in H^1(I), \quad \forall v_2 \in H^2(I),$$

$$\int_I \left(\left(v_2' + \frac{v_1}{R}\right)'\right)^2 \mathrm{d}s \geq \lambda_2 |v_2|_{2,I}^2 + \mu_2 |v_1|_{1,I}^2 + \nu_2 |v_1|_{0,I}^2. \tag{8.1.36}$$

If the constant μ_2 is positive, then inequality (8.1.34) is a direct consequence of inequalities (8.1.35) and (8.1.36). If $\mu_2 < 0$, let ϵ be so chosen that $0 < \epsilon < \min\{\lambda_1/|\mu_2|, 1\}$. Then

$$a(v, v) \geq \int_I \left(v_1' - \frac{v_2}{R}\right)^2 \mathrm{d}s + \epsilon \int_I \left(\left(v_2' + \frac{v_1}{R}\right)'\right)^2 \mathrm{d}s$$

$$\geq (\lambda_1 - \epsilon|\mu_2|)|v_1|_{1,I}^2 - \epsilon|\nu_2|\,|v_1|_{0,I}^2 + \mu_1|v_2|_{0,I}^2 + \epsilon\lambda_2|v_2|_{2,I}^2,$$

and inequality (8.1.34) is proved in all cases.

(ii) *The mapping*

$$v \to \sqrt{a(v, v)}$$

is a norm over the space V. Clearly it is a semi-norm, so it remains to prove that $a(v, v) = 0$ implies $v = 0$, i.e., we face the problem of solving, *in the sense of distributions*, the coupled system of differential equations:

$$\begin{cases} v_1' - \dfrac{v_2}{R} = 0 \quad \text{on} \quad I, \\ \left(v_2' + \dfrac{v_1}{R}\right)' = 0 \quad \text{on} \quad I, \end{cases} \tag{8.1.37}$$

along with the boundary conditions:

$$v_1(0) = v_2(0) = v_1(L) = v_2(L) = 0. \tag{8.1.38}$$

As suggested by the geometry of the problem (Fig. 8.1.3), let us introduce the angle θ between the vectors e^1 and a^1, so that the following relations hold:

$$\theta' = \frac{1}{R}, \tag{8.1.39}$$

$$\varphi_1' = \cos\theta, \quad \varphi_2' = \sin\theta. \tag{8.1.40}$$

We also introduce the Cartesian components V_1 and V_2 of the displacement. From the relations

$$\forall s \in I, \quad \boldsymbol{v}(s) = v_1(s)\boldsymbol{a}^1(s) + v_2(s)\boldsymbol{a}^2(s) = V_1(s)\boldsymbol{e}^1 + V_2(s)\boldsymbol{e}^2,$$

we deduce that

$$V_1 = v_1\cos\theta - v_2\sin\theta,$$

$$V_2 = v_1\sin\theta + v_2\cos\theta.$$

Since the functions v_1 and v_2 are both in the space $H^1(I)$ and since the function θ is in the space $\mathscr{C}^1(\bar{I})$, both functions V_1 and V_2 are in the space $H^1(I)$, and

$$\begin{cases} V_1' = (v_1' - v_2\theta')\cos\theta - (v_2' + v_1\theta')\sin\theta, \\ V_2' = (v_2' + v_1\theta')\cos\theta + (v_1' - v_2\theta')\sin\theta. \end{cases} \tag{8.1.41}$$

The second differential equation of (8.1.37) implies the existence of a constant a such that

$$v_2' + \frac{v_1}{R} = a, \tag{8.1.42}$$

so that we obtain, upon combining relations (8.1.37), (8.1.39), (8.1.40), (8.1.41) and (8.1.42):

$$V_1' + a\varphi_2' = 0,$$

$$V_2' - a\varphi_1' = 0.$$

Therefore, there exist constants b_1 and b_2 such that

$$V_1 = -a\varphi_2 + b_1,$$

$$V_2 = a\varphi_1 + b_2.$$

Let then $\boldsymbol{e}^3 = \boldsymbol{e}^1 \times \boldsymbol{e}^2$. We have proved that *the general solution of the differential system* (8.1.37) *is of the form*

$$\boldsymbol{v} = \boldsymbol{a} \times \boldsymbol{\varphi} + \boldsymbol{b}, \tag{8.1.43}$$

where the constant vectors $\boldsymbol{a}$ *and* $\boldsymbol{b}$ *are given by*

$$\boldsymbol{a} = a\boldsymbol{e}^3, \quad \boldsymbol{b} = b_1\boldsymbol{e}^1 + b_2\boldsymbol{e}^2 \tag{8.1.44}$$

(see Remark 8.1.1 for the interpretation of such a solution).

Finally, it is an easy matter to show that any solution of the form (8.1.43) necessarily vanishes when it is subjected to the boundary conditions (8.1.38), since $\varphi(0) \neq \varphi(L)$ by assumption.

(iii) *Using steps* (i) *and* (ii), *we are in a position to show the* $\boldsymbol{V}$-*ellipticity of the bilinear form*. If it were not $\boldsymbol{V}$-elliptic, there would exist a sequence $\boldsymbol{v}^k = (v_1^k, v_2^k) \in \boldsymbol{V}$ such that

$$\lim_{k\to\infty} a(\boldsymbol{v}^k, \boldsymbol{v}^k) = 0, \quad |\boldsymbol{v}^k| = 1 \quad \text{for all } k. \tag{8.1.45}$$

Since the sequence (v_1^k) is bounded in $H_0^1(I)$, there exists a subsequence, which we shall still denote by (v_1^k) for convenience, which converges weakly in $H_0^1(I)$ and converges strongly in $L^2(I)$ to the same limit v_1.

Likewise, the boundedness of the sequence (v_2^k) in $H^2(I) \cap H_0^1(I)$ implies that there exists a subsequence, still denoted by (v_2^k), which converges weakly in $H^2(I) \cap H_0^1(I)$ and converges strongly in $H^1(I)$ to the same limit v_2.

The function $\boldsymbol{v} \in \boldsymbol{V} \to a(\boldsymbol{v}, \boldsymbol{v})$ is continuous for the strong topology of the space $\boldsymbol{V}$ and it is a convex function (it is even a strictly convex function since its second derivative is positive definite, as was shown in step (ii)). Therefore, it is weakly lower semi-continuous. As a consequence, we have, setting $\boldsymbol{v} = (v_1, v_2)$,

$$a(\boldsymbol{v}, \boldsymbol{v}) \leq \liminf_{k\to\infty} a(\boldsymbol{v}^k, \boldsymbol{v}^k) = \lim_{k\to\infty} a(\boldsymbol{v}^k, \boldsymbol{v}^k) = 0,$$

and thus $\boldsymbol{v} = \boldsymbol{0}$ by step (ii). By step (i), we have for all k,

$$a(\boldsymbol{v}^k, \boldsymbol{v}^k) \geq \lambda |\boldsymbol{v}^k|^2 + \mu(|v_1^k|_{0,I}^2 + \|v_2^k\|_{1,I}^2)$$

(cf. (8.1.34)). We have therefore reached a contradiction, since

$$\lim_{k\to\infty} a(\boldsymbol{v}^k, \boldsymbol{v}^k) = 0 \quad (\text{cf. } (8.1.45)),$$

$$\lim_{k\to\infty} \{\lambda |\boldsymbol{v}^k|^2 + \mu(|v_1^k|_{0,I}^2 + \|v_2^k\|_{1,I}^2)\} = \lambda > 0. \qquad \square$$

Remark 8.1.1. By step (ii) in the preceding proof, any displacement $\boldsymbol{v} \in H^1(\Omega) \times H^2(\Omega)$ which satisfies $a(\boldsymbol{v}, \boldsymbol{v}) = 0$ is of the form (8.1.43), i.e., it corresponds to a *rigid body motion* in the plane of the arch. This condition is an instance of the *rigid displacement condition* that a mathematical model for an elastic system should be such that *the vanishing of the strain energy corresponds to rigid body motions* (similar conclusions hold for the system of linear elasticity; cf. Exercise 1.2.4).

In addition, this interpretation provides an approach for integrating in a simple way the differential system (8.1.37) by suggesting the introduction of the functions θ, V_1 and V_2. □

Exercises

8.1.1. Prove all the statements of Theorem 8.1.1.

8.1.2. Let $\bar{\Omega} = [\alpha, \beta] \times [-H, +H]$ and

$$\varphi(\xi) = R \cos \xi^1 e^1 + R \sin \xi^1 e^2 + \xi^2 e^3$$

so that the shell $\mathcal{S} = \varphi(\bar{\Omega})$ is a portion of a right circular cylinder (Fig. 8.1.4).

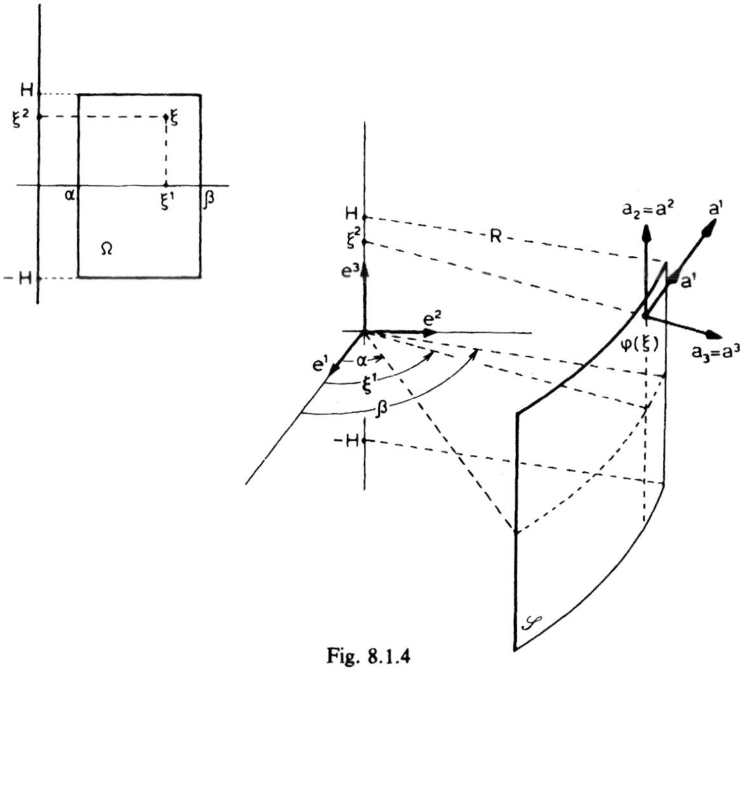

Fig. 8.1.4

Show that the energy of the shell $\mathscr{S}$ has the following expression:

$$J(\boldsymbol{v}) = \frac{Ee}{2(1-\sigma^2)} \int_\alpha^\beta \int_{-H}^{+H} \Big\{(1-\sigma)\Big(\frac{1}{R^4}(\partial_1 v_1 + R v_3)^2 + \frac{1}{2R^2}(\partial_2 v_1 + \partial_1 v_2)^2 + (\partial_2 v_2)^2\Big) + \sigma\Big(\frac{1}{R^2}(\partial_1 v_1 + R v_3)^2 + (\partial_2 v_2)^2\Big)\Big\} R \,\mathrm{d}\xi^1 \,\mathrm{d}\xi^2 + \frac{Ee^3}{24(1-\sigma^2)} \int_\alpha^\beta \int_{-H}^{+H} \Big\{(1-\sigma) \times \Big(\frac{1}{R^4}\Big(\partial_{11} v_3 - v_3 - \frac{2}{R}\partial_1 v_1\Big)^2 + \frac{2}{R^2}\Big(\partial_{12} v_3 - \frac{1}{R}\partial_2 v_1\Big)^2 + (\partial_{22} v_3)^2\Big) + \sigma\Big(\frac{1}{R^2}\Big(\partial_{11} v_3 - v_3 - \frac{2}{R}\partial_1 v_1\Big) + \partial_{22} v_3\Big)^2\Big\} R \,\mathrm{d}\xi^1 \,\mathrm{d}\xi^2 - \int_\alpha^\beta \int_{-H}^{+H} \{f^1 v_1 + f^2 v_2 + f^3 v_3\} R \,\mathrm{d}\xi^1 \,\mathrm{d}\xi^2.$$

8.1.3. Let $\bar{\Omega}$ be a rectangle with sides parallel to the coordinate axes and let $\boldsymbol{\varphi}(\xi) = \xi^1 \boldsymbol{e}^1 + \xi^2 \boldsymbol{e}^2$, i.e., the shell is a plate. Is the energy in this case identical to the energy of a plate as given in (1.2.46)?

8.1.4. In the case of a *clamped circular arch* (R = constant), one can give another proof of Theorem 8.1.2, along the following lines, suggested by C. Johnson.

(i) For any $\boldsymbol{v} = (v_1, v_2)$ in the space $\boldsymbol{V} = H_0^1(I) \times H_0^2(I)$, let

$$g_1 = v_1' - \frac{v_2}{R}, \quad g_2 = v_2'' + \frac{v_2'}{R}, \quad F = g_2 - \frac{g_1}{R},$$

and show that

$$\forall s \in I, \quad v_2(s) = \int_0^s R \sin\Big(\frac{1}{R}(s-t)\Big) F(t) \,\mathrm{d}t.$$

(ii) For all $\boldsymbol{v} \in \boldsymbol{V}$, deduce from (i) that $|v_2|_{0,I}^2 \leq c_1 a(\boldsymbol{v}, \boldsymbol{v})$, and then that $|v_2|_{2,I}^2 \leq c_2 a(\boldsymbol{v}, \boldsymbol{v})$ and finally that $|v_1|_{1,I}^2 \leq c_3 a(\boldsymbol{v}, \boldsymbol{v})$, for some constants c_1, c_2 and c_3 independent of $\boldsymbol{v} \in \boldsymbol{V}$.

8.1.5. Consider a *circular arch*, i.e., for which the radius of curvature is a constant. Assuming the solution $\boldsymbol{u} = (u_1, u_2)$ of the associated variational equations is smooth enough, derive the associated system of two differential equations and the boundary conditions corresponding to the choices $\boldsymbol{V} = H_0^1(I) \times H_0^2(I)$ and $\boldsymbol{V} = H_0^1(I) \times (H^2(I) \cap H_0^1(I))$.

8.2. Conforming methods

The discrete problem. Approximation of the geometry. Approximation of the displacement

We shall assume throughout this section that *the set $\bar{\Omega}$ is a polygon* (cf. Remark 8.2.1). Thus we may cover the set $\bar{\Omega}$ by triangulations $\mathcal{T}_h$ in such a way that $\bar{\Omega} = \bigcup_{K\in\mathcal{T}_h} K$, the sets K being the finite elements of the triangulation. *With such a triangulation are associated three finite element spaces Φ_h, V_h, W_h,* whose specific properties will be subsequently described (actually, the present analysis immediately extends to the case where the spaces Φ_h, V_h, W_h would be associated with different triangulations, an unrealistic case from a practical viewpoint, however).

The discrete problem then requires two approximations.

(i) *Approximation of the geometry of the surface:*
If Θ_h denotes the Φ_h-interpolation operator, then with the given mapping $\boldsymbol{\varphi} = \varphi_i \boldsymbol{e}^i$ is associated the *approximate mapping*

$$\boldsymbol{\varphi}_h = \varphi_{ih}\boldsymbol{e}^i, \quad \text{with} \quad \varphi_{ih} = \Theta_h\varphi_i, \quad 1 \leq i \leq 3. \tag{8.2.1}$$

Notice that if the finite elements of the space Φ_h are not of class $\mathscr{C}^0$, then the mapping $\boldsymbol{\varphi}_h$ is *a priori* defined only on the union $\bigcup_{K\in\mathcal{T}_h} \mathring{K}$ of the interiors of the finite elements.

(ii) *Approximation of the components of the displacement:* The approximations $u_{\alpha h}$ of the components u_α, $\alpha = 1, 2$, belong to the space V_h, while the approximation u_{3h} of the component u_3 belongs to the space W_h. Therefore the discrete solution $\boldsymbol{u}_h = (u_{1h}, u_{2h}, u_{3h})$ is in the space

$$\boldsymbol{V}_h = V_h \times V_h \times W_h. \tag{8.2.2}$$

The *discrete problem* is then defined as follows:
The discrete solution $\boldsymbol{u}_h \in \boldsymbol{V}_h$ is such that

$$\forall \boldsymbol{v}_h \in \boldsymbol{V}_h, \quad a_h(\boldsymbol{u}_h, \boldsymbol{v}_h) = f_h(\boldsymbol{v}_h), \tag{8.2.3}$$

where the *approximate bilinear form* $a_h(\cdot\,,\cdot)$ is given by

$$a_h(\boldsymbol{u}, \boldsymbol{v}) = \sum_{K\in\mathcal{T}_h} \int_{\mathring{K}} \sum_{I,J=1}^{12} A_{IJh}(\xi) U_I V_J \,\mathrm{d}\xi, \tag{8.2.4}$$

with

$$A_{IJh}(\xi) = f_{IJ}(\partial_\alpha\varphi_{ih}(\xi), \partial_{\alpha\beta}\varphi_{ih}(\xi), \partial_{\alpha\beta\gamma}\varphi_{ih}(\xi)), \tag{8.2.5}$$

the functions f_{IJ} being the same as in Theorem 8.1.1. In other words, *the coefficients A_{IJh} are expressed in terms of the partial derivatives of the approximate mapping φ_h exactly as the coefficients A_{IJ} are expressed in terms of the same partial derivatives of the mapping φ.*

In the same fashion, the *approximate linear form f_h* is given by

$$f_h(\boldsymbol{v}) = \sum_{K \in \mathcal{T}_h} \int_{\hat{K}} \boldsymbol{f} \cdot \boldsymbol{v} \sqrt{a_h} \, \mathrm{d}\xi, \tag{8.2.6}$$

where (compare with (8.1.27))

$$\sqrt{a_h} = \left(\left(\sum_{i=1}^{3} (\partial_1 \varphi_{ih})^2 \right) \left(\sum_{j=1}^{3} (\partial_2 \varphi_{jh})^2 \right) - \left(\sum_{i=1}^{3} \partial_1 \varphi_{ih} \partial_2 \varphi_{ih} \right)^2 \right)^{1/2}. \tag{8.2.7}$$

Notice that the replacement of the functions A_{IJ} by the functions A_{IJh} amounts to replacing each covariant derivative with respect to the surface $\mathcal{S}$ by the analogous covariant derivative with respect to the *approximate surface*

$$\mathcal{S}_h = \varphi_h(\bar{\Omega}). \tag{8.2.8}$$

Therefore *the approximate energy*

$$J_h(\boldsymbol{v}) = \tfrac{1}{2} a_h(\boldsymbol{v}, \boldsymbol{v}) - f_h(\boldsymbol{v}) \tag{8.2.9}$$

may be viewed either as an approximation of the energy of the shell $\mathcal{S}$ or *as the exact energy of the approximate shell $\mathcal{S}_h$.*

Remark 8.2.1. If the boundary Γ is *curved*, then a *third approximation* has to be taken into account, and similarly, a *fourth approximation has* to be considered in case *numerical quadrature schemes* are used for computing the coefficients of the resulting linear system. Taking these approximations into account requires an extension of the analysis made in Chapter 4. See BERNADOU (1976). □

Finite element methods conforming for the displacements

For the sake of definiteness, we shall assume in the sequel that we are considering the case of a *clamped shell*, i.e., the space $\boldsymbol{V}$ is given by

$$\boldsymbol{V} = H_0^1(\Omega) \times H_0^1(\Omega) \times H_0^2(\Omega), \tag{8.2.10}$$

but it should be clear that the subsequent analysis extends readily to other situations, such as that of a simply supported shell, etc...

We shall say that the discrete problem is *conforming for the dis-*

placements if the inclusions

$$V_h \subset H_0^1(\Omega), \quad W_h \subset H_0^2(\Omega), \quad \text{i.e., } \boldsymbol{V}_h \subset \boldsymbol{V},$$

hold (after we have established sufficient conditions for convergence in Theorem 8.2.4, we will also define what may be understood by a discrete problem which is "conforming for the geometry").

As regards the construction of the spaces $\boldsymbol{\Phi}_h$, V_h and W_h, let us consider one *example*, as described in ARGYRIS & LOCHNER (1972), ARGYRIS, HAASE & MALEJANNAKIS (1973). Let K be any triangle of the triangulation $\mathcal{T}_h$, with vertices a_i and mid-points b_i along the sides, and where the vectors ν_i denote the heights of the triangle (Fig. 8.2.1).

Then if the mapping $\boldsymbol{\varphi}: \bar{\Omega} \to \mathscr{E}^3$ is of class $\mathscr{C}^2$, there exists for each triangle $K \in \mathcal{T}_h$ a unique mapping $\boldsymbol{F}_K: K \to \mathscr{E}^3$ such that:

$$\begin{aligned}
&\boldsymbol{F}_K \in (P_5(K))^3,\\
&\boldsymbol{F}_K(a_i) = \boldsymbol{\varphi}(a_i), \quad 1 \leqslant i \leqslant 3,\\
&D\boldsymbol{F}_K(a_i)(a_{i-1} - a_i) = D\boldsymbol{\varphi}(a_i)(a_{i-1} - a_i), \quad 1 \leqslant i \leqslant 3,\\
&D\boldsymbol{F}_K(a_i)(a_{i+1} - a_i) = D\boldsymbol{\varphi}(a_i)(a_{i+1} - a_i), \quad 1 \leqslant i \leqslant 3,\\
&D\boldsymbol{F}_K(b_i)\nu_i = D\boldsymbol{\varphi}(b_i)\nu_i, \quad 1 \leqslant i \leqslant 3,\\
&D^2\boldsymbol{F}_K(a_i)(a_{j+1} - a_j)^2 = D^2\boldsymbol{\varphi}(a_i)(a_{j+1} - a_j)^2, \quad 1 \leqslant i, j \leqslant 3,
\end{aligned}$$

where the indices i and j are counted modulo 3, if necessary. We recognize here the *Argyris triangle*, which was introduced in Section 2.2, and whose interpolation properties were analyzed in Section 6.1.

We then choose the approximate mapping $\boldsymbol{\varphi}_h: \bar{\Omega} \to \mathscr{E}^3$ so that, for each

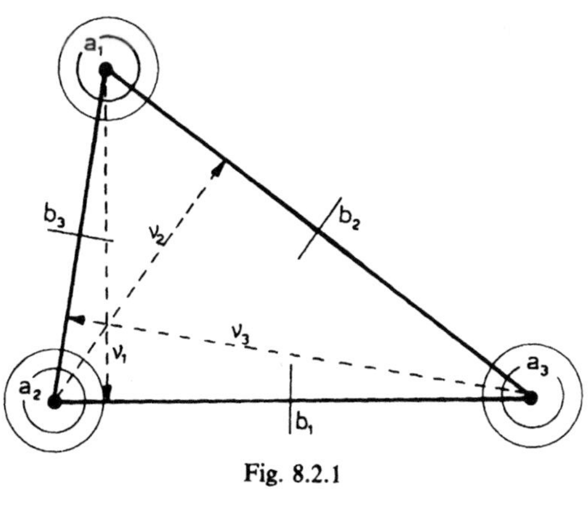

Fig. 8.2.1

$K \in \mathcal{T}_h$, its restriction to the set K coincides with the mapping F_K. Once the space Φ_h is defined in this fashion, we let V_h, and W_h, be the subspaces of Φ_h whose functions v_h and w_h satisfy the boundary condition $v_h = 0$ on Γ, and the boundary condition $w_h = \partial_\nu w_h = 0$ on Γ, respectively, i.e., with the notations of Chapter 2, we let $V_h = \Phi_{0h}$ and $W_h = \Phi_{00h}$. Since the inclusion $\Phi_h \subset \mathscr{C}^1(\bar{\Omega})$ holds, the inclusions $V_h \subset H_0^1(\Omega)$ and $W_h \subset H_0^2(\Omega)$ surely hold, and this method is therefore conforming for the displacements.

Actually, one has even "too much", in that the inclusion $V_h \subset H_0^1(\Omega) \cap H^2(\Omega)$ holds. However, it is clear that using basically a single finite element space has obvious advantages in terms of the actual numerical implementation of the method. This approach is also similar to that of DUPUIS & GOËL (1970) and DUPUIS (1971), who approximate the *Cartesian* components of the displacement $\boldsymbol{u}$, i.e., over the basis $(\boldsymbol{e}^i)$, so that all three components are in $H^2(\Omega)$, in general.

Let us next return to the general discussion. As far as the error analysis is concerned, it is clear that the ideal situation would correspond to the equality $\varphi = \varphi_h$ which implies the equalities $a_h(.,.) = a(.,.)$ and $f_h(.) = f(.)$. However it is equally clear that this is an exceptional situation. For instance, if we use the Argyris triangle, this would happen only if the restrictions $\varphi_{|K}$ belong to the spaces $(P_5(K))^3$ for all triangles $K \in \mathcal{T}_h$.

Nevertheless, we wish to emphasize the fact that *there are instances where this general approach would yield* $\varphi \neq \varphi_h$, *while the most straightforward approach yields the equalities* $a_h(.,.) = a(.,.)$ *and* $f_h(.) = f(.)$. To make this point clear, let us consider the case where the surface $\mathscr{S}$ is a portion of a right circular cylinder, whose energy was given in Exercise 8.1.2. In this case, the energy is expressed uniquely in terms of the functions v_i and their derivatives since the functions φ_i and their derivatives appear only as constants. Therefore the obvious discretization of this problem consists in minimizing the *same* energy over the space V_h. In this fashion, there is no approximation of the geometry so that one may consider that the approximated displacement are *piecewise polynomials in the curvilinear coordinates which define the surface* $\mathscr{S}$.

Let us assume on the other hand that we had applied the general approach to this particular case. Since the mapping φ is given by

$$\varphi(\xi) = R \cos \xi^1 \boldsymbol{e}^1 + R \sin \xi^1 \boldsymbol{e}^2 + \xi^2 \boldsymbol{e}^3 = \varphi_i(\xi)\boldsymbol{e}^i,$$

for $\xi = (\xi^1, \xi^2) \in [\alpha, \beta] \times [-H, +H]$, any standard finite element space Φ_h

(whose functions are essentially piecewise polynomials) would *not* contain the two functions φ_1 and φ_2, and thus, this approach would necessarily require an approximation of the energy.

Likewise, between several available mappings φ for a given shell $\mathcal{S}$, one should choose the "simplest" one. To illustrate this point, let us consider again the case of a portion of a right circular cylinder. With the same notations as in Fig. 8.1.4, assume that $0 < \alpha < \beta < \pi$, so that another possible mapping φ^* is given by

$$\varphi^*(\eta) = \eta^1 e^1 + \sqrt{R^2 - (\eta^1)^2} e^2 + \eta^2 e^3 = \varphi_i^*(\eta) e^i,$$

for $\eta = (\eta^1, \eta^2) \in [R \cos \alpha, R \cos \beta] \times [-H, +H]$. Then, had we chosen this mapping, some partial derivatives of the function φ_2^* would have resulted in non polynomial functions $A_{\beta\lambda}^*(\eta)$, and therefore the energy could not have been exactly reproduced in the subspace V_h.

Consistency error estimates

Let us turn to the estimation of the error. In the rest of this section, we shall assume that we are given *three families of finite element spaces* Φ_h, V_h *and* W_h. In order to avoid lengthy statements of theorems, we shall assume throughout this section that, whenever they are needed, hypotheses (H1), (H2), (H3) (cf. Section 3.2) or (H1*), (H2*) (cf. Section 6.1) are satisfied by any one of the above family of finite element spaces. However, we shall record the basic inclusions which govern the orders of convergence, as in (8.2.13), (8.2.25) through (8.2.27).

We denote by P_K, P_K', P_K'', the spaces spanned by the restrictions to a given finite element K of the functions in the space Φ_h, V_h, W_h respectively.

The error analysis depends essentially upon estimates of the expressions $|a(u, v) - a_h(u, v)|$ and $|f(v) - f_h(v)|$, which are derived in the following theorem.

We shall use the product norm

$$v \in H^1(\Omega) \times H^1(\Omega) \times H^2(\Omega) \to \|v\| = (\|v_1\|_{1,\Omega}^2 + \|v_2\|_{1,\Omega}^2 + \|v_3\|_{2,\Omega}^2)^{1/2}, \tag{8.2.11}$$

which, over the space V of (8.2.10) is equivalent to the semi-norm

$$v \to |v| = (|v_1|_{1,\Omega}^2 + |v_2|_{1,\Omega}^2 + |v_3|_{2,\Omega}^2)^{1/2}. \tag{8.2.12}$$

Theorem 8.2.1. *We assume that the spaces* Φ_h *are such that the in-*

clusions

$$\forall K \in \bigcup_h \mathcal{T}_h, \quad P_m(K) \subset P_K \subset \mathscr{C}^3(K) \tag{8.2.13}$$

hold, for a given integer $m \geq 3$. *Then if* h *is sufficiently small, the approximate bilinear form of* (8.2.4) *is also defined over the space* $H^1(\Omega) \times H^1(\Omega) \times H^2(\Omega)$ *and there exists a constant* C *independent of* h *such that, for all* $\boldsymbol{u}, \boldsymbol{v} \in H^1(\Omega) \times H^1(\Omega) \times H^2(\Omega)$,

$$|a(\boldsymbol{u}, \boldsymbol{v}) - a_h(\boldsymbol{u}, \boldsymbol{v})| \leq C\|\boldsymbol{u}\| \, \|\boldsymbol{v}\| h^{m-2}. \tag{8.2.14}$$

Similarly, if h *is sufficiently small, there exists a constant* C *independent of* h *such that*,

$$|f(\boldsymbol{v}) - f_h(\boldsymbol{v})| \leq C|\boldsymbol{f}|_{0,\Omega}\|\boldsymbol{v}\| h^m, \tag{8.2.15}$$

where f_h *denotes the approximate linear form of* (8.2.6), *and*

$$|\boldsymbol{f}|_{0,\Omega} = \left(\sum_{i=1}^{3} |f^i|_{0,\Omega}^2\right)^{1/2}.$$

Proof. In view of the assumption (8.2.13) made upon the spaces Φ_h, there exists a constant C independent of h such that

$$\begin{cases} \sup\limits_{\xi \in \bigcup \mathring{K}} |\partial_\alpha \varphi_i(\xi) - \partial_\alpha \varphi_{ih}(\xi)| \leq Ch^m |\varphi_i|_{m+1,\infty,\bar{\Omega}}, \\ \sup\limits_{\xi \in \bigcup \mathring{K}} |\partial_{\alpha\beta} \varphi_i(\xi) - \partial_{\alpha\beta} \varphi_{ih}(\xi)| \leq Ch^{m-1} |\varphi_i|_{m+1,\infty,\bar{\Omega}}, \\ \sup\limits_{\xi \in \bigcup \mathring{K}} |\partial_{\alpha\beta\gamma} \varphi_i(\xi) - \partial_{\alpha\beta\gamma} \varphi_{ih}(\xi)| \leq Ch^{m-2} |\varphi_i|_{m+1,\infty,\bar{\Omega}}, \end{cases} \tag{8.2.16}$$

for all $i \in \{1, 2, 3\}$, and for all $\alpha, \beta, \gamma \in \{1, 2\}$.

Let $\psi_{i\alpha}$, $i \in \{1, 2, 3\}$, $\alpha \in \{1, 2\}$, be real-valued functions defined over the union $\bigcup \mathring{K}$. With these functions, we associate the function

$$a(\psi_{i\alpha}) = \left(\sum_{i=1}^{3} (\psi_{i1})^2\right)\left(\sum_{j=1}^{3} (\psi_{j2})^2\right) - \left(\sum_{i=1}^{3} \psi_{i1}\psi_{i2}\right)^2,$$

i.e., the function constructed from the function $\psi_{i\alpha}$ exactly as the function a, and the approximate function a_h, are constructed from the functions $\partial_\alpha \varphi_i$, and $\partial_\alpha \varphi_{ih}$, respectively (cf. (8.1.27) and (8.2.7)). We then claim that there exist two constants δ and $a_0 > 0$ such that, for all functions $\psi_{i\alpha}$ which satisfy the uniform bound

$$\sup_{\xi \in \bigcup \mathring{K}} |\partial_\alpha \varphi_i(\xi) - \psi_{i\alpha}(\xi)| \leq \delta,$$

then

$$\inf_{\xi \in \bigcup \hat{K}} a(\psi_{i\alpha}(\xi)) \geq a_0 > 0. \tag{8.2.17}$$

To see this, we remark that $a(\psi_{i\alpha})$ is the square of the norm of the vector $(\psi_{i1}\boldsymbol{e}^i) \times (\psi_{i2}\boldsymbol{e}^i)$. Since the norms of the vectors $\boldsymbol{a}_\alpha(\xi) = \partial_\alpha\varphi_i(\xi)\boldsymbol{e}^i$ are bounded below by a strictly positive constant independent of the point $\xi \in \bar{\Omega}$, this property is also true of all corresponding vectors $\psi_{i\alpha}\boldsymbol{e}^i$ for a sufficiently small quantity $\sup_{\xi \in \bigcup \hat{K}} |\partial_\alpha\varphi_i(\xi) - \psi_{i\alpha}(\xi)|$. Likewise, since the cosine of the angle between the two vectors $\boldsymbol{a}_\alpha(\xi)$ is bounded away from 1, independently of $\xi \in \bar{\Omega}$, we deduce that, for a sufficiently small quantity $\sup_{\xi \in \bigcup \hat{K}} |\partial_\alpha\varphi_i(\xi) - \psi_{i\alpha}(\xi)|$, the cosine of the angle between all corresponding vectors $\psi_{i\alpha}\boldsymbol{e}^i$ has the same property. Thus the modulus of their vector product is certainly bounded below by a strictly positive constant independent of the point $\xi \in \bar{\Omega}$.

Let then h_0 be such that

$$\forall h \leq h_0, \quad \sup_{\xi \in \bigcup \hat{K}} |\partial_\alpha\varphi_i(\xi) - \partial_\alpha\varphi_{ih}(\xi)| \leq \delta, \tag{8.2.18}$$

which is certainly possible, in view of the first of the uniform bounds given in (8.2.16). Since the only denominators which may occur in the functions A_{IJ} (resp. the functions A_{IJh}) are integer powers of the function $\sqrt{a}$ (resp. the function $\sqrt{a_h}$), as was stated in Theorem 8.1.1, and since these same functions are otherwise regular, we deduce that the approximate bilinear form $a_h(.\,,.)$ is well defined over the space $\boldsymbol{V}$ for all $h \leq h_0$. To compare it with the bilinear form $a(.\,,.)$, we observe that

$$\forall \boldsymbol{u}, \boldsymbol{v} \in \boldsymbol{V}, \quad |a(\boldsymbol{u}, \boldsymbol{v}) - a_h(\boldsymbol{u}, \boldsymbol{v})| \leq \gamma_h \|\boldsymbol{u}\|\, \|\boldsymbol{v}\|, \tag{8.2.19}$$

where

$$\gamma_h = \sum_{I,J=1}^{12} \sup_{\xi \in \bigcup \hat{K}} |A_{IJ}(\xi) - A_{IJh}(\xi)|.$$

Using again Theorem 8.1.1, and the definition (8.2.5) of the functions A_{IJh}, we obtain

$$\begin{aligned} A_{IJ}(\xi) - A_{IJh}(\xi) = {} & f_{IJ}(\partial_\alpha\varphi_i(\xi), \partial_{\alpha\beta}\varphi_i(\xi), \partial_{\alpha\beta\gamma}\varphi_i(\xi)) \\ & - f_{IJ}(\partial_\alpha\varphi_{ih}(\xi), \partial_{\alpha\beta}\varphi_{ih}(\xi), \partial_{\alpha\beta\gamma}\varphi_{ih}(\xi)). \end{aligned}$$

Since all points $(\psi_{i\alpha}(\xi), \psi_{i\alpha\beta}(\xi), \psi_{i\alpha\beta\gamma}(\xi))$ of the segments joining the points $(\partial_\alpha\varphi_i(\xi), \partial_{\alpha\beta}\varphi_i(\xi), \partial_{\alpha\beta\gamma}\varphi_i(\xi))$ and $(\partial_\alpha\varphi_{ih}(\xi), \partial_{\alpha\beta}\varphi_{ih}(\xi), \partial_{\alpha\beta\gamma}\varphi_{ih}(\xi))$ are such that

$$\forall \xi \in \bigcup \mathring{K}, \quad |\psi_{i\alpha}(\xi) - \partial_\alpha\varphi_i(\xi)| \leq |\partial_\alpha\varphi_{ih}(\xi) - \partial_\alpha\varphi_i(\xi)| \leq \delta$$

for all $h \leq h_0$ by (8.2.18), it follows that the functions f_{IJ} are continuously differentiable along these segments for all ξ. Since, in addition, all these points are in a compact subset of $\mathbf{R}^{27}$ of the form (cf. (8.2.17))

$$\{(X_{i\alpha}, X_{i\alpha\beta}, X_{i\alpha\beta\gamma}) \in \mathbf{R}^{27};\ |X_{i\alpha}|, |X_{i\alpha\beta}|, |X_{i\alpha\beta\gamma}| \leq \rho,\ a(X_{i\alpha}) \geq a_0\},$$

it follows that, along these segments, all partial derivatives of the first order of the functions f_{IJ} (with respect to the arguments $X_{i\alpha}$, $X_{i\alpha\beta}$, $X_{i\alpha\beta\gamma}$) are bounded above in the norm $|.|_{0,\infty,\cup\mathring{K}}$ by some constants $M_{i\alpha}$, $M_{i\alpha\beta}$ and $M_{i\alpha\beta\gamma}$, respectively. Therefore, an application of Taylor's formula yields:

$$\begin{aligned}
\gamma_h &= \sum_{I,J=1}^{12} \sup_{\xi\in\cup\mathring{K}} |A_{IJ}(\xi) - A_{IJh}(\xi)| \\
&\leq 144\Big(\sum_{i,\alpha} M_{i\alpha} \sup_{\xi\in\cup\mathring{K}} |\partial_\alpha\varphi_i(\xi) - \partial_\alpha\varphi_{ih}(\xi)| \\
&\quad + \sum_{i,\alpha\beta} M_{i\alpha\beta} \sup_{\xi\in\cup\mathring{K}} |\partial_{\alpha\beta}\varphi_i(\xi) - \partial_{\alpha\beta}\varphi_{ih}(\xi)| \\
&\quad + \sum_{i,\alpha\beta\gamma} M_{i\alpha\beta\gamma} \sup_{\xi\in\cup\mathring{K}} |\partial_{\alpha\beta\gamma}\varphi_i(\xi) - \partial_{\alpha\beta\gamma}\varphi_{ih}(\xi)|\Big),
\end{aligned} \tag{8.2.20}$$

and the conclusion follows by combining inequalities (8.2.16), (8.2.19) and (8.2.20).

The difference $|f(v) - f_h(v)|$ is studied analogously. Since partial derivatives of the first order only (of the functions φ_i) appear in the surface element $dS = \sqrt{a}\, d\xi$, we are in this case led to the exponent m.

□

As was indicated in the previous section, the bilinear form $a(.,.)$ is V-elliptic, i.e., there exists a constant $\alpha > 0$ such that for all $v = (v_1, v_2, v_3) \in V = H_0^1(\Omega) \times H_0^1(\Omega) \times H_0^2(\Omega)$, one has

$$\alpha |v|^2 \leq a(v, v), \tag{8.2.21}$$

where the norm $|.|$ has been defined in (8.2.12).

Theorem 8.2.2. *Assume that the discrete problems are conforming for* *the displacements, and that the spaces Φ_h are such that the inclusions* (8.2.13) *are satisfied for all* $K \in \bigcup_h \mathcal{T}_h$.

Then, if h is sufficiently small, the approximate bilinear form $a_h(.,.)$ *is* V_h*-elliptic and therefore, the discrete problem has a unique solution.*

The bilinear forms $a_h(.,.)$ *are also V-elliptic and continuous over the space V, uniformly with respect to h, in that there exist two constants* $\tilde{\alpha} > 0$ *and* $\tilde{M}$ *such that for all h sufficiently small,*

$$\forall v \in V, \quad \tilde{\alpha}\|v\|^2 \leq a_h(v, v), \tag{8.2.22}$$

$$\forall u, v \in V, \quad |a_h(u, v)| \leq \tilde{M}\|u\| \|v\|. \tag{8.2.23}$$

Proof. Let C be the constant appearing in inequality (8.2.14). Then using inequality (8.2.21), we find that, for all $v \in V$,

$$\begin{aligned} a_h(v, v) &= a(v, v) + (a_h(v, v) - a(v, v)) \\ &\geq \alpha |v|^2 - Ch^{m-2}\|v\|^2, \end{aligned}$$

and thus there exists a constant $\tilde{\alpha} > 0$ such that inequalities (8.2.22) hold, provided h is sufficiently small. Likewise, the bilinear form $a(.,.)$ being continuous, there exists a constant M such that, for all $u, v \in V$,

$$\begin{aligned} |a_h(u, v)| &= |a(u, v) + (a_h(u, v) - a(u, v))| \\ &\leq (M + Ch^{m-2})\|u\| \|v\|, \end{aligned}$$

which proves the validity of inequality (8.2.23). □

Abstract error estimate

Thus we have another instance of a family of discrete problems for which the associated bilinear forms are *uniformly* V_h*-elliptic*. With this property as our main assumption, we first derive an abstract upper bound for the error. As usual, *consistency conditions* can be derived from inequality (8.2.24) below.

Theorem 8.2.3. *Given a family of discrete problems conforming for the displacements, for which the inequalities* (8.2.22) *and* (8.2.23) *hold for all h, there exists a constant C independent of h such that*

$$\begin{aligned} \|u - u_h\| \leq C\Big(\inf_{v_h \in V_h} \|u - v_h\| &+ \sup_{w_h \in V_h} \frac{|a(u, w_h) - a_h(u, w_h)|}{\|w_h\|} \\ &+ \sup_{w_h \in V_h} \frac{|f(w_h) - f_h(w_h)|}{\|w_h\|}\Big). \end{aligned} \tag{8.2.24}$$

Proof. Let $\boldsymbol{v}_h$ be an arbitrary element in the space $\boldsymbol{V}_h$. We may write

$$\begin{aligned}\tilde{\alpha}\|\boldsymbol{v}_h-\boldsymbol{u}_h\|^2 &\leqslant a_h(\boldsymbol{v}_h-\boldsymbol{u}_h,\boldsymbol{v}_h-\boldsymbol{u}_h)\\ &= a_h(\boldsymbol{v}_h-\boldsymbol{u},\boldsymbol{v}_h-\boldsymbol{u}_h)+\{a_h(\boldsymbol{u},\boldsymbol{v}_h-\boldsymbol{u}_h)-a(\boldsymbol{u},\boldsymbol{v}_h-\boldsymbol{u}_h)\}\\ &\quad+\{f(\boldsymbol{v}_h-\boldsymbol{u}_h)-f_h(\boldsymbol{v}_h-\boldsymbol{u}_h)\},\end{aligned}$$

from which we deduce

$$\begin{aligned}\tilde{\alpha}\|\boldsymbol{v}_h-\boldsymbol{u}_h\| &\leqslant \tilde{M}\|\boldsymbol{u}-\boldsymbol{v}_h\|+\frac{|a_h(\boldsymbol{u},\boldsymbol{v}_h-\boldsymbol{u}_h)-a(\boldsymbol{u},\boldsymbol{v}_h-\boldsymbol{u}_h)|}{\|\boldsymbol{v}_h-\boldsymbol{u}_h\|}\\ &\quad+\frac{|f(\boldsymbol{v}_h-\boldsymbol{u}_h)-f_h(\boldsymbol{v}_h-\boldsymbol{u}_h)|}{\|\boldsymbol{v}_h-\boldsymbol{u}_h\|}\\ &\leqslant \tilde{M}\|\boldsymbol{u}-\boldsymbol{v}_h\|-\sup_{\boldsymbol{w}_h\in\boldsymbol{V}_h}\frac{|a_h(\boldsymbol{u},\boldsymbol{w}_h)-a(\boldsymbol{u},\boldsymbol{w}_h)|}{\|\boldsymbol{w}_h\|}\\ &\quad+\sup_{\boldsymbol{w}_h\in\boldsymbol{V}_h}\frac{|f(\boldsymbol{w}_h)-f_h(\boldsymbol{w}_h)|}{\|\boldsymbol{w}_h\|},\end{aligned}$$

and the conclusion follows by combining the above inequality with the triangular inequality

$$\|\boldsymbol{u}-\boldsymbol{u}_h\|\leqslant\|\boldsymbol{u}-\boldsymbol{v}_h\|+\|\boldsymbol{v}_h-\boldsymbol{u}_h\|.$$ □

Estimate of the error $\left(\sum_{\alpha=1}^{2}\|u_\alpha-u_{\alpha h}\|_{1,\Omega}^2+\|u_3-u_{3h}\|_{2,\Omega}^2\right)^{1/2}$

We are now in a position to obtain sufficient conditions for convergence (to shorten the statement of the next theorem, it is to be implicitly understood that possible additional hypotheses upon the integers k and l may be needed so as to insure that the V_h-interpolation operator, or the W_h-interpolation operator, are well defined).

Theorem 8.2.4. *Assume that the discrete problems are conforming for the displacements and that the spaces Φ_h, V_h and W_h are such that, for all $\mathcal{T}_h$ and all $K\in\mathcal{T}_h$,*

$$P_m(K)\subset P_K\subset\mathscr{C}^3(K) \tag{8.2.25}$$

for some integer $m\geqslant 3$,

$$P_k(K)\subset P'_K\subset H^1(K) \tag{8.2.26}$$

for some integer $k \geqslant 1$,

$$P_l(K) \subset P_K'' \subset H^2(K) \tag{8.2.27}$$

for some integer $l \geqslant 2$, *respectively.*

Then if the solution $\boldsymbol{u} = (u_1, u_2, u_3)$ *belongs to the space*

$$H^{k+1}(\Omega) \times H^{k+1}(\Omega) \times H^{l+1}(\Omega), \tag{8.2.28}$$

there exists a constant C *independent of* h *such that*

$$\|\boldsymbol{u} - \boldsymbol{u}_h\| \leqslant Ch^{\min\{k, l-1, m-2\}}. \tag{8.2.29}$$

Proof. One has

$$\inf_{\boldsymbol{v}_h \in \boldsymbol{V}_h} \|\boldsymbol{u} - \boldsymbol{v}_h\| \leqslant \|\boldsymbol{u} - \boldsymbol{\Pi}_h \boldsymbol{u}\| = (\|u_1 - \Pi_h u_1\|_{1,\Omega}^2 + \|u_2 - \Pi_h u_2\|_{1,\Omega}^2 + \|u_3 - \Lambda_h u_3\|_{2,\Omega}^2)^{1/2},$$

where $\boldsymbol{\Pi}_h \boldsymbol{u} = (\Pi_h u_1, \Pi_h u_2, \Lambda_h u_3)$ is the $\boldsymbol{V}_h$-interpolant of the solution $\boldsymbol{u}$. Since it follows that $\Pi_h u_\alpha$ and $\Lambda_h u_3$ are the V_h-interpolants of the function u_α and the W_h-interpolant of the function u_3, respectively, an application of the standard error estimates shows that

$$\inf_{\boldsymbol{v}_h \in \boldsymbol{V}_h} \|\boldsymbol{u} - \boldsymbol{v}_h\| \leqslant C\{(|u_1|_{k+1,\Omega} + |u_2|_{k+1,\Omega})h^k + |u_3|_{l+1,\Omega} h^{l-1})\},$$

for some constant C independent of h.

From inequalities (8.2.14) and (8.2.15) of Theorem 8.2.1, we derive the *consistency error estimates*:

$$\sup_{\boldsymbol{w}_h \in \boldsymbol{V}_h} \frac{|a(\boldsymbol{u}, \boldsymbol{w}_h) - a_h(\boldsymbol{u}, \boldsymbol{w}_h)|}{\|\boldsymbol{w}_h\|} \leqslant C\|\boldsymbol{u}\| h^{m-2},$$

$$\sup_{\boldsymbol{w}_h \in \boldsymbol{V}_h} \frac{|f(\boldsymbol{w}_h) - f_h(\boldsymbol{w}_h)|}{\|\boldsymbol{w}_h\|} \leqslant C|f|_{0,\Omega} h^m,$$

and the conclusion follows by combining the last three inequalities and inequality (8.2.24) of Theorem 8.2.3. □

For instance, this result shows that the Argyris triangle yields an $O(h^3)$ convergence since it corresponds to the values $k = l = m = 5$. This is to be compared with the $O(h^4)$ convergence which it yields for plates: *the decrease of one in the order of convergence is due to the approximation of the geometry.*

Remark 8.2.2. In some shell models, partial derivatives of orders only 1 and 2 of the mapping $\tilde{\varphi}$ appear in the functions A_{IJ}. For such models, the analogues of Theorems 8.2.1 and 8.2.4 hold with the exponent $(m-1)$ instead of $(m-2)$. □

Finite element methods conforming for the geometry

In view of Theorem 8.2.4, we shall say that a finite element method is *conforming for the geometry if the inclusions*

$$\forall K \in \mathcal{T}_h, \quad P_m(K) \subset P_K \subset \mathscr{C}^3(K)$$

hold for some integer $m \geq 3$, so that we may obtain convergence with these sole conditions (as regards the geometry).

In this definition, it is unexpected that *no continuity is required across adjacent finite elements for the functions in the space* Φ_h, and this is a conclusion which differs from the requirement, usually found in the engineering literature, that the inclusion $\Phi_h \subset \mathscr{C}^1(\bar{\Omega})$ should hold. We believe that the origin of this difference is that there are essentially two points of view:

Either one can argue in terms of the approximate surface $\mathscr{S}_h = \varphi_h(\bar{\Omega})$ (cf. (8.2.8)) and, for physical reasons, this imposes some regularity requirements (such as $\mathscr{C}^1$-continuity) on the mapping φ_h. In this interpretation, one may think of the discrete solution $\boldsymbol{u}_h(\xi)$ as a displacement attached to the point $\varphi_h(\xi)$. Or one can consider that the main objective is to get a good approximation of the bilinear form $a(.,.)$: From this point of view, the proof of Theorem 8.2.1 shows that possible discontinuities of the approximate mapping φ_h along sides common to adjacent finite elements are irrelevant. What matters is only that sufficiently good "local" uniform approximations of the coefficients A_{IJ} can be obtained and this is exactly a consequence of the definition of a conforming method for the geometry. Let us add that in this second interpretation, which is chosen here, we think of the three functions u_{ih} *defined over the set* $\bar{\Omega}$ as approximations of the three functions u_i *also defined over the set* $\bar{\Omega}$, even if these functions, by means of the coordinate system $(\varphi(\xi), \boldsymbol{a}^i(\xi))$, allow to derive the displacement of the point $\varphi(\xi)$.

Conforming finite element methods for shells

In light of the preceding analysis, we shall say that *a finite element method for solving the shell problem is conforming* if it is both conforming for the displacement and conforming for the geometry, in the senses understood in this section. Consequently, *a finite element method for solving the shell problem will be called nonconforming* if it is not conforming in the previous sense.

8.3. A nonconforming method for the arch problem

The circular arch problem

Our purpose is to analyze a *nonconforming method* for solving the simplest problem similar to the shell problem: *the circular arch problem.*

We consider a circular arch of radius R (Fig. 8.3.1) and, for definiteness, we shall assume that the arch is clamped. Setting the physical constants EA and EI equal to one in the energy (8.1.30), the variational problem corresponds to the following data (notice the change of sign

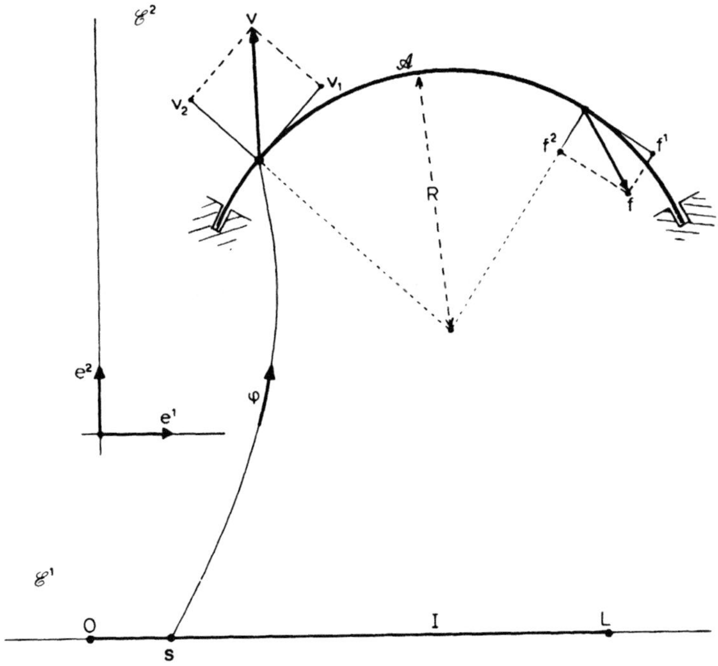

Fig. 8.3.1

because R is a positive constant in the present case):

$$\begin{cases} \boldsymbol{V} = H_0^1(I) \times H_0^2(I), \\ a(\boldsymbol{u}, \boldsymbol{v}) = \int_I \left\{ \left(u_1' + \frac{u_2}{R}\right)\left(v_1' + \frac{v_2}{R}\right) + \left(u_2'' - \frac{u_1'}{R}\right)\left(v_2'' - \frac{v_1'}{R}\right) \right\} \mathrm{d}s, \\ f(\boldsymbol{v}) = \int_I \boldsymbol{f} \cdot \boldsymbol{v} \, \mathrm{d}s = \int_I (f^1 v_1 + f^2 v_2) \, \mathrm{d}s, \end{cases} \tag{8.3.1}$$

where $\boldsymbol{u} = (u_1, u_2)$, $\boldsymbol{v} = (v_1, v_2)$ are functions of the curvilinear abscissa $s \in \bar{I} = [0, L]$. We recall that in Theorem 8.1.2, we have proved the $\boldsymbol{V}$-ellipticity of bilinear forms which contain that of (8.3.1) as a special case.

A natural finite element approximation

Let us first review what the most straightforward finite element method would be for solving this problem. Since all coefficients appearing in the bilinear form are constant, it is not necessary to approximate the geometry as we pointed out in Section 8.2. In other words, the discrete problem consists in letting $a_h(.\,,.) = a(.\,,.)$ and $f_h(.) = f(.)$, and then in looking for a discrete solution $\boldsymbol{u}_h = (u_{1h}, u_{2h}) \in \boldsymbol{V}_h = V_h \times W_h$, where V_h and W_h are subspaces of $H_0^1(I)$ and $H_0^2(I)$, respectively.

The simplest choices that can be made for these spaces are the following: Let

$$\bar{I} = [0, L] = \bigcup_{i=1}^{M} I_i, \quad \text{with} \quad I_i = [s_{i-1}, s_i],\, s_i = ih, \quad 1 \leq i \leq M, \tag{8.3.2}$$

be a *uniform partition* of the interval $\bar{I}$ associated with a mesh size $h = L/M$, M being a strictly positive integer.

We let V_h be the space of functions $v_{1h} \in \mathscr{C}^0(\bar{I})$ for which the restrictions $v_{1h}|_{I_i}$ span the space $P_1(I_i)$, $1 \leq i \leq M$, and which satisfy the boundary conditions $v_{1h}(0) = v_{2h}(0) = 0$, and we let W_h be the space of functions $v_{2h} \in \mathscr{C}^1(\bar{I})$ for which the restrictions $v_{2h}|_{I_i}$ span the space $P_3(I_i)$, $1 \leq i \leq M$, and which satisfy the boundary conditions $v_{2h}(0) = v_{2h}'(0) = v_{2h}(L) = v_{2h}'(L) = 0$. Therefore, the degrees of freedom of the space V_h are $v_{1h}(s_i)$, $1 \leq i \leq M-1$, and the degrees of freedom of the space W_h are $v_{2h}(s_i)$, $v_{2h}'(s_i)$, $1 \leq i \leq M-1$.

The discrete solution $\boldsymbol{u}_h$ satisfies the equations

$$\forall \boldsymbol{v}_h \in \boldsymbol{V}_h, \quad a(\boldsymbol{u}_h, \boldsymbol{v}_h) = f(\boldsymbol{v}_h), \tag{8.3.3}$$

and it is straightforward to show that if the solution $\boldsymbol{u} = (u_1, u_2)$ belongs to the space $\boldsymbol{V} \cap (H^2(I) \times H^3(I))$, there exists a constant C independent of h such that

$$|\boldsymbol{u} - \boldsymbol{u}_h| \leq Ch(|u_1|_{2,I}^2 + |u_2|_{3,I}^2)^{1/2}, \tag{8.3.4}$$

where

$$\boldsymbol{v} = (v_1, v_2) \in \boldsymbol{V} \to |\boldsymbol{v}| = (|v_1|_{1,I}^2 + |v_2|_{2,I}^2)^{1/2} \tag{8.3.5}$$

is a norm on the space V. We shall occasionally use the equivalent norm

$$\boldsymbol{v} = (v_1, v_2) \to \|\boldsymbol{v}\| = (\|v_1\|_{1,I}^2 + \|v_2\|_{2,I}^2)^{1/2}. \tag{8.3.6}$$

Finite element methods conforming for the geometry

Let us henceforth forget that we need not approximate the bilinear form of (8.3.1). If we follow the analysis made in Section 8.2, we are led to approximate the mapping φ by a mapping φ_h whose components lie in a finite element space Φ_h associated with the partition (8.3.2) of the interval $[0, L]$. We let P_{I_i} denote the spaces spanned by the restrictions to the sets I_i, $1 \leq i \leq M$, of the functions in the space Φ_h.

Since the third derivative of the mapping φ does not appear in the bilinear form (to see this, it suffices to choose any parametrization of the arch in which the derivative of the mapping are not constant), a method is *conforming for the geometry* provided the inclusions $P_2(I_i) \subset P_{I_i}$, $1 \leq i \leq M$, hold.

A finite element method which is not conforming for the geometry. Definition of the discrete problem

We shall now analyze a method for which the spaces P_{I_i} coincide with the spaces $P_1(I_i)$, $1 \leq i \leq M$, and which consequently is *not* conforming for the geometry. More precisely, let Φ_h denote the space of functions which are affine on each interval I_i, $1 \leq i \leq M$, and continuous over the interval $\bar{I}$. Then the approximate arch is defined by

$$\mathscr{A}_h = \varphi_h(\bar{I}), \tag{8.3.7}$$

where

$$\varphi_h = \varphi_{1h}\boldsymbol{e}^1 + \varphi_{2h}\boldsymbol{e}^2 \quad \text{with} \quad \varphi_{ih} \in \Phi_h, \quad i = 1, 2, \tag{8.3.8}$$

and the mapping φ_h is uniquely determined by the interpolation conditions

$$\varphi_h(s_i) = \varphi(s_i), \quad 0 \leqslant i \leqslant M. \tag{8.3.9}$$

Since the second derivative of the approximate mapping φ_h vanishes on each interval $\mathring{I}_i$, the corresponding "approximate" radius of curvature is infinite. Therefore, following the approach of Section 8.2 (cf. equations (8.2.4) and (8.2.5)) and the expression of the bilinear form as given in (8.3.1), we are led to the following approximate bilinear form:

$$a_h^*(u_h^*, v_h^*) = \sum_{i=1}^{M} \int_{I_i} (u_{1h}^{*\prime}(s) v_{1h}^{*\prime}(s) + \\ + u_{2h}^{*\prime\prime}(s) v_{2h}^{*\prime\prime}(s)) \sqrt{(\varphi_{1h}'(s))^2 + (\varphi_{2h}'(s))^2}\, \mathrm{d}s, \tag{8.3.10}$$

for functions u_{1h}^*, and u_{2h}^*, belonging to an appropriate finite element space contained in the space $\Pi_{i=1}^M H^1(I_i)$, and in the space $\Pi_{i=1}^M H^2(I_i)$, respectively.

The element of arc length along the approximate arch $\mathscr{A}_h$ is given by

$$\mathrm{d}\tilde{s} = \sqrt{(\varphi_{1h}')^2 + (\varphi_{2h}')^2}\, \mathrm{d}s = \frac{\sin\theta_h}{\theta_h}\, \mathrm{d}s, \tag{8.3.11}$$

where the angle θ_h is such that (Fig. 8.3.2)

$$h = 2R\theta_h. \tag{8.3.12}$$

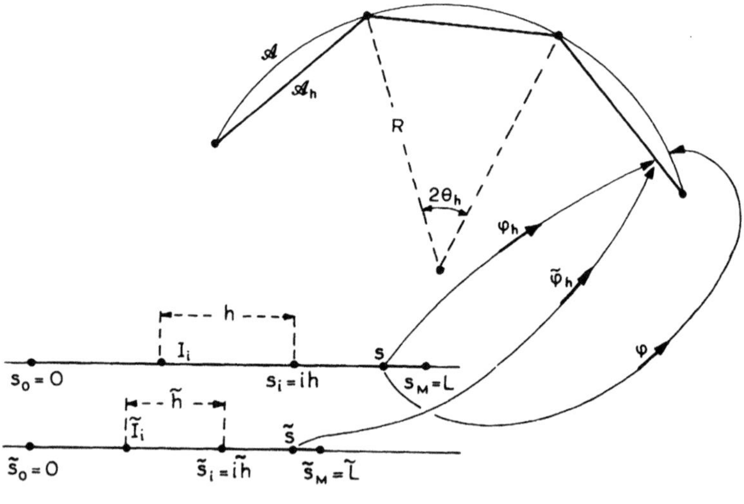

Fig. 8.3.2

Usually, the discrete problem is rather defined in terms of the abcissa $\tilde{s}$ along the approximate arch $\mathscr{A}_h$, which is given by

$$\tilde{s} = \frac{\sin \theta_h}{\theta_h} s, \quad 0 \leqslant s \leqslant L. \tag{8.3.13}$$

This being the case, we can also write (Fig. 8.3.2)

$$\mathscr{A}_h = \tilde{\varphi}_h(\bar{\tilde{I}}), \tag{8.3.14}$$

with

$$\bar{\tilde{I}} = [0, \tilde{L}], \quad \tilde{L} = \frac{\sin \theta_h}{\theta_h} L, \tag{8.3.15}$$

$$\tilde{\varphi}_h(\tilde{s}) = \varphi_h(s) \quad \text{for all } \tilde{s} = \frac{\sin \theta_h}{\theta_h} s, \quad 0 \leqslant s \leqslant L, \tag{8.3.16}$$

and we shall associate the uniform partition

$$\bar{\tilde{I}} = [0, \tilde{L}] = \bigcup_{i=1}^{M} \tilde{I}_i, \quad \text{with} \quad \tilde{I}_i = [\tilde{s}_{i-1}, \tilde{s}_i], \quad \tilde{s}_i = i\tilde{h},$$
$$1 \leqslant i \leqslant M, \tilde{h} = \frac{\sin \theta_h}{\theta_h} h, \tag{8.3.17}$$

with the partition of (8.3.2). Notice that

$$\tilde{h} = 2R \sin \theta_h. \tag{8.3.18}$$

Thus, rather than looking for the functions u_{1h}^*, u_{2h}^*: $s \in \bar{I} \to \boldsymbol{R}$, we shall look instead for functions $\tilde{u}_{1h}, \tilde{u}_{2h}$: $\tilde{s} \in \bar{\tilde{I}} \to \boldsymbol{R}$. To get the simplest correspondences, it suffices to let

$$\tilde{u}_{1h}(\tilde{s}) = \frac{\tilde{h}}{h} u_{1h}^*(s), \quad \tilde{u}_{2h}(\tilde{s}) = \left(\frac{\tilde{h}}{h}\right)^2 u_{2h}^*(s), \tag{8.3.19}$$

so that the strain energy (8.3.10) of the approximate arch takes the form

$$\tilde{a}_h(\tilde{\boldsymbol{u}}_h, \tilde{\boldsymbol{v}}_h) = a_h^*(\boldsymbol{u}_h^*, \boldsymbol{v}_h^*) = \sum_{i=1}^{M} \int_{\tilde{I}_i} (\tilde{u}_{1h}'(\tilde{s})\tilde{v}_{1h}'(\tilde{s}) + \tilde{u}_{2h}''(\tilde{s})\tilde{v}_{2h}''(\tilde{s}))\, d\tilde{s}, \tag{8.3.20}$$

i.e., *it is written as a sum of strain energies of "elementary" straight beams* $\varphi_h(I_i) = \tilde{\varphi}_h(\tilde{I}_i)$, $1 \leqslant i \leqslant M$, *which is indeed the main feature of such a method.*

We are therefore led to look for a discrete solution $\tilde{\boldsymbol{u}}_h = (\tilde{u}_{1h}, \tilde{u}_{2h})$ in a space $\tilde{V}_h$ whose elements $\tilde{\boldsymbol{v}}_h = (\tilde{v}_{1h}, \tilde{v}_{2h})$ are such that $\tilde{v}_{1h} \in \Pi_{i=1}^M H^1(\tilde{I}_i)$ and $\tilde{v}_{2h} \in \Pi_{i=1}^M H^2(\tilde{I}_i)$.

In view of the definitions of the spaces V_h and W_h, we shall assume for definiteness (leaving out for the time being the boundary conditions) that the restrictions $\tilde{v}_{1h|\tilde{I}_i}$ span the space $P_1(\tilde{I}_i)$, and that the restrictions $\tilde{v}_{2h|\tilde{I}_i}$ span the space $P_3(\tilde{I}_i)$, $1 \leq i \leq M$. In this fashion, a function $\tilde{\boldsymbol{v}}_h = (\tilde{v}_{1h}, \tilde{v}_{2h}) \in \tilde{V}_h$ is specified by the parameters

$$\tilde{v}_{1h}(\tilde{s}_{i-1}^+), \quad \tilde{v}_{1h}(\tilde{s}_i^-), \quad 1 \leq i \leq M,$$

and

$$\tilde{v}_{2h}(\tilde{s}_{i-1}^+), \quad \tilde{v}_{2h}'(\tilde{s}_{i-1}^+), \quad \tilde{v}_{2h}(\tilde{s}_i^-), \quad \tilde{v}_{2h}'(\tilde{s}_i^-), \quad 1 \leq i \leq M.$$

To find the compatibility relations between these parameters, it suffices to express that they correspond to a well-defined *displacement*

$$\boldsymbol{v}_h(s_i) = v_{1h}(s_i)\boldsymbol{a}_1(s_i) + v_{2h}(s_i)\boldsymbol{a}_2(s_i) \tag{8.3.21}$$

(Fig. 8.3.3) of the point $\boldsymbol{\varphi}(s_i)$, and to a well-defined *rotation*

$$\omega_h(s_i) = v_{2h}'(s_i) - \frac{1}{R} v_{1h}(s_i) \tag{8.3.22}$$

of the same point $\boldsymbol{\varphi}(s_i)$, for all $i = 0, 1, \ldots, M$.

With the self-explanatory notation of Fig. 8.3.3, we obtain

$$\begin{cases} \tilde{\boldsymbol{v}}_h(\tilde{s}_i^-) = \tilde{v}_{1h}(\tilde{s}_i^-)\boldsymbol{a}_1(\tilde{s}_i^-) + \tilde{v}_{2h}(\tilde{s}_i^-)\boldsymbol{a}_2(\tilde{s}_i^-), \quad 1 \leq i \leq M, \\ \tilde{\boldsymbol{v}}_h(\tilde{s}_i^+) = \tilde{v}_{1h}(\tilde{s}_i^+)\boldsymbol{a}_1(\tilde{s}_i^+) + \tilde{v}_{2h}(\tilde{s}_i^+)\boldsymbol{a}_2(\tilde{s}_i^+), \quad 0 \leq i \leq M-1, \\ \tilde{\omega}_h(\tilde{s}_i^-) = \tilde{v}_{2h}'(\tilde{s}_i^-), \quad 1 \leq i \leq M, \quad \tilde{\omega}_h(\tilde{s}_i^+) = \tilde{v}_{2h}'(\tilde{s}_i^+), \\ \qquad\qquad 0 \leq i \leq M-1, \end{cases} \tag{8.3.23}$$

and thus we must have

$$\begin{cases} \tilde{v}_{1h}(\tilde{s}_i^-) = \cos\theta_h v_{1h}(s_i) + \sin\theta_h v_{2h}(s_i), \\ \tilde{v}_{2h}(\tilde{s}_i^-) = -\sin\theta_h v_{1h}(s_i) + \cos\theta_h v_{2h}(s_i), \\ \tilde{v}_{1h}(\tilde{s}_i^+) = \cos\theta_h v_{1h}(s_i) - \sin\theta_h v_{2h}(s_i), \\ \tilde{v}_{2h}(\tilde{s}_i^+) = \sin\theta_h v_{1h}(s_i) + \cos\theta_h v_{2h}(s_i), \\ \tilde{v}_{2h}'(\tilde{s}_i^-) = \tilde{v}_{2h}'(\tilde{s}_i^+) = v_{2h}'(s_i) - \dfrac{1}{R} v_{1h}(s_i). \end{cases} \tag{8.3.24}$$

Using relations (8.3.21), (8.3.22), (8.3.23) and (8.3.24), we deduce the

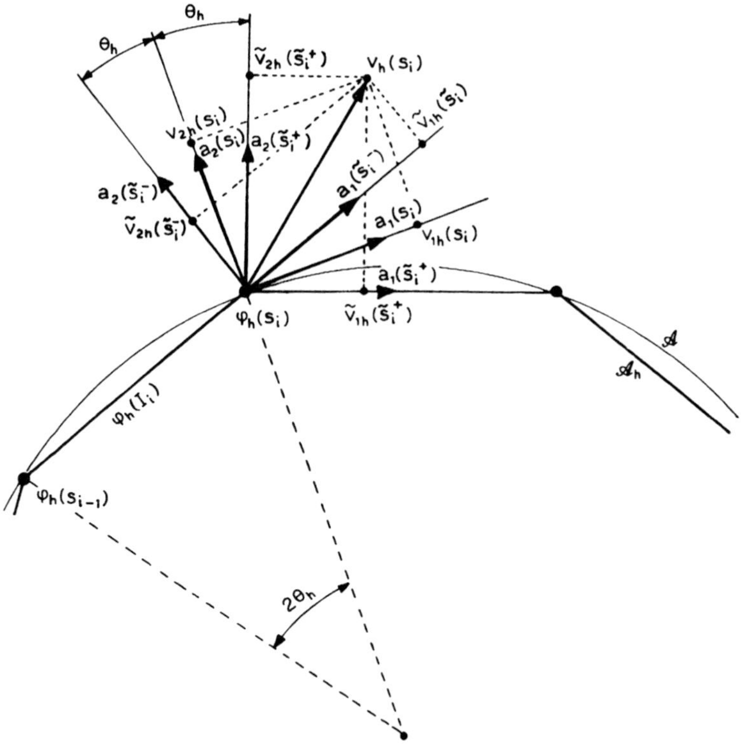

Fig. 8.3.3

compatibility relations which the functions $\tilde{v}_{1h}$ and $\tilde{v}_{2h}$ should satisfy:

$$\begin{cases} \cos\theta_h\tilde{v}_{1h}(\tilde{s}_i^-) - \sin\theta_h\tilde{v}_{2h}(\tilde{s}_i^-) = \cos\theta_h\tilde{v}_{1h}(\tilde{s}_i^+) + \sin\theta_h\tilde{v}_{2h}(\tilde{s}_i^+) \\ \qquad\qquad (= v_{1h}(s_i)), \\ \sin\theta_h\tilde{v}_{1h}(\tilde{s}_i^-) + \cos\theta_h\tilde{v}_{2h}(\tilde{s}_i^-) = -\sin\theta_h\tilde{v}_{1h}(\tilde{s}_i^+) + \cos\theta_h\tilde{v}_{2h}(\tilde{s}_i^+) \\ \qquad\qquad (= v_{2h}(s_i)), \\ \tilde{v}'_{2h}(\tilde{s}_i^-) = \tilde{v}'_{2h}(\tilde{s}_i^+) \quad (= v'_{2h}(s_i) - \frac{1}{R}v_{1h}(s_i)). \end{cases} \tag{8.3.25}$$

Thus at each point s_i, $0 \leq i \leq M$, one may consider that the *independent parameters* are the values $v_{1h}(s_i)$, $v_{2h}(s_i)$, $v'_{2h}(s_i)$, from which the parameters $\tilde{v}_{1h}(\tilde{s}_i^-)$, $\tilde{v}_{1h}(\tilde{s}_i^+)$, $\tilde{v}_{2h}(\tilde{s}_i^-)$, $\tilde{v}_{2h}(\tilde{s}_i^+)$, $\tilde{v}'_{2h}(\tilde{s}_i^-)$, $\tilde{v}'_{2h}(\tilde{s}_i^+)$ are derived through relations (8.3.24).

Finally, the following boundary conditions will be included in the definition of the spaces $\tilde{V}_h$:

$$\tilde{v}_{1h}(0^+) = \tilde{v}_{2h}(0^+) = \tilde{v}'_{2h}(0^+) = \tilde{v}_{1h}(L^-) = \tilde{v}_{2h}(L^-) = \tilde{v}'_{2h}(L^-) = 0, \tag{8.3.26}$$

or equivalently, using relations (8.3.25),

$$v_{1h}(0) = v_{2h}(0) = v'_{2h}(0) = v_{1h}(L) = v_{2h}(L) = v'_{2h}(L) = 0. \tag{8.3.27}$$

To sum up, *the space $\tilde{V}_h$ is completely defined*: it is composed of pairs $(\tilde{v}_{1h}, \tilde{v}_{2h})$

(i) whose restrictions $v_{1h|\tilde{I}_i}$ and $v_{2h|\tilde{I}_i}$ span the space $P_1(\tilde{I}_i)$ and the space $P_3(\tilde{I}_i)$, respectively,·

(ii) which satisfy the compatibility relations (8.3.25), and

(iii) which satisfy the boundary conditions (8.3.26).

By relations (8.3.24)–(8.3.25) and (8.3.26)–(8.3.27), there exists a bijection

$$(\tilde{v}_{1h}, \tilde{v}_{2h}) \in \tilde{\boldsymbol{V}}_h \to (v_{1h}, v_{2h}) \in \boldsymbol{V}_h = V_h \times W_h, \tag{8.3.28}$$

where $\boldsymbol{V}_h = V_h \times W_h$ is the "conforming" subspace introduced at the beginning of this section.

The *discrete problem* then consists in finding an element $\tilde{\boldsymbol{u}}_h \in \tilde{\boldsymbol{V}}_h$ such that (cf. (8.3.20))

$$\forall \tilde{\boldsymbol{v}}_h \in \tilde{\boldsymbol{V}}_h, \quad \tilde{a}_h(\tilde{\boldsymbol{u}}_h, \tilde{\boldsymbol{v}}_h) = h \sum_{i=1}^{M-1} (\boldsymbol{f} \cdot \boldsymbol{v}_h)(s_i), \tag{8.3.29}$$

or, more explicitly, such that

$$\sum_{i=1}^{M} \int_{\tilde{I}_i} (\tilde{u}'_{1h}(\tilde{s})\tilde{v}'_{1h}(\tilde{s}) + \tilde{u}''_{2h}(\tilde{s})\tilde{v}''_{2h}(\tilde{s}))\, \mathrm{d}\tilde{s} = \\ = h \sum_{i=1}^{M-1} \{f^1(s_i)v_{1h}(s_i) + f^2(s_i)v_{2h}(s_i)\}, \tag{8.3.30}$$

for all $\tilde{\boldsymbol{v}}_h \in \tilde{\boldsymbol{V}}_h$, where the functions $\tilde{\boldsymbol{v}}_h \in \tilde{\boldsymbol{V}}_h$ and $\boldsymbol{v}_h \in \boldsymbol{V}_h$ are in the correspondence (8.3.28).

Remark 8.3.1. In principle, the approximate linear form should also be given as a sum of integrals over the intervals $\tilde{I}_i$. However, the right-hand side of the discrete variational problem is usually defined as in (8.3.30) in the engineering literature, where it is considered that the applied force is approximated by a sum of concentrated forces, an obvious simplification for computational purposes. In addition, this simplification is theoretically justified, since it does not decrease the order of convergence, as we shall see. □

Notice that the discrete problem (8.3.29) can also be written as a problem posed *over the space* V_h: To find $u_h \in V_h$ such that

$$\forall v_h \in V_h, \quad a_h(u_h, v_h) = f_h(v_h), \tag{8.3.31}$$

where, *by definition*, the approximate bilinear form a_h is given by

$$a_h(u_h, v_h) = \tilde{a}_h(\tilde{u}_h, \tilde{v}_h), \tag{8.3.32}$$

for all $u_h, v_h \in V_h$ and $\tilde{u}_h, \tilde{v}_h \in \tilde{V}_h$ in the correspondence (8.3.28), and the approximate linear form is given by

$$f_h(v_h) = h \sum_{i=1}^{M-1} (f \cdot v_h)(s_i). \tag{8.3.33}$$

This is why our first task (Theorem 8.3.1) will be to explicitly compute the bilinear form $a_h(.,.)$. Since the space V_h is a subspace of the space V, we are exactly in the same abstract setting as we were when we studied the effect of numerical integration in Section 4.1. Accordingly, our objective is to be in a positition to apply the abstract error bound of Theorem 4.1.1. Therefore, we shall successively evaluate the quantities $|a_h(v_h, w_h) - a(v_h, w_h)|$ for $v_h, w_h \in V_h$ (Theorem 8.3.2) and $|f_h(w_h) - f(w_h)|$ for $w_h \in V_h$ (Theorem 8.3.3), before we combine them in our final result (Theorem 8.3.4).

Theorem 8.3.1. *Let* $\tilde{v}_h = (\tilde{v}_{1h}, \tilde{v}_{2h}) \in \tilde{V}_h$ *and* $v_h = (v_{1h}, v_{2h}) \in V_h$ *be in the correspondence* (8.3.28). *Then we have*

$$\tilde{a}_h(\tilde{v}_h, \tilde{v}_h) = a_h(v_h, v_h), \tag{8.3.34}$$

where

$$\begin{aligned} a_h(v_h, v_h) = {} & \sum_{i=1}^{M} \int_{I_i} \left(\frac{h}{\tilde{h}} \cos\theta_h v_{1h}'(s) + \frac{v_{2h}(s_{i-1}) + v_{2h}(s_i)}{2R} \right)^2 \frac{\tilde{h}}{h} \,\mathrm{d}s \\ & + \sum_{i=1}^{M} \int_{I_i} \left(\left(\frac{h}{\tilde{h}}\right)^2 \cos\theta_h v_{2h}''(s) - \frac{1}{R}\frac{h}{\tilde{h}} v_{1h}'(s) \right. \\ & + \left\{ \frac{1}{\tilde{h}} \left(1 - \cos\theta_h \frac{h}{\tilde{h}}\right) \right\} \left\{ v_{2h}'(s_{i-1}) \left(6\frac{(s - s_{i-1})}{h} - 4\right) \right. \\ & \left.\left. + v_{2h}'(s_i) \left(6\frac{(s - s_{i-1})}{h} - 2\right) \right\} \right) \frac{\tilde{h}}{h} \,\mathrm{d}s. \end{aligned} \tag{8.3.35}$$

Proof. Recall that (cf. (8.3.20))

$$\tilde{a}_h(\tilde{v}_h, \tilde{v}_h) = \sum_{i=1}^{M} \int_{\tilde{I}_i} (\tilde{v}'_{1h}(\tilde{s}))^2 \,\mathrm{d}\tilde{s} + \sum_{i=1}^{M} \int_{\tilde{I}_i} (\tilde{v}''_{2h}(\tilde{s}))^2 \,\mathrm{d}\tilde{s}.$$

To prove equality (8.3.35) we shall in fact prove more: With obvious notations, equality (8.3.35) can be written as

$$\sum_{i=1}^{M} \int_{\tilde{I}_i} \{(\tilde{\varphi}_i^1(\tilde{s}))^2 + (\tilde{\varphi}_i^2(\tilde{s}))^2\} \,\mathrm{d}\tilde{s} = \sum_{i=1}^{M} \int_{I_i} \{(\theta_i^1(s))^2 + (\theta_i^2(s))^2\} \frac{\tilde{h}}{h} \,\mathrm{d}s,$$

and we have

$$\sum_{i=1}^{M} \int_{\tilde{I}_i} \{(\tilde{\varphi}_i^1(\tilde{s}))^2 + (\tilde{\varphi}_i^2(\tilde{s}))^2\} \,\mathrm{d}\tilde{s} =$$

$$= \sum_{i=1}^{M} \int_{I_i} \left\{\left(\tilde{\varphi}_i^1\left(\frac{\tilde{h}}{h} s\right)\right)^2 + \left(\tilde{\varphi}_i^2\left(\frac{\tilde{h}}{h} s\right)\right)^2\right\} \frac{\tilde{h}}{h} \,\mathrm{d}s.$$

Then we shall derive the stronger equalities

$$\tilde{\varphi}_i^1\left(\frac{\tilde{h}}{h} s\right) = \theta_i^1(s) \text{ and } \tilde{\varphi}_i^2\left(\frac{\tilde{h}}{h} s\right) = \theta_i^2(s) \quad \text{for} \quad s \in I_i, \quad 1 \leqslant i \leqslant M.$$

Since, on each interval $\tilde{I}_i$, $\tilde{v}_{1h}$ is a polynomial of degree one, we can write, for all $\tilde{s} \in \tilde{I}_i$,

$$\tilde{v}'_{1h}(\tilde{s}) = \frac{\tilde{v}_{1h}(\tilde{s}_i^-) - \tilde{v}_{1h}(\tilde{s}_{i-1}^+)}{\tilde{h}}$$

$$= \frac{1}{\tilde{h}} \{\cos \theta_h (v_{1h}(s_i) - v_{1h}(s_{i-1})) + \sin \theta_h (v_{2h}(s_i) + v_{2h}(s_{i-1}))\},$$

by (8.3.24), and thus, since v_{1h} is also a polynomial of degree one in the variable s over I_i,

$$\tilde{v}'_{1h}(\tilde{s}) = \frac{h}{\tilde{h}} \cos \theta_h v'_{1h}(s) + \frac{v_{2h}(s_{i-1}) + v_{2h}(s_i)}{2R},$$

where we have also used the relation $2R \sin \theta_h = \tilde{h}$ ((cf. (8.3.18)). Therefore, the first equality is proved.

Since on each interval $\tilde{I}_i$, $\tilde{v}_{2h}$ is a polynomial of degree three, we can write, for all $\tilde{s} \in \tilde{I}_i$,

$$\tilde{v}''_{2h}(\tilde{s}) = \tilde{v}_{2h}(\tilde{s}_{i-1}^+)\left(\frac{12(\tilde{s} - \tilde{s}_i) + 6\tilde{h}}{\tilde{h}^3}\right) + \tilde{v}_{2h}(\tilde{s}_i^-)\left(\frac{-12(\tilde{s} - \tilde{s}_{i-1}) + 6\tilde{h}}{\tilde{h}^3}\right)$$

$$+ \tilde{v}'_{2h}(\tilde{s}_{i-1}^+)\left(\frac{6(\tilde{s} - \tilde{s}_i) + 2\tilde{h}}{\tilde{h}^2}\right) + \tilde{v}'_{2h}(s_i^-)\left(\frac{6(\tilde{s} - \tilde{s}_{i-1}) - 2\tilde{h}}{\tilde{h}^2}\right).$$

Using relations (8.3.24) and equalities $\tilde{s} = (\tilde{h}/h)s$, $\tilde{h} = 2R \sin \theta_h$ (cf. (8.3.13) and (8.3.18)), we obtain

$$\begin{aligned}
\tilde{v}''_{2h}(\tilde{s}) &= \left(\frac{h}{\tilde{h}}\right)^2 \left\{ \left(\sin \theta_h v_{1h}(s_{i-1}) + \cos \theta_h v_{2h}(s_{i-1}) \right) \left(\frac{12(s - s_i) + 6h}{h^3} \right) \right. \\
&\quad \left. + \left(- \sin \theta_h v_{1h}(s_i) + \cos \theta_h v_{2h}(s_i) \right) \left(\frac{-12(s - s_{i-1}) + 6h}{h^3} \right) \right\} \\
&\quad + \left(\frac{h}{\tilde{h}}\right) \left\{ \left(v'_{2h}(s_{i-1}) - \frac{1}{R} v_{1h}(s_{i-1}) \right) \left(\frac{6(s - s_i) + 2h}{h^2} \right) \right. \\
&\quad \left. + \left(v'_{2h}(s_i) - \frac{1}{R} v_{1h}(s_i) \right) \left(\frac{6(s - s_{i-1}) - 2h}{h^2} \right) \right\} \\
&= \left(\frac{h}{\tilde{h}}\right)^2 \cos \theta_h \left\{ v_{2h}(s_{i-1}) \left(\frac{12(s - s_i) + 6h}{h^3} \right) \right. \\
&\quad + v_{2h}(s_i) \left(\frac{-12(s - s_{i-1}) + 6h}{h^3} \right) + v'_{2h}(s_{i-1}) \left(\frac{6(s - s_i) + 2h}{h^2} \right) \\
&\quad \left. + v'_{2h}(s_i) \left(\frac{6(s - s_{i-1}) - 2h}{h^2} \right) \right\} + \frac{h}{\tilde{h}} \left(1 - \cos \theta_h \frac{h}{\tilde{h}} \right) \\
&\quad \times \left\{ v'_{2h}(s_{i-1}) \left(\frac{6(s - s_i) + 2h}{h^2} \right) + v'_{2h}(s_i) \left(\frac{6(s - s_{i-1}) - 2h}{h^2} \right) \right\} \\
&\quad + \frac{1}{R} \frac{h}{\tilde{h}} \left\{ \frac{v_{1h}(s_{i-1}) - v_{1h}(s_i)}{h} \right\} = \left(\frac{h}{\tilde{h}}\right)^2 \cos \theta_h v''_{2h}(s) \\
&\quad + \frac{h}{\tilde{h}} \left(1 - \cos \theta_h \frac{h}{\tilde{h}} \right) \left\{ v'_{2h}(s_{i-1}) \left(\frac{6(s - s_i) + 2h}{h^2} \right) \right. \\
&\quad \left. + v'_{2h}(s_i) \left(\frac{6(s - s_{i-1}) - 2h}{h^2} \right) \right\} - \frac{1}{R} \frac{h}{\tilde{h}} v'_{1h}(s),
\end{aligned}$$

where we have taken into account the fact that v_{1h}, and v_{2h}, are polynomials of degree one and three, respectively, in the variable s on each interval I_i. Thus the second equality is proved. □

Consistency error estimates

When this is not explicitly stated, it is understood in the remainder of this section that the letter C stands for any constant independent of h.

Theorem 8.3.2. *There exists a constant C independent of h such that,* *for all* $\boldsymbol{v}_h \in \boldsymbol{V}_h$, $\boldsymbol{w}_h \in \boldsymbol{V}_h$,

$$|a_h(\boldsymbol{v}_h, \boldsymbol{w}_h) - a(\boldsymbol{v}_h, \boldsymbol{w}_h)| \leq Ch\|\boldsymbol{v}_h\|\,\|\boldsymbol{w}_h\|, \tag{8.3.36}$$

where $\|.\|$ *is the norm defined in* (8.3.6).

Consequently, for h sufficiently small, the approximate bilinear forms $a_h(.\,,.)$ *are uniformly* $\boldsymbol{V}_h$*-elliptic.*

Proof. With self-explanatory notation, the bilinear forms $a(.\,,.)$ and $a_h(.\,,.)$ are of the following form, for all $\boldsymbol{v}, \boldsymbol{w} \in \boldsymbol{V}_h$ (cf. (8.3.1) and (8.3.35)):

$$a(\boldsymbol{v}, \boldsymbol{w}) = \sum_{i=1}^{M} \int_{I_i} A\boldsymbol{v}A\boldsymbol{w}\, \mathrm{d}s + \sum_{i=1}^{M} \int_{I_i} B\boldsymbol{v}B\boldsymbol{w}\, \mathrm{d}s,$$

$$a_h(\boldsymbol{v}, \boldsymbol{w}) = \sum_{i=1}^{M} \int_{I_i} A_h\boldsymbol{v}A_h\boldsymbol{w}\, \mathrm{d}s + \sum_{i=1}^{M} \int_{I_i} B_h\boldsymbol{v}B_h\boldsymbol{w}\, \mathrm{d}s$$

(notice that since the bilinear form $a_h(.\,,.)$ is symmetric, it sufficed to compute it on the diagonal $\boldsymbol{v} = \boldsymbol{w}$, as we did in Theorem 8.3.1). We shall use the inequalities

$$\left|\int_{I_i} A\boldsymbol{v}A\boldsymbol{w}\, \mathrm{d}s - \int_{I_i} A_h\boldsymbol{v}A_h\boldsymbol{w}\, \mathrm{d}s\right| \leq |A\boldsymbol{v} - A_h\boldsymbol{v}|_{0,I_i}|A\boldsymbol{w} - A_h\boldsymbol{w}|_{0,I_i} + \\ + |A\boldsymbol{v}|_{0,I_i}|A\boldsymbol{w} - A_h\boldsymbol{w}|_{0,I_i} + |A\boldsymbol{w}|_{0,I_i}|A\boldsymbol{v} - A_h\boldsymbol{v}|_{0,I_i}, \tag{8.3.37}$$

and similar inequalities for the other integrals. Since

$$A\boldsymbol{v} = v_1' + \frac{1}{R}v_2, \quad A_h\boldsymbol{v} = \sqrt{\frac{\tilde{h}}{h}}\left(\frac{h}{\tilde{h}}\cos\theta_h v_1' + \frac{1}{R}\frac{v_2(s_{i-1}) + v_2(s_i)}{2}\right),$$

we deduce

$$A\boldsymbol{v} - A_h\boldsymbol{v} = \left(1 - \sqrt{\frac{h}{\tilde{h}}}\cos\theta_h\right)v_1' + \frac{1}{R}\sqrt{\frac{\tilde{h}}{h}}\left(v_2 - \frac{v_2(s_{i-1}) + v_2(s_i)}{2}\right) \\ + \frac{1}{R}\left(1 - \sqrt{\frac{\tilde{h}}{h}}\right)v_2,$$

and consequently,

$$|A\boldsymbol{v} - A_h\boldsymbol{v}|_{0,I_i} \leq \left|1 - \sqrt{\frac{h}{\tilde{h}}}\cos\theta_h\right||v_1|_{1,I_i} \\ + \frac{1}{R}\sqrt{\frac{\tilde{h}}{h}}\left|v_2 - \frac{v_2(s_{i-1}) + v_2(s_i)}{2}\right|_{0,I_i} \\ + \frac{1}{R}\left|1 - \sqrt{\frac{\tilde{h}}{h}}\right||v_{2}|_{0,I_i}.$$

On the one hand, using (8.3.12) and (8.3.17), we obtain

$$\left|1-\sqrt{\frac{h}{\tilde{h}}}\cos\theta_h\right| = \mathrm{O}(h^2),\quad \sqrt{\frac{\tilde{h}}{h}} = 1+\mathrm{O}(h^2),\quad \left|1-\sqrt{\frac{\tilde{h}}{h}}\right| = \mathrm{O}(h^2),$$

and, on the other hand, there exists some constant C such that

$$\forall v_2 \in H^1(I_i),\quad \left|v_2 - \frac{v_2(s_{i-1})+v_2(s_i)}{2}\right|_{0,I_i} \leq Ch|v_2|_{1,I_i},$$

since the mapping

$$\Pi:\ v_2 \in H^1(I_i) \to \frac{v_2(s_{i-1})+v_2(s_i)}{2} \in L^2(I_i)$$

preserves polynomials of degree zero (Theorem 3.1.4). Since

$$|A\boldsymbol{v}|_{0,I_i} \leq |v_1|_{1,I_i} + \frac{1}{R}|v_2|_{0,I_i},$$

we eventually find, upon combining the above inequalities in inequality (8.3.37), that

$$\forall \boldsymbol{v}, \boldsymbol{w} \in \boldsymbol{V}_h,\quad \left|\int_{I_i} A\boldsymbol{v}A\boldsymbol{w}\ \mathrm{d}s - \int_{I_i} A_h\boldsymbol{v}A_h\boldsymbol{w}\ \mathrm{d}s\right| \leq Ch\|\boldsymbol{v}\|_{I_i}\|\boldsymbol{w}\|_{I_i},$$

where

$$\|\boldsymbol{v}\|_{I_i} = (\|v_1\|_{1,I_i}^2 + \|v_2\|_{2,I_i}^2)^{1/2},$$

and therefore,

$$\left|\sum_{i=1}^{M}\int_{I_i} A\boldsymbol{v}A\boldsymbol{w}\ \mathrm{d}s - \sum_{i=1}^{M}\int_{I_i} A_h\boldsymbol{v}A_h\boldsymbol{w}\ \mathrm{d}s\right| \leq Ch\|\boldsymbol{v}\|\,\|\boldsymbol{w}\|.$$

It remains to consider the analogous expression, where

$$B\boldsymbol{v} = v_2'' - \frac{1}{R}v_1',$$

$$B_h\boldsymbol{v} = \left(\frac{h}{\tilde{h}}\right)^{3/2}\cos\theta_h v_2'' - \frac{1}{R}\sqrt{\frac{h}{\tilde{h}}}\,v_1' +$$
$$+\frac{1}{\sqrt{h\tilde{h}}}\left(1-\cos\theta_h\frac{h}{\tilde{h}}\right)\left\{v_2'(s_{i-1})\left(6\frac{(s-s_{i-1})}{h}-4\right)\right.$$
$$\left.+\,v_2'(s_i)\left(6\frac{(s-s_{i-1})}{h}-2\right)\right\},$$

so that

$$Bv - B_h v = \left(1 - \left(\frac{h}{\bar{h}}\right)^{3/2} \cos\theta_h\right) v_2'' - \frac{1}{R}\left(1 - \sqrt{\frac{h}{\bar{h}}}\right) v_1' - \\ - \frac{1}{\sqrt{h\bar{h}}}\left(1 - \cos\theta_h \frac{h}{\bar{h}}\right)\left\{ v_2'(s_{i-1})\left(6\frac{(s - s_{i-1})}{h} - 4\right) \right. \\ \left. + v_2'(s_i)\left(\frac{6(s - s_{i-1})}{h} - 2\right)\right\}.$$

Using (8.3.12) and (8.3.17), it is first established that

$$\left|1 - \left(\frac{h}{\bar{h}}\right)^{3/2} \cos\theta_h\right| = O(h^2), \quad \left|1 - \sqrt{\frac{h}{\bar{h}}}\right| = O(h^2),$$

$$\frac{1}{\sqrt{h\bar{h}}}\left(1 - \cos\theta_h \frac{h}{\bar{h}}\right) = O(h).$$

Next, one has

$$\left|\left(6\frac{(\cdot - s_{i-1})}{h} - 4\right)\right|_{0,I_i} = O(h^{1/2}), \quad \left|\left(6\frac{(\cdot - s_{i-1})}{h} - 2\right)\right|_{0,I_i} = O(h^{1/2}),$$

and, finally, there exists a constant C such that

$$\forall v_2 \in H^2(I_i), \quad |v_2|_{1,\infty,I_i} \leqslant Ch^{-1/2}\|v_2\|_{2,I_i}.$$

Combining the above relations with an inequality similar to inequality (8.3.37), we find that

$$\forall \boldsymbol{v}, \boldsymbol{w} \in V_h, \quad \left|\int_{I_i} B\boldsymbol{v} B\boldsymbol{w}\, \mathrm{d}s - \int_{I_i} B_h\boldsymbol{v} B_h\boldsymbol{w}\, \mathrm{d}s\right| \leqslant Ch\|\boldsymbol{v}\|_{I_i}\|\boldsymbol{w}\|_{I_i},$$

and inequality (8.3.36) is proved.

The uniform V_h-ellipticity of the approximate bilinear forms is proved as in Theorem 8.2.2. □

Concerning the approximation of the linear form, we have the following result, whose proof is left to the reader (Exercise 8.3.1):

Theorem 8.3.3. *Assume that the functions f^1 and f^2 are Lipschitz-continuous on the interval $\bar{I}$. Then there exists a constant C independent of h (but dependent on the functions f^1, f^2) such that, for all $\boldsymbol{w}_h \in V_h$,*

$$|f(\boldsymbol{w}_h) - f_h(\boldsymbol{w}_h)| \leqslant Ch\|\boldsymbol{w}_h\|. \qquad \square \quad (8.3.38)$$

Estimate of the error $(|u_1 - u_{1h}|^2_{1,I} + |u_2 - u_{2h}|^2_{2,I})^{1/2}$

We are now in a position to prove the main result of this section.

Theorem 8.3.4. *Assume that the solution* $\boldsymbol{u} = (u_1, u_2)$ *belongs to the space* $V \cap (H^2(I) \times H^3(I))$, *and that the functions* f^1 *and* f^2 *are Lipschitz-continuous on the interval* I. *Then there exists a constant* C *independent of* h *(but dependent on the solution* $\boldsymbol{u}$*) such that*

$$|\boldsymbol{u} - \boldsymbol{u}_h| = (|u_1 - u_{1h}|^2_{1,I} + |u_2 - u_{2h}|^2_{2,I})^{1/2} \leqslant Ch, \tag{8.3.39}$$

where $\boldsymbol{u}_h$ is the solution of the discrete problem (8.3.31).

Proof. Since the approximate bilinear forms are uniformly V_h-elliptic for h sufficiently small (Theorem 8.3.2), we may apply Theorem 4.1.1: If we let $\boldsymbol{\Pi}_h \boldsymbol{u} = (\Pi_h u_1, \Lambda_h u_2)$ denote the V_h-interpolant of the solution $\boldsymbol{u}$, we have

$$\|\boldsymbol{u} - \boldsymbol{u}_h\| \leqslant C\Bigg(\|\boldsymbol{u} - \boldsymbol{\Pi}_h \boldsymbol{u}\| + \sup_{\boldsymbol{w}_h \in V_h} \frac{|a(\boldsymbol{\Pi}_h \boldsymbol{u}, \boldsymbol{w}_h) - a_h(\boldsymbol{\Pi}_h \boldsymbol{u}, \boldsymbol{w}_h)|}{\|\boldsymbol{w}_h\|} + \\ + \sup_{\boldsymbol{w}_h \in V_h} \frac{|f(\boldsymbol{w}_h) - f_h(\boldsymbol{w}_h)|}{\|\boldsymbol{w}_h\|}\Bigg).$$

Using the regularity assumption on the solution $\boldsymbol{u}$, we obtain,

$$\|\boldsymbol{u} - \boldsymbol{\Pi}_h \boldsymbol{u}\| \leqslant Ch(|u_1|^2_{2,I} + |u_2|^2_{3,I})^{1/2}.$$

Next, using Theorem 8.3.2, we find the *consistency error estimate*

$$\sup_{\boldsymbol{w}_h \in V_h} \frac{|a(\boldsymbol{\Pi}_h \boldsymbol{u}, \boldsymbol{w}_h) - a_h(\boldsymbol{\Pi}_h \boldsymbol{u}, \boldsymbol{w}_h)|}{\|\boldsymbol{w}_h\|} \leqslant Ch\|\boldsymbol{\Pi}_h \boldsymbol{u}\| \leqslant Ch\|\boldsymbol{u}\|,$$

where for the second inequality, we have used the fact that the operators Π_h and Λ_h preserve polynomials of degree zero and one, respectively.

Finally, we have by Theorem 8.3.3,

$$\sup_{\boldsymbol{w}_h \in V_h} \frac{|f(\boldsymbol{w}_h) - f_h(\boldsymbol{w}_h)|}{\|\boldsymbol{w}_h\|} \leqslant Ch,$$

and the proof is complete. □

We are now in a position to state the main conclusion of the present analysis: In order to compensate the discrepancies between the strain energies of the continuous and of the discrete problem, the functions in the spaces $\tilde{V}_h$ and $\tilde{W}_h$ must satisfy the compatibility relations (8.3.25) across the mesh points $\tilde{s}_i$. As a consequence, the inclusions $\tilde{V}_h \subset H^1(\tilde{I})$ and $\tilde{W}_h \subset H^2(\tilde{I})$ no longer hold.

In other words, *the fact that the method is not conforming for the geometry implies that it is not conforming also for the displacements.*

Exercise

8.3.1. Prove Theorem 8.3.3. Recall (cf. (8.3.1) and (8.3.33)) that for all $w \in V_h$, we have

$$f(w) - f_h(w) = \int_0^L (f^1(s)w_1(s) + f^2(s)w_2(s))\,ds - h\sum_{i=1}^{M-1} (f^1(s_i)w_1(s_i) + f^2(s_i)w_2(s_i)).$$

Bibliography and comments

8.1. There exist numerous references for the shell problem. The description of Koiter's model (including the nonlinear case) is found at various stages of its development in KOITER (1966, 1970), KOITER & SIMMONDS (1972). This model is based on certain physical hypotheses (essentially about the stress distribution across the thickness of the shell), which JOHN (1965) has theoretically justified.

A different model has been proposed by NAGHDI (1963, 1972). A simplified theory of the so-called "shallow" shells is presented in TIMOSHENKO & WOINOWSKY–KRIEGER (1959) and WASHIZU (1968).

For references in tensor calculus and differential geometry, the reader may consult GOUYON (1963), LELONG–FERRAND (1963), LICHNÉROWICZ (1967), VALIRON (1950, Chapters 12, 13, 14).

The ellipticity of Koiter's model in the linear case is proved in BERNADOU & CIARLET (1976), where the proof follows basically a method set up in CIARLET (1976a) for circular arches. Following CIARLET (1976c), we have presented here the extension of this method to the case of an arch of varying curvature. For the expression of the corresponding elastic energy, see for example MOAN (1974).

As regards the question of ellipticity for various shell models, we mention the works of ROUGÉE (1969) for "cylindrical" shells, COUTRIS (1973) for Naghdi's model, GORDEZIANI (1974) for the model of VEKUA (1965), SHOIKET (1974) for the model of NOVOZHILOV (1970). For cylindrical shells, see KOLAKOWSKI & DRYJA (1974), MIYOSHI (1973b).

8.2. The content of this section is essentially based on CIARLET (1976b). Conforming finite element methods of the type considered here are described in ARGYRIS & LOCHNER (1972), ARGYRIS, HAASE & MALEJANNAKIS (1973). Related methods are discussed in DUPUIS & GÖEL (1970b), DUPUIS (1971). These are only a few among the many papers which are concerned with the description of the application of finite element methods to shells and the various computational problems attached with them. In this direction, let us quote FRIED (1971b), GALLAGHER (1973). In the case of large deflections, see BATOZ, CHATTOPADHYAY & DHATT (1976), MATSUI & MATSUOKA (1976).

By contrast, there are very few papers that deal with the numerical analysis of such methods. MIYOSHI (1973b) has analyzed the convergence of a mixed finite element method for cylindrical shells and GELLERT & LAURSEN (1976) study a mixed method for arches. KIKUCHI & ANDO (1972d, 1973b) have described the application of a simplified hybrid method to shallow shells. A hybrid method for shells is also considered by STEPHAN & WEISSGERBER (1976). MOAN (1974) has examined the asymptotic rate of energy convergence for arches. KIKUCHI (1975a) has proposed a simplified method for thin "shallow" shells. Let us also mention the analysis of CLÉMENT & DESCLOUX (1972) regarding the validity of the rigid displacement condition (Remark 8.1.1) for the discrete problem.

The effect of curved boundaries and numerical integration (Remark 8.2.1) is analyzed in BERNADOU (1976) along the lines of the present treatment.

As was pointed out in the introduction, an elegant way of approximating shell problems consists in using finite elements directly derived from three-dimensional finite elements by reducing their thickness. See AHMAD, IRONS & ZIENKIEWICZ (1970), ZIENKIEWICZ, TAYLOR & TOO (1971). The corresponding numerical analysis is yet to be done. A related, and challenging, problem is to describe and analyze the "intermediate" finite elements which should be used at the junctions between two-dimensional or three-dimensional portions of a single mechanical structure.

8.3. The content of this section is based on a recent paper by JOHNSON

EPILOGUE:

Some "real-life" finite element model examples

GEOMETRY OF MIDDLE SURFACE

POSITION VECTOR

$$\mathbf{x} = \begin{bmatrix} r \cdot \cos \xi^1 \\ r \cdot \sin \xi^1 \\ \xi^2 \end{bmatrix}$$

WHERE

$$r = 24.85 \sqrt{1 + (\xi^2 / 64.62)^2}$$

PARAMETER DEFINITION $\xi^1 = \vartheta$, $\xi^2 = z$

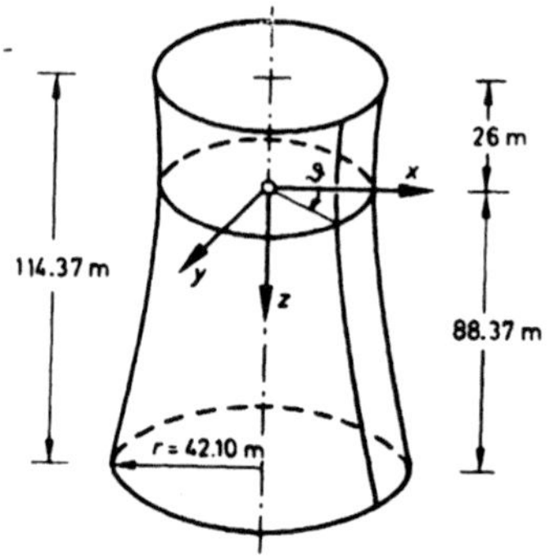

MATERIAL DATA

YOUNG'S MODULUS	$E = 3 \cdot 10^9$ kp/m^2	
POISSON'S RATIO	$\nu = 0.2$	
SPECIFIC GRAVITY	$\gamma = 1.0$ kp/m^3	(FOR BUCKLING ANALYSIS)
DENSITY	$\rho = 1.0$ kp s^2/m^4	(FOR VIBRATION ANALYSIS)

GEOMETRICAL DATA

SHELL THICKNESS	$t = 0.14$ m
STIFFENER AT THE TOP OF THE SHELL	$A = 0.4$ m^2
	$I_{ff} = 3.333 \cdot 10^{-2}$ m^4
	$I_{nn} = 5.333 \cdot 10^{-3}$ m^4
	$J = 1.597 \cdot 10^{-2}$ m^4

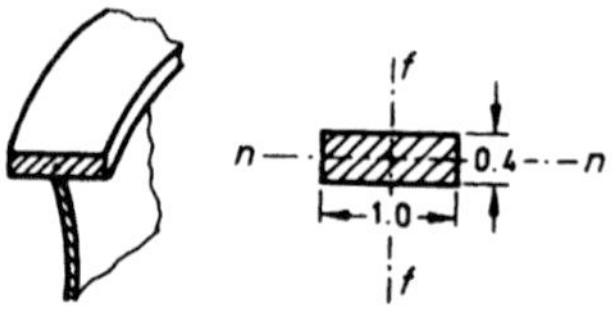

BOUNDARY CONDITIONS

$z = -26.0$ FREE EDGE

$z = 88.37$ BUILT-IN

WINDLOAD

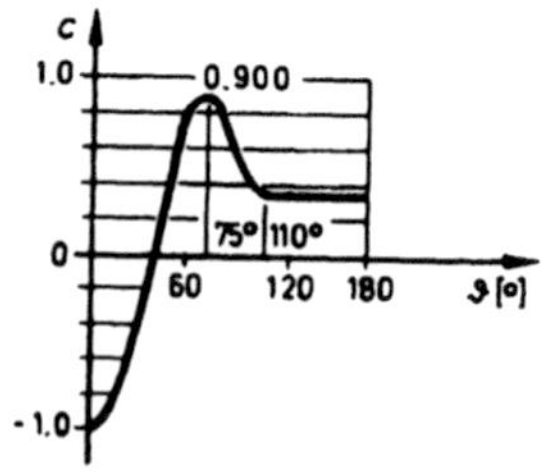

DISTRIBUTION OF WINDLOAD

$P_w = c(\vartheta) \cdot q(z)$

$q(z) = -100$ kp/m^2 = constant

Cooling tower: Geometry, dimensions and input data.
Reproduced by courtesy of Professor J.H. Argyris.

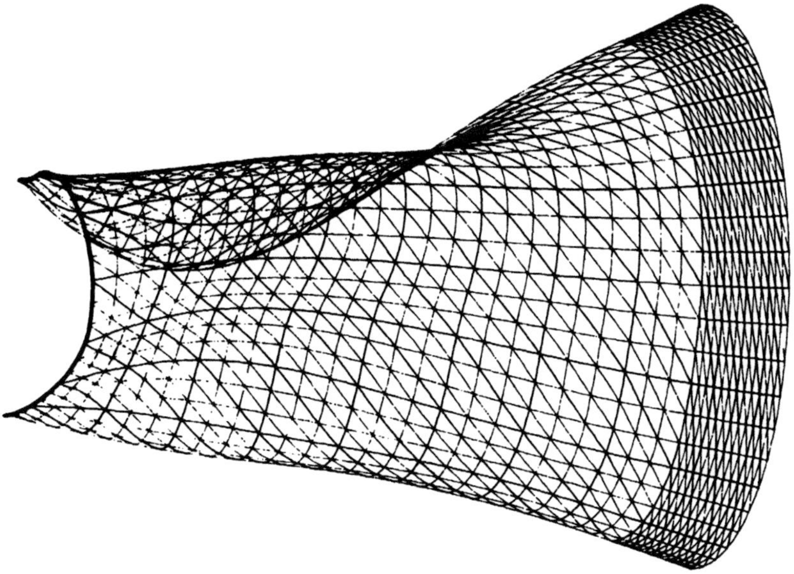

Cooling tower: Deformed structure under wind load.
Reproduced by courtesy of Professor J.H. Argyris.

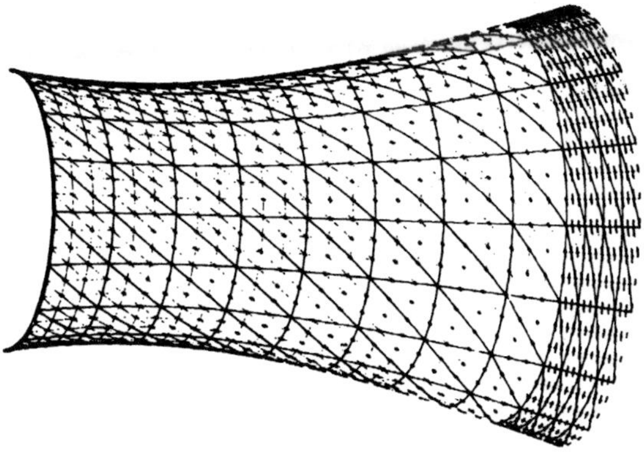

Cooling tower: Triangulation.
Reproduced by courtesy of Professor J.H. Argyris.

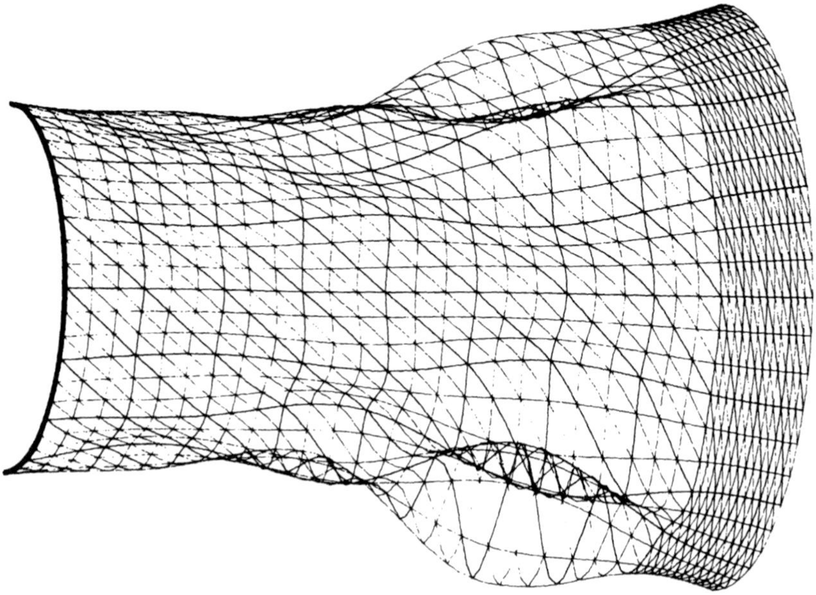

Cooling tower: First buckling mode under dead weight.
Reproduced by courtesy of Professor J.H. Argyris.

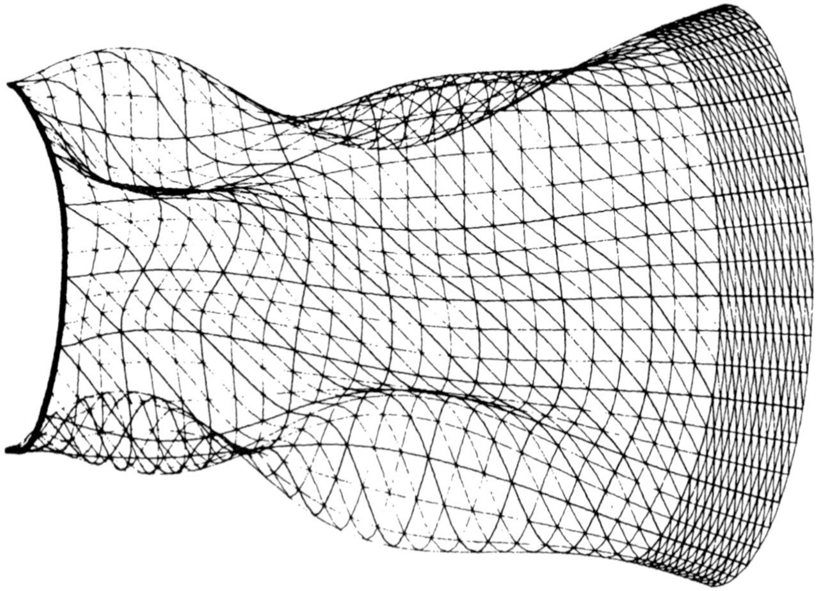

Cooling tower: First vibration mode.
Reproduced by courtesy of Professor J.H. Argyris.

Finite element stress analysis of complex tubular joints.
Reproduced by courtesy of Professor C.A. Felippa.

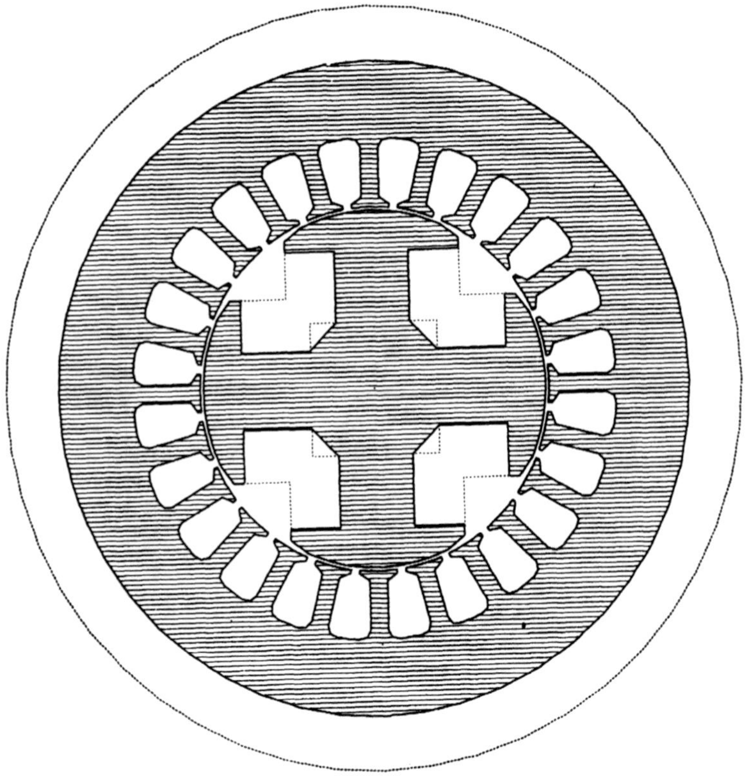

Tetrapolar alternator.
Reproduced by courtesy of Professor R. Glowinski and Mr. A. Marrocco.

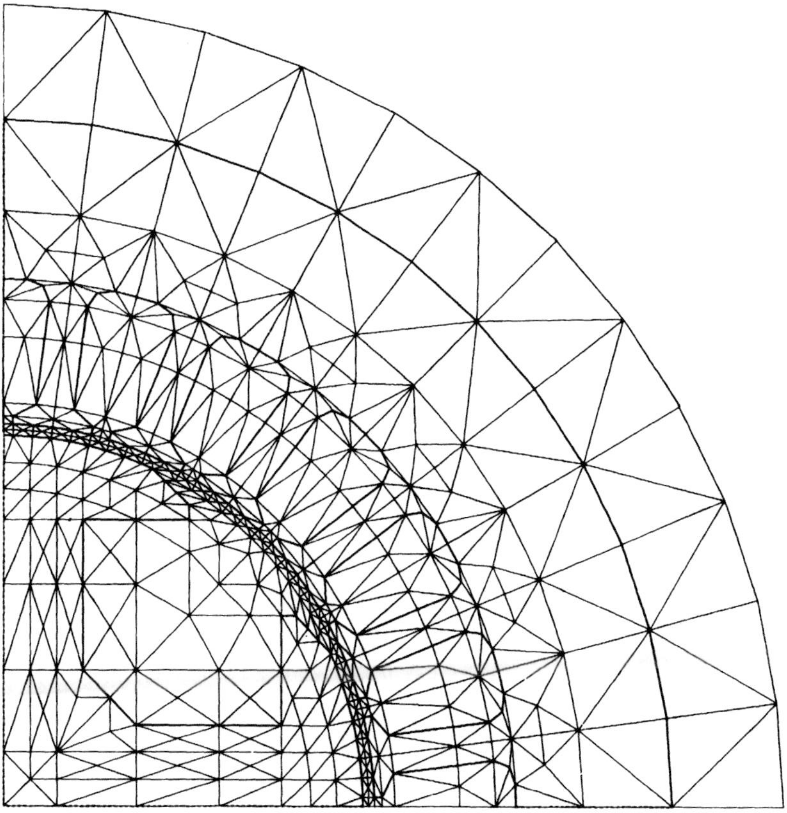

Tetrapolar alternator: Example of a triangulation.
Reproduced by courtesy of Professor R. Glowinski and Mr. A. Marrocco.

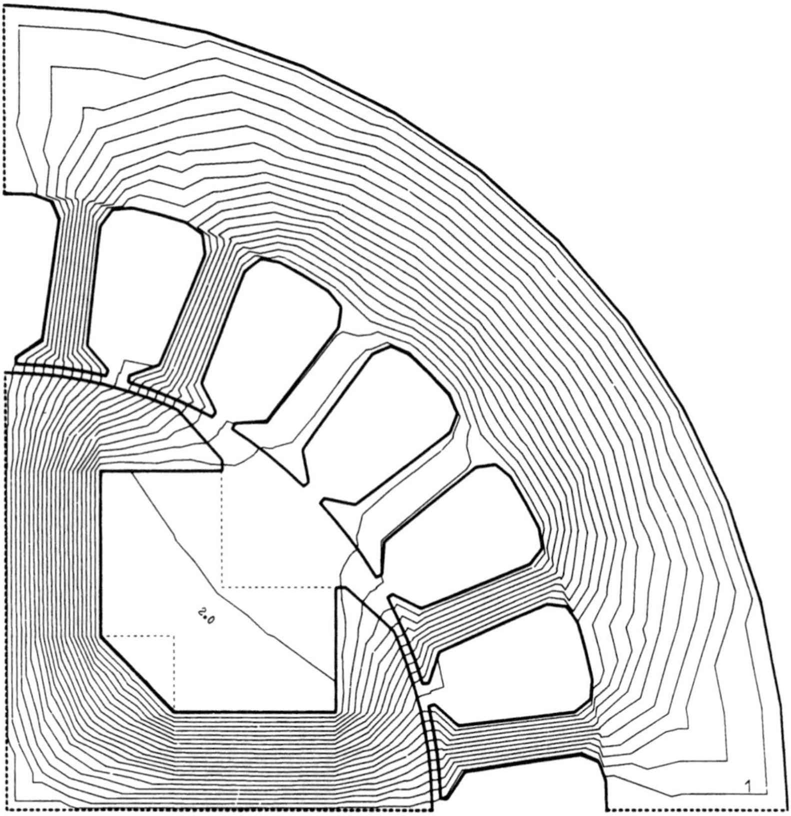

Tetrapolar alternator: Induction lines for $J = 2$.
(J: density of current)
Reproduced by courtesy of Professor R. Glowinski and Mr. A. Marrocco.

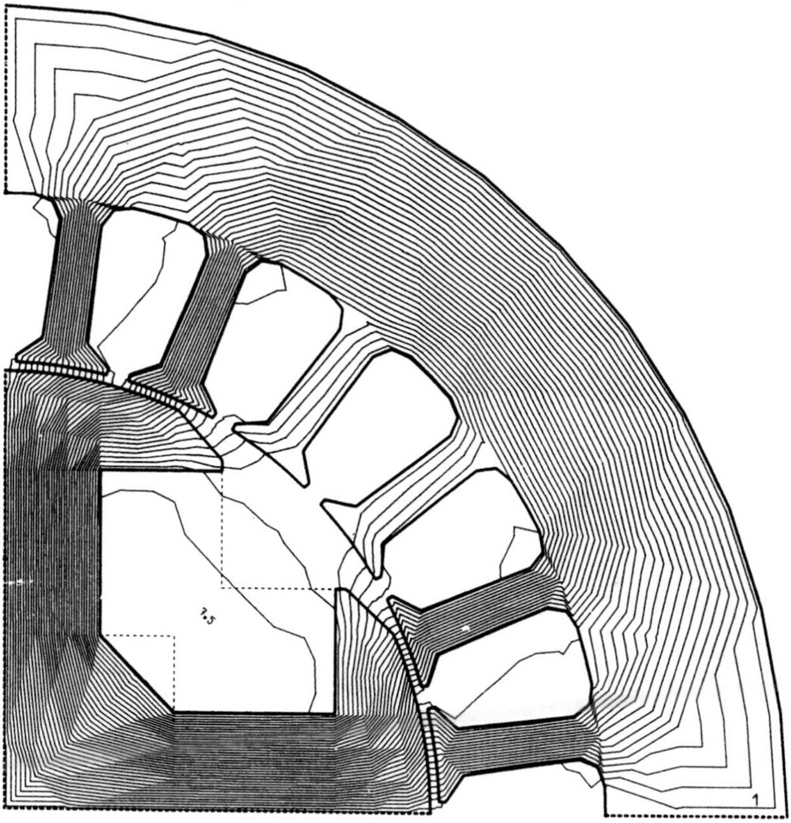

Tetrapolar alternator: Induction lines for $J = 7.5$.
(J: density of current)
Reproduced by courtesy of Professor R. Glowinski and Mr. A. Marrocco.

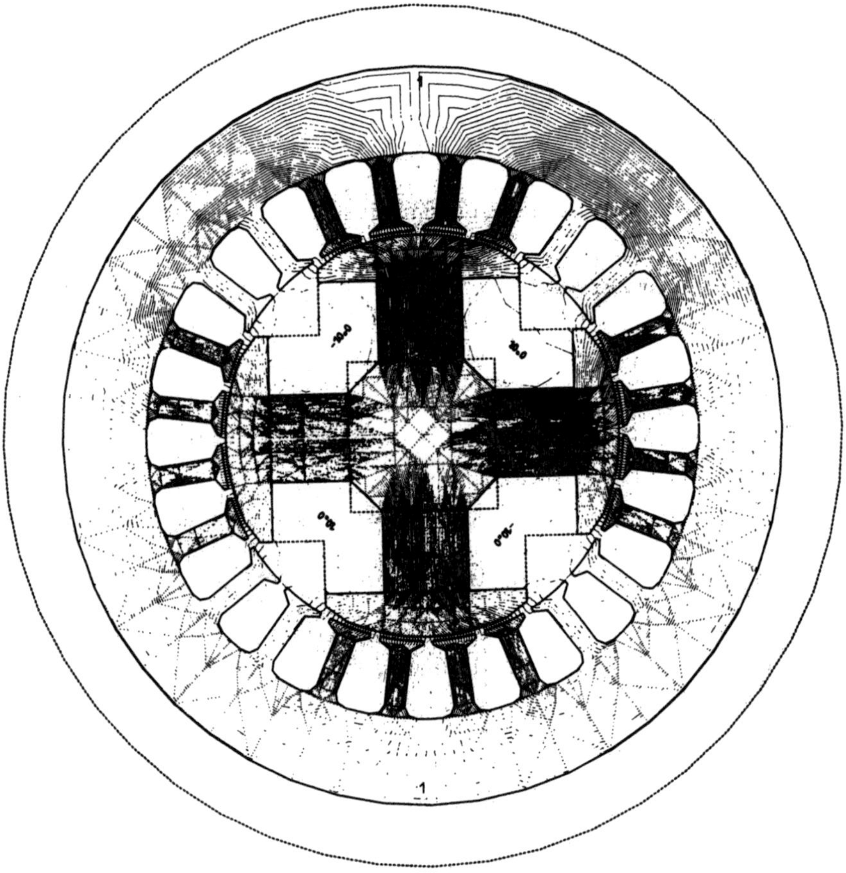

Tetrapolar alternator: Induction lines for $J = 10$.
(J: density of current)
Reproduced by courtesy of Professor R. Glowinski and Mr. A. Marrocco.

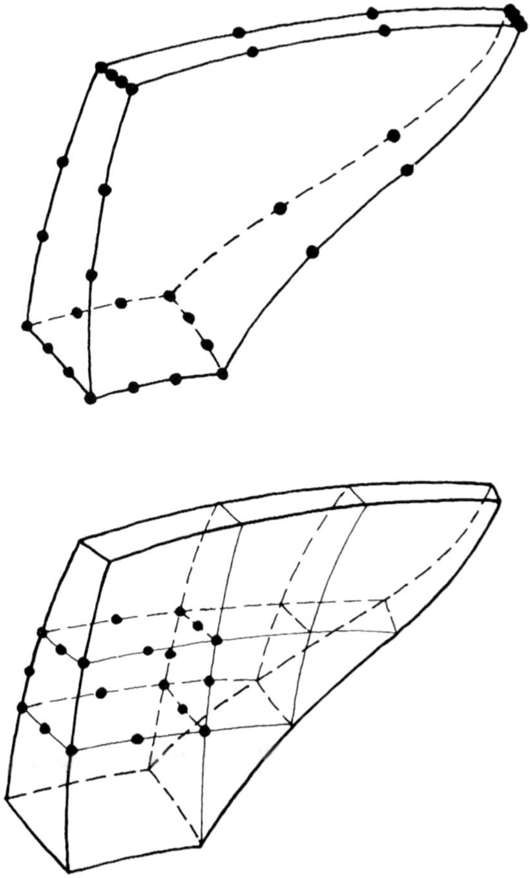

Arch dam in a rigid valley – Various element subdivisions.
Reproduced from Fig. 9.8 of Professor Zienkiewicz' book: "*The Finite Element Method in Engineering Science*", McGraw-Hill, London, 1971, by courtesy of Professor O.C. Zienkiewicz, and with permission of the Publisher.

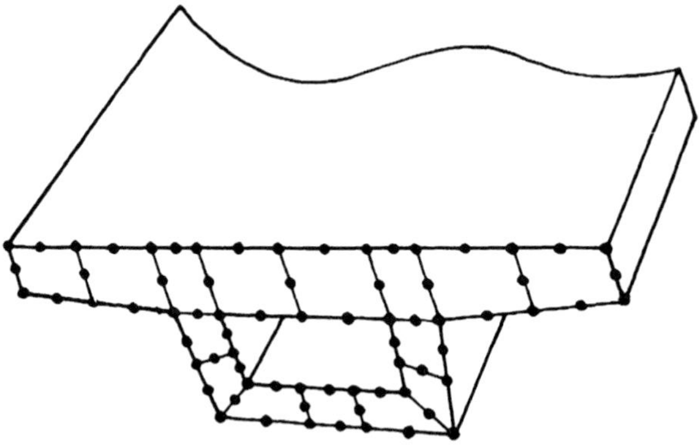

A thick box bridge reduced to a two-dimensional problem with isoparametric, quadratic, elements.
Reproduced from Fig. 13.2 of Professor Zienkiewicz' book: "*The Finite Element Method in Engineering Science*", McGraw-Hill, London, 1971, by courtesy of Professor O.C. Zienkiewicz, and with permission of the Publisher.

BIBLIOGRAPHY

The following list includes all titles referred to in the text. With minor exceptions, the journal abbreviations follow the usage of the Index of the *Mathematical Reviews*, Volume 51, No. 7 (June 1976).

ADAMS, R.A. (1975): *Sobolev Spaces*, Academic Press, New York.

ADINI, A.; CLOUGH, R.W. (1961): Analysis of plate bending by the finite element method, NSF report G. 7337.

AGMON, S. (1965): *Lectures on Elliptic Boundary Value Problems*, Van Nostrand, Princeton.

AHMAD, S.; IRONS, B.M.; ZIENKIEWICZ, O.C. (1970): Analysis of thick and thin shell structures by curved finite elements, Internat. J. Numer. Methods Engrg. **2**, 419–451.

ALLMAN, D.J. (1976): A simple cubic displacement element for plate bending, Internat. J. Numer. Methods Engrg. **10**, 263–281.

ARCANGELI, R.; GOUT, J.L. (1976): Sur l'évaluation de l'erreur d'interpolation de Lagrange dans un ouvert de R^n, Rev. Française Automat. Informat. Recherche Opérationnelle Sér. Rouge Anal. Numer. **10**, 5–27.

ARGYRIS, J.H. (1954–1955): Energy theorems and structural analysis, part I: General Theory, Aircraft Engineering **26**, 347–356, 383–387, 394; **27**, 42–58, 80–94, 125–134 (also published as a book, Butterworths Scientific Publications, London, 1960).

ARGYRIS, J.H.; FRIED, I. (1968): The LUMINA element for the matrix displacement method (Lagrangian interpolation). The Aeronautical Journal of the Royal Aeronautical Society **72**, 514–517.

ARGYRIS, J.H.; FRIED, I.; SCHARPF, D.W. (1968): The TUBA family of plate elements for the matrix displacement method, The Aeronautical Journal of the Royal Aeronautical Society **72**, 701–709.

ARGYRIS, J.H.; HAASE, M.; MALEJANNAKIS, G.A. (1973): Natural geometry of surfaces with specific reference to the matrix displacement analysis of shells. I, II and III, *Proceedings Koninklijke Nederlandse Akademie van Wetenschappen, Series B*, **76**, 361–410.

ARGYRIS, J.H.; LOCHNER, N. (1972): On the application of the SHEBA shell element, Comput. Methods Appl. Mech. Engrg. **1**, 317–347.

ATTEIA, M. (1975): Fonctions "spline" et méthode d'éléments finis, Rev. Française Automat. Informat. Recherche Opérationnelle Sér. Rouge Anal. Numér. **R-2**, 13–40.

ATTEIA, M. (1977): Evaluation de l'erreur dans la méthode des éléments finis, Numer. Math. **28**, 295–306.

AUBIN, J.P. (1967a): Approximation des Espaces de Distributions et des Opérateurs Différentiels, Mémoire 12, Bull. Soc. Math. France.

AUBIN, J.P. (1967b): Behavior of the error of the approximate solutions of boundary value problems for linear elliptic operators by Galerkin's and finite difference methods, Ann. Scuola Norm. Sup. Pisa **21**, 599–637.

AUBIN, J.P. (1968a): Evaluation des erreurs de troncature des approximations des espaces de Sobolev, J. Math. Anal. Appl. **21**, 356–368.

AUBIN, J.P. (1968b): Interpolation et approximations optimales et "spline functions", J. Math. Anal. Appl. **24**, 1–24.

AUBIN, J.P. (1969): Approximation des problèmes aux limites non homogènes et régularité de la convergence, Calcolo **6**, 117–139.

AUBIN, J.P. (1972): *Approximation of Elliptic Boundary-Value Problems*, Wiley-Interscience, New York.

AUBIN, J.P.; BURCHARD, H.G. (1971): Some aspects of the method of the hypercircle applied to elliptic variational problems, in *SYNSPADE 1970* (B. Hubbard, Editor), Academic Press, New York.

AUSLENDER, A. (1976): *Optimisation-Méthodes Numériques*, Masson, Paris.

BABUŠKA, I. (1963): The theory of small changes in the domain of existence in the theory of partial differential equations and its applications, in *Differential Equations and their Applications*, pp. 13–26, Academic Press, New York.

BABUŠKA, I. (1970): Approximation by hill functions, Comment Math. Univ. Carolinae **11**, 787–811.

BABUŠKA, I. (1971a): The rate of convergence for the finite element method, SIAM J. Numer. Anal. **8**, 304–315.

BABUŠKA, I. (1971b): Error-bounds for finite element method, Numer. Math. **16**, 322–333.

BABUŠKA, I. (1972a): A finite element scheme for domains with corners, Numer. Math. **20**, 1–21.

BABUŠKA, I. (1972b): Approximation by hill functions II, Comment. Math. Univ. Carolinae **13**, 1–22.

BABUŠKA, I. (1972c): The finite element method for infinite domains. I, Math. Comput. **26**, 1–11.

BABUŠKA, I. (1973a): The finite element method with Lagrangian multipliers, Numer. Math. **20**, 179–192.

BABUŠKA, I. (1973b): The finite element method with penalty, Math. Comput. **27**, 221–228.

BABUŠKA, I. (1974a): Method of weak elements, Technical Note BN-809, University of Maryland.

BABUŠKA, I. (1974b): Solution of problems with interfaces and singularities, in *Mathematical Aspects of Finite Elements in Partial Differential Equations* (C. de Boor, Editor), pp. 213–277, Academic Press, New York.

BABUŠKA, I. (1976): Singularities problem in the finite element method, Technical Note BN-835, Institute for Fluid Dynamics and Applied Mathematics, University of Maryland, College Park.

BABUŠKA, I., AZIZ, A.K. (1972): Survey Lectures on the Mathematical Foundations of the Finite Element Method, in *The Mathematical Foundations of the Finite Element Method with Applications to Partial Differential Equations* (A.K. Aziz, Editor), pp. 3–359, Academic Press, New York.

BABUŠKA, I.; AZIZ, A.K. (1976): On the angle condition in the finite element method, SIAM J. Numer. Anal. **13**, 214–226.

BABUŠKA, I.; KELLOGG, R.B. (1975): Nonuniform error estimates for the finite element method, SIAM J. Numer. Anal. **12**, 868–875.

BABUŠKA, I.; ODEN, J.T.; LEE, J.K. (1977): Mixed-hybrid finite element approximations of second-order elliptic boundary-value problems, Comput. Methods Appl. Mech. Engrg. **11**, 175–206.

BABUŠKA, I.; ROSENZWEIG, M.B. (1972): A finite element scheme for domains with corners, Numer. Math. **20**, 1–21.

BABUŠKA, I.; ZLÁMAL, M. (1973): Nonconforming elements in the finite element method with penalty, SIAM J. Numer. Anal. **10**, 863–875.

BAIOCCHI, C. (1971): Sur un problème à frontière libre traduisant le filtrage de liquides à travers des milieux poreux, C.R. Acad. Sci. Paris **273**, 1215–1217.

BAIOCCHI, C. (1972): Su un problema di frontiera libera connesso a questioni di idraulica, Ann. Mat. Pura Appl. **XCII**, 107–127.

BAIOCCHI, C. (1974): Problèmes à frontière libre en hydraulique, C.R. Acad. Sci. Paris **278**, 1201–1204.

BAIOCCHI, C. (1975): Free boundary problems in the theory of fluid flow through porous media (address given at the International Congress of Mathematicians, Vancouver, August 21–29, 1974), Publication No. 84, Laboratorio di Analisi Numerica del C.N.R., Pavia.

BAIOCCHI, C.; COMINCIOLI, V.; MAGENES, E.; POZZI, G.A. (1973): Free boundary problems in the theory of fluid flow through porous media: Existence and uniqueness theorems, Ann. Mat. Pura Appl. **XCVII**, 1–82.

BAKER, G.A. (1973): Simplified proofs of error estimates for the least squares method for Dirichlet's problem, Math. Comput. **27**, 229–235.

BARLOW, J. (1976): Optimal stress locations in finite element models, Internat. J. Numer. Methods Engrg. **10**, 243–251.

BARNHILL, R.E. (1976a): Blending function finite elements for curved boundaries, in The Mathematics of Finite Elements and Applications (J.R. Whiteman, Editor), pp. 67–76, Academic Press, London.

BARNHILL, R.E. (1976b): Blending function interpolation: a survey and some new results (to appear in the Proceedings of the Conference on Numerical Methods of Approximation Theory, Oberwolfach, May 29, 1975).

BARNHILL, R.E.; BIRKHOFF, G.; GORDON, W.J. (1973): Smooth interpolation in triangles, J. Approximation Theory **8**, 114–128.

BARNHILL, R.E.; BROWN, J.H. (1975): Curved nonconforming elements for plate problems, Report No. 8, University of Dundee.

BARNHILL, R.E.; BROWN, J.H.; MCQUEEN, N.; MITCHELL, A.R. (1976): Computable finite element error bounds for Poisson's equation, Internat. J. Numer. Methods Engrg. (to appear).

BARNHILL, R.E.; GREGORY, J.A. (1975a): Compatible smooth interpolation in triangles, J. Approximation Theory **15**, 214–225.

BARNHILL, R.E.; GREGORY, J.A. (1975b): Polynomial interpolation to boundary data on triangles, Math. Comput. **29**, 726–735.

BARNHILL, R.E.; GREGORY, J.A. (1976a): Sard kernel theorems on triangular domains with application to finite element error bounds, Numer. Math. **25**, 215–229.

BARNHILL, R.E.; GREGORY, J.A. (1976b): Interpolation remainder theory from Taylor expansions on triangles, Numer. Math. **25**, 401–408.

BARNHILL, R.E.; WHITEMAN, J.R. (1973): Error analysis of finite element methods with triangles for elliptic boundary value problems, in *the Mathematics of Finite Elements and Applications* (J.R. Whiteman, Editor), pp. 83–112, Academic Press, London.

BARNHILL, R.E.; WHITEMAN, J.R. (1975): Error analysis of Galerkin methods for Dirichlet problems containing boundary singularities, J. Inst. Math. Appl. **15** (1975), 121–125.

BARROS NETO, J. (1965): Inhomogeneous boundary value problems in a half space, Ann. Scuola Norm. Sup. Pisa **19**, 331–365.

BARSOUM, R.S. (1976): On the use of isoparametric finite elements in linear fracture mechanics, Internat. J. Numer Methods Engrg. **10**, 25–37.

BATHE, K.-J.; WILSON, E.L. (1973): Solution methods for eigenvalue problems in structural mechanics, Internat. J. Numer. Methods Engrg. **6**, 213–226.

BATOZ, J.L.; CHATTOPADHYAY, A.; DHATT, G. (1976): Finite element large deflection analysis of shallow shells, Internat. J. Numer. Methods Engrg. **10**, 39–58.

BAZELEY, G.P.; CHEUNG, Y.K.; IRONS, B.M.; ZIENKIEWICZ, O.C. (1965): Triangular elements in bending – conforming and nonconforming solutions, in *Proceedings of the Conference on Matrix Methods in Structural Mechanics*, Wright Patterson A.F.B., Ohio.

BEGIS, D.; GLOWINSKI, R. (1974): Application de la méthode des éléments finis à la résolution d'un problème de domaine optimal, in *Computing Methods in Applied Sciences and Engineering, Part 2* (R. Glowinski and J.L. Lions, Editors), pp. 403–434, Lecture Notes in Computer Science **11**, Springer-Verlag, Berlin.

BEGIS, D.; GLOWINSKI, R. (1975): Application de la méthode des éléments finis à l'approximation d'un problème de domaine optimal. Méthodes de résolution des problèmes approchés, Appl. Math. Optim. **2**, 130–169.

BELL, K. (1969): A refined triangular plate bending element, Internat. J. Numer. Methods Engrg. **1**, 101–122.

BENSOUSSAN, A.; LIONS, J.L. (1973): Nouvelle formulation de problèmes de contrôle impulsionnel et applications, C.R. Acad. Sci. Paris **276**, 1189–1192.

BENSOUSSAN, A.; LIONS, J.L. (1974): Sur l'approximation numérique d'inéquations quasi-variationnelles stationnaires, in *Computing Methods in Applied Sciences and Engineering, Part 2* (R. Glowinski and J.L. Lions, Editors), pp. 326–338, Lecture Notes in Computer Science, Vol. 11, Springer-Verlag, Berlin.

BERCOVIER, M. (1976): *Régularisation Duale des Problèmes Variationnels Mixtes. Application aux Eléments Finis Mixtes et Extension à Quelques Problèmes non Linéaires*, Doctoral Thesis, Université de Rouen.

BERCOVIER, M.; LIVNE, E. (1976): A 4 CST quadrilateral element for incompressible and nearly incompressible materials, Technical Note MB/76/3, Computation Center, Hebrew University, Jerusalem.

BERGAN, P.G.; CLOUGH, R.W. (1973): Large deflection analysis of plates and shallow shells using the finite element method, Internat. J. Numer. Methods Engrg. **5**, 543–556.

BERGER, A.E. (1973): L^2-error estimates for finite elements with interpolated boundary conditions, Numer. Math. **21**, 345–349.

BERGER, A.E. (1976): The truncation method for the solution of a class of variational inequalities, Rev. Française Automat. Informat. Recherche Opérationnelle Sér. Rouge Anal. Numér. **10**, 29–42.

BERGER, A.E.; SCOTT, R.; STRANG, G. (1972): Approximate boundary conditions in the finite element method, in *Symposia Mathematica*, Vol. 10, pp. 295–313, Academic Press, New York.

BERNADOU, M. (1976): Méthodes conformes d'éléments finis pour des problèmes de coques avec intégration numérique (to appear).

BERNADOU, M.; CIARLET, P.G. (1976): Sur l'ellipticité du modèle linéaire de coques de W.T. Koiter in *Computing Methods in Applied Sciences and Engineering* (R. Glowinski and J.L. Lions, Editors), pp. 89–136, Lecture Notes in Economics and Mathematical Systems, Vol. 134, Springer-Verlag, Berlin.

BERNADOU, M.; DUCATEL, Y. (1976): Méthodes conformes d'élements finis pour des problèmes elliptiques du quatrième ordre avec intégration numérique, Rapport de Recherche No. 195, I.R.I.A. Laboria, Rocquencourt.

BERS, L.; JOHN, F.; SCHECHTER, M. (1964): *Partial Differential Equations*, John Wiley, New York.

BIRKHOFF, G. (1969): Piecewise bicubic interpolation and approximation in polygons, in *Approximation with Special Emphasis on Spline Functions* (I.J. Schoenberg, Editor), pp. 185–121, Academic Press, New York.

BIRKHOFF, G. (1971): Tricubic polynomial interpolation, Proc. Nat. Acad. Sci. U.S.A. **68**, 1162–1164.

BIRKHOFF, G. (1972): Piecewise analytic interpolation and approximation in triangulated polygons, in *The Mathematical Foundations of the Finite Element Method with Applications to Partial Differential Equations* (A.K. Aziz, Editor), pp. 363–385, Academic Press, New York.

BIRKHOFF, G.; DE BOOR, C.; SWARTZ, B.; WENDROFF, B. (1966): Rayleigh–Ritz approximation by piecewise cubic polynomials, SIAM J. Numer. Anal. **3**, 188–203.

BIRKHOFF, G.; FIX, G.J. (1974): Higher-order linear finite element methods, Technical Report 1, Office of Naval Research, Arlington, Virginia.

BIRKHOFF, G.; GULATI, S. (1974): Optimal few-point discretizations of linear source problems, SIAM J. Numer. Anal. **11**, 700–728.

BIRKHOFF, G.; MANSFIELD, L. (1974): Compatible triangular finite elements, J. Math. Anal. Appl. **47**, 531–553.

BIRKHOFF, G.; SCHULTZ, M.H.; VARGA, R.S. (1968): Piecewise Hermite interpolation in one and two variables with applications to partial differential equations, Numer. Math. **11**, 232–256.

BLAIR, J.J. (1976): Higher order approximations to the boundary conditions for the finite element method, Math. Comput. **30**, 250–262.

BOGNER, F.K.; FOX, R.L.; SCHMIT, L.A. (1965): The generation of interelement compatible stiffness and mass matrices by the use of interpolation formulas, in *Proceedings of the Conference on Matrix Methods in Structural Mechanics*, Wright Patterson A.F.B., Ohio.

BOISSERIE, J.M.; PLANCHARD, J. (1971): Le problème de l'assemblage dans la méthode des éléments finis, Bull. Direction Etudes Recherches, Sér. C: Math.-Informat., 33–42.

BOSSAVIT, A. (1971): Une méthode de décomposition de l'opérateur biharmonique, Note HI 585/2, Electricité de France.

BOSSAVIT, A. (1973): Sur l'assemblage des éléments finis par la méthode frontale, Bull. Direction Etudes Recherches, Sér. C: Math.-Informat. **2**, 47–60.

BOSSAVIT, A.; FREMOND, M. (1976): The frontal method based on mechanics and dynamic programming, Comput. Methods Appl. Mech. Engrg. **8**, 153–178.

BOURGAT, J.F. (1976): Numerical study of a dual iterative method for solving a finite element approximation of the biharmonic equation, Comput. Methods Appl. Mech. **9**, 203–218.

BRAMBLE, J.H. (1970): *Variational Methods for the Numerical Solution of Elliptic Problems*, Chalmers Institute of Technology and the University of Göteborg.

BRAMBLE, J.H. (1972): On the approximation of eigenvalues of non-selfadjoint operators, in *Proceedings Equa Diff 3*, J.E. Purkyně University, Brno.

BRAMBLE, J.H. (1975): A survey of some finite element methods proposed for treating the Dirichlet problem, Advances in Math. **16**, 187–196.

BRAMBLE, J.H.; DUPONT, T.; THOMÉE, V. (1972): Projection methods for Dirichlet's problem in approximating polygonal domains with boundary-value corrections, Math. Comput. **26**, 869–879.

BRAMBLE, J.H.; HILBERT, S.R. (1970): Estimation of linear functionals on Sobolev spaces with application to Fourier transforms and spline interpolation, SIAM J. Numer. Anal. **7**, 113–124.

BRAMBLE, J.H.; HILBERT, S.R. (1971): Bounds for a class of linear functionals with applications to Hermite interpolation, Numer. Math. **16**, 362–369.

BRAMBLE, J.H.; NITSCHE, J.A. (1973): A generalized Ritz-least-squares method for Dirichlet problems, SIAM J. Numer. Anal. **10**, 81–93.

BRAMBLE, J.H.; NITSCHE, J.A.; SCHATZ, A.H. (1975): Maximum-norm interior estimates for Ritz–Galerkin methods, Math. Comput. **29**, 677–688.

BRAMBLE, J.H.; OSBORN, J.E. (1972): Approximation of Steklov eigenvalues of non-selfadjoint second order elliptic operators, in *The Mathematical Foundations of the Finite Element Method with Applications to Partial Differential Equations* (A.K. Aziz, Editor), pp. 387–408, Academic Press, New York.

BRAMBLE, J.H.; OSBORN, J.E. (1973): Rate of convergence estimates for nonselfadjoint eigenvalue approximations. Math. Comput. **27**, 525–549.

BRAMBLE, J.H.; SCHATZ, A.H. (1970): Rayleigh–Ritz–Galerkin methods for Dirichlet's problem using subspaces without boundary conditions, Comm. Pure Appl. Math. **23**, 653–675.

BRAMBLE, J.H.; SCHATZ, A.H. (1971): Least squares methods for 2mth order elliptic boundary-value problems, Math. Comput. **25**, 1–32.

BRAMBLE, J.H.; SCHATZ, A.H. (1974): Higher order local accuracy by averaging in the finite element method, in *Mathematical Aspects of Finite Elements in Partial Differential Equations* (C. de Boor, Editor), pp. 1–14, Academic Press, New York.

BRAMBLE, J.H.; SCHATZ, A.H. (1976): Estimates for spline projections, Rev. Française Automat. Informat. Recherche Opérationnelle Sér. Rouge Anal. Numér. **10**, 5–37.

BRAMBLE, J.H.; THOMÉE, V. (1974): Interior maximum norm estimates for some simple finite element methods, Rev. Française Automat. Informat. Recherche Opérationnelle Sér. Rouge Anal. Numér. **R-2**, 5–18.

BRAMBLE, J.H.; ZLÁMAL, M. (1970): Triangular elements in the finite element method, Math. Comput. **24**, 809–820.

BRAUCHLI, J.H.; ODEN, J.T. (1971): Conjugate approximation functions in finite element analysis, Quart. Appl. Math. **29**, 65–90.

BREZIS, H.; SIBONY, M. (1968): Méthodes d'approximation et d'itération pour les opérateurs monotones, Arch. Rational Mech. Anal. **1**, 59–82.

BREZIS, H.; STAMPACCHIA, G. (1968): Sur la régularité de la solution d'inéquations elliptiques, Bull. Soc. Math. France **96**, 153–180.

BREZIS, H.; STAMPACCHIA, G. (1973): Une nouvelle méthode pour l'étude d'écoulements stationnaires, C.R. Acad. Sci. Paris, Sér. A, **276**, 129–132.

BREZZI, F. (1974a): Sur une méthode hybride pour l'approximation du problème de la torsion d'une barre élastique, Ist. Lombardo Accad. Sci. Lett. Rend. A **108**, 274–300.

BREZZI, F. (1974b): On the existence, uniqueness and approximation of saddle-point problems arising from Lagrangian multipliers, Rev. Française Automat. Informat. Recherche Opérationnelle Sér. Rouge Anal. Numér. **R-2**, 129–151.

BREZZI, F. (1975): Sur la méthode des éléments finis hybrides pour le problème biharmonique, Numer. Math. **24**, 103–131.

BREZZI, F.; HAGER, W.W.; RAVIART, P.-A. (1977): Error estimates for the finite element solution of variational inequalities – Part 1: Primal theory, Numer. Math. **28**, 431–443.

BREZZI, F.; JOHNSON, C.; MERCIER, B. (1977): Analysis of a mixed finite element method for elasto-plastic plates, Math. Comput. (to appear).

BREZZI, F.; MARINI, L.D. (1975): On the numerical solution of plate bending problems by hybrid methods, Rev. Française Automat. Informat. Recherche Opérationnelle Sér. Rouge Anal. Numér. **R-3**, 5–50.

BREZZI, F.; RAVIART, P.-A. (1976): Mixed finite element methods for 4th order elliptic equations, Rapport interne No. 9, Centre de Mathématiques Appliquées, Ecole Polytechnique, Palaiseau.

BRISTEAU, M.-O. (1975): *Application de la Méthode des Eléments Finis à la Résolution Numérique d'Inéquations Variationnelles d'Evolution de Type Bingham*, Doctoral Thesis (3ème Cycle), Université Pierre et Marie Curie, Paris.

BROWDER, F.E. (1965): Existence and uniqueness theorems for solutions of nonlinear boundary value problems, Proc. Amer. Math. Soc., Symposia in Appl. Math. **17**, 24–49.

CAREY, G.F. (1974): A unified approach to three finite element theories for geometric nonlinearity, Comput. Methods Appl. Mech. Engrg. **4**, 69–79.

CARLSON, R.E.; HALL, C.A. (1971): Ritz approximations to two-dimensional boundary value problems, Numer. Math. **18**, 171–181.

CARLSON, R.E.; HALL, C.A. (1973): Error bounds for bicubic spline interpolation, J. Approximation Theory **4**, 41–47.

CARROLL, W.E.; BARKER, R.M. (1973): A theorem for optimum finite-element idealizations, Internat. J. Solids and Structures **9**, 883–895.

CARTAN, H. (1967): *Calcul Différentiel*, Hermann, Paris.

CAVENDISH, J.C.; GORDON, W.J.; HALL, C.A. (1976): Ritz–Galerkin approximations in blending function spaces, Numer. Math. **26**, 155–178.

CEA, J. (1964): Approximation variationnelle des problèmes aux limites, Ann. Inst. Fourier (Grenoble) **14**, 345–444.

CEA, J. (1971): *Optimisation: Théorie et Algorithmes*, Dunod, Paris.

CEA, J. (1976): Approximation variationnelle; Convergence des éléments finis; Un test, in *Journées Eléments Finis*, Université de Rennes, Rennes.

CEA, J.; GIOAN, A.; MICHEL, J. (1974): Adaptation de la méthode du gradient à un problème d'identification de domaine, in *Computing Methods in Applied Sciences and Engineering, Part 2* (R. Glowinski and J.L. Lions, Editors), pp. 391–402, Springer-Verlag, Berlin.

CHATELIN, F.; LEMORDANT, M.J. (1975): La méthode de Rayleigh–Ritz appliquée à des opérateurs différentiels elliptiques – Ordres de convergence des éléments propres, Numer. Math. **23**, 215–222.

CHERNUKA, M.W.; COWPER, G.R.; LINDBERG, G.M.; OLSON, M.D. (1972): Finite element analysis of plates with curved edges, Internat. J. Numer. Methods Engrg. **4**, 49–65.

CIARLET, P.G. (1966): *Variational Methods for Non-Linear Boundary-Value Problems*, Doctoral Thesis, Case Institute of Technology, Cleveland.

CIARLET, P.G. (1968): An $O(h^2)$ method for a non-smooth boundary value problem, Aequationes Math. **2**, 39–49.

CIARLET, P.G. (1970): Discrete variational Green's function. I, Aequationes Math. **4**, 74–82.

CIARLET, P.G. (1973): Orders of convergence in finite element methods, in *The Mathematics of Finite Elements and Applications* (J.R. Whiteman, Editor), pp. 113–129, Academic Press, London.

CIARLET, P.G. (1974a): Conforming and nonconforming finite element methods for solving the plate problem, in *Conference on the Numerical Solution of Differential Equations* (G.A. Watson, Editor), pp. 21–31, Lecture Notes in Mathematics, Vol. 363, Springer-Verlag, Berlin.

CIARLET, P.G. (1974b): Quelques méthodes d'éléments finis pour le problème d'une plaque encastrée, in *Computing Methods in Applied Sciences and Engineering, Part I* (R. Glowinski and J.L. Lions, Editors), pp. 156–176, Lecture Notes in Computer Science, Vol. 10, Springer-Verlag, Berlin.

CIARLET, P.G. (1974c): Sur l'élément de Clough et Tocher, Rev. Française Automat. Informat. Recherche Opérationnelle Sér. Rouge Anal. Numér. **R-2**, 19–27.

CIARLET, P.G. (1975): *Lectures on the Finite Element Method*, Tata Institute of Fundamental Research, Bombay.

CIARLET, P.G. (1976a): *Numerical Analysis of the Finite Element Method*, Séminaire de Mathématiques Supérieures, Presses de l'Université de Montréal.

CIARLET, P.G. (1976b): Conforming finite element methods for the shell problems, in The Mathematics of Finite Elements and Applications II (J.R. Whiteman, Editor), pp. 105–123, Academic Press, London.

CIARLET, P.G. (1976c): On questions of existence in shell theory, J. Indian Math. Soc. **40**, 131–143.

CIARLET, P.G.; GLOWINSKI, R. (1975): Dual iterative techniques for solving a finite element approximation of the biharmonic equation, Comput. Methods Appl. Mech. Engrg. **5**, 277–295.

CIARLET, P.G.; NATTERER, F.; VARGA, R.S. (1970): Numerical methods of high-order accuracy for singular nonlinear boundary value problems, Numer. Math. **15**, 87–99.

CIARLET, P.G.; RAVIART, P.-A. (1972a): General Lagrange and Hermite interpolation in R^n with applications to finite element methods, Arch. Rational Mech. Anal. **46**, 177–199.

CIARLET, P.G.; RAVIART, P.-A. (1972b): Interpolation theory over curved elements, with applications to finite element methods, Comput. Methods Appl. Mech. Engrg. **1**, 217–249.

CIARLET, P.G.; RAVIART, P.-A. (1972c): The combined effect of curved boundaries and numerical integration in isoparametric finite element methods, in *The Mathematical Foundations of the Finite Element Method with Applications to Partial Differential Equations* (A.K. Aziz, Editor), pp. 409–474, Academic Press, New York.

CIARLET, P.G.; RAVIART, P.-A. (1973): Maximum principle and uniform convergence for the finite element method, Comput. Methods Appl. Mech. Engrg. **2**, 17–31.

CIARLET, P.G.; RAVIART, P.-A. (1974): A mixed finite element method for the biharmonic equation, in *Mathematical Aspects of Finite Elements in Partial Differential Equations* (C. de Boor, Editor), pp. 125–145, Academic Press, New York.

CIARLET, P.G.; RAVIART, P.-A. (1975): L'effet de l'intégration numérique dans les méthodes d'éléments finis (to appear).

CIARLET, P.G.; SCHULTZ, M.H.; VARGA, R.S. (1967): Numerical methods of high-order accuracy for nonlinear boundary value problems. III. Eigenvalue problems, Numer. Math. **9**, 394–430.

CIARLET, P.G.; SCHULTZ, M.H.; VARGA, R.S. (1968a): Numerical methods of high-order accuracy for nonlinear boundary value problems. II. Eigenvalue problems, Numer. Math. **11**, 331–345.

CIARLET, P.G.; SCHULTZ, M.H.; VARGA, R.S. (1968b): Numerical methods of high-order accuracy for nonlinear boundary value problems. III. Eigenvalue problems, Numer. Math. **12**, 120–133.

CIARLET, P.G.; SCHULTZ, M.H.: VARGA, R.S. (1968c): Numerical methods of high-order accuracy for nonlinear boundary value problems. IV. Periodic boundary conditions, Numer. Math. **12**, 266–279.

CIARLET, P.G.; SCHULTZ, M.H.; VARGA, R.S. (1969): Numerical methods of high-order accuracy for nonlinear boundary value problems. V. Monotone operator theory, Numer. Math. **13**, 51–77.

CIARLET, P.G.; VARGA, R.S. (1970): Discrete variational Green's function. II. One dimensional problem, Numer. Math. **16**, 115–128.

CIARLET, P.G.; WAGSCHAL, C. (1971): Multipoint Taylor formulas and applications to the finite element method, Numer. Math. **17**, 84–100.

CIAVALDINI, J.F.; NEDELEC, J.C. (1974): Sur l'élément de Fraeijs de Veubeke et Sander, Rev. Française Automat. Informat. Recherche Opérationnelle Sér. Rouge Anal. Numér. **R-2**, 29–45.

CIAVALDINI, J.F.; TOURNEMINE, G. (1977): A finite element method to compute stationary steady state flows in the hodograph plane, J. Indian Math. Soc. (to appear).

CLEMENT, P. (1974): Méthode des éléments finis appliquée à des problèmes variationnels de type indéfini, Doctoral Thesis, Ecole Polytechnique Fédérale de Lausanne, Lausanne.

CLEMENT, P. (1975): Approximation by finite element functions using local regularization, Rev. Française Automat. Informat. Recherche Opérationnelle Sér. Rouge Anal. Numér. **R-2**, 77–84.

CLEMENT, P.; DESCLOUX, J. (1972): On the rigid displacement condition, Internat. J. Numer. Methods Engrg. **4**, 583–586.

CLOUGH, R.W. (1960): The finite element method in plane stress analysis, in Proceedings of the Second ASCE Conference on Electronic Computation, Pittsburg, Pennsylvania.

CLOUGH, R.W.; TOCHER, J.L. (1965): Finite element stiffness matrices for analysis of plates in bending, in *Proceedings of the Conference on Matrix Methods in Structural Mechanics*, Wright Patterson A.F.B. Ohio.

COATMELEC, C. (1966): Approximation et interpolation des fonctions différentiables de plusieurs variables, Ann. Sci. Ecole Norm. Sup. **83**, 271–341.

COURANT, R. (1943): Variational methods for the solution of problems of equilibrium and vibrations, Bull. Amer. Math. Soc. **49**, 1–23.

COURANT, R.; HILBERT, D. (1953): *Methods of Mathematical Physics, Vol. I*, Interscience, New York.

COURANT, R.; HILBERT, D. (1962): *Methods of Mathematical Physics, Vol. II*, Interscience, New York.

COUTRIS, N. (1973): Flexions élastique et élastoplastique d'une coque mince, J. Mécanique **12**, 463–475.

CROUZEIX, M.; LE ROUX, A.Y. (1976): Ecoulement d'une fluide irrotationnel, in *Journées Eléments Finis*, Université de Rennes, Rennes.

CROUZEIX, M.; RAVIART, P.-A. (1973): Conforming and nonconforming finite element methods for solving the stationary Stokes equations I, Rev. Française Automat. Informat. Recherche Opérationnelle Sér. Rouge Anal. Numér. **R-3**, 33–76.

CROUZEIX, M.; THOMAS, J.M. (1973): Eléments finis et problèmes elliptiques dégénérés, Rev. Française Automat. Informat. Recherche Opérationnelle Sér. Rouge Anal. Numér. **R-3**, 77–104.

DAILEY, J.W.; PIERCE, J.G. (1972): Error bounds for the Galerkin method applied to singular and nonsingular boundary value problems, Numer. Math. **19**, 266–282.

DAVIS, P.J.; RABINOWITZ, P. (1974): *Methods of Numerical Integration*, Academic Press, New York.

DAWE, D.J. (1972): Shell analysis using a simple facet element, Journal of Strain Analysis **7**, 266–270.

DELEZE, M.; GOEL, J.-J. (1976): Tétraèdre comme élément fini de Classe C^1, à seize paramètres, contenant les polynômes de degré deux, Report, Institut de Mathématiques, Université de Fribourg, Fribourg.

DENY, J.; LIONS, J.L. (1953–1954): Les espaces du type de Beppo Levi, Ann. Institut Fourier (Grenoble) **V**, 305–370.

DESCLOUX, J. (1972a): On finite element matrices, SIAM J. Numer. Anal. **9**, 260–265.

DESCLOUX, J. (1972b): Finite elements and numerical stability, in *Proceedings EquaDiff 3*, pp. 21–29, J.E. Purkyně University, Brno.

DESCLOUX, J. (1973): Two basic properties of finite elements, Report, Ecole Polytechnique Fédérale de Lausanne.

DESCLOUX, J. (1975): Interior regularity and local convergence of Galerkin finite element approximations for elliptic equations, in *Topics in Numerical Analysis II* (J.J.H. Miller, Editor), pp. 27–41, Academic Press, New York.

DESCLOUX, J.; NASSIF, N. (1977): Interior L^∞ estimates for finite element approximations of solutions of elliptic equations, Rev. Française Automat. Informat. Recherche Opérationnelle, Sér. Rouge Anal. Numér. (to appear).

DIEUDONNE, J. (1967): *Fondements de l'Analyse Moderne*, Paris, Gauthier-Villars.

DOUGLAS, J., Jr.; DUPONT, T. (1973): Superconvergence for Galerkin Methods for the two point boundary problem via local projections, Numer. Math. **21** (1973), 270–278.

DOUGLAS, J., Jr.; DUPONT, T. (1975): A Galerkin method for a nonlinear Dirichlet problem, Math. Comput. **29**, 689–696.

DOUGLAS, J., Jr.; DUPONT, T. (1976a): Interior penalty procedures for elliptic and parabolic Galerkin methods in *Computing Methods in Applied Sciences* (R. Glowinski and J.L. Lions, Editors), pp. 207–216, Lecture Notes in Physics, Vol. 58, Springer-Verlag, Berlin.

DOUGLAS, J., Jr.; DUPONT, T. (1976b): Galerkin approximations for the two point boundary problem using continuous, piecewise polynomial spaces (to appear).

DOUGLAS, J., Jr.; DUPONT, T.; PERCELL, P.; SCOTT, R. (1976): A family of C^1 finite elements with optimal approximation properties for various Galerkin methods for 2nd and 4th order problems (to appear).

DOUGLAS, J., Jr.; DUPONT, T.; WAHLBIN, L. (1975a): The stability in L^q of the L^2-projection into finite element function spaces, Numer. Math. **23**, 193–197.

DOUGLAS, J., Jr.; DUPONT, T.; WAHLBIN, L. (1975b): Optimal L_∞ error estimates for Galerkin approximations to solutions of two point boundary value problems, Math. Comput. **29**, 475–483.

DOUGLAS, J., Jr.; DUPONT, T.; WHEELER, M.F. (1974a): A Galerkin procedure for approximating the flux on the boundary for elliptic and parabolic boundary value problems, Rev. Française Automat. Informat. Recherche Opérationnelle Sér. Rouge Anal. Numér. **R-2**, 47–59.

DOUGLAS, J., Jr.; DUPONT, T.; WHEELER, M.F. (1974b): An L^∞ estimate and a superconvergence result for a Galerkin method for elliptic equations based on tensor products of piecewise polynomials, Rev. Française Automat. Informat. Recherche Opérationnelle Sér. Rouge Anal. Numér. **R-2**, 61–66.

DUCHON, J. (1976a): Interpolation des fonctions de deux variables suivant le principe de la flexion des plaques minces, Rev. Française Automat. Informat. Recherche Opérationnelle Sér. Rouge Anal. Numér. **10**, No. 12, 5–12.

DUCHON, J. (1976b): Splines minimizing rotation – invariant semi-norms in Sobolev spaces (to appear in the Proceedings of Mehrdimensionale konstructive Functiontheorie, Oberwolfach, 1976).

DUPUIS, G. (1971): Application of Ritz method to thin elastic shell analysis, J. Appl. Mech., 1–9.

DUPUIS, G.; GOËL, J.-J. (1970a): Finite elements with a high degree of regularity, Internat. J. Numer. Methods Engrg. **2**, 563–577.

DUPUIS, G.; GOËL, J.-J. (1970b): A curved finite element for thin elastic shells, Internat. J. Solids and Structures **6**, 1413–1428.

DUVAUT, G.; LIONS, J.L. (1972): *Les Inéquations en Mécanique et en Physique*, Dunod, Paris.

EHRLICH, L.W. (1971): Solving the biharmonic equation as coupled finite difference equations, SIAM J. Numer. Anal. **8**, 278–287.

EKELAND, I.; TEMAM, R. (1974): *Analyse Convexe et Problèmes Variationnels*, Dunod, Paris.

ERGATOUDIS, I.; IRONS, B.M.; ZIENKIEWICZ, O.C. (1968): Curved, isoparametric, "quadrilateral" elements for finite element analysis, Internat. J. Solids and Structures **4**, 31–42.

FALK, R.S. (1974): Error estimates for the approximation of a class of variational inequalities, Math. Comput. **28**, 963–971.

FALK, R.S. (1975): Approximation of an elliptic boundary value problem with unilateral constraints, Revue Française Automat. Informat. Recherche Opérationnelle Sér. Rouge Anal. Numér. **R-2**, 5–12.

FALK, R.S. (1976a): An analysis of the finite element method using Lagrange multipliers for the stationary Stokes equation, Math. Comput. **30**, 241–249.

FALK, R.S. (1976b): A Ritz method based on a complementary variational principle, Rev. Française Automat. Informat. Recherche Opérationnelle Sér. Rouge Anal. Numér. **10**, 39–48.

FALK, R.S. (1976c): A finite element method for the stationary Stokes equation using trial functions which do *not* have to satisfy div $v = 0$, Math. Comput. **30**, 698–702.

FALK, R.S. (1976d): Approximation of the biharmonic equation by a mixed finite element method, SIAM J. Numer. Anal. (to appear).

FALK, R.S.; KING, J.T. (1976): A penalty and extrapolation method for the stationary Stokes equations, SIAM J. Numer. Anal. **13**, 814, 829.

FALK, R.S.; MERCIER, B. (1977): Error estimates for elasto-plastic problems, Rev. Française Automat. Informat. Recherche Opérationnelle Sér. Rouge Anal. Numér. **11**, 135–144.

FELIPPA, C.A. (1966): *Refined Finite Element Analysis of Linear and Nonlinear Two-Dimensional Structures*, Doctoral Thesis, University of California, Berkeley.

FELIPPA, C.A.; CLOUGH, R.W. (1970): The finite element method in solid mechanics, in *Numerical Solution of Field Problems in Continuum Mechanics* (G. Birkhoff & R.S. Varga, Editors), pp. 210–252, American Mathematical Society, Providence.

FICHERA, G. (1972): Existence theorems in elasticity-Boundary value problems of elasticity with unilateral constraints, in *Encyclopedia of Physics* (S. Flügge, Chief Editor), Vol. VIa/2: *Mechanics of Solids* II (C. Truesdell, Editor), pp. 347–424, Springer-Verlag, Berlin.

FIX, G.J. (1969): Higher-order Rayleigh-Ritz approximations, Journal of Mathematics and Mechanics **18**, 645–658.

FIX, G.J. (1972a): On the effects of quadrature in the finite element method, in *Advances in Computational Methods in Structural Mechanics and Design* (J.T. Oden, R.W. Clough, Y. Yamamoto, Editors), pp. 55–68, The University of Alabama Press, Huntsville.

FIX, G.J. (1972b): Effects of quadrature errors in finite element approximation of steady state, eigenvalue and parabolic problems, in *The Mathematical Foundations of the Finite Element Method with Applications to Partial Differential Equations* (A.K. Aziz, Editor), pp. 525–556, Academic Press, New York.

FIX, G.J. (1973): Eigenvalue approximation by the finite element method, Advances in Math. **10**, 300–316.

FIX, G.J. (1976): Hybrid finite element methods, SIAM Rev. **18**, 460–484.

FIX, G.J.; GULATI, S.; WAKOFF, G.I. (1973): On the use of singular functions with finite element approximations, J. Computational Phys. **13**, 209–228.

FIX, G.J.; LARSEN, K. (1971): On the convergence of SOR iterations for finite element approximations to elliptic boundary value problems, SIAM J. Numer. Anal. **8**, 536–547.

FIX, G.J.; STRANG, G. (1969): Fourier analysis of the finite element method in Ritz-Galerkin theory, Studies in Appl. Math. **48**, 265–273.

FORTIN, M. (1972a): *Calcul Numérique des Ecoulements des Fluides de Bingham et des Fluides Newtoniens Incompressibles par des Méthodes d'Eléments Finis*, Doctoral Thesis, Université de Paris VI.

FORTIN, M. (1972b): Résolution des équations des fluides incompressibles par la méthode des éléments finis, in *Proceedings of the Third International Conference on the Numerical Methods in Fluid Mechanics* (Paris, July 03–07, 1972). Springer-Verlag.

FORTIN, M. (1976): Résolution numérique des équations de Navier–Stokes par des éléments finis de type mixte, in *Journées Eléments Finis*, Université de Rennes, Rennes.

FRAEIJS DE VEUBEKE, B. (1965a): Bending and stretching of plates, in *Proceedings of the*

Conference on Matrix Methods in Structural Mechanics, Wright Patterson A.F.B., Ohio.

FRAEIJS DE VEUBEKE, B. (1965b): Displacement and equilibrium models in the finite element method, in *Stress Analysis* (O.C. Zienkiewicz & G.S. Holister, Editors), pp. 145–197, Wiley, New York.

FRAEIJS DE VEUBEKE, B. (1968): A conforming finite element for plate bending, Internat. J. Solids and Structures **4**, 95–108.

FRAEIJS DE VEUBEKE, B. (1973): Diffusive equilibrium models, Lecture Notes, University of Calgary, Calgary.

FRAEIJS DE VEUBEKE, B. (1974): Variational principles and the patch test, Internat. J. Numer. Methods Engrg. **8**, 783–801.

FREHSE, J. (1976): Eine gleichmäßige asymptotische Fehlerabschätzung zur Methode der finiten Elemente bei quasilinearen elliptischen Randwertproblemen, in *Theory of Nonlinear Operators. Constructive Aspects*, Tagungsband der Akademie der Wissenschaften, Berlin.

FREHSE, J.; RANNACHER, R. (1976): Optimal uniform convergence for the finite element approximation of a quasilinear elliptic boundary value problem (to appear in Proceedings of the U.S.-Germany Symposium "Formulations and Computational Algorithms in Finite Element Analysis", M.I.T., 1976).

FREHSE, J.; RANNACHER, R. (1977): Asymptotic L^∞-error estimates for linear finite element approximations of quasilinear boundary value problems, SIAM J. Numer. Anal. (to appear).

FREMOND, M. (1971a): Etude de Structures Visco-Elastiques Stratifiées Soumises à des Charges Harmoniques, et de Solides Elastiques reposant sur ces Structures, Doctoral Thesis, Université Pierre et Marie Curie (Paris VI), Paris.

FREMOND, M. (1971b): Formulations duales des énergies potentielles et complémentaires. Application à la méthode des éléments finis, C.R. Acad. Sci. Paris, Sér. A, **273**, 775–777.

FREMOND, M. (1972): Utilisation de la dualité en élasticité. Compléments sur les énergies de Reissner. Equilibre d'une dalle élastique reposant sur une structure stratifiée, Annales de l'Institut Technique du Bâtiment et des Travaux Publics, Supplément au No. 294, 54–66.

FREMOND, M. (1973): Dual formulations for potential and complementary energies. Unilateral boundary conditions. Applications to the finite element method, in *The Mathematics of Finite Elements and Applications* (J.R. Whiteman, Editor), pp. 175–188, Academic Press, London.

FREMOND, M. (1974): La méthode frontale pour la résolution des systèmes linéaires, in *International Computing Symposium 1973* (A. Günther *et al.*, Editors), pp. 337–343, North-Holland, Amsterdam.

FRIED, I. (1971a): Discretization and round-off errors in the finite element analysis of elliptic boundary value problems, Doctoral Thesis, Massachusetts Institute of Technology, Cambridge.

FRIED, I. (1971b): Basic computational problems in the finite element analysis of shells, Internat. J. Solids and Structures **7**, 1705–1715.

FRIED, I. (1973a): Bounds on the spectral and maximum norms of the finite element stiffness, flexibility and mass matrices, Internat. J. Solids and Structures **9** (1973), 1013–1034.

FRIED, I. (1973b): The l_2 and l_∞ condition numbers of the finite element stiffness and mass matrices, and the pointwise convergence of the method, in *The Mathematics of Finite Elements and Applications* (J.R. Whiteman, Editor), pp. 163–174, Academic Press, London.

FRIED, I. (1973c): Shear in C^0 and C^1 bending finite elements, Internat. J. Solids and Structures **9**, 449–460.

FRIED, I.; YANG, S.K. (1972): Best finite elements distribution around a singularity, AIAA J. **10**, 1244–1246.

FRIED, J.; YANG, S.K. (1973): Triangular, nine-degrees-of-freedom, C^0 plate bending element of quadratic accuracy, Quart. Appl. Math. XX, 303–312.

FRIEDRICHS, K.O. (1962): A finite-difference scheme for the Neumann and the Dirichlet problem, A.E.C. Research and Development Report, Institute of Mathematical Sciences, New York University, New York.

FUTAGAMI, T. (1976): *Several Mathematical Methods in Water Pollution Control: The Finite Element & Linear Programming Method*, Doctoral Thesis, Kyoto University, Kyoto.

GABAY, D.; MERCIER, B. (1976): A dual algorithm for the solution of nonlinear variational problems via finite element approximation, Comput. Math. Appl. (to appear).

GALLAGHER, R.H. (1973): The finite element method in shell stability analysis, Computers & Structures **3**, 543–557.

GARTLING, D.K.; BECKER, E.B. (1976): Finite element analysis of viscous, incompressible fluid flow. Part 1: Basic methodology, Comput. Methods Appl. Mech. Engrg. **8**, 51–60.

GELLERT, M.; LAURSEN, M.E. (1976): Formulation and convergence of a mixed finite element method applied to elastic arches of arbitrary geometry and loading, Comput. Methods Appl. Mech. Engrg. **7**, 285–302.

GIRAULT, V. (1976a): Nonelliptic approximation of a class of partial differential equations with Neumann boundary conditions, Math. Comput. **30**, 68–91.

GIRAULT, V. (1976b): A combined finite element and Markes and Cell method for solving Navier–Stokes equations, Numer. Math. **26**, 39–59.

GIRAULT, V. (1976c): A mixed finite element method for the stationary Stokes equations (to appear).

GLOWINSKI, R. (1973): Approximations externes, par éléments finis de Lagrange d'ordre un et deux, du problème de Dirichlet pour l'opérateur biharmonique. Méthodes itératives de résolution des problèmes approchés, in *Topics in Numerical Analysis* (J.J.H. Miller, Editor), pp. 123–171, Academic Press, London.

GLOWINSKI, R. (1975): Analyse numérique d'inéquations variationnelles d'ordre 4, Rapport No. 75002, Laboratoire d'Analyse Numérique, Université Pierre et Marie Curie, Paris.

GLOWINSKI, R. (1976a): Sur l'approximation d'une inéquation variationnelle elliptique de type Bingham, Rev. Française Automat. Informat. Recherche Opérationnelle Sér. Rouge Anal. Numér. **10**, No. 12, 13–30.

GLOWINSKI, R. (1976b): *Numerical Analysis of some Nonlinear Elliptic Boundary Value Problems*, Presses de l'Université de Montréal (to appear).

GLOWINSKI, R.; LIONS, J.-L.; TREMOLIERES, R. (1976a): *Analyse Numérique des Inéquations Variationnelles*, Vol. 1: *Théorie Générale, Premières Applications*, Dunod, Paris.

GLOWINSKI, R.; LIONS, J.-L.; TREMOLIERES, R. (1976b): *Analyse Numérique des In-*

équations Variationnelles, Vol. 2: *Applications aux Phénomènes Stationnaires et d'Evolution*, Dunod, Paris.

GLOWINSKI, R.; MARROCCO, A. (1974): Analyse numérique du champ magnétique d'un alternateur par éléments finis et sur-relaxation ponctuelle non linéaire, Comput. Methods Appl. Mech. Engrg. **3**, 55–85.

GLOWINSKI, R.; MARROCCO, A. (1975): Sur l'approximation par éléments finis d'ordre un, et la résolution par pénalisation-dualité, d'une classe de problèmes de Dirichlet non linéaires, Rev. Française Automat. Informat. Recherche Opérationnelle Sér. Rouge Anal. Numér. **R-2**, 41–76.

GLOWINSKI, R., PIRONNEAU, O. (1976a): Sur la résolution numérique du problème de Dirichlet pour l'opérateur biharmonique par une méthode "quasi-directe", C.R. Acad. Sci. Paris, Sér. A, **282**, 223–226.

GLOWINSKI, R.; PIRONNEAU, O. (1976b): Sur la résolution numérique du problème de Dirichlet pour l'opérateur biharmonique par la méthode du gradient conjugué. Applications, C.R. Acad. Sci. Paris, Sér. A, **282**, 1315–1318.

GLOWINSKI, R.; PIRONNEAU, O. (1976c): Sur la résolution par une méthode "quasi-directe", et par diverses méthodes itératives, d'une approximation par éléments finis mixtes du problème de Dirichlet pour l'opérateur biharmonique, Rapport No. 76010, Laboratoire d'Analyse Numérique, Université Pierre et Marie Curie, Paris.

GLOWINSKI, R.; PIRONNEAU, O. (1976d): Sur la résolution, via une approximation par éléments finis mixtes, du problème de Dirichlet pour l'opérateur biharmonique, par une méthode "quasi-directe" et diverses méthodes itératives, Rapport LABORIA No. 197, IRIA-Laboria, Le Chesnay.

GOËL, J.-J. (1968): *Utilisation Numérique de la Méthode de Ritz, Application au Calcul de Plaque*, Doctoral Thesis, Ecole Polytechnique de l'Université de Lausanne, Lausanne.

GOËL, J.-J. (1968b): Construction of basic functions for numerical utilisation of Ritz's method, Numer. Math. **12**, 435–447.

GOPALACHARYULU, S. (1973): A higher order conforming, rectangular plate element, Internat. J. Numer. Methods Engrg. **6**, 305–309.

GOPALACHARYULU, S. (1976): Author's reply to the discussion by Watkins, Internat. J. Numer. Methods Engrg. **10**, 472–474.

GORDEZIANI, D.G. (1974): On the solvability of some boundary value problems for a variant of the theory of thin shells, Dokl. Akad. Nauk SSSR **215**, No. 6, 677–680.

GORDON, W.J.; HALL, C.A. (1973): Transfinite element methods: blending-function interpolation over arbitrary curved element domains, Numer. Math. **21**, 109–129.

GOUT, J.L. (1976): Estimation de l'erreur d'interpolation d'Hermite dans R^n (to appear).

GOUYON, R. (1963): *Calcul Tensoriel*, Vuibert, Paris.

GREGOIRE, J.P.; NEDELEC, J.C.; PLANCHARD, J. (1976): A method of finding the eigenvalues and eigenfunctions of self-adjoint elliptic operators, Comput. Methods Appl. Mech. Engrg. **8**, 201–214.

GRISVARD, P. (1976): Behavior of the solutions of an elliptic boundary value problem in a polygonal or polyhedral domain, in *Numerical Solution of Partial Differential Equations*. III (SYNSPADE 1975) (B. Hubbard, Editor), pp. 207–274, Academic Press, New York.

GUGLIELMO, F. Di (1970): Méthode des éléments finis: Une famille d'approximation des espaces de Sobolev par les translatées de p fonctions, Calcolo **7**, 185–234.

GUGLIELMO, F. Di (1971): Résolution approchée de problèmes aux limites elliptiques par des schémas aux éléments finis à plusieurs fonctions arbitraires, Calcolo **8**, 185–213.

HABER, S. (1970): Numerical evaluation of multiple integrals, SIAM Rev. **12**, 481–526.

HARVEY, J.W.; KELSEY, S. (1971): Triangular plate bending elements with enforced compatibility, AIAA J. **9**, 1023–1026.

HASLINGER, J.; HLAVÁČEK, I. (1975): Curved elements in a mixed finite element method close to the equilibrium model, Apl. Mat. **20**, 233–252.

HASLINGER, J.; HLAVÁČEK, I. (1976a): A mixed finite element method close to the equilibrium model, Numer. Math. **26**, 85–97.

HASLINGER, J.; HLAVÁČEK, I. (1976b): A mixed finite element method close to the equilibrium model applied to plane elastostatics, Apl. Mat. **21**, 28–42.

HEDSTROM, G.W.; VARGA, R.S. (1971): Application of Besov spaces to spline approximation, J. Approximation Theory **4**, 295–327.

HELFRICH, H.-P. (1976): Charakterisierung des K-Funktionales zwischen Hilberträumen und nichtuniforme Fehlerschranken, Bonn. Math. Schr. **89**, 31–41.

HENNART, J.P.; MUND, E.H. (1976): Singularities in the finite element approximation of two dimensional diffusion problems, Nuclear Science and Engineering (to appear).

HENSHELL, R.D. (1973): On hybrid finite elements, in *The Mathematics of Finite Elements and Applications* (J.R. Whiteman, Editor), pp. 299–311, Academic Press, London.

HENSHELL, R.D.; SHAW, K.G. (1975): Crack tip finite elements are unnecessary, Internat. J. Numer. Methods Engrg. **9**, 495–507.

HERBOLD, R.J. (1968): *Consistent Quadrature Schemes for the Numerical Solution of Boundary Value Problems by Variational Techniques*, Doctoral Thesis, Case Western Reserve University, Cleveland.

HERBOLD, R.J.; SCHULTZ, M.H.; VARGA, R.S. (1969): The effect of quadrature errors in the numerical solution of boundary value problems by variational techniques, Aequationes Math. **3**, 247–270.

HERBOLD, R.J.; VARGA, R.S. (1972): The effect of quadrature errors in the numerical solution of two-dimensional boundary value problems by variational techniques, Aequationes Math. **7**, 36–58.

HERMANN, L. (1967): Finite element bending analysis for plates, Journal of Mechanics Division, ASCE, **93**, EM5.

HESS, J.L. (1975a): Review of integral-equation techniques for solving potential-flow problems with emphasis on the surface-source method, Comput. Methods Appl. Mech. Engrg. **5**, 145–196.

HESS, J.L. (1975b): Improved solution for potential flow about arbitrary axisymmetric bodies by the use of a higher-order surface source method, Comput. Methods Appl. Mech. Engrg. **5**, 297–308.

HILBERT, S. (1973): A mollifier useful for approximations in Sobolev spaces and some applications to approximating solutions of differential equations, Math. Comput. **27**, 81–89.

HINATA, M.; SHIMASAKI, M.; KIYONO, T. (1974): Numerical solution of Plateau's problem by a finite element method, Math. Comput. **28**, 45–60.

HINTON, E.; CAMPBELL, J.S. (1974): Local and global smoothing of discontinuous finite element functions using a least squares method, Internat. J. Numer. Methods Engrg. **8**, 461–480.

HLAVÁČEK, I.; NEČAS, J. (1970): On inequalities of Korn's type. I. Boundary-value problems for elliptic systems of partial differential equations, Arch. Rational Mech. Anal. **36**, 305–311.

HOPPE, V. (1973): Finite elements with harmonic interpolation functions, in *The Mathematics of Finite Elements and Applications* (J.R. Whiteman, Editor), pp. 131–142, Academic Press, London.

HSIAO, G.C.; WENDLAND, W. (1976): A finite element method for some integral equations of the first kind, J. Math. Anal. Appl. (to appear).

IRONS, B.M. (1974a): Un nouvel élément de coques générales – "semiloof", in *Computing Methods in Applied Sciences and Engineering, Part 1* (R. Glowinski and J.L. Lions, Editors), pp. 177–192, Springer-Verlag, New York.

IRONS, B.M. (1974b): A technique for degenerating brick-type isoparametric elements using hierarchical midside nodes, Internat. J. Numer. Methods Engrg. **8**, 203–209.

IRONS, B.M.; RAZZAQUE, A. (1972a): Experience with the patch test for convergence of finite elements, in *The Mathematical Foundations of the Finite Element Method with Applications to Partial Differential Equations* (A.K. Aziz, Editor), pp. 557–587, Academic Press, New York.

IRONS, B.M.; RAZZAQUE, A. (1972b): Shape function formulations for elements other than displacement models, paper presented at the International Conference on Variational Methods in Engineering, Southampton, 1972.

JAMET, P. (1976a): Estimations d'erreur pour des éléments finis droits presque dégénérés, Rev. Française Automat. Informat. Recherche Opérationnelle Sér. Rouge Anal. Numér. **10**, 43–61.

JAMET, P. (1976b): Estimation de l'erreur d'interpolation dans un domaine variable et application aux éléments finis quadrilatéraux dégénérés, in *Méthodes Numériques en Mathématiques Appliquées*, pp. 55–100, Presses de l'Université de Montréal.

JAMET, P.; RAVIART, P.-A. (1974): Numerical solution of the stationary Navier–Stokes equations by finite element methods, in *Computing Methods in Applied Sciences and Engineering, Part 1* (R. Glowinski and J.-L. Lions, Editors), Lecture Notes in Computer Science, Vol. 10, Springer-Verlag.

JENKINS, H.; SERRIN, J. (1968): The Dirichlet problem for the minimal surface equation in higher dimensions, J. Reine Angew. Math. **229**, 170–187.

JOHN, F. (1965): Estimates for the derivatives of the stresses in a thin shell and interior shell equations, Comm. Pure Appl. Math. **18**, 235–267.

JOHNSON, C. (1972): Convergence of another mixed finite-element method for plate bending problems, Report No. 1972–27, Department of Mathematics, Chalmers Institute of Technology and the University of Göteborg, Göteborg.

JOHNSON, C. (1973): On the convergence of a mixed finite-element method for plate bending problems. Numer. Math. **21**, 43–62.

JOHNSON, C. (1975): On finite element methods for curved shells using flat elements, in *Numerische Behandlung von Differentialgleichungen*, pp. 147–154, International Series of Numerical Mathematics, Vol. 27, Birkhäuser Verlag, Basel and Stuttgart.

JOHNSON, C.; THOMEE, V. (1975): Error estimates for a finite element approximation of a minimal surface, Math. Comput. **29**, 343–349.

JOHNSON, M.W.; Jr.; MCLAY, R.W. (1968): Convergence of the finite element method in the theory of elasticity, J. Appl. Mech., Ser. E, **35**, 274–278.

JONES, E. (1964): A generalization of the direct-stiffness method of structure analysis, AIAA J., **2**, 821–826.

JOURON, C. (1975): Résolution numérique du problème des surfaces minima, Arch. Rational Mech. Anal. **59**, 311–341.

KELLOGG, R.B.; OSBORN, J.E. (1976): A regularity result for the Stokes problem in a convex polygon, J. Functional Analysis **21**, 397–431.

KESAVAN, S.; VANNINATHAN, M. (1977): Sur une méthode des éléments finis mixtes pour l'équation biharmonique, Rev. Française Automat. Informat. Recherche Opérationnelle Sér. Rouge Anal. Numér. **11**, 255–270.

KIKUCHI, F. (1973): Some considerations of the convergence of hybrid stress methods, in *Theory and Practice in Finite Element Structural Analysis*, pp. 25–42, University of Tokyo Press, Tokyo.

KIKUCHI, F. (1975a): On the validity of an approximation available in the finite element shell analysis, Computers & Structures **5**, 1–8.

KIKUCHI, F. (1975b): On the validity of the finite element analysis of circular arches represented by an assemblage of beam elements, Comput. Methods Appl. Mech. Engrg. **5**, 253–276.

KIKUCHI, F. (1975c): Approximation in finite element models, Report No. 531, Institute of Space and Aeronautical Science, University of Tokyo.

KIKUCHI, F. (1975d): Convergence of the ACM finite element scheme for plate bending problems, Publ. Res. Inst. Math. Sci., Kyoto University, **11**, 247–265.

KIKUCHI, F. (1975e): On a finite element scheme based on the discrete Kirchhoff assumption, Numer. Math. **24**, 211–231.

KIKUCHI, F. (1976a): Theory and examples of partial approximation in the finite element method, Internat. J. Numer. Methods Engrg. **10**, 115–122.

KIKUCHI, F. (1976b): An iterative finite element scheme for bifurcation analysis of semi-linear elliptic equations, Report No. 542, Institute of Space and Aeronautical Science, University of Tokyo.

KIKUCHI, F.; ANDO, Y. (1972a): Rectangular finite element for plate bending analysis based on Hellinger–Reissner's variational principle, J. Nuclear Sci. and Tech. **9**, 28–35.

KIKUCHI, F.; ANDO, Y. (1972b): Convergence of simplified hybrid displacement method for plate bending, J. Fac. Engrg. Univ. Tokyo Ser. B **XXXI**, 693–713.

KIKUCHI, F.; ANDO, Y. (1972c): Some finite element solutions for plate bending problems by simplified hybrid displacement method, Nuclear Engineering and Design **23**, 155–178.

KIKUCHI, F.; ANDO, Y. (1972d): A new variational functional for the finite-element method and its application to plate and shell problems, Nuclear Engineering and Design **21**, 95–113.

KIKUCHI, F.; ANDO, Y. (1973a): On the convergence of a mixed finite element scheme for plate bending, Nuclear Engineering and Design **24**, 357–373.

KIKUCHI, F.; ANDO, Y. (1973b): Application of simplified hybrid displacement method to large deflection analysis of elastic-plastic plates and shells, J. Fac. Engrg. Univ. Tokyo Ser. B **XXXII**, 117–135.

KING, J.T. (1974): New error bounds for the penalty method and extrapolation, Numer. Math. **23**, 153–165.

KOITER, W. T. (1966): On the nonlinear theory of thin elastic shells, Proceedings Koninklijke Nederlandse Akademie van Wetenschappen, Series B, **69**, 1–54.

KOITER, W.T. (1970): On the foundations of the linear theory of thin elastic shells. I, II,

Proceedings Koninklijke Nederlandse Akademie van Wetenschappen, Series B, **73**, 169–195.

KOITER, W. T.; SIMMONDS, J. C. (1972): Foundations of shell theory, in *Proceedings of the Thirteenth International Congress of Theoretical and Applied Mechanics*, Moscow, pp. 150–176.

KOLAKOWSKI, H.; DRYJA, M. (1974): A boundary value problem for equations of elastic cylindrical shell, Bull. Acad. Polon. Sci. Sér. Sci. Tech. **XXII**, 37–42.

KONDRAT'EV, V.A. (1967): Boundary value problems for elliptic equations in domains with conical or angular points, Trudy Moskov. Mat. Obšč. **16**, 209–292.

KOUKAL, S. (1973): Piecewise polynomial interpolations in the finite element method, Apl. Mat. **18**, 146–160.

LADYŽENSKAJA, O.A. (1963): *The Mathematical Theory of Viscous Incompressible Flow*, Gordon and Breach, New York.

LADYŽENSKAJA, O.A.; URAL'CEVA, N.N. (1968): *Linear and Quasilinear Elliptic Equations*, Academic Press, New York.

LANCHON, H. (1972): Torsion élastoplastique d'un arbre cylindrique de section simplement ou multiplement connexe, Doctoral Thesis, Université Pierre et Marie Curie (Paris VI).

LANDAU, L.; LIFCHITZ, E. (1967): *Théorie de l'Elasticité*, Mir, Moscou.

LASCAUX, P.; LESAINT, P. (1975): Some nonconforming finite elements for the plate bending problem, Rev. Française Automat. Informat. Recherche Opérationnelle Sér. Rouge Anal. Numér **R-1**, 9–53.

LAURENT, P.J. (1972): *Approximation et Optimisation*, Hermann, Paris.

LAX, P.D.; MILGRAM, A.N. (1954): Parabolic equations, Annals of Mathematics Studies No. 33, 167–190, Princeton University Press, Princeton.

LEAF, G.K.; KAPER, H.G. (1974): L^∞-error bounds for multivariate Lagrange approximation, SIAM J. Numer. Anal. **11**, 363–381.

LEAF, G.K.; KAPER, H.G.; LINDEMAN, A.J. (1976): Interpolation and approximation properties of rational coordinates over quadrilaterals, J. Approximation Theory **16**, 1–15.

LELONG-FERRAND, J. (1963): *Géométrie Différentielle*, Masson, Paris.

LERAY, J.; LIONS, J.L. (1965): Quelques résultats de Visik sur les problèmes elliptiques non linéaires par les méthodes de Minty-Browder, Bull. Soc. Math. France **93**, 97–107.

LE ROUX, M.N. (1974): Equations intégrales pour le problème du potentiel électrique dans le plan, C.R. Acad. Sci. Paris, Sér. A, **278**, 541–544.

LE ROUX, M.N. (1977): Méthode d'éléments finis pour la résolution numérique de problèmes extérieurs en dimension deux, Rev. Française Automat. Informat. Recherche Opérationnelle Sér. Rouge Anal. Numér. **11**, 27–60.

LESAINT, P. (1973): Finite element methods for symmetric hyperbolic equations, Numer. Math. **21**, 244–255.

LESAINT, P. (1975): *Sur la Résolution des Systèmes Hyperboliques du Premier Ordre par des Méthodes d'Eléments Finis*, Doctoral Thesis, Université Pierre et Marie Curie, Paris.

LESAINT, P. (1976): On the convergence of Wilson's nonconforming element for solving the elastic problem, Comput. Methods Appl. Mech. Engrg. **7**, 1–16.

LEWY, H.; STAMPACCHIA, G. (1969): On the regularity of the solution of a variational inequality, Comm. Pure Appl. Math. **22**, 153–188.

LICHNEROWICZ, A. (1967): *Eléments de Calcul Tensoriel*, Armand Colin, Paris (8th edition).

LICHNEWSKY, A. (1974a): Principe du maximum local et solutions généralisées de problèmes du type hypersurfaces minimales, Bull. Soc. Math. France **102**, 417–434.

LICHNEWSKY, A. (1974b): Sur le comportement au bord des solutions généralisées du problème non paramétrique des surfaces minimales, J. Math. Pures Appl. **53**, 397–425.

LINDBERG, G.M.; OLSON, M.D. (1970): Convergence studies of eigenvalue solutions using two finite plate bending elements, Internat. J. Numer. Methods Engrg. **2**, 99–116.

LIONS, J.L. (1962): *Problèmes aux Limites dans les Equations aux Dérivées Partielles*, Presses de l'Université de Montréal, Montréal.

LIONS, J.L. (1968), *Contrôle Optimal des Systèmes Gouvernés par des Equations aux Dérivées Partielles*, Dunod–Gauthier–Villars, Paris.

LIONS, J.L. (1969): *Quelques Méthodes de Résolution des Problèmes aux Limites Non Linéaires*, Dunod, Paris.

LIONS, J.L. (1975a): Sur la théorie du contrôle (address given at the International Congress of Mathematicians, Vancouver, August 21–29, 1974), in *Actes du Congrès International des Mathématiciens*, pp. 139–154, Canadian Mathematical Congress.

LIONS, J.L. (1975b): On free surface problems: Methods of variational and quasi variational inequalities, in *Computational Mechanics* (J.T. Oden, Editor), pp. 129–148, Lecture Notes in Mathematics, Vol. 461, Springer-Verlag, Berlin.

LIONS, J.L. (1976): *Sur Quelques Questions d'Analyse, de Mécanique et de Contrôle Optimal*, Presses de l'Université de Montréal.

LIONS, J.L.; MAGENES, E. (1968): *Problèmes aux Limites non Homogènes et Applications, Vol. 1*, Dunod, Paris.

LIONS, J.L.; STAMPACCHIA, G. (1967): Variational inequalities, Comm. Pure Appl. Math. **20**, 493–519.

LOUIS, A. (1976): *Fehlerabschätzungen für Lösungen quasilinearer elliptischer Differentialgleichungen mittels Finiter Elemente*, Doctoral Thesis, Universität in Mainz.

LUKÁŠ, I.L. (1974): Curved boundary elements - General forms of polynomial mappings, in *Computational Methods in Nonlinear Mechanics* (J.T. Oden *et al.*, Editors), pp. 37–46, The Texas Institute for Computational Mechanics, Austin.

MANSFIELD, L.E. (1971): On the optimal approximation of linear functionals in spaces of bivariate functions, SIAM J. Numer. Anal. **8**, 115–126.

MANSFIELD, L.E. (1972a): Optimal approximation and error bounds in spaces of bivariate functions, J. Approximation Theory **5**, 77–96.

MANSFIELD, L.E. (1972b): On the variational characterization and convergence of bivariate splines, Numer. Math. **20**, 99–114.

MANSFIELD, L.E. (1974): Higher order compatible triangular finite elements, Numer. Math. **22**, 89–97.

MANSFIELD, L.E. (1976a): Mixed finite element methods for elliptic equations, Report No. 76–24, Institute for Computer Applications in Science and Engineering, NASA Langley Research Center, Hampton, Virginia.

MANSFIELD, L.E. (1976b): Approximation of the boundary in the finite element solution of fourth order problems (to appear).

MANSFIELD, L.E. (1976c): Interpolation to boundary data in tetrahedra with applications to compatible finite elements, J. Math. Anal. Appl. **56**, 137–164.

MATSUI, T.; MATSUOKA, O. (1976): A new finite element scheme for instability analysis of thin shells, Internat. J. Numer. Methods Engrg. **10**, 145–170.

MCLAURIN, J.W. (1974): A general coupled equation approach for solving the biharmonic boundary value problem, SIAM J. Numer. Anal. **11**, 14–33.

MCLAY, R.W. (1963): *An Investigation into the Theory of the Displacement Method of Analysis for Linear Elasticity*, Doctoral Thesis, University of Wisconsin, Madison.

MCLEOD, R.; MITCHELL, A.R. (1972): The construction of basis functions for curved elements in the finite element method, J. Inst. Math. Appl. **10**, 382–393.

MCLEOD, R.; MITCHELL, A.R. (1975): The use of parabolic arcs in matching curved boundaries in the finite element method, J. Inst. Math. Appl. **16**, 239–246.

MCNEICE, G.M.; MARCAL, P.V. (1973): Optimization of finite element grids based on minimum potential energy, Transactions of the Amer. Soc. Mech. Engrs., 95 Ser. B, No. 1, 186–190.

MEINGUET, J. (1975): Realistic estimates for generic constants in multivariate pointwise approximation, in *Topics in Numerical Analysis II* (J.J.H. Miller, Editor), pp. 89–107, Academic Press, New York.

MEINGUET, J.; DESCLOUX, J. (1977): An operator-theoretical approach to error estimation, Numer. Math. **27**, 307–326.

MELKES, F. (1970): The finite element method for non-linear problems, Apl. Mat. **15**, 177–189.

MELOSH, R.J. (1963): Basis of derivation of matrices for the direct stiffness method, AIAA J. **1**, 1631–1637.

MERCIER, B. (1974): Numerical solution of the biharmonic problem by mixed finite elements of class C^0, Boll. Un. Mat. Ital. **10**, 133–149.

MERCIER, B. (1975a): Approximation par éléments finis et résolution, par un algorithme de pénalisation-dualité, d'un problème d'élasto-plasticité, C.R. Acad. Sci. Paris, Sér. A, **280**, 287–290.

MERCIER, B. (1975b): Une méthode de résolution du problème des charges limites utilisant les fluides de Bingham, C.R. Acad. Sci. Paris, Sér. A, **281**, 525–527.

MIKHLIN, S.G. (1964): *Variational Methods in Mathematical Physics*, Pergamon, Oxford (original Russian edition: 1957).

MIKHLIN, S.G. (1971): *The Numerical Performance of Variational Methods*, Wolters-Noordhoff, Groningen.

MIRANDA, C. (1970): *Partial Differential Equations of Elliptic Type*, Springer-Verlag, New York.

MITCHELL, A.R. (1976): Basis functions for curved elements in the mathematical theory of finite elements in *The Mathematics of Finite Elements and Applications II* (J.R. Whiteman, Editor), pp. 43–58, Academic Press, London.

MITCHELL, A.R.; MARSHALL, J.A. (1975): Matching of essential boundary conditions in the finite element method, in *Topics in Numerical Analysis II* (J.J.H. Miller, Editor), pp. 109–120, Academic Press, New York.

MITTELMANN, H.D. (1977): On pointwise estimates for a finite element solution of nonlinear boundary value problems, SIAM J. Numer. Anal. **14**, 773–778.

MIYOSHI, T. (1972): Convergence of finite elements solutions represented by a non-conforming basis, Kumamoto J. Sci. (Math.) **9**, 11–20.

MIYOSHI, T. (1973a): A finite element method for the solutions of fourth order partial differential equations, Kumamoto J. Sci. (Math.) **9**, 87–116.

MIYOSHI, T. (1973b): Finite element method of mixed type and its convergence in linear shell problems, Kumamoto J. Sci. (Math), **10**, 35–58.

MIYOSHI, T. (1976a): A mixed finite element method for the solution of the von Karman Equations, Numer. Math. **26**, 255–269.

MIYOSHI, T. (1976b): Lumped mass approximation to the nonlinear bending of elastic plates (to appear).

MIYOSHI, T. (1976c): Some aspects of a mixed finite element method applied to fourth order partial differential equations in *Computing Methods in Applied Sciences* (R. Glowinski and J.L. Lions, Editors), pp. 237–256, Lecture Notes in Physics, Vol. 58, Springer-Verlag, Berlin.

MIYOSHI, T. (1977): Application of a mixed finite element method to a nonlinear problem of elasticity (to appear in Proceedings of the Symposium on the Mathematical Aspects of the Finite Element Methods, Rome, December, 1975).

MOAN, T. (1974): A note on the convergence of finite element approximations for problems formulated in curvilinear coordinate systems, Comput. Methods Appl. Mech. Engrg. **3**, 209–235.

MOCK, M.S. (1975): A global *a posteriori* error estimate for quasilinear elliptic problems, Numer. Math. **24**, 53–61.

MOCK, M.S. (1976): Projection methods with different trial and test spaces, Math. Comput. **30**, 400–416.

MORGAN, J.; SCOTT, R. (1975): A nodal basis for C^1 piecewise polynomials of degree $n \geq 5$, Math. Comput. **29**, 736–740.

MORGAN, J.; SCOTT, R. (1976): The dimension of the space of C^1 piecewise polynomials (to appear).

MORLEY, L.S.D. (1968): The triangular equilibrium element in the solution of plate bending problems, Aero. Quart. **19**, 149–169.

MORREY, Jr., C.B. (1966): *Multiple Integrals in the Calculus of Variations*, Die Grundlehren der math. Wissenschaften, Band 130, Springer-Verlag, New York.

MOSCO, U.; SCARPINI, F. (1975): Complementarity systems and approximations of variational inequalities, Rev. Française Automat. Informat. Recherche Opérationnelle Sér. Rouge Anal. Numér. **R-1**, 5–8.

MOSCO, U.; STRANG, G. (1974): One-sided approximation and variational inequalities, Bull. Amer. Math. Soc. **80** (1974), 308–312.

NAGHDI, P.M. (1963): Foundations of elastic shell theory, in *Progress in Solid Mechanics*, Vol. 4, pp. 1–90, North-Holland, Amsterdam.

NAGHDI, P.M. (1972): *The Theory of Shells and Plates, Handbuch der Physik, Vol. VI* No. a-2, pp. 425–640, Springer-Verlag, Berlin.

NATTERER, F. (1975a): Über die punktweise Konvergenz finiter Elemente, Numer. Math. **25**, 67–77.

NATTERER, F. (1975b): Berechenbare Fehlerschranken für die Methode der Finiten Elemente, pp. 109–121, International Series of Numerical Mathematics, Vol. 28, Birkhäuser Verlag, Basel.

NATTERER, F. (1976): Optimale L_2-Konvergenz finiter Elemente bei Variationsungleichungen, Bonn. Math. Schr. **89**, 1–12.

NATTERER, F. (1977): Uniform convergence of Galerkin's method for splines on highly nonuniform meshes (to appear).

NEČAS, J. (1967): *Les Méthodes Directes en Théorie des Equations Elliptiques*, Masson, Paris.

NEDELEC, J.C. (1976): Curved finite element methods for the solution of singular integral equations on surfaces in R^3, Comput. Methods Appl. Mech. Engrg. **8**, 61–80.

NEDELEC, J.C.; PLANCHARD, J. (1973): Une méthode variationnelle d'éléments finis pour la résolution numérique d'un problème extérieur dans R^3, Rev. Française Automat. Informat. Recherche Opérationnelle Sér. Rouge Anal. Numér. **R-3**, 105–129.

NICOLAIDES, R.A. (1972): On a class of finite elements generated by Lagrange interpolation, SIAM J. Numer. Anal. **9**, 435–445.

NICOLAIDES, R.A. (1973): On a class of finite elements generated by Lagrange interpolation. II, SIAM J. Numer. Anal. **10**, 182–189.

NIELSON, G.M. (1973): Bivariate spline functions and the approximation of linear functionals, Numer. Math. **21**, 138–160.

NITSCHE, J.A. (1968): Ein kriterium für die quasi-optimalitat des Ritzchen Verfahrens, Numer. Math. **11**, 346–348.

NITSCHE, J.A. (1969): Orthogonalreihenentwicklung nach linearen Spline-Funktionen, J. Approximation Theory **2**, 66–78.

NITSCHE, J.A. (1970): Linear Spline-Funktionen und die Methoden von Ritz für elliptische Randwertprobleme, Arch. Rational Mech. Anal. **36**, 348–355.

NITSCHE, J.A. (1971): Über ein Variationsprinzip zur Lösung von Dirichlet-Problemen bei Verwendung von Teilräumen, die keinen Randbedingungen unterworfen sind, Abh. Math. Sem. Univ. Hamburg **36**, 9–15.

NITSCHE, J.A. (1972a): Interior error estimates of projection methods, in *Proceedings EquaDiff 3*, pp. 235–239, J.E. Purkyně University, Brno.

NITSCHE, J.A. (1972b): On Dirichlet problems using subspaces with nearly zero boundary conditions, in *The Mathematical Foundations of the Finite Element Method with Applications to Partial Differential Equations* (A.K. Aziz, Editor), pp. 603–627, Academic Press, New York.

NITSCHE, J.A. (1974): Convergence of nonconforming methods, in *Mathematical Aspects of Finite Elements in Partial Differential Equations* (C. de Boor, Editor), pp. 15–53, Academic Press, New York.

NITSCHE, J.A. (1975): L_∞-convergence of finite element approximation, Second Conference on Finite Elements, Rennes.

NITSCHE, J.A. (1976a): Der Einfluss von Randsingularitäten beim Ritzschen Verfahren, Numer. Math. **25**, 263–278.

NITSCHE, J.A. (1976b): Über L_∞-Abschätzungen von Projektionen auf finite Elemente, Bonn. Math. Schr. **89**, 13–30.

NITSCHE, J.A. (1977): L_∞-convergence of finite element approximations (to appear in Proceedings of the Symposium on the Mathematical Aspects of the Finite Element Methods, Rome, December, 1975).

NITSCHE, J.A. (1976c): On L_∞-convergence of finite element approximations to the solution of a nonlinear boundary value problem (to appear in Proceedings of the Conference on Numerical Analysis, Dublin, August 16–20, 1976).

NITSCHE, J.A.; SCHATZ, A.H. (1974): Interior estimates for Ritz–Galerkin methods, Math. Comput. **28**, 937–958.

NITSCHE, J.C.C. (1975): *Vorlesungen über Minimalflächen*, Die Grundlehren der Mathematischen Wissenschaften in Einzeldarstellungen, Band 199, Springer-Verlag, Berlin.

NOOR, M.A.; WHITEMAN, J.R. (1976): Error bounds for finite element solutions of mildly nonlinear elliptic boundary value problems, Numer. Math. **26**, 107–116.

NOVOZHILOV, V.V. (1970): *Thin Shell Theory*, Wolters-Noordhoff, Groningen.

ODEN, J.T. (1972a): *Finite Elements of Nonlinear Continua*, McGraw-Hill, New York.

ODEN, J.T. (1972b): Generalized conjugate functions for mixed finite element approximations of boundary value problems, in *The Mathematical Foundations of the Finite Element Method with Applications to Partial Differential Equations* (A.K. Aziz, Editor), pp. 629–669, Academic Press, New York.

ODEN, J.T. (1973a): Theory of conjugate projections in finite element analysis, in *Lectures on Finite Element Methods in Continuum Mechanics* (J.T. Oden & E.R.A. Oliveira, Editors), pp. 41–75, The University of Alabama at Huntsville Press.

ODEN, J.T. (1973b): Approximations and numerical analysis of finite deformations of elastic solids, in *Nonlinear Elasticity*, pp. 175–228, Academic Press, New York.

ODEN, J.T. (1973c): Some contributions to the mathematical theory of mixed finite element approximation, in *Theory and Practice in Finite Element Structural Analysis*, pp. 3–23, University of Tokyo Press, Tokyo.

ODEN, J.T. (1975) Mathematical aspects of finite-element approximations in continuum mechanics, in *Mechanics Today*, Vol. II (S. Nemat-Nasser, Editor), pp. 159–250, Pergamon Press, New York.

ODEN, J.T. (1976a): Some new results on the theory of hybrid finite element methods (to appear in Proceedings of the Symposium on the Mathematical Aspects of the Finite Element Methods, Rome, December, 1975).

ODEN, J.T. (1976b): Galerkin approximations of a class of nonlinear boundary-value problems and evolution problems in elasticity, in *Computing Methods in Applied Sciences* (R. Glowinski and J.L. Lions, Editors), pp. 175–192, Lecture Notes in Physics, Vol. 58, Springer-Verlag, Berlin.

ODEN, J.T.; LEE, J.K. (1975): Theory of mixed and hybrid finite-element approximations in linear elasticity (IUTAM/IUM Symposium on Applications of Methods of Functional Analysis to Problems of Mechanics), Lecture Notes in Mathematics, Springer-Verlag, Berlin.

ODEN, J.T.; LEE, J.K. (1977): Dual-mixed-hybrid finite element method for second-order elliptic problems (to appear).

ODEN, J.T.; REDDY, J.N. (1974): On dual-complementary variational principles in mathematical physics, Internat. J. Engrg. Sci. **12**, 1–29.

ODEN, J.T.; REDDY, J.N. (1975): Some observations on properties of certain mixed finite element approximations, Internat. J. Numer. Methods. Engrg. **9**, 933–938.

ODEN, J.T.; REDDY, J.N. (1976a): *An Introduction to the Mathematical Theory of Finite Elements*, Wiley Interscience, New York.

ODEN, J.T.; REDDY, J.N. (1976b): *Variational Methods in Theoretical Mechanics*, Springer-Verlag, Heidelberg.

ODEN, J.T.; REDDY, J.N. (1976c): On mixed finite element approximations, SIAM J. Numer. Anal. **13**, 393–404.

OGANESJAN, L.A.; RUKHOVETS, P.A. (1969): Investigation of the convergence rate of variational-difference schemes for elliptic second order equations in a two-dimensional domain with a smooth boundary, Ž. Vyčisl. Mat. i Mat. Fyz. **9**, 1102–1120.

OLIVEIRA, E.R. DE ARANTES E (1968): Theoretical foundations of the finite element method, Internat. J. Solids and Structures **4**, 929–952.

OLIVEIRA, E.R. DE ARANTES E (1969): Completeness and convergence in the finite element method, in *Proceedings of the Second Conference on Matrix Methods in Structural Mechanics*, pp. 1061–1090, Wright-Patterson AFB, Ohio.

OLIVEIRA, E.R. DE ARANTES E (1976): The patch test and the general convergence criteria of the finite element method, Internat. J. Solids and Structures (to appear).

OSBORN, J.E. (1974): Spectral approximation for compact operators, Technical Report 74–26, Department of Mathematics, University of Maryland, College Park.

OSBORN, J.E. (1976a): Approximation of the eigenvalues of a nonselfadjoint operator arising in the study of the stability of stationary solutions of the Navier–Stokes equations, SIAM J. Numer. Anal. **13**, 185–197.

OSBORN, J.E. (1976b): Regularity of solutions of the Stokes problem in a polygonal domain, in *Numerical Solution of Partial Differential Equations – III* (SYNSPADE 1975) (B. Hubbard, Editor), pp. 393–411, Academic Press, New York.

PAYNE, L.E. (1970): Some isoperimetric inequalities for harmonic functions, SIAM J. Math. Anal. **1**, 354–359.

PELISSIER, M.C. (1975): *Sur Quelques Problèmes Non Linéaires en Glaciologie*, Doctoral Thesis, Université Paris XI, Orsay.

PERCELL, P. (1976): On cubic and quartic Clough–Tocher finite elements, SIAM J. Numer. Anal. **13**, 100–103.

PETROVSKY, I.G. (1954): *Lectures on Partial Differential Equations*, Interscience, New York.

PIAN, T.H.H. (1971): Formulations of finite element methods for solid continua, in *Recent Advances in Matrix Methods* (R.H. Gallagher, Y. Yamada, J.T. Oden, Editors), pp. 49–83, The University of Alabama Press, Huntsville.

PIAN, T.H.H. (1972): Finite element formulation by variational principles with relaxed continuity requirements, in *The Mathematical Foundations of the Finite Element Method* (A.K. Aziz, Editor), pp. 671–687, Academic Press, New York.

PIAN, T.H.H.; TONG, P. (1969a): Basis of finite element methods for solid continua, Internat. J. Numer. Methods Engrg. **1**, 3–28.

PIAN, T.H.H.; TONG, P. (1969b): A variational principle and the convergence of a finite element method based on assumed stress distribution, International J. Solids and Structures **5**, 463–472.

PIERCE, J.G.; VARGA, R.S. (1972a): Higher order convergence results for the Rayleigh–Ritz method applied to eigenvalue problems. 1: Estimates relating Rayleigh–Ritz and Galerkin approximations to eigenfunctions, SIAM J. Numer. Anal. **9**, 137–151.

PIERCE, J.G.; VARGA, R.S. (1972b): Higher order convergence results for the Rayleigh–Ritz method applied to eigenvalue problems: 2. Improved error bounds for eigenfunctions, Numer. Math. **19**, 155–169.

PINI, F. (1974): Approximation by finite element functions using global regularization, Report, Département de Mathématiques, Ecole Polytechnique Fédérale de Lausanne.

POCESKI, A. (1975): A mixed finite element method for bending of plates, Internat. J. Numer. Methods Engrg. **9**, 3–15.

PÓLYA, G. (1952): Sur une interprétation de la méthode des différences finies qui peut fournir des bornes supérieures ou inférieures, C.R. Acad. Sci. Paris **235**, 995–997.

PRAGER, W. (1975): A note on the optimal choice of finite element grids, Comput. Methods Appl. Mech. Engrg. **6**, 363–366.

RABIER, P. (1977): Interpolation harmonique, Rev. Française Automat. Informat. Recherche Opérationnelle Sér. Rouge Anal. Numér. **11**, 159–180.

RACHFORD, H.H., Jr.; WHEELER, M.F. (1974): An H^{-1}-Galerkin procedure for the two-point boundary value problem, in *Mathematical Aspects of Finite Elements in*

Partial Differential Equations (C. de Boor, Editor), pp. 353–382, Academic Press, New York.

RADÓ, T. (1930): The problem of the least area and the problem of Plateau, Math. Z. **32**, 763–796.

RAJAGOPALAN, K. (1976): Comment on: A note on the optimal choice of finite element grids, Comput. Methods Appl. Mech. Engrg. **8**, 361–362.

RANNACHER, R. (1976a): Punktweise Konvergenz der Methode der finiten Elemente beim Plattenproblem, Manuscripta Math. **19**, 401–416.

RANNACHER, R. (1976b): Zur L^∞-Konvergenz linearer finiter Elemente beim Dirichlet problem, Math. Z. **149**, 69–77.

RANNACHER, R. (1977): Some asymptotic error estimates for finite element approximation of minimal surfaces, Rev. Française Automat. Informat. Recherche Opérationnelle Sér. Rouge Anal. Numér. **11**, 181–196.

RAPPAZ, J. (1976): Approximation par la Méthode des Eléments Finis du Spectre d'un Opérateur Non Compact Donné par la Stabilité Magnétohydrodynamique d'un Plasma, Doctoral Thesis, Ecole Polytechnique Fédérale, Lausanne.

RAPPAZ, J. (1977): Approximation of the spectrum of a non-compact operator given by the magnetohydrodynamic stability of a plasma, Numer. Math. **28**, 15–24.

RAVIART, P.-A. (1972): Méthode des Eléments Finis, Lecture Notes (D.E.A. Analyse Numérique), Laboratoire d'Analyse Numérique, Université Pierre et Marie Curie (Paris VI).

RAVIART, P.-A. (1975): Hybrid finite element methods for solving 2nd order elliptic equations, in *Topics in Numerical Analysis. II* (J.J.H. Miller, Editor), pp. 141–155, Academic Press, New York.

RAVIART, P.-A. (1976): On some applications of mixed finite element methods (to appear in Proceedings of the "Colloque Franco-Brésilien sur les Méthodes Numériques de l'Ingénieur", August 1976, Rio de Janeiro).

RAVIART, P.-A.; THOMAS, J.M. (1977a): A mixed finite element method for 2nd order elliptic problems (to appear in Proceedings of the Symposium on the Mathematical Aspects of the Finite Element Methods, Rome, December, 1975).

RAVIART, P.-A.; THOMAS, J.M. (1977b): Primal hybrid finite element methods for 2nd order elliptic equations, Math. Comput. **31**, 391–413.

RAZZAQUE, A. (1973): Program for triangular bending elements with derivatives smoothing, Internat. J. Numer. Methods Engrg. **6**, 333–343.

REDDY, J.N. (1973): *A Mathematical Theory of Complementary-Dual Variational Principles and Mixed Finite-Element Approximations of Linear Boundary-Value Problems in Continuum Mechanics*, Doctoral Thesis (TICOM Report 73-7), The University of Texas at Austin, Austin.

REDDY, J.N. (1976): On mixed-hybrid finite element approximations of the biharmonic equation (to appear in Proceedings of the Second SIAM-SIGNUM 1975 Fall meeting).

REDDY, J.N.; ODEN, J.T. (1973): Convergence of mixed finite-element approximations of a class of linear boundary-value problems, J. Struct. Mech. **2**, 83–108.

RHAM, G. DE (1955): *Variétés Differentiables*, Hermann, Paris.

RIESZ, F.; NAGY, B.SZ. (1952): *Leçons d'Analyse Fonctionnelle*, Budapest, Akadémiai Kiadó.

ROBINSON, J. (1973): Basis for isoparametric stress elements, Comput. Methods Appl. Mech. Engrg. **2**, 43–63.

ROSE, M.E. (1975): Weak-element approximations to elliptic differential equations, Numer. Math. **24**, 185–204.

ROUGEE, P. (1969): *Equilibre des Coques Elastiques Minces Inhomogènes en Théorie non Linéaire*, Doctoral Thesis, Université de Paris.

ROUX, J. (1976): Résolution numérique d'un problème d'écoulement subsonique de fluides compressibles, Rev. Française Automat. Informat. Recherche Opérationnelle Sér. Rouge Anal. Numér. **10**, No. 12, 31–50.

SAMUELSSON, (1973): Mixed finite element methods in theory and application, in *Proceedings of the Finite Element Course* (Tirrenia), Istituto di elaborazione della informazione, Pisa.

SANDER, G. (1964): Bornes supérieures et inférieures dans l'analyse matricielle des plaques en flexion-torsion, Bull. Soc. Roy. Sci. Liège **33**, 456–494.

SARD, A. (1963), *Linear Approximation*, Math. Survey 9, American Mathematical Society, Providence, R.I.

SCARPINI, F.; VIVALDI, M.A. (1977): Error estimates for the approximation of some unilateral problems, Rev. Française Automat. Informat. Recherche Opérationnelle Sér. Rouge Anal. Numér. **11**, 197–208.

SCHATZ, A.H. (1974): An observation concerning Ritz–Galerkin methods with indefinite bilinear forms, Math. Comput. **28**, 959–962.

SCHATZ, A.H.; WAHLBIN, L.B. (1976): Maximum norm error estimates in the finite element method for Poisson equation on plane domains with corners (to appear).

SCHATZ, A.H.; WAHLBIN, L.B. (1977): Interior maximum norm estimates for finite element methods, Math. Comput. **31**, 414–442.

SCHEURER, B. (1977): Existence et approximation de point-selle pour certains problèmes non linéaires (to appear).

SCHOLZ, R. (1976): Approximation von Sattelpunkten mit finiten Elementen, Bonn. Math. Schr. **89**, 53–66.

SCHOLZ, R. (1977): L_∞-convergence of saddle-point approximations for second order problems, Rev. Française Automat. Informat. Recherche Opérationnelle Sér. Rouge Anal. Numér. **11**, 209–216.

SCHULTZ, M.H. (1969a): Error bounds for the Rayleigh–Ritz–Galerkin Method, J. Math. Anal. Appl. **27**, 524–533.

SCHULTZ, M.H. (1969b): L^∞-multivariate approximation theory, SIAM J. Numer. Anal. **6**, 161–183.

SCHULTZ, M.H. (1971): L^2 error bounds for the Rayleigh–Ritz–Galerkin method, SIAM J. Numer. Anal. **8**, 737–748.

SCHULTZ, M.H. (1972): Quadrature–Galerkin approximations to solutions of elliptic differential equations, Proc. Amer. Math. Soc. **33**, 511–515.

SCHULTZ, M.H. (1973): Error bounds for a bivariate interpolation scheme, J. Approximation Theory **8**, 189–194.

SCHWARTZ, L. (1966): *Théorie des Distributions*, Hermann, Paris.

SCHWARTZ, L. (1967): *Cours d'Analyse*, Hermann, Paris.

SCOTT, R. (1973a): *Finite Element Techniques for Curved Boundaries*, Doctoral Thesis, Massachusetts Institute of Technology, Cambridge.

SCOTT, R. (1973b): Finite element convergence for singular data, Numer. Math. **21**, 317–327.

SCOTT, R. (1974): C^1 continuity via constraints for 4th order problems, in *Mathematical Aspects of Finite Elements in Partial Differential Equations* (C. de Boor, Editor), pp. 171–193, Academic Press, New York.

SCOTT, R. (1975): Interpolated boundary conditions in the finite element method, SIAM J. Numer. Anal. **12**, 404–427.

SCOTT, R. (1976a): Optimal L^∞ estimates for the finite element method on irregular meshes, Math. Comput. **30**, 681–697.

SCOTT, R. (1976b): A survey of displacement methods for the plate bending problem (to appear in Proceedings of the U.S.-Germany Symposium on Formulations and Computational Algorithms in Finite Element Analysis, Massachusetts Institute of Technology, Cambridge, August 09–13, 1976).

SHAH, J.M. (1970): Two-dimensional polynomial splines, Numer. Math. **15**, 1–14.

SHILOV, G.E. (1968): *Generalized Functions and Partial Differential Equations*, Gordon and Breach, New York.

SHOIKET, B.A. (1974): On existence theorems in linear shell theory, J. Appl. Math. Mech. **38**, 527–531.

SILVESTER, P.; HSIEH, M.S. (1971): Finite-element solution of 2-dimensional exterior-field problems, Proc. Inst. Elec. Engrs. **118**, 1743–1748.

SMITH, J. (1968): The coupled equation approach to the numerical solution of the biharmonic equation by finite differences. I, SIAM J. Numer. Anal. **5**, 323–339.

SMITH, J. (1973): On the approximate solution of the first boundary value problem for $\nabla^4 u = f$, SIAM J. Numer. Anal. **10**, 967–982.

SMITH, K.T. (1961): Inequalities for formally positive integro-differential forms, Bull. Amer. Math. Soc. **67**, 368–370.

SOBOLEV, S.L. (1950): *Application of Functional Analysis in Mathematical Physics*, Leningrad (English translation: American Mathematical Society, Providence, 1963).

STAKGOLD, I. (1968): *Boundary Value Problems of Mathematical Physics*, Vol. II, The MacMillan Company, New York.

STAMPACCHIA, G. (1964): Formes bilinéaires coercitives sur les ensembles convexes, C.R. Acad. Sci. Paris Sér. A, **258**, 4413–4416.

STEIN, E.; AHMAD, R. (1974): On the stress computation in finite element models based upon displacement approximations, Comput. Methods Appl. Mech. Engrg. **4**, 81–96.

STEPHAN, E.; WEISSGERBER, V. (1976): Zur Approximation von Schalen mit Hybriden Elementen, Preprint Nr. 300, Fachbereich Mathematik, Technische Hochschule Darmstadt, Darmstadt.

STRANG, G. (1971): The finite element method and approximation theory, in *Numerical Solutions of Partial Differential Equations II* (B.E. Hubbard, Editor), pp. 547–583. Academic Press, New York.

STRANG, G. (1972a): Approximation in the finite element method, Numer. Math. **19**, 81–98.

STRANG, G. (1972b): Variational crimes in the finite element method, in *The Mathematical Foundations of the Finite Element Method with Applications to Partial Differential Equations* (A.K. Aziz, Editor), pp. 689–710, Academic Press, New York.

STRANG, G. (1973): Piecewise polynomials and the finite element method, Bull. Amer. Math. Soc. **79**, 1128–1137.

STRANG, G. (1974a): The dimension of piecewise polynomials, and one-sided approximation, in *Conference on the Numerical Solution of Differential Equations* (G.A. Watson, Editor), pp. 144–152, Lecture Notes in Mathematics, Vol. 363, Springer-Verlag, New York, 1974.

STRANG, G. (1974b): The finite element method-Linear and nonlinear applications (address given at the International Congress of Mathematicians, Vancouver, August 21–29, 1974).

STRANG, G.; BERGER, A. (1971): The change in solution due to change in domain, in *Proceedings of the A.M.S. Symposium on Partial Differential Equations* (Berkeley, 1971), Academic Press, New York.

STRANG, G.; FIX, G.J. (1971): A Fourier analysis of the finite element method, in *Proceedings of the CIME Summer School*, Crimonese, Rome.

STRANG, G.; FIX, G.J. (1973): *An Analysis of the Finite Element Method*, Prentice-Hall, Englewood Cliffs.

STRICKLIN, J.A.; HAISLER, W.E.; TISDALE, P.R.; GUNDERSON, R. (1969): A rapidly converging triangular plate element, AIAA J. **7**, 180–181.

STROUD, A.H. (1971): *Approximate Calculation of Multiple Integrals*, Prentice Hall, Englewood Cliffs.

SYNGE, J.L. (1957): *The Hypercircle in Mathematical Physics*, Cambridge University Press.

TAYLOR, C.; HOOD, P. (1973): A numerical solution of the Navier–Stokes equations using the finite element technique, Computers and Fluids **1**, 73–100.

TEMAM, R. (1971): Solutions généralisées de certaines équations du type hypersurfaces minima, Arch. Rational. Mech. Anal. **44**, 121–156.

TEMAM, R. (1973): *On the Theory and Numerical Analysis of the Navier–Stokes Equations*, Lecture Notes #9, Department of Mathematics, University of Maryland, College Park.

TEMAM, R. (1977): *Navier Stokes Equations*, North-Holland, Amsterdam.

TEMAM, R.; THOMASSET, F. (1976): Numerical solution of Navier–Stokes equations by a finite element method (to appear in Proceedings of the Conference on Numerical Methods in Fluid Mechanics, Rapallo 1976).

THATCHER, R.W. (1976): The use of infinite grid refinements at singularities in the solution of Laplace's equation, Numer. Math. **25**, 163–178.

THOMAS, J.M. (1975): Méthode des éléments finis équilibres, in *Journées Eléments Finis 1975*, Université de Rennes.

THOMAS, J.M. (1976): Méthode des éléments finis hybrides duaux pour les problèmes elliptiques du second ordre, Rev. Française Automat. Informat. Recherche Opérationnelle Sér. Rouge Anal. Numér. **10**, No. 12, 51–79.

THOMAS, J.M. (1977): *Sur l'Analyse Numérique des Méthodes d'Eléments Finis Hybrides et Mixtes*, Doctoral Thesis, Université Pierre et Marie Curie.

THOMASSET, F. (1974): *Etude d'une Méthode d'Eléments Finis de Degré 5; Application aux Problèmes de Plaques et d'Ecoulement de Fluides*, Doctoral Thesis (3ème cycle), Université de Paris-Sud, Orsay.

THOMÉE, V. (1973a): Approximate solution of Dirichlet's problem using approximating polygonal domain, in *Topics in Numerical Analysis* (J.J.H. Miller, Editor), pp. 311–328, Academic Press, New York.

THOMÉE, V. (1973b): Polygonal domain approximation in Dirichlet's problem, J. Inst. Math. Appl. **11**, 33–44.

TIMOSHENKO, S.; WOINOWSKY-KREIGER, S. (1959): *Theory of Plates and Shells*, McGraw-Hill.

TOMLIN, G.R. (1972): An optimal successive overrelaxation technique for solving second order finite difference equations for triangular meshes, Internat. J. Numer. Methods Engrg. **5**, 25–39.

TONTI, E. (1970): On the formal structure of continuum mechanics, part I: Deformation theory, Meccanica **V**, n° 1.

TORBE, I.; CHURCH, K. (1975): A general quadrilateral plate element, Internat. J. Numer. Methods Engrg. **9**, 855–868.

TREVES, F. (1967): *Topological Vector Spaces, Distributions and Kernels*, Academic Press, New York.

TURCKE, D.J.; MCNEICE, G.M. (1972): A variational approach to grid optimisation in the finite element method, in *Proceedings International Conference on Variational Methods in Engineering, Southampton, Vol. 1*, pp. 4/114–4/130.

TURNER, M.J.; CLOUGH, R.W.; MARTIN, H.C.; TOPP, L.J. (1956): Stiffness and deflection analysis of complex structures, J. Aero. Sci. **23**, 805–823.

VALIRON, G. (1950): *Equations Fonctionnelles. Applications*, Masson, Paris (2nd edition).

VARGA, R.S. (1966): Hermite interpolation-type Ritz methods for two-point boundary value problems, in *Numerical Solution of Partial Differential Equations* (J.H. Bramble, Editor), pp. 365–373, Academic Press, New York.

VARGA, R.S. (1971): *Functional Analysis and Approximation Theory in Numerical Analysis*, Regional Conference Series in Applied Mathematics, SIAM, Philadelphia, Pennsylvania.

VEIDINGER, L. (1972): On the order of convergence of the Rayleigh–Ritz method with piecewise linear trial functions, Acta Math. Acad. Sci. Hungar. **23**, 507–517.

VEIDINGER, L. (1974): On the order of convergence of a finite element scheme, Acta Math. Acad. Sci. Hungar. **25**, 401–412.

VEIDINGER, L. (1975): On the order of convergence of a finite element method in regions with curved boundaries, Acta Math. Acad. Sci. Hungar. **26**, 419–431.

VEKUA, I.N. (1965): Theory of thin shallow shells of variable thickness, Akad. Nauk Gruzin. SSR Trudy Tbilissi Mat. Inst. Razmadze **30**, 3–103.

VO-KHAC KHOAN (1972a): *Distributions, Analyse de Fourier, Opérateurs aux Dérivées Partielles, Vol. 1*, Vuibert, Paris.

VO-KHAC KHOAN (1972b): *Distributions, Analyse de Fourier, Opérateurs aux Dérivées Partielles, Vol. 2*, Vuibert, Paris.

WACHSPRESS, E.L. (1971): A rational basis for function approximation, J. Inst. Math. Appl. **8**, 57–68.

WACHSPRESS, E.L. (1973): A rational basis for function approximation, II. Curved sides, J. Inst. Math. Appl. **11**, 83–104.

WACHSPRESS, E.L. (1975): *A Rational Finite Element Basis*, Academic Press, New York.

WAIT, R.; MITCHELL, A.R. (1971): Corner singularities in elliptic problems by finite element methods, J. Computational Phys. **8**, 45–52.

WALSH, J. (1971): Finite-difference and finite-element methods of approximation, Proc. Roy. Soc. London Ser. A **323**, 155–165.

WASHIZU, K. (1968): *Variational Methods in Elasticity and Plasticity*, Pergamon, Oxford.

WHITEMAN, J.R. (1975): *A Bibliography for Finite Elements*, Academic Press, New York.

WILSON, E.L.; TAYLOR, R.L. (1971): Incompatible displacement models, in *Proceedings of the Symposium on Numerical and Computer Methods in Structural Engineering* (O.N.R.; University of Illinois, September, 1971).

WOLF, J.P. (1975): Alternate hybrid stress finite element models, Internat. J. Numer. Methods Engrg. **9**, 601–615.

YAMAMOTO, Y.; SUMI, Y. (1976): Stress intensity factors for three-dimensional cracks (to appear in the Proceedings of the 14th International Congress on Theoretical and Applied Mechanics, Delft, 1976).

YAMAMOTO, Y.; TOKUDA, N. (1973): Determination of stress intensity factors in cracked plates by the finite element method, Internat. J. Numer. Methods Engrg. **6**, 427–439.

ŽENÍŠEK, A. (1970): Interpolation polynomials on the triangle, Numer. Math. **15**, 283–296.

ŽENÍŠEK, A. (1972): Hermite interpolation on simplexes in the finite element method, in *Proceedings EquaDiff 3*, pp. 271–277, J.E. Purkyně University, Brno.

ŽENÍŠEK, A. (1973): Polynomial approximation on tetrahedrons in the finite element method, J. Approximation Theory **7**, 334–351.

ŽENÍŠEK, A. (1974): A general theorem on triangular finite $C^{(m)}$-elements, Rev. Française Automat. Informat. Recherche Opérationnelle Sér. Rouge Anal. Numér. **R-2**, 119–127.

ZIENKIEWICZ, O.C. (1971): *The Finite Element Method in Engineering Science*, McGraw-Hill, London.

ZIENKIEWICZ, O.C. (1973): Finite elements. The background story, in *The Mathematics of Finite Elements and Applications* (J.R. Whiteman, Editor), pp. 1–35, Academic Press, London.

ZIENKIEWICZ, O.C. (1974): Constrained variational principles and penalty function methods in finite element analysis, in *Conference on the Numerical Solution of Differential Equations* (G.A. Watson, Editor), pp. 207–214, Lecture Notes in Mathematics, Vol. 363, Springer-Verlag, New York.

ZIENKIEWICZ, O.C.; TAYLOR, R.L.; TOO, J.M. (1971): Reduced integration technique in general analysis of plates and shells, Internat. J. Numer. Methods Engrg. **3**, 275–290.

ZLÁMAL, M. (1968): On the finite element method, Numer. Math. **12**, 394–409.

ZLÁMAL, M. (1970): A finite element procedure of the second order of accuracy, Numer. Math. **16**, 394–402.

ZLÁMAL, M. (1973a): The finite element method in domains with curved boundaries, Internat. J. Numer. Methods Engrg. **5**, 367–373.

ZLÁMAL, M. (1973b): Curved elements in the finite element method. I. SIAM J. Numer. Anal. **10**, 229–240.

ZLÁMAL, M. (1973c): Some recent advances in the mathematics of finite elements, in *The Mathematics of Finite Elements and Applications* (J.R. Whiteman, Editor), pp. 59–81, Academic Press, London.

ZLÁMAL, M. (1973d): A remark on the "Serendipity family", Internat. J. Numer. Methods Engrg **7**, 98–100.

ZLÁMAL, M. (1974): Curved elements in the finite element method. II, SIAM J. Numer. Anal. **11**, 347–362.

GLOSSARY OF SYMBOLS

General notation

$v(\cdot)$, $v(\cdot,\cdot)$, etc...: function v of one variable, two variables, etc...
$v(\cdot, b)$: partial mapping $x \to v(x, b)$.
$\operatorname{supp} v = \{x \in X;\ v(x) \neq 0\}^-$: support of a function v.
$\operatorname{osc}(v; A) = \sup_{x,y \in A} |v(x) - v(y)|$.
v_A or $v_{|A}$: restriction of a function v to the set A.
$P(A) = \{P_{|A};\ \forall p \in P\}$, where P is any space of functions defined over a domain which contains the set A.
$\operatorname{tr} v$, or simply v: trace of a function v.
$R(v) = \dfrac{a(v, v)}{(v, v)}$: Rayleigh quotient.
$C(a)$, $C(a, b)$, etc... : any "constant" which depends solely on a, a and b, etc...
$\mathring{A}$: interior of a set A.
∂A: boundary of a set A.
$\bar{A}$ or A^-: closure of a set A.
card A: number of elements of a set A.
diam A: diameter of a set A.
$\complement A$, or $\complement_X A$, or $X - A$: Complement set of the subset A of the set X.
$\Rightarrow$: implies.

Derivatives and differential calculus

$Dv(a)$, or $v'(a)$: first (Fréchet) derivative of a function v at a point a.
$D^2v(a)$, or $v''(a)$: second (Fréchet) derivative of a function v at a.
$D^kv(a)$: k-th (Fréchet) derivative of a function v at a point a.
$D^kv(a)h^k = D^kv(a)(h_1, h_2, \ldots, h_k)$ if $h_1 = h_2 = \cdots = h_k = h$.
$$\mathcal{R}_k(v; b, a) = v(b) - \left\{v(a) + Dv(a)(b-a) + \cdots + \frac{1}{k!} D^kv(a)(b-a)^k\right\}.$$

$$\left.\begin{aligned}\partial_i v(A) &= Dv(a)e_i,\\ \partial_{ij} v(a) &= D^2 v(a)(e_i, e_j),\\ \partial_{ijk} v(a) &= D^3 v(a)(e_i, e_j, e_k).\end{aligned}\right\}\quad \text{(also used for vector-valued functions)}$$

$J_F(\hat{x}) = \det(\partial_j F_i(\hat{x}))$ = Jacobian of a mapping $F: \hat{x} \in \mathbf{R}^n \to F(\hat{x}) = (F_i(x))_{i=1}^n \in \mathbf{R}^n$.

$\operatorname{div} \boldsymbol{v} = \sum_{i=1}^n \partial_i v$.

$\nabla v(a) = (\partial_i v)_{i=1}^n$, also denoted $\boldsymbol{\nabla} v(a)$, grad $v(a)$.

$\Delta v = \sum_{i=1}^n \partial_{ii} v$, $\Delta \boldsymbol{v} = (\Delta v_i)_{i=1}^n$.

$|\alpha| = \sum_{i=1}^n \alpha_i$, for a multi-index $\alpha = (\alpha_1, \ldots, \alpha_n) \in N^n$.

$\partial^\alpha v(a) = D^{|\alpha|} v(a)(\overbrace{e_1, \ldots, e_1}^{\alpha_1 \text{ times}}, \overbrace{e_2, \ldots, e_2}^{\alpha_2 \text{ times}}, \ldots, \overbrace{e_n, \ldots, e_n}^{\alpha_n \text{ times}})$.

$\nu = (\nu_1, \nu_2, \ldots, \nu_n)$: unit outer normal vector.

$\partial_\nu = \sum_{i=1}^n \nu_i \partial_i$: (outer) normal derivative operator.

$\tau = (\tau_1, \tau_2)$: unit tangential vector along the boundary of a plane domain.

$\partial_\tau v(a) = Dv(a)\tau = \sum_{i=1}^2 \tau_i \partial_i v(a)$.

$\partial_{\nu\tau} v(a) = D^2 v(a)(\nu, \tau) = \sum_{i,j=1}^2 \nu_i \tau_j \partial_{ij} v(a)$.

$\partial_{\tau\tau} v(a) = D^2 v(a)(\tau, \tau) = \sum_{i,j=1}^2 \tau_i \tau_j \partial_{ij} v(a)$.

$(V_I)_{I=1}^{12} = \{\partial^\alpha v_\beta, |\alpha| \leq 1,\ \beta = 1, 2,\ \partial^\alpha v_3,\ |\alpha| \leq 2\}$ (notation for admissible displacements $\boldsymbol{v} = (v_1, v_2, v_3)$ in shell theory).

Differential geometry

$(a_{\alpha\beta})$: first fundamental form of a surface.
$a = \det(a_{\alpha\beta})$.
$(b_{\alpha\beta})$: second fundamental form of a surface.
$(c_{\alpha\beta})$: third fundamental form of a surface.
$\Gamma^\alpha_{\beta\gamma}$: Christoffel symbols.
$v_{|\beta}, v_{|\alpha\beta}, \ldots$: covariant derivatives along a surface.
$ds = \sqrt{a}\, d\xi$: surface element.
$\frac{1}{R}$: curvature of a plane curve.

General notation for vector spaces.

$B(a; r) = \{x \in X;\ \|x - a\| \leq r\}$.
$\mathscr{L}(X; Y)$: space of continuous linear mappings from X into Y.
$\mathscr{L}(X) = \mathscr{L}(X; X)$.
$\mathscr{L}_k(X; Y)$: space of continuous k-linear mappings from X^k into Y.
$\mathscr{L}_2(X_1 \times X_2; Y)$: space of continuous bilinear mappings from $X_1 \times X_2$ into Y.
X': dual of a space X.
$\|\cdot\|^*$: norm in the space X'.
$\langle \cdot, \cdot \rangle$: duality pairing between a space and its dual.
$x + Y = \{x + y;\ y \in Y\}$.
$X + Y = \{x + y;\ x \in X, y \in Y\}$.
$X \oplus Y = \{x + y;\ x \in X, y \in Y\}$ when $X \cap Y = \{0\}$.
X/Y: quotient space of X by Y.
$\bigvee \{e_\lambda, \lambda \in \Lambda\}$: vector space spanned by the vectors e_λ, $\lambda \in \Lambda$.
I: identity mapping.
$\hookrightarrow$: inclusion with continuous injection.
$\overset{c}{\subset}$: inclusion with compact injection.
$\dim X$: dimension of the space X.
$\ker A = \{x \in X;\ Ax = 0\}$.

Notation for specific vector spaces

$(u, v) = \int_\Omega uv \, \mathrm{d}x$ (inner product in $\mathrm{L}^2(\Omega)$).

$(\boldsymbol{u}, \boldsymbol{v}) = \int_\Omega \boldsymbol{u} \cdot \boldsymbol{v} \, \mathrm{d}x$ (inner product in $(\mathrm{L}^2(\Omega))^n$).

$\mathscr{C}^m(A)$: space of functions m times continuously differentiable on a subset A of $\mathbf{R}^n$.

$\mathscr{C}^\infty(A) = \bigcap_{m=0}^{\infty} \mathscr{C}^m(A)$.

$$\mathscr{C}^{m,\alpha}(A) = \{v \in C^m(\bar{\Omega});\ \forall \beta, |\beta| = m, \exists \Gamma_\beta, \forall x, y \in A,$$
$$|\partial^\beta v(x) - \partial^\beta v(y)| \leq \Gamma_\beta \|x - y\|^\alpha\}.$$

$$\|v\|_{\mathscr{C}^{m,\alpha}(A)} = \|v\|_{m,\infty,A} + \max_{|\beta|=m} \sup_{\substack{x,y \in A \\ x \neq y}} \frac{|\partial^\beta v(x) - \partial^\beta v(y)|}{\|x - y\|^\alpha}.$$

$\mathscr{D}(\Omega) = \{v \in \mathscr{C}^\infty(\Omega);$ supp v is a compact subset of $\Omega\}$.
$\mathscr{D}'(\Omega)$: space of distributions over Ω.
$H^m(\Omega) = \{v \in \mathrm{L}^2(\Omega);\ \forall \alpha, |\alpha| \leq m, \partial^\alpha v \in \mathrm{L}^2(\Omega)\}$.
$H_0^m(\Omega) =$ closure of $\mathscr{D}(\Omega)$ in $H^m(\Omega)$.

$\|v\|_{m,\Omega} = \left(\sum_{|\alpha|\leq m} \int_\Omega |\partial^\alpha v|^2 \, dx\right)^{1/2}$.

$|v|_{m,\Omega} = \left(\sum_{|\alpha|=m} \int_\Omega |\partial^\alpha v|^2 \, dx\right)^{1/2}$.

$\|\boldsymbol{v}\|_{m,\Omega} = \left(\sum_{i=1}^n \|v_i\|_{m,\Omega}^2\right)^{1/2}$ (for functions $\boldsymbol{v} = (v_i)_{i=1}^n$ in $(H^m(\Omega))^n$).

$|\boldsymbol{v}|_{m,\Omega} = \left(\sum_{i=1}^n |v_i|_{m,\Omega}^2\right)^{1/2}$ (for functions $\boldsymbol{v} = (v_i)_{i=1}^n$ in $(H^m(\Omega))^n$).

$W^{m,p}(\Omega) = \{v \in L^p(\Omega);\ \forall\alpha,\ |\alpha| \leq m,\ \partial^\alpha v \in L^p(\Omega)\}$.

$W_0^{m,p}(\Omega)$ = closure of $\mathcal{D}(\Omega)$ in $W^{m,p}(\Omega)$.

$\|v\|_{m,p,\Omega} = \left(\sum_{|\alpha|\leq m} \int_\Omega |\partial^\alpha v|^p \, dx\right)^{1/p}, \quad 1 \leq p < \infty$.

$\|v\|_{m,\infty,\Omega} = \max_{|\alpha|\leq m} \left\{\text{ess.}\sup_{x\in\Omega} |\partial^\alpha v(x)|\right\}$ (also used to denote the norm in $C^m(\bar{\Omega})$).

$\|v\|_{m,p,\Omega}^*$ = norm in the dual space of $W^{m,p}(\Omega)$.

$|v|_{m,p,\Omega} = \left(\sum_{|\alpha|=m} \int_\Omega |\partial^\alpha v|^p \, dx\right)^{1/p}, \quad 1 \leq p < \infty$.

$|v|_{m,\infty,\Omega} = \max_{|\alpha|=m} \left\{\text{ess.}\sup_{x\in\Omega} |\partial^\alpha v(x)|\right\}$.

$\dot{v} = \{w \in W^{k+1,p}(\Omega);\ (w - v) \in P_k(\Omega)\}$,
$\|\dot{v}\|_{k+1,p,\Omega} = \inf_{p\in P_k(\Omega)} \|v + p\|_{k+1,p,\Omega},\ v \in \dot{v}$,
$|\dot{v}|_{k+1,p,\Omega} = |v|_{k+1,p,\Omega},\ v \in \dot{v}$.
} notation in the quotient space $W^{k+1,p}(\Omega)/P_k(\Omega)$

$[v]_{m,\Omega} = \left(\sum_{i=1}^n \int_\Omega |D^m v(x)(e_i^m)|^2 \, dx\right)^{1/2}$.

$[v]_{m,p,\Omega} = \left(\sum_{i=1}^n \int_\Omega |D^m v(x)(e_i^m)|^p \, dx\right)^{1/p}$.

$|v|_{\varphi;m,\Omega} = \left\{\int_\Omega \varphi \sum_{|\beta|=m} |\partial^\beta v|^2 \, dx\right\}^{1/2}$ (weighted semi-norm).

$|v|_{m,\infty,K} = \sup_{x\in K} \|D^m v(x)\|_{\mathcal{L}_m(\mathbf{R}^n;\mathbf{R})}$ (for $v\colon K \subset \mathbf{R}^n \to \mathbf{R}$).

$|F|_{m,\infty,\hat{K}} = \sup_{\hat{x}\in\hat{K}} \|D^m F(\hat{x})\|_{\mathcal{L}_m(\mathbf{R}^n;\mathbf{R}^n)}$ (for $F\colon \hat{K} \subset \mathbf{R}^n \to \mathbf{R}^n$).

$[\![F]\!]_{m,\infty,\hat{K}} = \max_{1\leq i\leq n} \sup_{\hat{x}\in\hat{K}} \|D^m F(\hat{x})(e_i)^m\|$ (for $F\colon \hat{K} \subset \mathbf{R}^n \to \mathbf{R}^n$).

$H^m(K)$, $W^{m,p}(K)$, $\|\cdot\|_{m,p,K}$, etc. . . : alternate notation for $H^m(\mathring{K})$, $W^{m,p}(\mathring{K})$, $\|\cdot\|_{m,p,\mathring{K}}$, etc. . . ($K$: a subset of $\mathbf{R}^n$ with interior $\mathring{K}$).

$H^{1/2}(\Gamma) = \{r \in L^2(\Gamma);\ \exists v \in H^1(\Omega);\ \text{tr}\, v = r \text{ on } \Gamma\}$.

$\|r\|_{H^{1/2}(\Gamma)} = \inf\{\|v\|_{1,\Omega};\ v \in H^1(\Omega),\ \mathrm{tr}\ v = r \text{ on } \Gamma\}$.
$H^{-1/2}(\Gamma)$: dual space of $H^{1/2}(\Gamma)$.
$\|\cdot\|_{H^{-1/2}(\Gamma)}$: norm of $H^{-1/2}(\Gamma)$.
$\langle\cdot,\cdot\rangle_\Gamma$: duality pairing between the spaces $H^{-1/2}(\Gamma)$ and $H^{1/2}(\Gamma)$.
$W_0^1(\mathbf{R}^3)$ = completion of $\mathcal{D}(\mathbf{R}^3)$ with respect to the norm $|\cdot|_{1,\mathbf{R}^3}$.
$H(\mathrm{div};\Omega) = \{\boldsymbol{q} \in (\mathrm{L}^2(\Omega))^n;\ \mathrm{div}\ \boldsymbol{q} \in \mathrm{L}^2(\Omega)\}$.
$\|\boldsymbol{q}\|_{H(\mathrm{div};\Omega)} = (|\boldsymbol{q}|_{0,\Omega}^2 + |\mathrm{div}\ \boldsymbol{q}|_{0,\Omega}^2)^{1/2}$.

Elasticity

λ, μ: Lamé's coefficient of a material.
$E = \dfrac{\mu(3\lambda + 2\mu)}{\lambda + \mu}$: Young's modulus.
$\sigma = \dfrac{\lambda}{2(\lambda + \mu)}$: Poisson's coefficient.
$\epsilon_{ij}(\boldsymbol{v}) = \frac{1}{2}(\partial_j v_i + \partial_i v_j)$: components of the (linearized) strain tensor.
σ_{ij}: components of the stress tensor.
e: thickness of a plate, or a shell.
A: area of a cross section of an arch.
I: moment of inertia of a cross-section of an arch.
$(\gamma_{\alpha\beta})$: strain tensor (of the middle surface of a shell).
$(\bar{\rho}_{\alpha\beta})$: change of curvature tensor (of the middle surface of a shell).

Some spaces of polynomials

P_k: space of all polynomials in $x_1, \ldots, x_n$ of degree $\leqslant k$.
$P_3' = \{p \in P_3;\quad \phi_{ijk}(p) = 0,\quad 1 \leqslant i < j < k \leqslant n+1\}$, with $\phi_{ijk}(p) = 12p(a_{ijk}) + 2\sum_{l=i,j,k} p(a_l) - 3\sum_{\substack{l,m=i,j,k\\ l\neq m}} p(a_{llm})$ (cf. the n-simplex of type (3')).
$P_3'' = \{p \in P_3;\quad \psi_{ijk}(p) = 0,\quad 1 \leqslant i < j < k \leqslant n+1\}$, with $\psi_{ijk}(p) = 6p(a_{ijk}) - 2\sum_{l=i,j,k} p(a_l) - \sum_{l=i,j,k} Dp(a_l)(a_l - a_{ijk})$ (cf. the Hermite n-simplex of type (3')).
$P_5'(K) = \{p \in P_5(K);\ \partial_\nu p \in P_3(K') \text{ for each side } K' \text{ of } K\}$
$= \{p \in P_5(K);\quad \chi_{ij}(\partial_\nu p) = 0,\quad 1 \leqslant i < j \leqslant 3\}$, with $\chi_{ij}(v) = 4(v(a_i) + v(a_j)) - 8v(a_{ij}) + Dv(a_i)(a_j - a_{ij}) + Dv(a_j)(a_j - a_{ij})$ (cf. the Bell triangle).
Q_k: space of all polynomials in $x_1, \ldots, x_n$, of degree $\leqslant k$ with respect to each variable x_i, $1 \leqslant i \leqslant n$.

$Q_2' = \{p \in Q_2;\ 4p(a_9) + \sum_{i=1}^{4} p(a_i) - 2\sum_{i=5}^{8} p(a_i) = 0\}$ (cf. the rectangle of type (2')).

$Q_3' = \{p \in Q_3;\ \psi_i(p) = 0,\ 1 \leqslant i \leqslant 4\}$, with $\psi_1(p) = 9p(a_{13}) + 4p(a_1) + 2p(a_2) + p(a_3) + 2p(a_4) - 6p(a_5) - 3p(a_6) - 3p(a_{11}) - 6p(a_{12})$, etc... (cf. the rectangle of type (3')).

$T_3(K)$: space of tricubic polynomials (i.e., whose restrictions along any parallel to any side of a triangle K are polynomials of degree $\leqslant 3$ in one variable).

Notation special to $\mathbf{R}^n$

e_i, $1 \leqslant i \leqslant n$: canonical basis of $\mathbf{R}^n$, also denoted e^i, for $n = 3$.

$\|v\| = \left(\sum_{i=1}^{n} |v_i^2|\right)^{1/2}$: Euclidean norm of the vector $v = (v_i)_{i=1}^n$.

$\|B\| = \sup_{v \in \mathbf{R}^n} \frac{\|Bv\|}{\|v\|}$: norm of the matrix B, induced by the Euclidean vector norm.

$\boldsymbol{a} \cdot \boldsymbol{b}$: Euclidean scalar product in $\mathbf{R}^n$ of the vectors $\boldsymbol{a}$ and $\boldsymbol{b}$.

$\boldsymbol{a} \times \boldsymbol{b}$: vector product of the vectors $\boldsymbol{a}$ and $\boldsymbol{b}$.

$\det B$: determinant of a square matrix B.

$\operatorname{meas}(A) = \mathrm{d}x$-measure of a set $A \subset \mathbf{R}^n \left(= \int_A \mathrm{d}x\right)$.

$\mathrm{d}\gamma$ = superficial measure along a Lipschitz-continuous boundary of an open subset of $\mathbf{R}^n$.

$\lambda_j = \lambda_j(x)$: barycentric coordinates of a point $x \in \mathbf{R}^n$, $1 \leqslant j \leqslant n+1$.

$a_{ij} = \dfrac{a_i + a_j}{2}$, $i < j$.

$a_{iij} = \dfrac{2a_i + a_j}{3}$, $i \neq j$.

$a_{ijk} = \dfrac{a_i + a_j + a_k}{3}$, $i \neq j$, $j \neq k$, $k \neq i$.

$L_k(K) = \left\{x = \sum_{j=1}^{n+1} \lambda_j a_j;\ \sum_{j=1}^{n+1} \lambda_j = 1,\ \lambda_j \in \left\{0, \frac{1}{k}, \ldots, \frac{k-1}{k}, 1\right\},\ 1 \leqslant j \leqslant n+1\right\}$.

$\hat{M}_k = \left\{x = \left(\frac{i_1}{k}, \frac{i_2}{k}, \ldots, \frac{i_n}{k}\right) \in \mathbf{R}^n;\ i_j \in \{0, 1, \ldots, k\},\ 1 \leqslant j \leqslant n\right\}$.

$M_k(K) = F_K(M_k)$, $F_K: x \to F_K(x) = B_K x + b_K$, B_K: diagonal matrix.

Finite Elements (most common notation)

(K, P, Σ) or (K, P_K, Σ_K): finite element.

$P = P_K$: space of functions p, or $p_K: K \to \mathbf{R}$.
$\Sigma = \Sigma_K$: set of degrees of freedom of a finite element.
$\varphi_i = \varphi_{i,K}$, $1 \leq i \leq N$: degrees of freedom of a finite element.
$p_i = p_{i,K}$, $1 \leq i \leq N$: basis functions of a finite element.
$\mathcal{N}_K$: set of nodes of a finite element.
$s = s_K$: maximal order of directional derivatives found in the set Σ.
$\Pi v = \Pi_K v = P$-, or P_K-, interpolant of a function v.
dom $\Pi = \mathscr{C}^s(K)$.
$h_K = \operatorname{diam}(K)$.
$\rho_K = \sup\{\operatorname{diam}(S);\ S$ is a ball contained in $K\}$.
$\hat{x} \in \hat{K} \to x = F(\hat{x}) \in K$: bijection between points of $\hat{K}$ and $K = F(\hat{K})$ (F: bijection).
$\hat{v}: \hat{K} \to \mathbf{R} \to v = \hat{v} \cdot F^{-1}$: $K \to \mathbf{R}$: bijection between functions defined over $\hat{K}$ and $K = F(\hat{K})$ (F: bijection).
$F \in (\hat{P})^n \Leftrightarrow F_i \in \hat{P}$, $1 \leq i \leq n$, with $\hat{P}$: space of functions $\hat{p}: \hat{K} \subset \mathbf{R}^n \to \mathbf{R}$.
$\tilde{K} = \tilde{F}(\hat{K})$, where $\tilde{F} \in (P_1(\hat{K}))^n$ and $\tilde{F}(\hat{a}_i) = a_i$, $1 \leq i \leq n+1$,
$h_K = \operatorname{diam}(\tilde{K})$,
$\rho_K =$ diameter of the sphere inscribed in $\tilde{K}$.
} for isoparametric simplicial elements
$\int_K \varphi(x)\,dx \sim \sum_{l=1}^{L} \omega_l \varphi(b_l)$: quadrature formula with weights ω_l and nodes b_l.
$\hat{E}(\hat{\varphi}) = \int_{\hat{K}} \hat{\varphi}(\hat{x})\,d\hat{x} - \sum_{l=1}^{L} \hat{\omega}_l \hat{\varphi}(\hat{b}_l)$: quadrature error functional on $\hat{K}$.
$E_K(\varphi) = \int_K \varphi(x)\,dx - \sum_{l=1}^{L} \omega_{l,K} \varphi(b_{l,K})$: quadrature error functional on $K = F_K(\hat{K})$, with $\omega_{l,K} = \hat{\omega}_l J_{F_K}(\hat{b}_l)$, $b_{l,K} = F_K(\hat{b}_l)$.

Finite element spaces (most common notation)

$\mathscr{T}_h$: triangulation of a set $\bar{\Omega}$.
X_h: finite element space without boundary conditions.
$X_{0h} = \{v_h \in X_h;\ v_h = 0$ on $\Gamma\}$.
$X_{00h} = \{v_h \in X_h;\ v_h = \partial_\nu v_h = 0$ on $\Gamma\}$.
V_h: finite element space with boundary conditions.
$\Sigma_h =$ set of degrees of freedom of a finite element space X_h.
φ_h or φ_{kh}, $1 \leq k \leq M$: degrees of freedom of a finite element space X_h.
$(w_k)_{k=1}^M$: basis in a finite element space X_h or V_h.
$\mathcal{N}_h$: set of nodes of a finite element space X_h.
$\Pi_h v$: X_h-interpolant of a function v.
dom $\Pi_h = \mathscr{C}^s(\bar{\Omega})$, $s = \max_{K \in \mathscr{T}_h} s_K$.

Various sets of hypotheses concerning the finite element method

(FEM 1): Existence of a triangulation.

(FEM 2): The spaces P_K, $K \in \mathcal{T}_h$, contain polynomials or "nearly polynomials".

(FEM 3): There exists a basis in the finite element space V_h whose functions have "small" support.

($\mathcal{T}_h$1): $\bar{\Omega} = \bigcup_{K \in \mathcal{T}_h} K$.

($\mathcal{T}_h$2): $\forall K \in \mathcal{T}_h$, $\mathring{K} \neq \emptyset$.

($\mathcal{T}_h$3): $K_1 \neq K_2 \Rightarrow \mathring{K}_1 \cap \mathring{K}_2 = \emptyset$.

($\mathcal{T}_h$4): For all $K \in \mathcal{T}_h$, the boundary ∂K is Lipschitz-continuous.

($\mathcal{T}_h$5): Condition on adjacent finite elements.

(H1): Regularity of a family of triangulations.

(H2): All finite elements (K, P_K, Σ_K), $K \in \bigcup_h \mathcal{T}_h$, are affine-equivalent to a single reference finite element.

(H3): All finite elements (K, P_K, Σ_K), $K \in \bigcup_h \mathcal{T}_h$, are of class $\mathscr{C}^0$.

(H4): The family of triangulations satisfies an inverse assumption.

(H1*): The family (K, P_K, Σ_K), $K \in \bigcup_h \mathcal{T}_h$, is almost affine.

(H2*): All finite elements (K, P_K, Σ_K), $K \in \bigcup_h \mathcal{T}_h$, are of class $\mathscr{C}^1$.

INDEX

Note: An asterisk in the left margin indicates a specific finite element.